Coal Combustion Byproducts and Environmental Issues

Kenneth S. Sajwan
Irena Twardowska
Tracy Punshon
Ashok K. Alva

Editors

Coal Combustion Byproducts and Environmental Issues

Springer

Kenneth S. Sajwan
Department of Natural Sciences and Mathematics
Savannah State University
Savannah, GA 31404
USA

Irena Twardowska
Institute of Environmental Engineering
Polish Academy of Sciences
41-819 Zabrze
Poland

Tracy Punshon
Savannah River Ecology Laboratory
University of Georgia
Aiken, SC 29802
USA

and

Environmental and Occupational Health Sciences Institute
Rutgers University
Piscataway, NJ 08854
USA

Ashok K. Alva
Vegetable and Forage Crops Research Unit
USDA-ARS, Pacific West Area
Prosser, WA 99350
USA

Library of Congress Control Number: 2005930800

ISBN-10: 0-387-25865-5
ISBN-13: 978-0387-25865-2

Printed on acid-free paper.

Printed in the United States of America. (TB/IBT)

9 8 7 6 5 4 3 2 1 SPIN 11423706

springeronline.com

IN GRATITUDE TO

My family, Maria, Mia, and Joseph Sajwan
Who've been through it all before

and

Dr. George Gobran
Swedish University of Agricultural Sciences, Uppsala, Sweden
For his major contribution in organizing the Seventh International Conference on the Biogeochemistry of Trace Elements, of which the present work is a part.

Preface

The massive accumulation of coal fly ash generated by electric power plants during fossil fuel combustion has become a major environmental health concern in the United States. Even though, stringent environmental regulations have been mandated by the Environmental Protection Agency through the Clean Air Act, Clean Water Act, and the Resource Conservation and Recovery Act, coal combustion products continue to pose serious environmental concerns due to our major reliance upon fossil fuels to meet ever increasing demands for energy production within the United States. The concentrations of trace elements in coal residues are extremely variable and depend upon the composition of the original coal, conditions experienced during combustion, the efficiency of emission control devices, storage and handling procedures, and overall climatic conditions.

The research papers carefully selected for publication within this book were originally presented as a part of the Seventh International Conference on the Biogeochemistry of Trace Elements held at the Swedish University of Agricultural Sciences, Uppsala, Sweden, from June 15–19, 2003. This conference offered the unique opportunity for international scientists and scholars to share the most current knowledge concerning the source, pathways, behavior, interactions, and effects of trace elements in soils, water, plants, and animals. Notably, this book also contains the invited research papers from leading scientists who have conducted significant research within the area of coal and coal combustion products. All of the research papers presented herein have been subjected to a peer review process. The editors have arranged the articles systematically by topic, beginning with introductory chapter entitled "Production of Coal Combustion Products and Their Potential Uses" followed by the sections on Environmental Impact of Coal Combustion Residues, Trace Elements in Fly Ashes, Transport and Leachability of Metals from Coal and Ash Piles, and the Use of Coal Ash as an Agricultural Soil Amendment.

This book addresses the major implications and critical issues surrounding coal combustion products and their impact upon the environment. It provides invaluable information particularly to scientists specializing and conducting research in coal and coal combustion products. Even so, it will certainly appeal beyond this initial target audience to serve a wide variety of scientists, scholars, academicians, and professionals within the fields of public health, engineering, energy industry, and a wide realm of environmental science related disciplines. It is our sincere endeavor that this volume of work will serve as a valuable resource tool for those engaged in fossil fuel energy research to benefit both our economy and environment to assure quality of life for future generations.

The editors wish to generously express their genuine appreciation and gratitude to all of the contributing authors for their professional insight and scientific contributions to this body of knowledge, along with their diligence and patience throughout the publication process.

Kenneth S. Sajwan
Irena Twardowska
Tracy Punshon
Ashok K. Alva

Contributors

S. A. Aburime
Department of Engineering, Clark Atlanta University, 233 James P. Brawley Drive, Atlanta, GA 30314, USA.

Domy C. Adriano
The University of Georgia, Savannah River Ecology Laboratory, Aiken, SC 29802, USA.

J. Afolabi
Department of Natural Sciences and Mathematics, Savannah State University, Savannah, GA 31404, USA.

A. K. Alva
USDA-ARS, Pacific West Area, Vegetable and Forage Crops Research Unit, 24106 N. Bunn Rd., Prosser, WA 99350, USA.

Christopher Barton
Department of Forestry, University of Kentucky, Thomas Poe Cooper Bldg., Lexington, KY 40546-0073, USA.

Nanthi Bolan
Institute of Natural Resources, Massy University, Palmerston North, New Zealand.

Joanna Burger
The University of Georgia, Savannah River Ecology Laboratory, Aiken, SC 29802, USA.

D. Chaudhuri
Department of Applied Science, College of Engineering, University of California, Davis, Livermore, CA 94550, USA.

Jianjun Chen
Mid-Florida Research and Education Center, IFAS, University of Florida, Apopka, FL 32703, USA.

E. M. D'Angelo
Department of Agronomy, University of Kentucky, N-122K Ag. Science-North, Lexington, KY 40546-0091, USA.

Holger Ecke
Division of Waste Science & Technology, Lulea University of Technology, SE-97187, Lulea, Sweden.

X. Feng
Ferro Corporation, Cleveland, OH 44114, USA.

G. E. Fryxell
Pacific Northwest National Laboratory, Richland, WA 99352, USA.

G. S. Ghuman
Department of Natural Sciences and Mathematics, Savannah State University, Savannah, GA 31404, USA.

B. R. Hart
Department of Earth Sciences, University of Western Ontario, London, N6A 5B7, Canada.

Mike Hedley
Institute of Natural Resources, Massy University, Palmerston North, New Zealand.

Dave Horne
Institute of Natural Resources, Massey University, Palmerston North, New Zealand.

J. M. Hutchison
Advanced Analytical Center for Environmental Sciences, Savannah River Ecology Laboratory, The University of Georgia, Aiken, SC 29802, USA.

Brian P. Jackson
The University of Georgia, Savannah River Ecology Laboratory, Aiken, SC 29802, USA.

A. D. Karathanasis
Department of Agronomy, University of Kentucky, N-122K Ag. Science-North, Lexington, KY 40546-0091, USA.

U. M. Khodke
College of Agricultural Engineering, Marathwada Agricultural University, Parbhani, Maharashtra, 431 402, India.

Ryunosuke Kikuchi
Department of Basic Science and Environment, CERNAS; ESAC—Polytechnic Institute of Coimbra, Bencanta, 3040-316, Coimbra, Portugal.

Waldemar Klassen
Tropical Plant Research and Education Center, Institute of Food and Agricultural Sciences, University of Florida, Homestead, FL 33031, USA.

Anna S. Knox
Savannah River National Laboratory, Westinghouse Savannah River Company, 773-42A, Savannah River Site, Aiken, SC 29802. USA.

John D. Knox
Columbia County Board of Education, Appling, GA 30802, USA.

Bon-Jun Koo
Department of Forestry, University of Kentucky, Thomas Poe Cooper Bldg., Lexington, KY 40546-0073, USA.

Jurate Kumpiene
Division of Waste Science & Technology, Lulea University of Technology, SE-97187, Lulea, Sweden.

Yuncong Li
Tropical Plant Research and Education Center, IFAS, University of Florida, Homestead, FL 33031, USA.

Sally Maharaj
Department of Forestry, University of Kentucky, Thomas Poe Cooper Bldg., Lexington, KY 40546-0073, USA.

S. V. Mattigod
Pacific Northwest National Laboratory, Richland, WA 99352, USA.

Christian Maurice
Division of Water Science & Technology, Lulea University of Technology, SE-97187, Lulea, Sweden.

William P. Miller
Department of Crop and Soil Sciences, University of Georgia, Athens, GA 30602-7272, USA.

Lee Newman
University of South Carolina, Arnold School of Public Health, 800 Sumpter St., Columbia, SC 29208, USA.

R. K. Panda
Department of Agricultural and Food Engineering, Indian Institute of Technology, Kharagpur, 721 302, West Bengal, India.

S. Paramasivam
Department of Natural Sciences and Mathematics, Savannah State University, Savannah, GA 31404, USA.

K. E. Parker
Pacific Northwest National Laboratory, Richland, WA 99352, USA.

E. M. Piers
Pacific Northwest National Laboratory, Richland, WA 99352, USA.

M.C. Potts
Department of Natural Sciences and Mathematics, Savannah State University, Savannah, GA 31404, USA.

M. A. Powell
Department of Earth Sciences, University of Western Ontario, London, N6A 5B7, Canada.

T. Prahraj
Department of Earth Sciences, University of Ottawa, 140, Louis Pasteur, Ottawa, KIN 6N5, Canada.

Tracy Punshon
Savannah River Ecology Laboratory, The University of Georgia, Aiken, SC 29802, USA. /Consortium for Risk Evaluation with Stakeholder Participation, Environmental and Occupational Health Sciences Institute, Division of Life Sciences, Rutgers University, 604 Allison Road, Piscataway, NJ 08854, USA.

J. E. Robinson
Department of Agronomy, University of Kentucky, N-122K Ag. Science-North, Lexington, KY 40546-0091, USA.

Kenneth S. Sajwan
Department of Natural Sciences and Mathematics, Savannah State University, Savannah, GA 31404, USA.

Maxim Schlossberg
Department of Crop and Soil Sciences, University of Georgia, Athens, GA 30602-7272, USA.

John C. Seaman
The University of Georgia, Savannah River Ecology Laboratory, Aiken SC 29802, USA.

Sebastian Stefaniak
Polish Academy of Sciences, Institute of Environmental Engineering, 34 M. Sklodowska-Curie St., 41-819 Zabrze, Poland.

S. Tripathy
Department of Geology & Geophysics, Indian Institute of Technology, Kharagpur, 721 302, West Bengal, India.

Irena Twardowska
Polish Academy of Sciences, Institute of Environmental Engineering, 34 M. Sklodowska-Curie St., 41-819 Zabrze, Poland.

H. Veeresh
Department of Soil Science and Agricultural Chemistry, U.A.S., G.K.V.K., Banglore, Karnataka, 560 065, India.

F. Clint Waltz, Jr.
Department of Crop and Soil Sciences, Georgia Agricultural Experiment Station, 119 Redding Building, Experiment, GA 30212, USA.

Hailong Wang
Forest Research, Private Bag 3020, Rotorua, New Zealand.

Qingren Wang
Tropical Plant Research and Education Center, Institute of Food and Agricultural Sciences, Homestead, FL 33031, USA.

Paul F. Ziemkiewicz
West Virginia Water Research Institute, West Virginia University, Morgantown, WV 26506, USA.

Contents

Preface vii

Contributors ix

Part I Introduction

1 Production of Coal Combustion Products and Their Potential Uses 3
K.S. Sajwan, T. Punshon, and J.C. Seaman

Part II Environmental Impact of Coal Combustion Residues

2 Coal and Coal Combustion Products: Prospects for Future and Environmental Issues 13
Irena Twardowska and Sebastian Stefaniak

3 Alternative By-Products of Coal Combustion and Simultaneous $SO_2/SO_3/NO_x$ Treatment of Coal-Fired Flue Gas: Approach to Environmentally Friendly Use of Low-Rank Coal 21
Ryunosuke Kikuchi

4 Fly Ash as a Sealing Material for Spontaneous Combustion and Acid Rock Drainage Prevention and Control 33
Irena Twardowska and Sebastian Stefaniak

5 Delineation of Water and Sediment Contamination in River Near a Coal Ash Pond in Orissa, India 41
S. Tripathy, and T. Praharaj

6 Prediction of Coal Ash Leaching Behavior in Acid Mine Water: Comparison of Laboratory and Field Studies 50
Paul F. Ziemkiewicz and Anna S. Knox

Part III Trace Elements in Flyashes

7 Occurrence and Sorption of Radionuclides Onto Coal-Fired Power Plant Combustion Waste 61
Irena Twardowska and Sebastian Stefaniak

8 Heavy Metals Adsorption and Their Distribution in Three Soil Types of India: Effect of Coal Fly Ash and Sewage Sludge Amendment 66
S. Tripathy, H. Veeresh, D. Chaudhuri, M.A. Powell, and B.R. Hart

Part IV Transport and Leachability of Metals from Coal and Ash Piles

9 Impact of Grassed Swales on the Fate of Metals Leached from Roads Built with Municipal Solid Waste Incineration Bottom Ashes 87
Jurate Kumpiene, Holger Ecke, and Christian Maurice

10 Removal of Mercury from Aqueous Streams of Fossil Fuel Power Plants Using Novel Functionalized Nonoporous Sorbents 99
S.V. Mattigod, G.E. Fryxell, X. Feng, K.E. Parker, and E.M. Piers

11 Leachability of Trace Metals from Sandy or Rocky Soils Amended with Coal Fly Ash 105
Yuncong Li and Jianjun Chen

12 Arsenic and Selenium Speciation in Aged Flue Gas Desulfurization Amended Soil 114
Tracy Punshon, Brian P. Jackson, John C. Seaman, Domy C. Adriano, and Joanna Burger

13 Trace Element Transport in Putting Green Root Mixes Amended by Coal Combustion Products (CCP) 124
Maxim J. Schlossberg and William P. Miller

14 Solute Leaching from Fly Ash Amended Soil Under Varying Degrees of Saturation 134
J.M. Hutchison, J.C. Seaman, B.P. Jackson, and S.A. Aburime

15 Solution Geochemistry Gradients in an Acid Mine Drainage Wetland Substrate 142
A.D. Karathanasis, and J.E. Robinson, and E.M. D'Angelo

16 Removal of Trace Elements from Aqueous System: Comparison of Two Fly Ash Materials 150
K.S. Sajwan, S. Paramasivam, M.C. Potts, G.S. Ghuman, J. Afolabi, and A.K. Alva

Part V Use of Coal Ash as Agriculture Soil Amendment

17 Transport and Plant Uptake of Zn in an Oxyaquic Haplustalf Amended with Coal Ash and Sewage Sludge: A Field Study 159
U.M. Khodke, S. Tripathy, R.K. Panda, H. Veeresh, and K.S. Sajwan

18 Influence of Fly Ash and Other Amendments on Leaching of Soil Nutrients and Metals from a Calcareous Soil 171
Qingren Wang, Yuncong Li, and Waldemar Klassen

19 Amendment of Fly Ash to Container Substrates for Ornamental Plant Production 177
Jianjun Chen and Yuncong Li

20 Influence and Coal Combustion Flue Gas Desulfurization Waste on Element Uptake by Maize (*Zea Mays L.*) 184
Anna S. Knox, John D. Knox, Domy C. Adriano, and Kenneth S. Sajwan

21 Amelioration of Soil Acidity with a Class-C Fly Ash: A Field Study 190
Maxim J. Schlossberg, F. Clint Waltz Jr., and William P. Miller

22 Phytoavailability of Trace Elements from a Landfill Containing Coal Combustion Waste 195
Sally Maharaj, Christopher Barton, Bon-Jun Koo, and Lee Newman

23 Potential Uses of Fluidised Bed Boiler Ash (FBA) as a Liming Material, Soil Conditioner and Sulphur Fertilizer 202
Hailong Wang, Nanthi Bolan, Mike Hedley, and Dave Horne

24 Potential of Fly Ash and Organic Wastes for Uses as Amendments to Agricultural Soils: A Review 216
G.S. Ghuman, K.S. Sajwan, and S. Paramasivam

25 Evaluation of Bahiagrass (*Paspalum notatum*) as a Vegetative Cover for a Landfill Containing Coal Combustion Waste 225
Bob-Jun Koo, Christopher Barton, and Domy Adriano

About the Editors 232

Index 235

I
Introduction

1
Production of Coal Combustion Products and Their Potential Uses

K.S. Sajwan[1], T. Punshon[2], and J.C. Seaman[2]
[1]*Department of Natural Sciennces and Mathematics, Savannah State University, Savannah, GA 31404, USA*
[2]*Savannah River Ecology Laboratory, The University of Georgia, Aiken SC 29802, USA*

Abstract

Coal Combustion Products (CCPs) arise from the combustion of coal for energy, with fly ash (FA), bottom ash (BA) and flue-gas desulfurization residues (FGD) the most abundant. Our reliance on fossil fuel for energy is set to continue into the 21st century, and research into the environmental safety of beneficial re-use options, as well as novel re-use options, must continue. Since previous editions of collected CCP research[1], significant changes have been made to both the New Source Review and the Clean Air Act that directly impact CCP production figures. New techniques such as x-ray absorption spectroscopy are increasingly being used to reveal micron-scale elemental characteristics of CCPs, and aid our understanding of the distribution and chemical form of the metallic constituents. This chapter summarizes production and use covering the period 2001–2003, new trends in reuse applications and discusses new research on the environmental safety of CCP re-use.

Introduction

Coal Combustion Products (CCPs) predominantly consist of fly ash (FA), bottom ash (BA), boiler slag (BS) and flue gas desulfurization residue (FGD or synthetic gypsum). In 2003, approximately 110.4 million metric tons (Mt) of CCPs were produced[2], an increase of about 8% on the previous year. Continued disposal of material on this scale is no longer considered feasible, and beneficial re-use is essential. Re-use in various construction applications, by far the most common avenue, now stands at 38.1% of the total CCP produced, although clearly the scope of CCP re-use can be expanded in the future. The focus of this volume is to collate information that will promote environmentally safe CCP re-use and foster this expansion.

The environmental hazards associated with CCPs are posed by the content of potentially toxic trace metals and metalloids which readily leach out when they enter soils[3–5]. While the metal (loid) content of CCPs reflects that of the parent coal, the most commonly found elements of concern are boron (B)[6], molybdenum (Mo)[7,8], arsenic (As)[9–11] and Selenium (Se)[12,13]. A variety of other metals have also been reported in CCPs, such as nickel (Ni), cadmium (Cd), mercury (Hg) and lead (Pb)[12]. The waste from coal and lignite burning power stations can also be enriched with radionuclides, such as uranium (^{238}U), radon (^{226}Ra), lead (^{210}Pb), thorium (^{232}Th) and potassium (^{40}K)[14]. Lasting environmental damage has been attributed, at least in part, to Se from CCP release in to settling lagoons, because this metalloid has a tendency to be transferred through the food chain[15,16]. For instance, developmental abnormalities in the mouthparts of amphibians living in FA disposal ponds are thought to be due to elevated Se[17], and there is also evidence of direct toxicity[6,18,19].

It is the major elemental properties of FA and FGD residues, such as the presence of calcium (Ca), potassium (K), sodium (Na), and sulfur (S), that has led to their application to soil in the hope that they can be safely used as amendments for various soil problems. Adding FA and FGD to nutrient poor soils has been reported to increase short term crop yield[20], correct nutrient deficiencies[21], and change the physical structure to alleviate compaction[22]. Nonetheless, the presence of metals and metalloids in CCPs is a significant impediment to their agronomic use and it is now believed that CCP application to soil should only occur after rigorous elemental analysis of the CCP, the soil and the crop requirement[21].

In addition, the input of mercury (Hg) into the atmosphere and aquatic ecosystems from coal combustion is currently receiving attention. Coal combustion is one of the most significant sources of Hg input into the biosphere[12,23–26], and Hg is currently unregulated in the U.S. The recent fish consumption advisory by the U.S. Environmental Protection Agency[27] has raised the profile of atmospheric Hg, and has prompted the formulation of a regulatory framework to address monitoring and safe limits[28].

Current CCP re-use rates are at their highest ever, although there is much research needed to address the safe exploitation of CCPs. Our understanding of metal(loid) bioavailability is still developing, and is directly applicable to CCP disposal and re-use issues. This volume brings together key biogeochemical studies, using novel techniques to directly address long-term toxicity and bioavailability. They include analysis of varying types of vegetation cover on the physical mobility of potentially toxic metals and metalloids leaching from CCPs, novel sorbents for Hg removal from aqueous CCPs, soil sorption characteristics when CCPs are combined with organic waste materials such as sewage sludge, the influence of CCPs on plant growth and elemental composition and the long term bioavailability and speciation of elements of concern in the soil following CCP application.

Coal Combustion Products

Fly Ash

Fly ash is a fine powder made up of hollow ferroaluminosilicate particles enriched with Ca, K and Na[29–31], and is collected by mechanical filters or electrostatic precipitators from the flue gas during coal combustion. Typical FA particle sizes are within 0.1–1.0 μm, and electron microscopy has revealed particles with rough surfaces covered with smaller adhering spherical particles[31]. Composite FA[32] comprises several types of particles, including true hollow particles, smaller aggregations known as microspheres, and opaque magnetite spheres[33]. Trace elements, including potentially toxic metals and metalloids, condense upon the surface of FA particles during combustion[29,33,34]. Fly ash is pozzolanic in nature; a siliceous (or combination siliceous and aluminous) material that forms cementitious compounds when in the presence of moisture[35]. Using x-ray absorption fine structure spectroscopy (XAFS), Shoji *et al.*[11] showed S to be present predominantly as sulfate, with some thiophene and sulfite in larger particle size fractions (>2.5 μm in diameter). X-ray absorption near-edge structure spectroscopy (XANES) showed Cr to be present as the toxic Cr^{6+} valence state in 10–30% of western U.S. coal fly ashes, but only Cr^{3+} was detected within the eastern U.S. bituminous coal FA. They identified As in all FA as As^{5+}, although there were small spectral differences in As speciation between eastern and western coals, which were not clarified. Significant variation in Zn speciation was observed between different FA phases of eastern and western coals, with $ZnFe_2O_4$ the principal form in eastern coal fly ashes. Struis *et al.*[36] found that 60% of the Zn in raw FA was hydrozincite ($Zn_5(CO_3)_2(OH)_6$) and the remaining 40% was inert forms such as willemite (Zn_2SiO_4) and gahnite ($ZnAl_2O_4$). Pires and Querol[37] investigated the composition of Brazilian fly ashes using ICP-MS, ICP-AES, x-ray diffraction (XRD) and scanning electron microscopy, finding that in the leachable fraction the metal classification was B (40–50%)>Mo>Cu>Ge= lithium (Li) =Zn=As>, Ni, tin (Sb), thallium (Tl), U > barium (Ba), Cd, strontium (Sr), vanadium (V)(0.3–2%). Utilization of micron-scale spatial metal analysis techniques to determine metal(loid) species within coal and resultant combustion products[38–40] is a significant development within CCP research; it will allow engineers to adjust the combustion conditions for parent coal type so that toxic species of metal(loid)s do not predominate in the resultant FA, and will allow the potential environmental hazards as a result of FA use to be more fully understood, by understanding the bioavailability and distribution of metal(loid) species within the soil and biota.

Bottom Ash

Bottom ash is uncombusted material that settles to the bottom of the boiler; boiler slag is formed when operating temperatures exceed ash fusion temperature and the slag remains molten until drained from the bottom of the combustion chamber[41]. Bottom ash is granular and is similar to concrete sand[42]. Boiler slag is a shiny, black granular material that has abrasive properties, and is used as grit for snow and ice control, structural embankments, aggregate and as road base material (Table 1). The re-use potential of BA is influenced by its physical characteristics, such as grain-size distribution, staining potential and color[43], which are typically variable properties[44]. In the scientific literature, the BA derived from coal and municipal solid waste (MSW) are frequently confused; and some workers suggest that these materials have considerable similarities[26], or are similar in nature to FA. However, in the present volume, which focuses on the chemical properties of the CCPs, distinctions are made between those arising from coal, and those from the combustion of other solid materials.

Flue Gas Desulfurization Residues

The Clean Air Act Amendments of 1990 (CAAA '90 Public Law 101-549) placed stringent restrictions on the release of sulfur oxide (SO_x) from coal-fired power plants, with a two phase implementation plan, requiring electric utility companies to reduce SO_2 emissions, in an effort to reduce atmospheric pollution and acid rain[45]. The majority of utility companies previously used high-sulfur bituminous coal, which was thought to have significantly contributed to incidences of acid rain in North America. Following the instatement of the act, many companies switched to low-sulfur coal or fuel oil for partial and rapid compliance with regulations, although retrofitting power plants with flue-gas scrubbing systems was ultimately necessary to fully comply. This change effectively resulted in the creation of a new waste stream, termed flue-gas desulfurization residue (FGD).

Flue gas desulfurization residue is the alkaline material produced when SO_x is extracted from coal-fired power plant flue-gases[46,47]. There are several technologies currently in use, differentiated by the type of sorbent (e.g. lime or dolomitic lime) used and the method of SO_xextraction[48], however, FGD typically consists of calcium sulfite ($CaSO_3$), calcium sulfate ($CaSO_4$), unreacted sorbent, and FA particles. Other types of FGD can include magnesium, ammonium or sodium sulfites and sulfates.

TABLE 1. Production and use of category I (dry) and II (ponded) coal combustion products (CCPs) in the United States during the period of 2001–2003 (metric tons).

	2001			2002			2003		
	FA	BA	FGD	FA	BA	FGD	FA	BA	FGD
PRODUCED	*61,841,517*	*17,044,190*	*25,839,153*	*69,399,630*	*17,962,257*	*10,341,906*	*63,639,007*	*16,420,043*	*10,795,498*
		TOTAL	**104,724,860**		**TOTAL**	**97,703,792**		**TOTAL**	**90,854,548**
USED									
Cement/Concrete/Grout	11,213,022	707,170	443,170	11,411,600	368,548	54,981	11,126,774	270,505	59,505
Raw Feed for Clinker	937,470	147,408	27,946	1,739,699	531,138	275,609	2,744,170	447,936	381,057
Flowable Fill Material	729,107	6,749	0	412,785	0	0	123,938	18,440	0
Structural Fill/Embankments	2,911,617	1,052,572	172,330	3,811,067	1,856,594	0	4,986,747	2,216,439	0
Road Base etc.	931,516	553,257	35,474	695,976	1,335,640	0	447,684	1,032,468	0
Soil Modification/Stabilization	668,582	103,022	0	820,771	89,366	0	467,674	61,687	0
Mineral Filler in Ashphalt	96,651	7,423	1,181	93,597	87,287	0	47,725	0	0
Snow/Ice Control	0	774,212	0	2,400	696,223	0	1,749	620,112	0
Blasting Grit/Roofing Granules	0	36,368	0	56,213	124,697	0	0	38,650	0
Mining Applications	743,518	107,452	127,215	1,713,540	728,089	0	620,446	1,074,948	0
Wallboard	0	0	5,647,109	0	0	6,575,144	0	0	7,058,719
Waste Stabilization/Solidification	1,305,808	62,532	42,872	2,891,899	17,319	0	3,556,071	27,676	0
Agriculture	18,603	20,057	103,949	0	6,235	70,488	11,013	3,206	29,500
Aggregate	*na*	*na*	*na*	0	615,170	5,639	124,439	465,176	0
Miscellaneous	406,665	1,603,978	278,387	507767.6059	519,569	66,964	359381.2189	1204557.123	0
TOTAL	19,962,559	5,182,200	6,879,631	24,157,313	6,975,877	7,048,825	24,617,812	7,481,800	7,528,780
% USE	**32.3**	**30.4**	**26.6**	**34.8**	**38.8**	**68.2**	**38.7**	**45.6**	**69.7**

FGD residues are a rapidly changing group of CCPs; research is continually underway to increase scrubbing efficiency[49] resulting in a higher sulfur content of the final product. In common with other CCPs, the quality of the product also depends on the characteristics of the parent coal, the type of scrubbing system used (i.e., wet or dry[47]) and the handling and stabilization procedures. Stabilization usually takes the form of mixing the FGD with FA[20], and this often changes the re-use options of the stabilized material. Fly ash and additional quicklime are usually added to stabilize FGD filter cake prior to landfilling[50].

Production of Coal Combustion Byproducts

The American Coal Ash Association (ACAA) reports the production of 107 million metric tons (Mt) of CCPs in 2001[51]; an increase of 8.4% from the previous year. However, in 2003, CCP production was seven million Mt less than the previous year (Table 1)[2,52]. In February 2002, the U.S. Administration proposed significant changes to the New Source Review (NSR), and the Clean Air Act[53], known as the 'Clear Skies' initiative. The NSR is the section of the Clean Air Act which requires industrial organizations and factories to install modern pollution control devices (such as the flue-gas scrubbers which produce FGD) whenever they make changes in their activities or output that result in an increase in pollution. These changes have been perceived by environmental groups as a weakening of the NSR; widening pre-existing loopholes and allowing some facilities to bypass pollution control installment or upgrade, and in effect increase atmospheric pollution, rather than decrease it. These changes can be clearly seen in the figures reported by the ACAA in Table 1, which shows production and re-use of all classes (wet and dry) of FA, BA and FGD (million Mt)[2,51,52]. The production of FGD in particular has shown a dramatic fall; down by 60% in 2002 and 58% in 2003, despite the previous trend of increasing production concurrent with the total production of CCPs[48] and the increasing consumption of coal which has steadily increased from 961 million Mt in 2001 to 992 million Mt in 2003[54,55].

Potential Uses of Coal Combustion Products

The percentage of FA re-used in the United States has steadily increased from 7% in 1966 to 38.7% in 2003; an average increase of approximately 0.5 million Mt per year. On average the production of FA has increased approximately 4.5% per year between 1966 and the present, or approximately 1.3 million Mt a year[48]. Although re-use rates of FGD residues appear to have increased dramatically (Table 1), this is due to the changes in FGD production.

Fly Ash

Application of FA alone to agricultural land does not meet all crop requirements for essential nutrients such as N and P, but can enhance K, Ca, S, B, Mo levels, as well as other essential micronutrients such as Zn in the short term. Alkaline FA can be effective in neutralizing soil acidity[56], and when mixed with other coal refuse, has effectively controlled acid mine drainage[4]. The presence of potentially toxic metal(loids) in FA limit its potential use for land application[21]. Historically, the use of FA in agriculture has been based on its liming potential and supply of essential elements such as Ca, B, S, and Mo[57], although the metal(loid) enrichment has diverted FA research toward determining leaching potential and minimizing environmental risks of the materials. Many agricultural studies conclude that FA may only be of use in situations where the plants are tolerant to the salinity[58], metal contamination[59] and are nitrogen fixers[59].

The advantages and disadvantages of FA application to agricultural land are well documented, and includes metal(loid) enrichment and toxicity, plant nutrient imbalance such as P deficiency from soils treated with alkaline FA[60], and antagonistic interactions among elements due to of excessive Ca, K, and S[57,61,62]. Research into the agronomic use of CCPs continues, however, although many now focus on using CCPs in a specifically formulated mix with organic matter. For example, Schumann and Sumner[21] used nutrient availability data and linear programming to formulate mixtures of FA and biosolid (sewage sludge and animal manure) to successfully avoid FA-related issues such as B toxicity, excessive As levels and overliming, and derive environmentally safe FA formulations. In addition, Schlossberg *et al.*[63] adopted a similar technique of mixing FA with an organic waste product to successfully establish and manage bermudagrass sod production.

Successful, and perhaps more appropriate use of FA however, has been in the remediation of severely eroded lands[64,65], where FA is mixed with an organic waste material such as poultry litter, to supply nitrogen and phosphorus to plants. In terms of restoration, the combination of FA with other industrial by-products such as sewage sludge, can result in a high quality restoration material [22,66], and applied to eroding soils that require physical stabilization in addition to chemical improvement. Sajwan *et al.*[65] combined FA with sewage sludge (SS) and applied various mixture (ratios of SS:FA mixtures of 4:1, 4:2, 4:3, and 4:4) to *Sorghum vulgaris* var. Sudanese Hitche ("sorgrass") and found stimulating in biomass at rates of 50–100 tons acre-1 of all ratios of SS:FA mixtures.

Flue-Gas Desulfurization Residues

Flue-gas desulfurization residue is increasingly being used in the production of wallboard material, with over 7 million Mt re-used in 2003, amounting to 93% of the FGD re-used for that year. Due to their alkaline nature, FGD residues have potential value as neutralizing agents[50,67] for agricultural soils which suffer from excessive acidity[68], or for the alleviation of excessive sodicity [69,70]. In comparison with FA, considerably more FGD residue is used in agricultural applications; in 2001, 103,949 Mt of FGD was used in agricultural applications compared to 18,603 Mt FA. The use of FGD in agriculture, however, has steadily decreased during the 2001–2003 period; from 0.1 to 0.03 million Mt.

New research into the agronomic application of FGD has similarly applied the material in a mixture with an organic waste material, such as dairy, swine or broiler litter manures; Zhang *et al.*[71] found that co-application of FGD with organic waste reduced the availability of P, from the water-soluble to the bicarbonate extractable, which retained its availability for plant uptake, while reducing the likelihood of environmental losses through leaching. Stout *et al.*[72,73] also used FGD to reduce the bioavailability of P in high-P soils, reducing surface P runoff effectively for a period of several years. This reduction can be attributed to the amount of Ca supplied from both the FGD and organic amendment. High-P soils pose environmental problems from P run-off, which damages water quality by causing algal blooms through eutrophication.

Using an FGD residue without co-application of an organic waste product, Clark and Baligar[74,75] compared its effect on growth[75] and mineral composition[74] of plants grown on an acidic soils (pH 4) to that $CaCO_3$, $CaSO_4$ and $CaSO_3$. They reported growth enhancement of *Medicago sativa*, *Trifolium repens* and *Festuca arundinacea* as a result of FGD addition, and in particular when magnesium (Mg) was co-applied. In this study, only FGD materials high in B and lower in $CaSO_3$ were found to be detrimental to plant growth, and overall increased the growth responses of plants in an otherwise infertile acidic soil. Sakai *et al.*[76] used FGD residues to restore pH balance, without a detectable increase in the metal(loid) concentration (in comparison with FGD-free controls) of the plant material grown on amended soil. In this case, their product was a combination of wet and semi-dry desulfurization.

Conclusion

Although the beneficial re-use of CCPs such as FA and FGD have been impeded in the past by the presence of potentially toxic metal(loids), research is now moving ahead to better understand the distribution and chemical speciation of metal(loids) in parent coals, combustion products and environmental matrices using analytical techniques which have a high resolution, and capabilities for micron-scale spatial analysis. The future of CCP research into finding new and safer re-use applications depends on the information these techniques can provide. Agronomic use of these materials has been advanced a great deal in recent years by incorporation with organic wastes, and by using these more balanced mixtures to fertilize non-food source crops.

Acknowledgments. This work was supported by the Environmental Remediation Science Division of the Office of Biological and Environmental Research, U.S. DOE, through the financial assistance award number DE-FC09-96SR18546 to the University of Georgia Research Foundation.

References

1. Sajwan, K.S., Alva, A.K. and Keefer, R.F. eds. ed. Chemistry of Trace Elements in Fly Ash. 2003. Kluwer Academic/Plenum Publishers: New York. 346.
2. ACAA 2003. 2003 Coal Combustion Product (CCP) Production and Use Survey. http://www.acaa-usa.org/CCPSurveyShort.htm.
3. Choi, S.-K., Lee, S., Song, Y.-K. and Moon, H.-S. 2002. Leaching characteristics of selected Korean fly ashes and its implications for the groundwater composition near the ash disposal mound. *Fuel* 81: 1083–1090.
4. Stewart, B.R., Daniels, W.L. and Jackson, M.L. 1997. Evaluation of leachate quality from codisposed coal fly ash and coal refuse. *J. Environ Qual.* 26: 1417–1424.
5. Ugurlu, A. 2004. Leaching characteristics of fly ash. *Environmental Geology* 46(6–7): 890–895.
6. Pavlovic, P., Mitrovic, M. and Djurdjevic, L. 2004. An ecophysiological study of plants growing on the fly ash deposits from the "Nikola tesla-A" thermal power station in Serbia. *Environmental Management* 33(5): 654–663.
7. Furr, A.K., Parkinson, T.F., Gutenmann, W.H., Pakkala, I.S. and Lisk, D.I. 1978. Elemental content of Vegetables, Grains and Forages Field-Grown on Fly-Ash Amended Soil. *J. Agric. Food Chem.* 26(2): 357–359.
8. Elseewi, A.A. and Page, A.L. 1984. Molybdenum enrichment of plants grown on fly-ash amended soils. *J. Environ Qual.* 13: 394–398.
9. Jackson, B.P. and Miller, W.P. 1998. Arsenic and selenium speciation in coal fly ash extracts by ion chromatography-inductively coupled plasma mass spectrometry. *Journal of Analytical Atomic Spectrometry* 13(10): 1107–1112.
10. Jackson, B.P. and Miller, W.P. 1999. Soluble arsenic and selenium species in fly ash organic waste-amended soils using ion chromatography inductively coupled plasma mass spectrometry. *Environmental Science & Technology* 33(2): 270–275.
11. Shoji, T., Huggins, F.E., Huffman, G.P., Linak, W.P. and Miller, C.A. 2002. XAFS spectroscopy analysis of selected elements in fine particulate matter derived from coal combustion. *Energy & Fuels* 16(2): 325–329.
12. Huang, Y.J., Jin, B.S., Zhong, Z.P., Xiao, R., Tang, Z.Y. and Ren, H.F. 2004. Trace elements (Mn, Cr, Pb, Se, Zn, Cd and Hg) in emissions from a pulverized coal boiler. *Fuel Processing Technology* 86(1): 23–32.
13. Moscoso-Perez, C., Moreda-Pineiro, J., Lopez-Mahia, P., Muniategui-Lorenzo, S., Fernandez-Fernandez, E. and Prada-Rodriguez, D. 2004. As, Bi, Se(IV), and Te(IV) determination in acid extracts of raw materials and by-products from coal-fired power plants by hydride generation-atomic fluorescence spectrometry. *Atomic Spectroscopy* 25(5): 211–216.
14. Karangelos, D.J., Petropoulos, N.P., Anagnostakis, M.J., Hinis, E.P. and Simopoulos, S.E. 2004. Radiological characteristics and investigation of the radioactive equilibrium in the ashes

produced in lignite-fired power plants. *Journal of Environmental Radioactivity* 77(3): 233–246.
15. Jackson, B.P., Allen, P.L.S., Hopkins, W.A. and Bertsch, P.M. 2002. Trace element speciation in largemouth bass (*Micropterus salmoides*) from a fly ash settling basin by liquid chromatography-ICP-MS. *Analytical and Bioanalytical Chemistry* 374(2): 203–211.
16. Wallschlager, D. and Carlton, R.G. Selenium speciation and cycling in fly ash ponds. Proceedings of *6th International Conference on the biogeochemistry of Trace Elements*. 2001. Guelph, Canada.
17. Jackson, B.P., Hopkins, W.A., Unrine, J., Bainonno, J. and Punshon, T. Selenium speciation in amphibian larvae developing in a coal fly ash settling basin. Proceedings of *9th International Conference on Plasma Source Mass Spectrometry*. 2004. Durham, UK: University of Durham.
18. Hopkins, W.A., Staub, B.P., Snodgrass, J.W., DeBiase, A., Taylor, B., Roe, J.H. and Jackson, B.P. 2004. Responses of benthic fish exposed to contaminants in outdoor mesocosms: Examining the ecological relevance of previous laboratory toxicity tests. *Aquatic Toxicology* 68(1): 1–12.
19. Staub, B.P., Hopkins, W.A., Novak, J. and Congdon, J.D. 2004. Respiratory and reproductive characteristics of eastern mosquitofish (Gambusia holbrooki) inhabiting a coal ash settling basin. *Archives of Environmental Contamination and Toxicology* 46(1): 96–101.
20. Punshon, T., Adriano, D.C. and Weber, J.T. 2001. Effect of Flue Gas Desulfurization Residue on Plant Establishment and Soil and Leachate Quality. *Journal of Environmental Quality* 30(3): 1071–1080.
21. Schumann, A.W. and Sumner, M.E. 2004. Formulation of environmentally sound waste mixtures for land application. *Water Air and Soil Pollution* 152(1–4): 195–217.
22. Chang, A.C., Lund, L.J., Page, A.L. and Warneke, J.E. 1977. Physical properties of fly-ash amended soils. *Journal of Environmental Quality* 6: 267–270.
23. Billings, C.E. and Matson, W.R. 1972. Mercury Emissions from Coal Combustion. *Science* 176(4040): 1232–&.
24. Fthenakis, V.M., Lipfert, F.W., Moskowitz, P.D. and Saroff, L. 1995. An Assessment of Mercury Emissions and Health Risks from a Coal-Fired Power-Plant. *Journal of Hazardous Materials* 44(2–3): 267–283.
25. Huang, H.S., Wu, J.A.M. and Livengood, C.D. 1996. Development of dry control technology for emissions of mercury in flue gas. *Hazardous Waste & Hazardous Materials* 13(1): 107–119.
26. Goodwin, R.W. and Schuetzenduebel, W.G. Residues from mass burn systems: testing, disposal and Utilization issues. Proceedings of *New York State Legislative Commission of Solid Waste Management and Materials Policy Conference*. 1988. New York.
27. US-FDA and US-EPA, 2004. What you need to know about mercury in fish and shellfish. United States Food and Drug Administration and United States Environmental Protection Agency, Washington, DC., EPA-823-F-04-009 2.
28. Mason, R.P., Abbott, M.L., Bodaly, R.A., Bullock Jr, O.R., Driscoll, C.T., Evers, D., Lindberg, S.E., Murray, M. and Swain, E.B. 2004. Monitoring the response to changing mercury deposition. *Environmental Science and Technology* 39(1): 15A–22A.
29. Page, A.L., Elseewi, A.A. and Straughan, I. 1979. Physical and chemical properties of fly ash from coal-fired power plants with references to environmental impacts. *Residue Rev.* 71: 83–120.
30. El-Mogazi, D.D., Lisk, D.J. and Weinstein, L.H. 1988. A review of physical, chemical and biological properties of fly ash and effects on environmental ecosystems. *Science of the Total Environment* 74: 1–37.
31. Peng, M., Ruan, X.G., Chen, X.M., Xu, J.W. and Jiang, Z.C. 2004. Study on both shape and chemical composition at the surface of fly ash by scanning electron microscope, focused ion beam, and field emission-scanning electron microscope. *Chinese Journal of Analytical Chemistry* 32(9): 1196–1198.
32. Fisher, G.L. and Natusch, D.F.S. 1979. Size dependence of the physical and chemical properties of coal fly ash. in *Analytical methods for coal and coal products.*, C. Karr Jr., Editor. Academic Press: New York. 489–541.
33. Mattigod, S.V., Rai, D., Eary, L.E. and Ainsworth, C.C. 1990. Geochemical Factors Controlling the Mobilization of Inorganic Constituents from Fossil Fuel Combustion Residues: I. Review of the major elements. *J. Environ Qual.* 19: 188–201.
34. US-EPA, 1988. Wastes from the combustion of coal by electricity power plants. U.S. Environmental Protection Agency, Washington DC.
35. Manz, O.E. 1999. Coal fly ash: a retrospective and future look. *Fuel* 78: 133–136.
36. Struis, R.P.W.J., Ludwig, C., Lutz, H. and Scheidegger, A.M. 2004. Speciation of zinc in municipal solid waste incineration fly ash after heat treatment: An X-ray absorption spectroscopy study. *Environmental Science & Technology* 38(13): 3760–3767.
37. Pires, M. and Querol, X. 2004. Characterization of Candiota (South Brazil) coal and combustion by-product. *International Journal of Coal Geology* 60(1): 57–72.
38. Camerani, M.C., Golosio, B., Somogyi, A., Simionovici, A.S., Steenari, B.M. and Panas, I. 2004. X-ray fluorescence tomography of individual municipal solid waste and biomass fly ash particles. *Analytical Chemistry* 76(6): 1586–1595.
39. Huggins, F.E., Huffman, G.P., Linak, W.P. and Miller, C.A. 2004. Quantifying hazardous species in particulate matter derived from fossil-fuel combustion. *Environmental Science & Technology* 38(6): 1836–1842.
40. Galbreath, K.C. and Zygarlicke, C.J. 2004. Formation and chemical speciation of arsenic-, chromium-, and nickel-bearing coal combustion PM2.5. *Fuel Processing Technology* 85(6-7): 701–726.
41. Bluedorn II, D.C. Recent environmental regulation of coal combustion wastes—revised. Proceedings of *2001 Conference on Unburned Carbon (UBC) on Utility Fly Ash.* 2001: National Energy Technology Laboratory.
42. Keefer, R.F. 1993. Coal ashes—Industrial wastes or beneficial byproducts? in *Trace Elements in Coal and Coal Combustion Residues.*, R.F. Keefer and K.S. Sajwan, Editors. Lewis Publishers: Ann Arbor. 3–9.
43. Kula, I., Olgun, A., Sevine, V. and Erdogan, Y. 2002. An investigation on the use of tincal ore waste, fly ash and coal bottom ash as Portland cement replacement materials. *Cement and Concrete Research* 32: 227–232.
44. Bethanis, S., Cheeseman, C.R. and Sollars, C.J. 2004. Effect of sintering temperature on the properties and leaching of

incinerator bottom ash. *Waste Management & Research* 22(4): 255–264.
45. Kalyoncu, R. 1998.Coal Combustion Products. http://minerals.usgs.gov/minerals/pubs/commodity/coal/874498.pdf,
46. Punshon, T., Knox, A.S., Adriano, D.C., Seaman, J.C. and Weber, T.J. 1999. Flue Gas Desulfurization residue (FGD): Potential Applications and Environmental Issues. in *Biochemistry of Trace Elements in Coal and Coal Combustion Byproducts.*, K.S. Sajwan and R.F. Keefer, Editors. Lewis Publishers: Boca Raton, FL. 7–28.
47. Punshon, T., Seaman, J.C. and Adriano, D.C. 2002. The effect of flue gas desulfurization residue on corn (*Zea mays* L.) growth and leachate salinity: Multiple season data from amended mesocosms. in *Chemistry of Trace Elements in Fly Ash*, K.S. Sajwan, A.K. Alva, and R.F. Keefer, Editors. Kluwer Academic/Plenum Press: New York, NY. In press.
48. Punshon, T., Seaman, J.C. and Sajwan, K.S. 2003. The production and use of coal combustion products. in *Chemistry of Trace Elements in Fly Ash*, K.S. Sajwan, A.K. Alva, and R.F. Keefer, Editors. KluwerAcademic/Plenum Publishers: New York. 1–11.
49. Baege, R. and Sauer, H. 2000. Recent developments in CFB-FGD technology. *VGB Powertech* 80(2): 57–60.
50. Chen, L., Dick, W.A. and Nelson, S. 2001. Flue gas desulfurization addition to acid soil: alfalfa productivity and environmental quality. *Environmental Pollution* 114(2): 161–168.
51. ACAA 2001. 2001 Coal Combustion Product (CCP) Production and Use (Short Tons). http://www.acaa-usa.org/CCP SurveyShort.htm,
52. ACAA 2002. 2002 Coal Combustion Product (CCP) Production and Use Survey. http://www.acaa-usa.org/CCPSurvey Short.htm,
53. Cooney, C.M. 2002. 'Clear skies' may be ahead for electric power plants. *Environmental Science and Technology* 136(9): 181A–182A.
54. EIA, 2001. Annual Coal Report. Energy Information Agency, Washington, DC, DOE/EIA-0584 (2001) 73.
55. EIA, 2003. Annual Coal Report. Energy Information Administration, Washington, DC, DOE/EIA-0584 (2003) 78.
56. Matsi, T. and Keramidas, V.Z. 1999. Fly ash application on two acid soils and its effect on soil salinity, pH, B, P and on ryegrass growth and composition. *Environmental Pollution.* 104: 107–112.
57. Adriano, D.C., Page, A.L., Elseewi, A.A., Chang, A.C. and Straughan, I. 1980. Utilization and disposal of fly ash and other coal residues in terrestrial ecosystems. A review. *Journal of Environmental Quality* 9: 333–334.
58. Adriano, D.C., Webber, J., Bolan, N.S., Paramasivam, S., Koo, B.-J. and Sajwan, K.S. 2002. Effects of high rates of coal fly ash on soil, turfgrass and groundwater quality. *Water Air and Soil Pollution* 139: 365–385.
59. Gupta, D.K., Rai, U.N., Tripathi, R.D. and Inouhe, M. 2002. Impacts of fly-ash on soil and plant responses. *Journal of Plant Research* 115(1122): 401–409.
60. Kukier, U. and Sumner, M.E. 1996. Boron availability to plants from coal combustion by-products. *Water, Air & Soil Pollution* 87(1–4): 93–110.
61. Carlson, C.L. and Adriano, D.C. 1993. Environmental impacts of coal combustion residues. *Journal of Environmental Quality* 22(2): 227–247.
62. Gary, C.A. and Schwab, P. 1993. Phosphorus fixing ability of high pH, high calcium, coal combustion waste material. *Water Air and Soil Pollution* 69: 309–320.
63. Schlossberg, M.J., Vanags, C.P. and Miller, W.P. 2004. Bermudagrass sod growth and metal uptake in coal combustion by-product-amended media. *Journal of Environmental Quality* 33(2): 740–748.
64. Punshon, T., Adriano, D.C. and Weber, J.T., 1999. Restoration of Eroded Land Using Coal Fly Ash and Biosolids. Electrical Power Research Institute, Pal Alto, CA., TR-113940.
65. Sajwan, K.S., Ornes, W.H. and Youngblood, T. 1995. The effect of fly ash/sewage sludge mixtures and application rates on biomass production. *Journal of Environmental Science and Health* A30(6): 1327–1337.
66. Sloan, J.J. and Cawthorn, D. 2003. Mine soil remediation using coal ash and compost mixtures. in *Chemistry of Trace Elements in Fly Ash*, K.S. Sajwan, A.K. Alva, and R.F. Keefer, Editors. Kluwer Academic/Plenum Publishers: New York. 309–318.
67. Schlossberg, M.J., Sumner, M., Miller, W.P. and Dudka, S. Utilization of coal combustion by-products (CBP) in horticulture and turfgrass industries: technical and environmental feasibility studies. Proceedings of *6th International Conference on the Biogeochemistry of Trace Elements.* 2001. Guelph, Canada.
68. Stehouwer, R.C., Sutton, P. and Dick, W.A. 1996. Transport and plant uptake of soil-applied dry flue gas desulfurization by products. *Soil Science* 161(9): 562–574.
69. Clark, R.B., Ritchey, K.D. and Baligar, V.C. 2001. Benefits and constraints for use of FGD products on agricultural land. *Fuel* 80(6): 821–828.
70. Chun, S., Nishiyama, M. and Matsumoto, S. 2001. Sodic soils reclaimed with by-product from flue-gas desulfurization: corn production and soil quality. *Environmental Pollution* 114(3): 453–459.
71. Zhang, G.Y., Dou, Z., Toth, J.D. and Ferguson, J. 2004. Use of flyash as environmental and agronomic amendments. *Environmental Geochemistry and Health* 26(2): 129–134.
72. Stout, W.L., Sharpley, A.N. and Pionke, H.B. 1998. Reducing soil phosphorus solubility with coal combustion by products. *J. Environ Qual.* 27: 111–118.
73. Stout, W.L., Sharpley, A.N. and Weaver, S.R. 2003. Effect of amending high phosphorus soils with flue-gas desulfurization gypsum on plant uptake and soil fractions of phosphorus. *Nutrient Cycling in Agroecosystems* 67(1): 21–29.
74. Clark, R.B. and Baligar, V.C. 2003. Mineral concentrations of forage legumes and grasses grown in acidic soil amended with flue gas desulfurization products. *Communications in Soil Science and Plant Analysis* 34(11–12): 1681–1707.
75. Clark, R.B. and Baligar, V.C. 2003. Growth of forage legumes and grasses in acidic soil amended with flue gas desulfurization products. *Communications in Soil Science and Plant Analysis* 34(1–2): 157–180.
76. Sakai, Y., Matsumoto, S. and Sadakata, M. 2004. Alkali soil reclamation with flue gas desulfurization gypsum in China and assessment of metal content in corn grains. *Soil & Sediment Contamination* 13(1): 65–80.

II
Environmental Impact of Coal Combustion Residues

2
Coal and Coal Combustion Products: Prospects for Future and Environmental Issues

Irena Twardowska and Sebastian Stefaniak
Polish Academy of Sciences, Institute of Environmental Engineering, 34, M. Sklodowska-Curie St., 41-819 Zabrze, Poland

Abstract

At the background of the overview of prospects for coal-based electricity production, the present and projected status and trends of coal combustion products (CCP) management was discussed. According to the latest projections, coal-based electricity share will remain the largest in world's energy balance and is set to double in the first three decades of 21st century. The majority of this growth is forecasted for developing Asian economies, but also for some developed countries not bound by the Kyoto Protocol (USA, Australia). Coal combustion products (CCP) are one of the most abundant high-volume waste materials worldwide that are partially being reused in a number of commercially and environmentally proven applications—both traditional and advancing, some are in the stage of extensive studies. Despite of numerous beneficial properties, their reuse rates are still far from being satisfactory; also statistics on CCP generation is fragmentary. There is also different approach to CCP in terms of its legal definition: ACAA (American Coal Ash Association) considers CCP as a "product" and a mineral commodity equivalent to natural materials; also some other countries—large coal producers and CCP generators adopted this approach (e.g. India). European legislation treats CCP as a waste, though it does not mean that the material is a waste in all circumstances, but only where the definition of waste is met. In view of anticipated CCP generation and growth in the future, a crucial task is to identify and remove obstacles and barriers in the way of increasing CCP utilization, along with developing new application fields. Based on the analysis of regulatory instruments and enforcement procedures compared to the utilization effects, it has been stated that the pro-environmental and pro-recovery CCP recycling policy must be based on the term "waste" and the "polluter pays" principle. The rationale of this statement is that the legal definition in no case should absolve the producer or the holder from the responsibility for the generated waste until it is transformed into an environmentally safe product. The exemplified practice confirms that the efficient regulatory and enforcement mechanisms should comprise a well-balanced system of precepts, prohibitions and charges for CCP disposal (fees, penalties) that would encourage power plants as waste generators to support financially the environmentally safe utilization of CCP by the waste reuse industry on a cost-benefit basis, in order to reduce charges for the disposal and to assure competitiveness of these products in the market.

1. Introduction

At the beginning of the third Millennium coal remains the primary fuel used in generation electricity worldwide—in 2002 coal generated 39.0% of the world's electricity. Total global hard coal production in 2003 has been estimated for 4037.5 Mt and increased 3.3% over the previous year (in 2002 it accounted for 3909.9 Mt). In the past 25 years it showed over 46% growth.

Coal domination in electricity generation is based on coal abundant and widely dispersed resources compared with oil and gas that is also a matter of energy security and minimizes the risk of energy supply disruption. Coal deposits occur in about 70 countries. At present extraction level, the proven coal reserves are estimated to last over 190 years. In contrast, oil and gas reserves are estimated to last for 41 and 67 years, respectively; besides, over 69% of oil and 67% of gas reserves are concentrated in the Middle East and Russia, which makes these resources highly insecure and endangered by supply disruption[1]. This strongly justify use coal incineration in power plants for electricity generation. This process causes formation of large amounts of coal combustion residues that are environmentally problematic, but at the same time display a number of beneficial properties, that might make these materials recyclable in a wide number of efficient and cost-effective applications. Up to now, though, not all the beneficial properties of these materials are well recognized and adequately utilized. At the same time, it should be taken into consideration that these materials are not environmentally safe and might adversely affect the ambient environment, if improperly handled. This chapter is focused on discussion of beneficial and adverse properties of fly ash (FA) that is the most abundant coal incineration residue in a context of potential applications, as well as on regulatory aspects that might optimize its management.

2. Coal as A Source of Electricity

The countries that are major producers are also major consumers of hard coal for electricity generation (Table 1). A number of countries use imported coal as a significant component in a balanced energy mix (Japan, Republic of Korea, Chinese Taipei, Germany and UK). The major steam coal exporters are Australia, PR China, Indonesia, S. Africa, and in lesser amounts (<50 Mt) Russia and Colombia; USA and Poland exported below 20 Mt.

Data presented in Table 1 reflect both temporary fluctuations in coal generation, but also permanent trends connected with growth of electrification rate, particularly strong in Asian/Pacific region. A dynamic increase of coal generation is observed in China, Indonesia and India; The EU and OECD countries show stagnation or even decreasing trend, which greatly depends on the implementation of the Kyoto Protocol to the United Nations Framework Convention on Climate Change (UNFCC) that sets mandatory targets on greenhouse gas (GHG) emissions for the parties that ratified the Protocol. The Protocol was first negotiated at COP3 in Japan in 1997 and will finally enter into force on 16 February 2005, when it will be legally binding on its 128 parties, 30 of whom are industrialized EU and OECD countries with emissions targets. Refusal to ratify Protocol by the USA that is the world's largest emitter and Australia, as well as the absence of targets for developing countries reduces Kyoto Protocol implementation to countries responsible for only 32% of global emissions[3]. It should be though noted that the US declared the alternative "voluntary targets" program that rely on recommendations outlined by the Council of Economic Advisors.[4]

According to the latest edition of the IEA—International Energy Agency's World Energy Outlook 2004[5,6], coal-based electricity share will remain the largest in world's energy balance. Between 2002 and 2030, coal demand is projected to grow annually by 1.4% and by 2030 will reach the amount of over 7 billion tons that is almost 50% higher than at present. In 2030 coal will meet 22% of energy needs, similarly to the present level (23%). Asian countries will show continuous increase in demand for coal; China and India are estimated to participate for 68% in this growth.

Coal-based electricity is set to double in the first three decades of 21st century, from 16,074 TWh in 2002 to 31,657 TWh in 2030 and will continue to play a key role in world electricity generation. Despite high nominal growth of coal-based electricity generation, its share in total electricity generation (in %) will remain in 2003 close to the current level showing only slight decrease from 39 to 38%. The strongest increase of coal-based electricity gross generation is projected in developing countries, in particular in China, where a quarter of the total growth will occur[5,6]. By 2030 developing countries will reach almost half of total energy demand. Though the majority of this growth is forecasted for developing Asian economies (China, ASEAN, India), also in some developed countries not bound by the Kyoto Protocol (USA, Australia), increasing competition will favour low cost coal-based electricity generation[7]. The EU and OECD countries with emission targets that ratified Kyoto protocol are tending to reduce share of this kind of energy, mainly by increasing gas use and developing alternative renewable electricity sources (Table 2). Nevertheless, having in mind almost double increase of electricity generation, similar or even reduced share of coal as a source of electricity, means significant nominal increase of incinerated coal.

In the cited outlook, IEA for the first time has used an Energy Development Index that is a composite measure reflecting commercial energy consumption per capita, share of commercial energy in total energy use and share of population with

TABLE 1. Coal production and major producers in 2001 (after WCI—World Coal Institute, 2004[1,2]).

	Production					
	Mt		% of Total		Coal-based electricity % of total	
Hard coal	2002	2003e	2002	2003e	2002p	2003p
World	**3837.0***	**4037.5**	**100**	**100**	**38.7**	**39.0**
China	1326.0	1502.4	34.5	37.2	76.2 (2001 data)	77.5
USA	916.7	891.9	23.9	22.1	49.9	52.2
India	333.7	340.4	8.7	8.4	78.3 (2001 data)	70.1
Australia	276.0	274.1	7.2	6.8	76.9	76.9
S. Africa	223.0	239.3	5.8	5.9	93.0	92.2 (2002 data)
Russia	163.6	188.4	3.5	4.7		
Indonesia	101.2	120.1	2.6	3.0		39.7
Poland	102.6	100.4	2.7	2.5	94.8	94.7
Kazakhstan	70.6	74.8	1.8	1.9		69.9 (2002 data)
Ukraine	82.9	56.8	2.2	1.4		

*Data from 2003 edition of Coal Facts; after 2004 edition the world coal production in 2002 was 3909.9 Mt; e—estimated; p—preliminary;

TABLE 2. Projections of world's electricity generation trends over the period of 2002–2030[6,7]

	Share in electricity generation (%)					
	OECD		Transition economics		Developing countries	
Electricity source	2002	2030	2002	2030	2002	2030
Coal	**38**	**33**	**22**	**16**	**45**	**47**
Oil	6	2	4	2	12	5
Gas	18	29	37	54	17	26
Nuclear	23	15	18	11	2	3
Hydro	13	11	19	15	23	16
Other renewables*	3	10	0	2	1	3

*Solar, wind, biomass, waste incineration etc

access to electricity. According to IEA projection, electrification rates in developed countries will increase from 66% in 2002 to 78% in 2030; the total number of people without electricity will account for 1.4 billions in 2030, while electrification rates in developing countries vary significantly—from over 98% in China to an average 23% in Sub-Saharan Africa, where there is a number of countries with electrification rate <5% (Figure 1).

This shows that in the future (after 2030) further increase of coal use as a source of energy can be anticipated unless novel energy sources are developed by that time.

3. Coal Combustion Products (CCP) Generation, Use and Disposal

Coal combustion products (CCP) are one of the most abundant high-volume waste materials worldwide. Their proportion in the total waste stream highly depends upon the role of coal in power production, and is as a rule the highest in coal producing countries. Despite the omnipresence of CCP, the statistical data concerning its generation and managing in the different countries of the world are fragmentary[9].

American Coal Ash Association (ACAA) regularly publishes in its web site CCP generation and use annual surveys based on a representative sampling of several hundreds of coal-fueled power plants in the U.S.[10,11,12] (e.g. 2002 data are estimates based on a sampling survey of nearly 600 utilities)[10]. These data show distinct increase of CCP generation and use in the last eight years covered by ACAA reports (1996–2003) (Table 3). The last available data for 2003 estimate total CCB amount for 110.44 Mt. that means an extrapolated increase of about 19.5% compared to 1996. It should be noted that total CCP generation can vary distinctly from tear to year depending on the amount and ash content of coal burned, though the general trend is distinctly increasing since 1996[10,11,12]. Its category structure in the last reported years 1996–2003 remains practically stable and consists predominantly of fly ash (ca. 58%), bottom ash (ca.15%), boiler slag (1.5–2.5%), FGD solids (23–25%) and FBC Ash (<1%, 0.7% in 2003), fly as being invariably the largest by mass component of CCP. Though in 1966–1996, along with growth of CCP generation, also its use was growing, an overall CCP utilization for this period remained at almost unchanged level at around 25%. An observed significant and regular upward trend of overall CCP utilization rate at 13.4% in 1996–2003 in parallel with an increase of CCP generation evidences an unquestionable success of utilization industry in the field of coal combustion products usage.

The growth of CCP use under the conditions of a slow economy and a relatively flat construction industry ACAA tentatively explains by possible local material/mineral

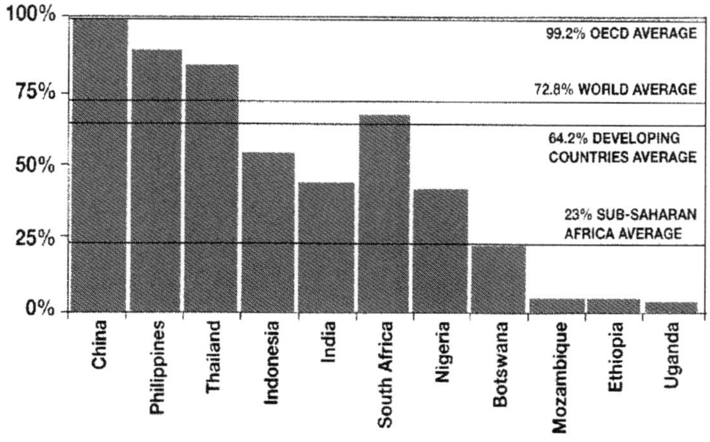

FIGURE 1. Electrification rates for selected developing world countries (after World Energy Outlook, IEA)[8].

TABLE 3. Coal Combustion Product (CCP) generation and use in the USA in 1996–2003 (ACAA survey)[10,11,12] (Mt—million metric tons, %)

Year	Generation/Use	FA	BA	BS	FGDR	FBCA	CCP total
	Generation Category Totals, Mt	53.88	14.58	2.33	21.65	nd	92.45
1996	Used Category Totals, Mt	14.74	4.42	2.18	1.50		22.84
	Category Use % of Total Generated	**24.7**	**30.3**	**93.3**	**6.9**		**24.7**
	Generation Category Totals, Mt	57.14	15.38	2.43	23.28	nd	98.22
2000	Used Category totals, Mt	18.407	4.483	2.108	4.380		29.19
	Category Use % of Total Generated	**31.9**	**29.2**	**86.5**	**18.8**		**29.7**
	Generation Category Totals, Mt	61.80	17.04	2.30	24.93	nd	106.98
2001	Used Category Totals, Mt	19.96	5.18	1.65	6.88		33.67
	Category Use % of Total Generated	**32.3**	**30.4**	**71.7**	**26.6**		**31.5**
	Generation Category Totals, Mt	69.40	17.96	1.74	26.52	1.13	116.76
2002	Used Category Totals, Mt	24.16	6.97	1.41	7.89	0.86	41.30
	Category Use % of Total Generated	**34.81**	**38.84**	**80.75**	**29.76**	**76.36**	**35.37**
	Generation Category Totals, Mt	63.64	16.42	1.66	28.00	0.72	110.44
2003	Used Category Totals, Mt	24.62	7.48	1.59	8.30	0.24	42.08
	Category Use % of Total Generated	**38.68**	**45.57**	**95.63**	**29.64**	**33.09**	**38.10**

FA—Fly Ash; BA—Bottom Ash; BS—Boiler Slag; FGDR—Flue Gas Desulfurization Residues, comprises FGD Gypsum (39.0%), FGD Material Wet Scrubbers (57.8%), FGD Material Dry Scrubbers (3.2%) and other FGD residues (0%)—2003 data[10]; FBCA—Fluidized Bed Combustion Ash.

shortages, improved fly ash quality due to beneficiation technologies, larger percentages of fly ash (FA) used in mix-designs, increased usage of fly ash in soil projects and waste stabilization or a combination of these factors.

Comparison and analysis of data from sufficiently remote year with the current status (Table 4) shows that in CCP management the most substantial role in growth of CCP utilization rate plays use of FA and BA as a substitute for sand and gravel in structural fills and embankment construction that cause also adequate increase of overall CCP use in these applications. Another dynamically developing field of CCP use is synthetic FGD gypsum for wallboard production. This trend is expected to continue in parallel with the developing new flue gas scrubbing systems in power plants. Distinct growth that has a visible effect upon the overall CCP usage has been observed in mining applications of FA that are particularly strong in such categories as bottom ash BA, and FGD residues. Mining applications of FA remain in mass units practically at the same level that results in the decrease of a percentage of FA use in 2003 compared to 1996.

TABLE 4. Changes in CCP utilization in 2003 compared to 1996 (based on ACAA data[10])

	FA		BA		FGDR		CCP total	
CCP generation	1996	2003	1996	2003	1996	2003	1996	2003
Generation Category Totals, Mt	53.88	63.64	14.58	16.42	21.65	28.0	92.45	110.44
% of Total CCP Generated	58.3	57.6	15.8	14.9	23.4	25.3	100.0	100.0
CCP Use, metric tons, Mt	14.74	24.62	4.42	7.48	1.50	8.30	22.84	42.08
CCP Use, % Total Used	64.5	58.5	19.3	17.8	**6.6**	**19.7**	100.0	100.0
CCP Use, % Category Used	100.0	100.0	100.0	100.0	100.0	100.0	100.0	100.0
Concrete/Concrete Prod./Grou	49.39	45.20	15.61	3.62	0.04	1.11	35.20	27.33
Cement/Raw Feed for Clinker		11.15		5.99		4.88		8.56
Flowable Fill	1.90	0.50	0.90	0.25	0.67	0.10	1.44	0.36
Structural fills/Embankments	**13.36**	**20.26**	**13.80**	**29.62**	3.33	2.63	**11.69**	**17.65**
Road Base/Sub-base/Pavement	4.61	1.82	14.93	13.80	7.33	0.000	6.30	3.58
Soil Modification/Stabilization		1.90		0.82		0.009		1.67
Mineral Filler in Asphalt	1.02	0.19	0.68	0.000	0.000	0.000	0.96	0.18
Snow and Ice Control	0.000	0.007	13.80	8.29	0.000	0.000	3.11	1.70
Blasting Grit/Roofing Granules	0.000	0.000	3.39	0.52	0.000	0.000	9.32	3.23
Mining Applications	4.68	2.52	**1.36**	**14.37**	**0.02**	**4.35**	**3.37**	**5.15**
Wallboard	0.003	0.000	0.000	0.000	**52.67**	**86.64**	**3.55**	**16.77**
Waste Stabilizat./Solidification	**0.53**	**14.44**	5.20	0.37	3.33	0.000	8.93	8.62
Agriculture	**0.003**	**0.045**	2.26	0.043	**0.000**	**0.39**	**0.09**	**0.11**
Aggregate	0.58	0.50	30.32	6.22	26.67	0.07	16.02	1.48
Miscellaneous/Other		1.45		16.10		0.000		3.75

The aforementioned three categories of extensively developing fields of application result also in the overall growth trend of CCP use that exceeds the growth rate of FA generation. An increase of FGDR and FA use in agriculture also occurs, though this application is marginal and does not influence significantly the FA utilization rate. In view of observed adverse transformations of pure FA in time due to weathering processes restricts uncontrollable high-rate use of FA as acidic soil improver/amendment[13,14,15] or acidic mining waste reclamation[16]. FGD residues appear to be a more promising material for soil amendment. An overview of studies on CCP application to nutrient deficient, disturbed, or eroded soils showed that careful use of CCP, also in co-application with biosolids may distinctly improve soil quality[11,17]. However, the authors stress the need of determination safe rates prior to use, in order to avoid potential health risks from enrichment of various potentially toxic trace elements in amended soils. Considering time-delayed adverse environmental and health effect of both CCP and biosolids, the application of these products to the soil, also in the field of land reclamation, remediation and stabilization should be preceded by the reliable long-term environmental impact assessment that takes into consideration CCP properties and long-term environmental behavior in the prospective area of application.

Concrete/cement and construction materials production remains the major field of CCP use; the CCP rate used for this activity has been practically unchanged, which means that the use of CCP for concrete/cement production grows in parallel with the increase of CCP generation.

High increase of FA use for waste stabilization/solidification has been substantially reduced by the declining of this application to bottom ash and FGD residues; the overall application rate has thus remained at the practically unchanged level.

A number of traditional; fields of CCP usage show distinct declining in time; this downward trend show use CCP for flowable fill, road base/sub-base, as mineral filler in asphalt, for snow and ice control and for production of blasting grit/roofing granules. These areas are though of a minor importance to the overall CCP use balance and trends in the USA, which resulted in a recent growth of CCP utilization in a multitude of commercially and environmentally friendly applications (Table 4). Still, though over 38% of CCP in 2003 was used, major part at about 62% was disposed in the landfills thus contributing to the huge and continuously growing amount of CCP laying at disposal sites all over the country (It should be noted that the utilization rate of dry, moisture conditioned and compacted CCP is considerably higher than that of the ponded CCP, while in the majority of developing countries the predominant way of CCP disposal is ponding). In several prospective CCP applications, this material can successfully compete with natural mineral commodities or even display better properties. Nevertheless the CCP use is still inadequate worldwide.

In India, around 75 Mt of CCP was estimated to be generated annually in 1996–97, and a further growth up to 290 Mt in 2011–2012 has been anticipated[18]. According to incomplete data, 15 EC countries including new member and associated states but without Spain, France, Italy, Luxembourg, Austria, Sweden, Island, Norway, Switzerland, Cyprus, and Baltic States generated a total of 96.404 Mt of waste from all the sources of power generation, and of this the contributions of Germany (26%), Poland (17%) and the UK (13%) amounted to 56% of the total[9]. In Poland, 17.3 Mt of CCP was generated in 2002[19].

The beneficial properties of CCP give rise of their using in a wide array of commercially and technically proven applications. The traditional leading markets for CCP use has been concrete, cement and concrete production, structural fills and embankments, waste stabilization/solidification, blasting grit/roofing granules as well as road base and sub-base. Other markets for CCP as engineering materials are advancing, with a goal of full use of these materials in a technically sound, commercially effective and environmentally safe way. Among the emerging markets, the most recently developing ones are wallboard production from FGD gypsum, as a sealing material for fire control and mining applications. A prospective field of FA application at the surface is use for sealing sulfidic mine waste dumps to prevent ARD generation and high-metal leachate. Another very promising but not yet fully utilized area of using excellent sealing properties of FA is fire interception in coalfields and coal mining waste dumping sites. Dense FA: water mixtures appeared to be also an effective sink of high radioactivity loads from waters of the Ra-Ba type to contents below MCL for drinking water. An attractive field of high-volume FA utilization is proven bulk use in deep mines in the form of dense mixtures with mine water for methane, fire and subsidence control, for backfilling, reduction of saline mine waters discharge at the surface, reducing greenhouse effect and. for sealing mined out and abandoned workings[20]. A brief review of other actual fields and research on the beneficial reuse and recycling of CCP shows its great potential and viability.

Nevertheless, the utilization of CCP is still far from achieving target of its full use. The major limiting factors include: inadequate legislation, discrepancy between the CCP generation rates and demand for the end product and availability of competing materials for lower costs. As a result, a large portion of CCP stream is still being disposed of in surface ponds or landfills. The rate of CCB use and disposal is a resultant of the joint effect of these factors in different countries.

With 67.7% of world's coal output currently devolved to China, the USA and India, and heavy dependence of these countries on coal for power generation, a low rate of CCP use (33% (estimate), 38% and 2–5%, respectively) reflects the scale of its disposal[2]. In Poland, which holds eighth place among the largest worlds' hard coal producers, and first place in Europe, 76.1% of CCP was utilized and 23.9% disposed of in 2002; this rate of CCP reuse is the highest among the countries—large coal producers. The total amount of CCP stored in the disposal sites was evaluated for 247.9 Mt[19]. Huge amounts of CCP stored in disposal sites worldwide are growing continuously with the intensity depending upon its utilization rate. Only few European countries—small CCB generators—utilize it in 90–100%. Anticipated the highest world's energy growth based on coal in Asia in the first decades of the Millennium indicates that this is the most critical area with respect to CCB management and its environmental consequences in the near future.

These data show that despite all the efforts of reuse proponents to change the way CCP are classified in the legislative arena and to consider ash, slag and FGD residues entirely as beneficial raw materials and produced mineral commodities[12], disposed CCP (mainly fly ash FA) must be regarded as waste. The occurrence of highly leachable macro-components, high hydraulic conductivity, about tenfold higher concentration of trace elements compared to coal/lithosphere and observed their mobilization as a result of acidification of the most abundant low-alkaline fly ash in a long-range time due to devitrification of glass phase, generation of secondary minerals and regeneration of primary minerals from their amorphous relics are the major factors that lead to the conclusion that the potential risk to the ground water from disposed FA should be evaluated as high[13,14]. The nature and extent of the risk varies at the different stages of weathering. This also shows that uncontrolled use of FA in agriculture as soil amendment, particularly attractive to developing countries as a low-tech sink of bulk amounts of CCP, should be treated with a great caution due to possibility of soil contamination with mobile potentially hazardous trace elements. Besides, due to frequently elevated radioactivity, FA cannot be used unrestrictedly for a production of construction materials and should be tested for the level of α- and γ-radiation before use[21].

4. Regulatory Instruments

To facilitate CCP disposal and utilization in the environmentally safe way and to prioritize its use, countries—high CCP generators need a regulatory framework and adequate enforcement procedures that would make uncontrollable CCP disposal highly unprofitable. The everyday practice shows that efficient regulatory framework should be based on consideration CCP as a "waste", and not a "product" or "a mineral commodity". The pro-environmental and pro-recovery CCP recycling policy must be based on the term "waste" and the "polluter pays" principle. The rationale of this approach is that waste, which is technically suitable for use does not automatically become a raw material if there is no market for it, or its use is not commercially effective and, hence, they should be disposed off. The majority of disposed wastes, including CCP, are not geochemically stable and environmentally safe.

CCP despite of all its beneficial properties is not produced in purpose and in required amounts, but much in excess; it is also not a neutral material and may pose a hazard to the environment in the place of its disposal due to high contaminants contents and their leachability, as well as due to aforementioned time-delayed adverse transformation of its properties. CCP falls perfectly within the definition of waste in accordance with Council Directive 75/442/EEC on waste[22], which states that *"waste" is any substance or object..., which the holder discards or intends or is required to discard"*. CCP categories are listed also in the EC single list of wastes[23] under code 10 "Waste from thermal processes",

10 01: Waste from power stations and other combustion plants, 10 01 01—10 01 99.
10 01 01—Bottom ash, slag and boiler dust;
10 01 02—Coal fly ash;
10 01 05—Calcium-based reaction wastes from flue-gas desulfurization in solid form;
10 01 07—as above, in sludge form;

The EC list of wastes contains an important statement that *"The inclusion of a material in the list does not mean that the material is a waste in all circumstances. Materials are considered to be waste only* ***where the definition of waste is met****"*.

In practice, this means that a power plant must pay charges for every ton of discarded CCP that are particularly high for recyclable waste. This coerces power plants as recyclable waste holders into seeking cooperation with adequate waste management companies to viable reuse their CCP in the adequate beneficial application in the environmentally safe way that is a prerequisite of every reuse application. To make an application profitable for such waste management enterprise and competitive with natural products in the market, the power plant subsidizes the company for the recycling its CCP that results in the development of CCP use and its successful competitiveness in the market. At the same time, the power plant benefits from the difference between a high charge for disposal of waste and a lower subsidy for CCP reuse. This simple mechanism based on cost-benefit calculations appears to be very efficient with respect to CCP use giving to Poland the leading position in CCP utilization among the countries—large CCP generators (Figures 2, 3).

The legal definition in no case should absolve the producer or the holder from the responsibility for the generated waste since the moment of generation until it is transformed into an environmentally safe product. An example of the successful implementation of EU Directive on waste in Poland with respect to CCP confirms that the efficient regulatory enforcement procedures should comprise a well-balanced system of precepts, prohibitions and charges for disposal (fees, penalties) that would encourage power plants as waste generators to support financially the environmentally safe utilization of CCP by the waste reuse industry on a cost-benefit basis, in order to reduce charges for the disposal and to assure competitiveness of these products in the market.

The current practice in India that adopted ACAA approach with respect to CCP in the activity of Indian Fly Ash Mission

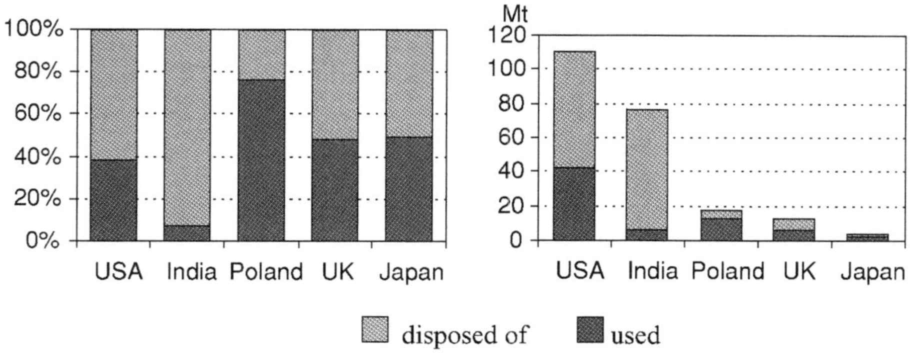

FIGURE 2. Examples of annual coal combustion waste generation and use in selected countries—large CCP generators (Data for the USA-2003, Poland-2002, India-2002 estimate, UK and Japan-1996/1998).

shows that under the conditions of the lack of the Resource Conservation Act and of financial incentives for CCP end users from waste generators, their products cannot compete financially with natural raw materials, and a fast growth of FA market remains a wishful thinking.

The lack of properly working system of finance transfer from waste generators to utilization industries seems to be a serious obstacle for growing of CCW market also in the USA, despite of an adequate regulatory framework and efforts of ACAA (whose purpose is to advance the safe and economic utilization of CCPs), along with the cooperation of many companies in the US with the Coal Combustion Products Partnership (C2P2) sponsored by the US EPA and the Department of Energy (DOE). It appears that the governmental sponsorship without the regulatory induced financial support of the power plants as CCP generators is not enough to reduce financial barriers associated with increased CCP utilization.

5. Conclusions

The major conclusions that could be derived from the state-of the-art in the area of Coal Combustion Products (CCP) management can be summarized as follows:

1. The coal-based power generation in the next three decades of 21st century is set to double; the majority of this growth is forecasted for developing Asian economies, but also for some developed countries not bound by the Kyoto Protocol (USA, Australia). This will result in the adequate growth of Coal Combustion Products (CCP) generation.
2. Waste that is technically suitable for use does not automatically become a raw material if there is no market for it, or its use is not commercially effective and, hence, they should be disposed off. This statement is thoroughly applicable also to CCP.
3. CCPs as the majority of disposed wastes are not geochemically stable and environmentally safe.
4. The legal definition in no case should absolve the producer or the holder from the responsibility for the generated waste since the moment of generation until it is transformed into an environmentally safe product; the pro-environmental and pro-recovery CCP recycling policy must thus be based on the term "waste" and the "polluter pays" principle.
5. A regulatory framework and adequate enforcement procedures should make CCP disposal highly unprofitable. Well-balanced system of precepts, prohibitions and charges for disposal (fees, penalties) should encourage power plants as waste generators to support financially the environmentally safe use of CCP by the waste reuse industry on a cost-benefit basis, in order to reduce charges for the disposal and assure competitiveness of these products in the market.

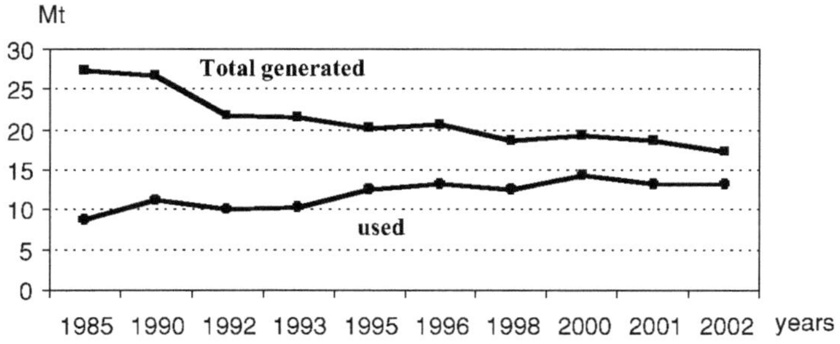

FIGURE 3. Generation and use of CCP in Poland, 1985–2002.

References

1. WCI—World Coal Institute. Key coal statistics for 2004, *Ecoal,* 51, 8, October 2004a.
2. WCI—World Coal Institute. Key coal statistics for 2002, *Ecoal,* 48, 8, January 2004b.
3. WCI—World Coal Institute. Kyoto is coming, *Ecoal,* 52, 1–3, January 2005.
4. WCI—World Coal Institute. The US alternative, *Ecoal,* 41, 5–6, March 2002.
5. WCI—World Coal Institute. World Energy Outlook, 2004. *Ecoal,* 52, 7, January 2005.
6. IEA—International Energy Agency. World Energy Outlook 2004. http://www.worldenergyoutlook.org
7. ABARE—Australian Bureau of Agricultural and Resource Economics (2002) *Global Coal Markets: Prospects for 2010.* http://www.abareconomics.com
8. IEA—World Energy Outlook 2002. http://www.worldenergyoutlook.org
9. EUROSTAT, *Environment Statistics Yearbook,* 2001 Edition, Luxembourg.
10. ACAA—American Coal Ash Association. 2001, 2002, 2003 Coal Combustion Product (CCP) Production and Use Survey. http://www.acaa-usa.org
11. Punshon T., Seaman J.C. and Sajwan K.S. The production and use of coal combustion products, in *Chemistry of Trace Elements,* K.S. Sajwan, A.K. Alva and R.F. Keefer, Editors, 2003, Kluwer Academic/Plenum Publishers, New York, Boston, Dordrecht, London, Moscow, 1–11.
12. Stewart B.S. Coal combustion product (CCP) production and use: survey results, in *Biogeochemistry of Trace Elements in Coal and Coal Combustion Byproducts,* K.S. Sajwan, A.K. Alwa and R.F. Keefer, Editors, 1999, Kluwer Academic/Plenum Publishers, New York, Boston, Dordrecht, London, Moscow, 1–6.
13. Twardowska I. and Szczepanska J. Solid waste: terminological and long-term environmental risk assessment problems exemplified in a power plant fly ash study, *Sci. Total Environ.*285, 29–51, 2002.
14. Twardowska I. and Szczepanska J. Coal combustion waste, in *Solid Waste: Assessment, Monitoring and remediation,* I. Twardowska, H.E. Allen, A.A.F. Kettrup and W.J. Lacy, Editors, 2004, Elsevier B.V., Amsterdam, Boston, Heidelberg, London, New York, Oxford, Paris, San Diego, San Francisco, Singapore, Sydney, Tokyo, 387–449.
15. Twardowska I., Tripathi P.S.M., Singh G. and Kyziol J. Trace elements and their mobility in coal ash/fly ash from Indian power plants in view of its disposal and bulk use in agriculture, in *Chemistry of Trace Elements,* K.S. Sajwan, A.K. Alva and R.F. Keefer, Editors, 2003, Kluwer Academic/Plenum Publishers, New York, Boston, Dordrecht, London, Moscow, 25–44.
16. Dunker R., Adriano D.C, Barton, C. and Punshon T. Revegetation of a coal fly ash—reject landfill, in *ICOBTE 2001 Conference Proceedings,* 2001, University of Guelph, Guelph, Canada, 671.
17. Use of coal ash as agricultural soil amendment. Chapter in *Chemistry of Trace Elements,* K.S. Sajwan, A.K. Alva and R.F. Keefer, Editors, 2003, Kluwer Academic/Plenum Publishers, New York, Boston, Dordrecht, London, Moscow, 251–334.
18. Prasad B., Bose J.M and Dubey A.K. Present situation of fly ash disposal and utilization in India: an appraisal, in *Indo-Polish Workshop on Fly Ash Management,* Calcutta, February 2000, R.P. Das, Editor, 2000, RRL Bhubaneswar, CMRI and CFRI Dhanbad, CGCRI Calcutta, 7.1–7.10.
19. ITB—Institute of Construction Techniques. Guidelines for testing natural radioactivity of raw materials and construction materials. *Instruction No. 234/95,* 1995, Warsaw (in Polish).
20. Central Statistical Office, *Environment Protection 2003. Information and Statistical Papers,* 2003, GUS, Warsaw (in Polish)
21. Twardowska I., Bulk use of power plant fly ash in deep mines and at the surface for contaminant and fire control, in *Solid Waste: Assessment, Monitoring and remediation,* I. Twardowska, H.E. Allen, A.A.F. Kettrup and W.J. Lacy, Editors, 2004, Elsevier B.V., Amsterdam, Boston, Heidelberg, London, New York, Oxford, Paris, San Diego, San Francisco, Singapore, Sydney, Tokyo, 949–1002.
22. Council Directive 75/442/EEC of 15 July 1975 on waste. *OJ L* 194, 25.07.1975. Amended by Council Directive 91/156/EEC of 26 March 1992, *OJ L* 078, 26.03.1991, and other amendments: *OJ L* 377, 31.12.1991; *OJ L* 001, 03.01.1994; *OJ L* 135, 06.06.1996; *OJ L* 332, 24.09.1996.
23. Commission Decision 2000/532/EC of 3 May 2000 replacing Decision 94/3/EC establishing a list of wastes pursuant to Article 1(a) of Council Directive 75/442/EEC on waste and Council Decision 94/904/EC establishing a list of hazardous waste pursuant to article 1(4) of Council Directive 91/689/EEC on hazardous waste (notified under document number C(2000)1147) (text with EEA relevance). *OJ L* 236, 06.09.2000; amended by 2001/118/EC as regards the list of wastes, *OJ L* 047, 16.02.2001, and by 2001/119/EC, *OJ L* 047, 16.02.2001.

3
Alternative By-Products of Coal Combustion and Simultaneous $SO_2/SO_3/NO_x$ Treatment of Coal-Fired Flue Gas: Approach to Environmentally Friendly Use of Low-Rank Coal

Ryunosuke Kikuchi
Department of Basic Science and Environment, CERNAS; ESAC—Polytechnic Institute of Coimbra Bencanta, 3040-316 Coimbra, Portugal

Abstract

The most widely used type of flue gas desulfurization (FGD) system fitted to coal-fired plants is the wet scrubber process using limestone slurry. The utilization rate of the by-product (gypsum) produced from the wet limestone FGD system is much lower than that of ash. There is also a possibility that the disposal by-product may cause environmental problems such as free-lime leachate and dust dispersion. Low-rank coal such as lignite accounts for about 50% of total coal reserves. SOx (SO_2 and SO_3) content is generally high in flue gas originating from lignite combustion. The wet limestone FGD unit does not treat SO_3 effectively, and it discharges the formed fine sulfuric acid mist to the atmosphere. To respond to the demand for processes that have low power consumption, a semi-pilot scale project (800 Nm^3/h) intended to demonstrate the adaptability of the electron beam process to flue gas having a high SOx concentration was initiated, and it proved that the process is capable of transforming SO_2, SO_3 and NOx into quality fertilizer. An evaluation based on the initial cost and marketability of the by-product fertilizer indicated that industrial application of this process would be economically feasible. In addition to the technical perspective regarding the reuse of CO_2 contained in coal-fired flue gas, this paper presents a possible approach to environmentally friendly use of low-rank coal.

1. Introduction

Effort has been made to promote the use of renewable energy (wind power, solar power, etc.), but this type of energy accounts for only 5% of energy use in Europe, with its share expected to rise to about 10% by 2020.[1] The remaining 90% of required energy will have to come from conventional energy sources such as coal and petroleum. Rapid economic growth has given rise to increasing demand for modern energy in developing countries, but nearly 2 billion people globally still have no access to modern energy.[2] Coal reserves are distributed worldwide and the amounts are vast (about 10 trillion tons),[3] making them easily accessible to people everywhere. Therefore, it is considered that coal has a major role in the energy sector. However, low-rank coal (low calorie and high sulfur content) such as brown coal and lignite accounts for about 50% of the total reserves.[3] It is necessary to discuss how to use coal wisely because of increasing pressure throughout the world for protection of the environment.

Though it is obvious that coal cannot become a renewable energy source, it is possible that coal can become an environmentally friendly energy source. It is important to utilize low-rank coal in such a way that it provides clean energy. The purpose of this study is not to describe recycling technology in great detail. The main purpose is to clarify actual and potential problems related to coal combustion by-products and then to encourage a new concept for coal use. The background of the presented study is described first, followed by a description of the actual technologies for utilizing coal as a clean energy and the associated problems.

2. Coal Combustion and Byproducts

To utilize the energy in coal, it is necessary to bum coal directly in air or first gasify it and then combust the clean fuel gas. In both processes the incombustible components in the coal remain as residues.

In a typical coal-fired power station, coal is injected into the furnace and ignited while in suspension. Combustion usually takes place in furnaces operating at temperatures above 1500°C.[4] The hot gases generate steam in the boiler tubes and then release heat to reheaters, superheaters, economizers and air pre-heaters depending on the boiler design. Figure 1 depicts the scheme of a typical combustion system for pulverized coal and the type of byproducts.

When coal is burnt, inorganic material is decomposed. A portion of the non-combustible material remains in the furnace. In most systems, the majority of the inorganic material is carried out of the combustion furnace with the flue gases, and is collected downstream by high efficiency filters, such

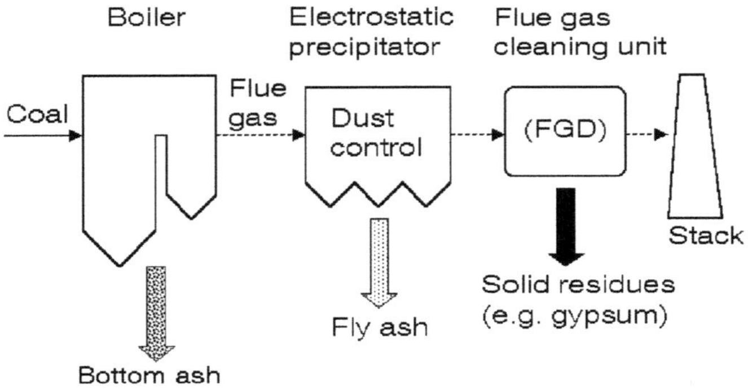

FIGURE 1. Typical system and by-products in pulverized coal combustion.

as electrostatic precipitators or bag houses. Pulverized coal-fired boilers can be described as either "dry-bottom" or "wet-bottom". In dry-bottom boilers, the bottom ash is removed from the bottom of the furnace as discrete, non-molten particles or clinkers. In these systems, most of the residues are collected as fly ash (about 80–90%).[4] In wet-bottom furnaces, most of the residues are collected as bottom ash (about 40–80%),[4] which is removed as a molten slag. After passing through the particulate control devices, gases are normally further treated in downstream gas cleaning devices such as a flue gas desulfurization (FGD) system that produces solid material as a by-product in the removal of sulfur dioxide from the utility boiler flue gas.[5]

From the aforementioned viewpoint, coal combustion by-products may be roughly classified into two types: ash (bottom ash and fly ash) and FGD by-product material (e.g. gypsum).

3. Utilization of Coal Combustion By-Products

In the past, coal combustion by-products (CCBs) were treated largely as waste. However, since the 1960's, the applications of ash have multiplied. CCBs are now used, for example, as a substitute for portland cement, in cement manufacturing, and in roofing tiles, structural fills, sheet rock, etc.[6] By contrast, FGD by-product material is becoming a serious problem from the viewpoint of the environmental impact of disposal.

3.1. Fly Ash and Bottom Ash

Both fly ash and bottom ash contain basic mineral elements which make them similar to the earth's crust: silicon dioxide, aluminum oxide, iron oxide, calcium oxide and trace amounts of sulfur trioxide, sodium oxide and potassium oxide. Thus, they are an excellent replacement for natural materials.[6]

Fly ash is the very small particle mineral residue that results from the burning of powdered coal in utility boilers. The individual particles are very small, like talcum powder, and are carried up and out of the boiler in the flow of flue gases leaving the boiler after the coal is consumed. The fly ash particles are removed from the flue gases using electrostatic precipitators, FGD systems or bag houses and are collected and stored dry for recycling. Fly ash is a pozzolan: a silica, alumina, and calcium based material which, in the presence of water, will chemically combine with the free lime contained in the fly ash and produce a cementitious material with excellent structural properties.[5,6] Some fly ash contains enough calcium compounds to be self-hardening when mixed with water.[6] All fly ash can be used as a direct replacement for portland cement in making concrete, in addition to many other applications such as absorbent, artificial aggregate, filter media, etc.[5]

Bottom ash is the coarse, solid mineral residue that results from the burning of coal in utility boilers.[5] The individual particles are much larger than fly ash particles and fall down through the air flow to the bottom of the boiler. The material is removed from the bottom of the boiler either in a wet or dry state and is transported to handling areas by conveyor or pipe. Bottom ash has a similar chemical composition to flu ash, but is produced in sizes ranging from fine sand to medium gravel. Because of its larger size, it does not have any cementitious properties. It is utilized in a variety of markets as an aggregate or filler material in construction projects, building products, etc. One other product, boiler slag, is bottom ash produced by some older boilers, where the material is actually melted in the boiler, quenched in the bottom hoppers, and becomes hard and glassy. This material has found an excellent market as a low silica blasting grit.[6]

The quantity of coal ash produced worldwide probably exceeds 550 million tons/year.[5] The utilization rate (1992) of coal ash by country is shown in figure 2.[5]

It follows from figure 2 that the utilization rate of coal ash varies between countries, from 25% in the USA to 99% in the Netherlands. In European countries, the cement and concrete industries use a high rate of coal ash.[5] In the USA, applications that involve little or no processing are important, and about 25% of the total ash is used in structural fill and road construction.

Fly ash and bottom ash are emitted mainly in the winter months while thermal coal-fired power stations are operating at full capacity. On the other hand, the construction industry

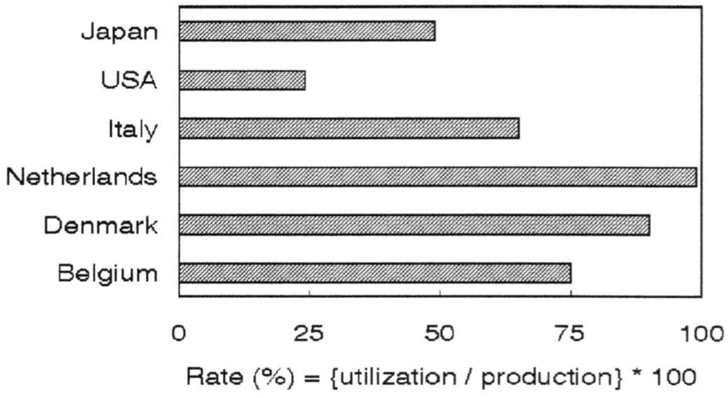

FIGURE 2. Utilization rate of coal ash in selected countries.

needs coal ash during the building season in the summer.[7] The distance between the construction sites and the power stations is another factor emphasizing the importance of interim storage facilities.[7] The rate of ash utilization will be further improved when solutions are found to the problems of uneconomic transport and lack of storage facilities.

3.2. FGD By-Product

Flue gas desulfurization (FGD) has been applied worldwide to remove sulfur dioxide (SO_2) from coal-fired flue gas, and the wet limestone process is now one of the leading methods of flue gas desulfurization, as shown in figure 3.[8]

The wet scrubber process (also known as the wet limestone process) generally uses limestone slurry as an absorbent.[4] SO_2 is transformed into calcium sulfate ($CaSO_4$, equivalent to natural gypsum). Limestone is abundant all over the world, easily handleable and cheap, and utilization of limestone has popularized the process.[9] However, it is difficult to imagine that calcium sulfate is commercially valuable because it is not generally in demand. Though the FGD by-product gypsum is utilized in various sectors (e.g. wallboard, road base, cement, etc.), the quantity of FGD gypsum used is very small and its utilization rate (=utilization/production) is low, about 3% on average.[10] The USA has a large number of limestone based wet FGD units,[10] but the FGD by-product gypsum has shown limited market potential and is usually disposed of.[6] According to other publications,[11,12] 20 million tons of by-product gypsum in Bulgaria and 2 million tons in Finland have been disposed of in landfills. The following reasons may be considered:[10] (1) there are already a number of supply sources of gypsum—natural mineral deposits and by-product gypsum produced from the phosphoric acid industry; therefore, the market must be shared with these existing sources; (2) processing of FGD gypsum to meet user specifications may incur additional operating costs such as dying for moisture reduction (about US$2/ton) and agglomeration (about US$2/ton); (3) gypsum is a low value commodity that generally cannot be shipped economically over a long distance; (4) it is often necessary to provide intermediate storage (about US$4/ton) because

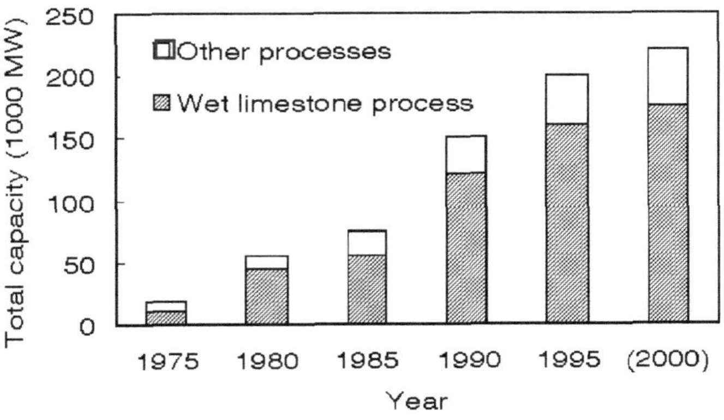

FIGURE 3. Trend in FGD unit installation throughout the world.

FIGURE 4. Photograph 1. Landscape degraded by disposal gypsum (Devnia in Bulgaria).

the supply and demand cycles for gypsum are not synchronized; (5) the wallboard industry is a major consumer of gypsum, but the ash content in FGD gypsum (about 2%) may cause discoloration of wallboard; (6) if an FGD unit is to produce by-product gypsum suitable for use in wallboard production, it requires a higher capital input at the beginning of the project, and it is judged that sales of the material justify the additional investment; and (7) a producer of FGD gypsum is not generally interested in its sale because the market price for FGD gypsum is comparatively low and ranges from 4 to 10 US dollars/ton in Europe.

Photograph 1 (Devnia in Bulgaria) shows the landscape at a site where unusable gypsum is dumped. This disposal site resembles the lunar landscape.

Environmental problems may arise with the disposal of by-product gypsum due to the high content of free lime and the alkalinity of any leachate.[10,13] Dust problems may occur during the handling of these dry residues with a high CaO content.[10,13] There can be violent exothermic reactions on addition of water to CaO-rich residues, followed by solidification and hardening.[10] Solidification has the advantage of reducing the amount of leachate formed, but can cause serious handling problems.[10]

The US Environmental Protection Agency reports that the limestone FGD process emits CO_2 gas to the air when limestone ($CaCO_3$) reacts with SO_2 gas and that its CO_2 emission amount was about 2.2 million tons in the USA in 1994.[14] This description is expressed by the following chemical reaction formula:

$$SO_2 + CaCO_3 \cdot aq(\text{limestone slurry}) + 1/2O_2 \\ \rightarrow CaSO_4 \cdot aq(\text{gypsum slurry}) + CO_2 \Uparrow$$

The above-mentioned wet limestone FGD unit uses slurry, so it is necessary to consider how to treat wastewater discharged from the scrubber. Heavy metals and fluoric compounds are removed from the FGD wastewater by the sedimentation

TABLE 1. Typical pollutants in coal-fired flue gas (measurements at filter exit)

Coal-fired plant	Coal type	SO_2	SO_3	NOx
Marista II TPP unit 8 (Bulgaria)[19]	Lignite	5600	140	390
Prunerov TPP (Czech)[20]	Lignite	3000	75	340
Ljubliana TPP unit 3 (Slovenia)[20]	Lignite	4900	120	350
Typical value in Japanese TPP[11]	Hard coal	1200	<10	140

Unit = ppm (vol./vol.); TPP = thermal power plant

process, and thionic acid ($HO_3S\text{-}S_n\text{-}SO_3H$, $n \geq 0$) is adsorbed in an ion exchanger and then neutralized.[15] In most cases, sludge from the wastewater of the various FGD systems applied in Europe is classified as hazardous waste.[16]

Though the by-product gypsum is recoverable, it is usually mixed with fly ash (and/or bottom ash) at a thermal power station and then deposited in depleted coal mines because of the lack of demand.[6] In order to preserve the landscape and fix CO_2, tree-planting campaigns (recultivation; e.g. bush, grass) have been started in depleted open mines in Europe.[17] A coal mine planted with trees is not a landfill. Disposal of the by-product gypsum in coal mines is unconscionable.

Common to all residues from the various wet FGD systems applied in Europe is the fact that they are highly contaminated and in most cases classified as hazardous.[15] Residues that have been treated further are also very soluble and therefore difficult to store in landfill.[16]

From the environmental standpoint, a process that reduces one pollutant but emits another (secondary pollution) should be avoided.

4. Reconsideration of The Most Widely Used FGD System

Acid rain is one of the most serious environmental problems facing the world today, and coal-fired flue gas contains sulfur dioxide which causes acid rain.[18] FGD systems improve the environmental performance of coal-fired plants by reducing emissions to the atmosphere. The most widely used type of FGD system fitted to coal-fired plants is the wet scrubber process using limestone slurry (see figure 3). This type of FGD system presents problems because of the large amounts of by-product requiring disposal and its CO_2 emissions (see section 3.2). It is now necessary to assess and reconsider the most widely used FGD system (wet limestone process), from the viewpoint of its by-product and technical performance.

4.1. FGD Performance and Flue Gas Generated by Burning Low-Rank Coal

As stated in the introduction, low-rank coal (e.g. lignite) accounts for about 50% of total coal reserves and is distributed worldwide:[3] Europe, Asia, Australia, Russia, North America and South America. Such type of coal is characterized by high moisture (30–50%),[3] low calorific value (below 4610 kcal/kg, ASTM 388) and high sulfur content (e.g. 5.9% in Leuna,[3] 3,0% in North Dakota,[3] 2,8% in Marista Iztok[19]). The properties of the coal influence the chemical composition of the flue gas. Table 1 shows the main pollutants contained in coal-fired flue gas. These concentrations were measured at the exit of the dust filter.

Most sulfur content in fossil fuel is converted to sulfur dioxide (SO_2), but a small percentage of sulfur is oxidized into sulfur trioxides (SO_3) by a two-step reaction.[21]

$$S + O_2 \rightarrow SO_2 \text{ (1st step); and } SO_2 + 1/2O_2 \rightarrow SO_3 \text{ (2nd step)}$$

SO_3 concentration might be negligible in a small incinerator, but it increases as the temperature of the fire chamber rises and the excess air factor increases in a large one. Its concentration increases considerably at 800°C or higher, and reaches a maximum at about 1600°C. It is known that reducing the excess air factor (ratio of supply air and theoretical air) is effective in controlling SO_3 during the combustion process. An acceptable factor is generally 1.2 to 1.3 for coal combustion.[21] Combustion of low-grade coal has a tendency to require a relatively higher factor, and SO_3 emission from burning coal with high sulfur content cannot be controlled effectively by this method. It has been reported that SO_3 concentration seldom exceeds 35 ppm.[21] However, table 1 shows a high SO_3 concentration (75–140 ppm) in flue gas generated by lignite combustion. There has been little discussion about the effects of SO_3 on the environment. SO_3 has toxicity over 10 times that of SO_2 and is highly corrosive.[22,23] Although there are significant differences between the two, SO_2 and SO_3 are often confused and expressed by the same chemical formula, SOx.[23] In this paper, sulfur dioxide and sulfur trioxide are expressed by SO_2 and SO_3 respectively, and a mixture of SO_2 and SO_3 is expressed by SOx.

The SO_3 released into the air reacts with vapor (water) to form sulfuric acid mist. One mechanism by which the acid mist may produce strong toxicity is by changing extracellular and intracellular pH. There is evidence that pH plays a critical role in growth control and cell differentiation, and that disrupting the control exerted by pH may lead to adverse effects.[24] A common continuous gas analyzer using infrared spectrometry is available for detecting SO_2 but not SO_3.[23] Ordinary FGD such as the wet limestone process can remove

TABLE 2. Example of pollutant concentration measured by antipollution equipment

Pollutants in flue gas	Inlet of FGD	Outlet of FGD	Outlet of wet-type ESP
SO_2, ppm (vol./vol.)	1000	80	80
SO_3, ppm (vol./vol.)	40	20	2
NOx, ppm (vol./vol.)	150	150	150
Fly ash, mg/Nm^3	300	60	10

Flow of flue gas = (ESP) → FGD → Wet-type ESP; FGD = wet limestone FGD; ESP = electrostatic precipitator.

SO_2 effectively (90% or more), but its removal efficiency for SO_3 is low and varies from 15% to 50%.[9,25] Coal-fired flue gas treated in a dust collector is cooled and fed to a scrubber because the reaction ratio of SO_2 with limestone slurry is comparatively low at high temperatures. That is, it is necessary to lower the flue gas temperature in order to improve the reaction rate (i.e. removal efficiency of SO_2). However, SO_3 contained in the cooled flue gas condenses into fine mist ($\leq 1\,\mu m$) of sulfuric acid in an environment of high moisture and low temperature.[9] As this fine mist cannot be trapped by the sprayed slurry,[9] flue gas containing sulfuric acid mist is discharged from the stack into the atmosphere. Furthermore, it promotes corrosion the plant's materials. This mist sometimes moistens the ash and causes clogging in the ducts and heat exchanger. It may be possible to remove it by injecting fly ash into the FGD unit because SO_3 reacts with CaO contained in fly ash. However, this solution exerts greater load on the fans, wears out the equipment and ducts due to the friction with the ash, and intermixes fly ash with FGD by-product. This approach shortens the life of the plant, increases energy consumption and lowers the commercial value of the FGD by-product. Ultimately, this approach is nonsensical because fly ash is again injected into the flue gas from which the fly ash has already been removed by dust collection. This controversial fine mist can be removed effectively ($\geq$90%) by a wet-type electrostatic precipitator using water on collecting electrodes.[9] Measurements of pollutant concentration by antipollution equipment are summarized in table 2 (modified from the published data[9]).

It follows from table 2 that each pollutant needs a specific antipollution device. If a wet limestone FGD unit only is used to remove SO_2 from lignite-fired flue gas containing SO_3, the resulting sulfuric acid mist is not removed effectively and is discharged to the atmosphere. If a wet-type electrostatic precipitator only is used to remove SO_3, the untreated SO_2 is discharged to the atmosphere. That is, both units are necessary to utilize low-rank coal such as lignite in an environmentally friendly way. Though this solution is ideal, the initial cost and the operating cost come to a large sum.

4.2. FGD By-Product Options

FGD by-product options were assessed by commission of the European Bank for Reconstruction and Development (EBRD) at Marista II thermal power station (TPP) in Bulgaria.[26] The situation in Bulgaria is exactly as described above; that is, there is little demand for by-product gypsum (see section 3.2.) and lignite is commonly used (refer to table 1). The assessment focused on Marista II TPP unit 8 which burns domestic lignite having a heat value of 1650 kcal/kg and a sulfur content of 2.8%, and which has an output capacity of 250 MW, flow rate of 1,500,000 Nm^3/h and the flow gas properties shown in table 1. Theoretically, SO_2 removed from flue gas by the FGD system can be transformed into any of several other sulfur compounds such as gypsum, sulfuric acid, elemental sulfur, ammonium sulfate (fertilizer), etc. Based on the possible amounts of by-product under the above-mentioned flue gas conditions at Marista II TPP and both national and international markets (estimate product price), each product was assessed assuming the following conditions: removal efficiency of SO_2 = 96%; and operation rate of FGD unit = 6400 hours/year. Figure 4 summarizes the economic results.[26]

It follows from figure 4 that by-product gypsum has a negative economic value, and ammonium sulfate (fertilizer) has the greatest economic value among the other options. These results can be interpreted as follows: there is no demand for by-product gypsum, and expense is therefore incurred in its disposal. Ammonium sulfate (fertilizer) may have a good market in the agriculture sector. As stated in section 3.2, serious environmental problems are being caused by disposal gypsum discharged from the most widely used FGD system (wet limestone process). The EBRD report indicates that there are other possible FGD by-products besides gypsum,[26] that is, it

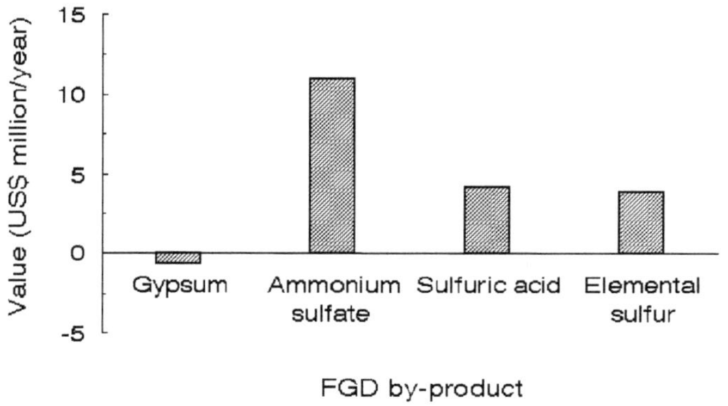

FIGURE 4. Economic evaluation of FGD by-product in the case of Marista II TPP.

proposes the transformation of SOx into ammonium sulfate (fertilizer).

5. Ammonium Sulfate and Processes for Making IT

It can be concluded here that it is desirable to convert the SO_2 contained in coal-fired flue gas into ammonium sulfate from the viewpoint of utilizing FGD by-product.

5.1. Function of Ammonium Sulfate as Fertilizer

Balanced nutrition is the key to profitable production of agricultural crops;[27] sulfur (S) along with nitrogen (N), phosphorus (P_2O_5) and potassium (K_2O) is a major plant nutrient needed for crop production. However, deficiencies of sulfur have been observed in most of the world's agricultural regions and this situation is receiving worldwide attention.[27] The so-called "hidden hunger" stage refers to a condition where there is a shortage of sulfur within the plant and this shortage depresses yields but does not produce visible symptoms of a sulfur deficiency.[28]

Ammonium sulfate contains 21% N and 24% S, and can be used as an ingredient in a compound fertilizer or blended with other materials because of its low hydroscopicity, easy storage and long-lastingness.[28] Since ammonium sulfate is long lasting, it is considered that positively charged ammonium ions attach to the soil particles, remaining in the root zone until needed by the plant, rarely leaching into the ground water.[28] This is an important factor from the environmental viewpoint.

As is customary in the trade, ammonium sulfate is commonly classified as nitrogen fertilizer. According to a report by the sulfur network of the Food and Agriculture Organization (FAO),[27] ammonium sulfate (regarded as a binary N+S fertilizer) can ideally meet the requirements of a plant for nitrogen and sulfur.

5.2. Existing Processes for Transforming SO_2 into Ammonium Sulfate

The following methods for making ammonium sulfate from flue gas are presented as proven commercial processes:[26] *General Electric Environmental Services* (*GEESI*) wet ammonia process, *Krupps Kopper Walther* process, and dry regenerable sodium process.

In the *GEESI* process, SO_2 contained in flue gas reacts with ammonium sulfite solution ($(NH_4)_2SO_3$) to form ammonium bisulfite solution (NH_4HSO_3) in the scrubbing step. This ammonium bisulfite is fed into an absorbing step and reacts with ammonia to form ammonium sulfite solution ($(NH_4)_2SO_3$). This is a so-called wet-type process that is equipped with a scrubber that uses water. There is therefore a possibility that the treated flue gas will contain ammonium sulfate mist. If so, a great amount of white smoke is generated from the top of the stack. Furthermore, this process discharges wastewater.

The *Krupps Kopper Walther* process is also a wet-type process and is similar to the *GEESI* process that recovers ammonium sulfate crystals from solution. Therefore, the same disadvantages as those of the *GEESI* process can be considered. Flue gas containing high moisture would make this process unfeasible.[26]

The dry regenerable sodium process can remove sulfur dioxides (SOx) from hot flue gas ($\geq 150°C$) by injecting finely ground sodium bicarbonate ($NaHCO_3$).[26] The sodium bicarbonate can be regenerated, and ammonium sulfate ($(NH_4)_2SO_4$) is formed by reaction with carbon dioxide and ammonia. The problem lies in the great amount of heat energy and concentrated CO_2 gas required for this regeneration.

6. Semi-Pilot Scale Test of The Electron Beam Process

As shown in table 1, flue gas originating from lignite combustion contains a high concentration of SOx (both SO_2 and SO_3). A project intended to demonstrate the adaptability of the electron beam process to flue gas having a high SOx concentration was initiated on a semi-pilot scale of 800 Nm^3/h gas flow in accordance with the following criteria: (1) formation of ammonium sulfate from the flue gas under the same conditions as those at Marista II TPP as shown in table 1; (2) effective treatment of SOx contained in the flue gas; (3) no discharge of wastewater to the surrounding area; (4) low power consumption; and (5) tentative removal of nitrogen oxides (NOx). Additional explanation is given regarding criterion (5). As stated in section 4.1, each pollutant needs a specific antipollution device in the flue gas treatment: an FGD unit for SO_2 removal and a wet-type ESP for SO_3 removal. Selective catalytic removal (SCR) units for treating NOx have been introduced worldwide.[9] The electron beam process theoretically makes it possible to treat SO_2, SO_3 and NOx simultaneously.[29] The concentration of NOx in the Marista II flue gas is lower than the national standard limit, and NOx removal was therefore included in the project only as a reference parameter.

6.1. Mechanism of Electron Beam Flue Gas Treatment

Electron beam processing is a method for producing functional materials by chemical reactions induced by fast electrons that originate from a cathode and then are accelerated in a vacuum acceleration tube.[30]

Under electron beam irradiation, molecules (e.g. N_2, O_2, H_2O) in the flue gas are converted into free radicals (e.g. N, O, OH, HO_2) with high oxidation potential.

$$N_2, O_2, H_2O \rightarrow (+e^-) \rightarrow N, O, OH, HO_2 \text{ (free radicals)}$$

These free radicals oxidize NOx and SO_2 in the flue gas to produce sulfuric acid and nitric acid as intermediate products.

$$SO_2 + OH \rightarrow HSO_3 + OH \rightarrow H_2SO_4$$
$$SO_2 + O \rightarrow SO_3 + H_2O \rightarrow H_2SO_4$$
$$NO + OH \rightarrow HNO_2 + O \rightarrow HNO_3$$
$$NO_2 + OH \rightarrow HNO_3$$

The sulfuric acid (H_2SO_4) and nitric acid (HNO_3) are immediately neutralized by ammonia that is injected to the reaction vessel beforehand, and then form powders of ammonium sulfate ($(NH_4)_2SO_4$) and ammonia nitrate (NH_4NO_3).

$$H_2SO_4 + 2NH_3 \rightarrow (NH_4)_2SO_4$$
$$HNO_3 + NH_3 \rightarrow NH_4NO_3$$

The fertilizing effect of ammonium sulfate is stated in section 5.1, and ammonia nitrate is also a fertilizer suitable for production of crops such as maize, cotton, wheat and vegetables.[31]

Without the effect of free radicals, SO_3 contained in the flue gas directly reacts with ammonia to form ammonium sulfate.

$$SO_3 + 2NH_3 + H2O \rightarrow (NH_4)_2SO_4$$

In the case that unreacted SO_2 and NH_3 are present at the vessel outlet, they are mostly consumed in the formation of ammonium sulfate by thermal reactions on the surface of the by-product powder in a collector.

$$SO_2 + 2NH_3 + H_2O + O_2 \rightarrow (NH_4)_2SO_4$$

6.2. Materials and Methods

The flue gas was produced in an oil-fired boiler, and was mixed with water vapor and SO_2 in order to simulate the flue gas conditions at Marista II TPP (see table 1). The prepared flue gas of 800 Nm^3/h was supplied to the test facility shown in figure 5.

The flue gas was cooled from 140°C to about 70°C in a spray cooler (diameter 1.2m and height 10m). The sprayed water in the cooler was completely evaporated by the heat of the introduced flue gas, so that no wastewater was produced. Prior to electron beam irradiation in a process vessel, the cooled flue gas was injected with a stoichiometric amount of gaseous ammonia. The flue gas with ammonia was fed to the process vessel (diameter 1.1m and length 2.2m) where the electron beam was irradiated (5 to 10kGy). Pulverous by-products readily formed and were collected in an electrostatic precipitator (50kV, 30mA, parallel plane electrode). The flue gas was separated from the by-products by an induced draft fan (1,500 Nm^3/h and 350 mmH_2O), and then released to the atmosphere through a stack.

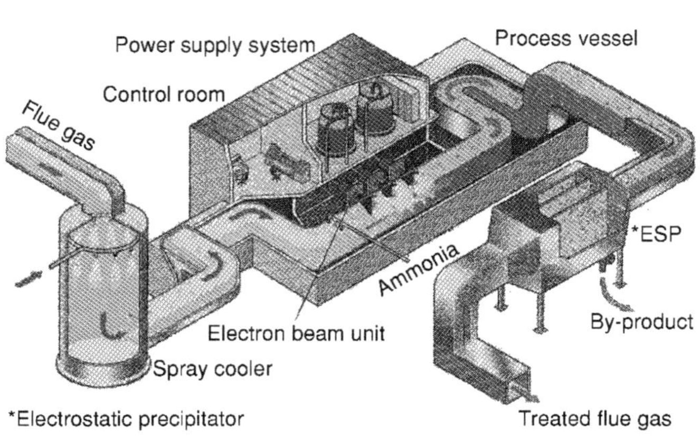

FIGURE 5. Schematic diagram of experimental system using electron beam process.

6.3. Results and Interpretation

Taking power consumption into account, a test using an electron beam dose of 5kGy (=5 kJ/kg of flue gas $\approx$ 0.36W/(Nm^3/h) of flue gas) was conducted over 2 months to treat flue gas under the same conditions as at Marista II (high SOx and high moisture). Under such conditions, it is likely that not only the target by-product but also undesirable impurities will be formed. NOx removal efficiency was determinate as a reference (see section 6). This test proved that the electron beam process affords the following: (1) proper treatment of high SOx flue gas; (2) no dust clogging; (3) no corrosion in the process vessel or the electron beam equipment (especially windows); (4) stable gas cooling to a target temperature; and (5) no leakage of the injected ammonia. Table 3 summarizes the removal efficiencies calculated according to the obtained data.

SOx removal efficiency and NOx removal efficiency increase with an increase in the dose of the electron beam.[29] It is therefore technically easy to improve the efficiencies shown in table 3 by increasing the dose, but this approach will cause an increase in the plant's power consumption. It was pointed out that the technology of the electron beam process suffered problems due to its high consumption of energy.[32] The energy consumption of the electron beam process at a dose of 5kGy is almost equal to that of a wet limestone FGD unit. At the same power consumption, a limestone FGD unit treats only SO_2. In contrast, an electron beam unit treats SO_2, SO_3 and NOx, and then produces by-product fertilizer consisting of 98.0% ammonium sulfate, 1.5% ammonium nitrate and 0.5% fly ash.

A vegetable (a kind of cabbage; *Brassica campestris*) growth test using the by-product fertilizer was conducted to verify the agricultural effect of the fertilizer. The results are shown in table 4.

The chief ingredient of the fertilizer is ammonium sulfate (98.0%). The fertilizer also contains a small amount of fly ash (0.5%). Though the fly ash normally consists of silicon oxides and aluminum oxide that make it similar to the earth's crust,[6] it also contains heavy metals. The heavy metal content of the fly ash was determined: 0.00003% As, 0,00002% Cr, 0.0001% Ni and 0.0017% Ti by weight. The amount of each heavy metal is so low that it may not influence vegetable growth. As evidence of this, it follows from table 4 that the quality (contribution to vegetable growth) of the by-product fertilizer is comparable to that of other commercial fertilizers such as urea and ammonium

TABLE 3. Removal performance obtained from the test

Flue gas component	Inlet of process vessel	Outlet of process vessel	Outlet of ESP (cleaned flue gas)	Removal efficiency
N_2	Balance	Balance	Balance	—
O_2	7.5%	7.5%	7.5%	—
CO_2	12.0%	12.0%	12.0%	—
SO_2	5500 ppm	803 ppm	795 ppm	85.5%
SO_3	140 ppm	21 ppm	20 ppm	85.7%
NOx	390 ppm	156 ppm	155 ppm	60.3%
Moisture	25.8%	27.5%	27.3%	—
Dust	140.0 mg/Nm^3	140.0 mg/Nm^3	0.2 mg/Nm^3	99.8%
Ammonium sulfate	0 mg/Nm^3	26.8 g/Nm^3 (21.5kg/h)	0.008 g/Nm^3	*99.9%
Ammonium nitrate	0 mg/Nm^3	0.4 g/Nm^3 (0.3kg/h)	0.004 g/Nm^3	*99.9%

Flue gas temperature at the vessel inlet = 65°C; dose of electron beam = 5kGy; ESP = electrostatic precipitator for by-product collection; *efficiency in by-product collection.

sulfate. After taking plant physiology and phytotoxicology into account, this vegetable test concludes that the obtained by-product has sufficient effect as a fertilizer, and the negative impact of the fly ash and other likely impurities is negligible.

7. Discussion

The main point at issue is how to promote environmentally friendly use of coal. The utilization rate of FGD by-product is much lower than that of ash in many parts of the world. This situation presents an obstacle to efforts to promote wiser use of coal. The most widely used type of FGD system fitted to coal-fired plants is the wet scrubber process using limestone slurry,[6,8] but this system has environmental problems because of the large amounts of by-product (gypsum) disposal;[11,12,13] hence, nitrogen fertilizer is proposed in this paper as an alternative by-product. To realize this proposal, a test of the electron beam process was conducted. However, there is still doubt as to whether such by-product fertilizer (a type of nitrogen fertilizer, abbreviated as N-fertilizer) is in demand.

7.1. Demand for By-Product N-Fertilizer

In 1996, 53 million tons of fertilizer were used in developed countries and 78 million tons of fertilizer were used in developing countries.[33] The annual growth rate of fertilizer use was +5.3% in developing countries during 1985-1997.[34] In 1999, Africa imported 2 million tons of fertilizer, Latin America imported 8 million tons and Asia imported 24 million tons.[33] Potash fertilizer and phosphate fertilizer are not completely absorbed into the plants in one cropping, and their fertilizing effects are expected to last through the next cropping too. That is, they have residual fertilization effects.[31] However, N-fertilizer should be applied each year, so its consumption is greater than that of other fertilizers. The annual growth rate of N-fertilizer use worldwide was about +8% during 1985–1998.[35] In spite of this, global production of fertilizer declined steadily during this time, with much of the decline occurring in the former USSR and Eastern Europe. The major reason for this decline was the economic chaos in these regions.[35] The decrease in production and the increase in use have caused a serious fertilizer deficit, especially in Asia,[36] and the gap between consumption and production has been widening since 1993 despite the recovery of production.[35]

To clarify this description, the Asian case (balance between consumption and supply) is given in figure 6.[36] A remarkable

TABLE 4. Vegetable culture test of the by-product fertilizer

Parameter\ Fertilizer type	Non-nitrogen	Ammonium sulfate	Urea	*By-product
Leaf height (cm)	19.8	28.2	28.5	28.5
Leaf width (cm)	7.1	11.4	11.8	11.7
Number of leaves	5.0	8.0	7.0	8.0
Weight (g)	20.6	71.5	63.1	78.2

* By-product obtained through the conducted test

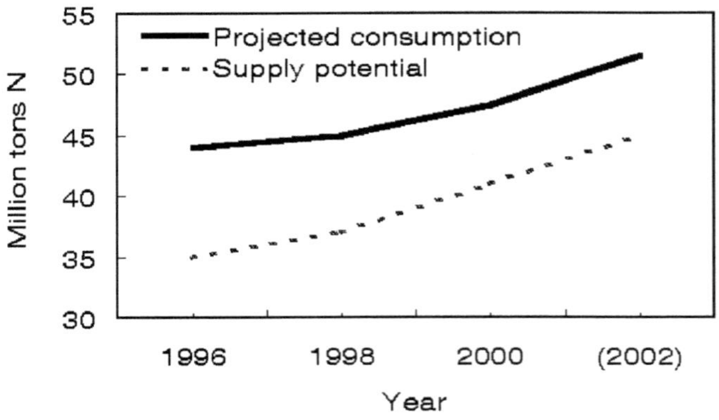

FIGURE 6. N-fertilizer consumption and supply in Asia.

increase of about 20% from 44 million tons in 1996 to 52 million tons in 2002 is shown in figure 6. The Asian deficit of N-fertilizer was projected to be about 7 million tons of N in 2002. In order to make up this deficiency, about 9 million tons of SO_2 accounting for 100% of the EU's total annual emission of SO_2 and about 4 million tons of NO_2 accounting for 50% of the EU's total annual emission of NO_2 would need to be transformed into 18.4 million tons of ammonium sulfate (21% N) and 6.2 million tons of ammonium nitrate (31% N) respectively; therefore, it can be concluded that by-product N-fertilizer is in great demand.

7.2. Acceptance of The Concept of Making By-Product Fertilizer From Coal-Fired Flue Gas

More widespread use of industrial processes for making N-fertilizer from coal-fired flue gas would help reduce the disposal amount of coal combustion by-products, make up the serious fertilizer deficit and cover the cost of reducing pollutants through sales of the fertilizer. However, these processes are not yet popular worldwide. On the other hand, the wet limestone FGD process is now very popular, perhaps as the result of misunderstanding, prejudice and a narrow outlook on the part of combustion plant owners. The energy sector is a major consumer of coal and a contributor to SOx and NOx emissions. Coal consumers such as thermal power stations, which attempt to deal with the environmental problem they create by installing air cleaning systems, generally have no connection with agriculture, food production or fertilizers. Hence they do not have any idea about making fertilizer from their own air pollutants. A local environmental administrator will call for a reduction in air pollution but may not point out the problem of disposing of the by-product from the air cleaning facility. The polluter (e.g. the owner of the coal combustion plant) is interested in the local cost for pollution abatement but is not interested in the profit from recycling the by-product on a global scale.

Furthermore, at the SO_2 control symposium,[37] it was pointed out that processes for making N-fertilizer would be expensive for FGD. At the time this was reported, these processes were still in the R&D stage (on a pilot scale). Potential users therefore gained the impression that processes for making N-fertilizer were expensive.

Recently the concept of linking a SOx/NOx emission source to fertilizer production has slowly penetrated the attitude of the energy sector in some parts of the world. A few projects employing the electron beam process are currently under way (private communication with Energostal, Polish Chamber of Power Industry and Ebara Corporation): A capacity plant of 100,000 Nm^3/h flue gas treatment is projected at Slavyanskaya coal-fired power station in Ukraine; a 540,000 Nm^3/h capacity plant (270,000 $Nm^3/h \times 2$ units) is under construction at Pomorzany coal-fired power station in Poland; a 620,000 Nm^3/h capacity plant is under construction at Chubu power station in Japan; and a 300,000 Nm^3/h capacity plant is under operation at Sichuan coal-fired power station in China.

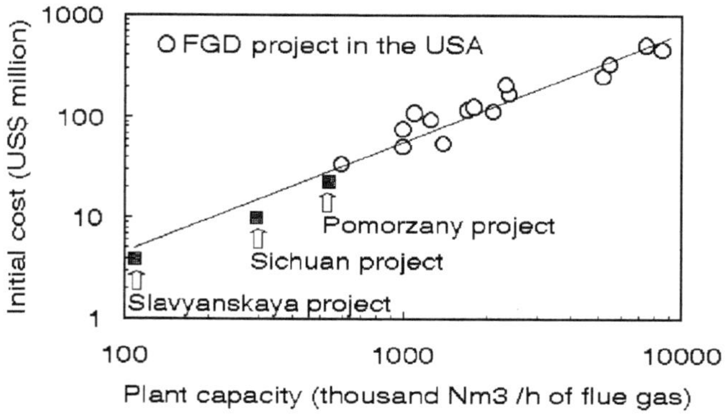

FIGURE 7. Comparison of initial costs: electron beam process *vs.* ordinary FGD.

An economic comparison of industrial (or semi-industrial) electron beam plants with wet limestone FGD plants is considered. Figure 7 shows the initial costs of several of the above-mentioned projects (Slavyanskaya, Pomorzany and Sichuan) plotted on a graph obtained from published cost data of FGD projects.[38]

Considering the initial cost and plant capacity, the projects for building an electron beam plant are cheaper than those for constructing a wet limestone FGD plant. However, the cost-capacity line drawn in figure 7 is based on economic data obtained from projects in the USA, and the plant costs of the electron beam process are based on the Polish project, the Ukrainian project and the Chinese project. Therefore, the costs cannot be directly compared. Taking into account the economic difference between the USA and the countries in which these projects are being carried out, it is considered that the construction cost of an electron beam plant for flue gas treatment is about the same as that of a wet limestone FGD plant.

7.3. Treatment of Gaseous By-Products in Coal-Fired Flue Gas

In order to ensure environmentally friendly use of coal, it is necessary to discuss not only solid by-products but also gaseous by-products originating from coal combustion. Special mention should be made of emission of greenhouse gas: CO_2 emission from coal combustion is 1.7 times greater than that from oil combustion and 2.3 times greater than that from natural gas combustion.[15]

A process of CO_2 recovery from coal-fired flue gas was attempted in Lobbock, Texas (USA) in the early 1980s, but this process was shut down within a year and dismantled because recovery of flue gas CO_2 was judged not to be feasible.[39] However, the CO_2 recovery technique has been advanced during the past 20 years, and it is in industrial operation in Hungary,

Turkey, Norway, Japan, Malaysia and other countries.[39] Some typical recovery processes are described: (1) PTSA and PSA.[40] CO_2 is separated from gas containing carbon dioxide by means of the pressure swing adsorption (PSA) process or the pressure and temperature swing adsorption (PTSA) process; and (2) CO_2 scrubbing with monoethanolamine solution (abbreviated as amine).[39,41] The cooled flue gas is scrubbed with amine solution (20–30%) in an adsorption tower, and then the solution containing CO_2 is fed to a regeneration tower. After CO_2 is desorbed from the amine solution in a regeneration tower by heat (110–120°C), CO_2 is supplied as product gas (CO_2 purity of over 90%) and the regenerated solution is recycled to the adsorption tower.[41]

How to apply the recovered CO_2 becomes a matter of great concern because the amount will be enormous. CO_2 is now mainly used for casting, welding, control of water pH, steal refining, production of refreshing beverages, chemical production (urea, soda ash, methanol, insecticides etc.), medicine (local anesthesia) and fire extinguishers,[42] but the amount of CO_2 emission is far greater than the current industrial demand for CO_2.

Enhanced oil recovery (EOR) and enhanced gas recovery (EGR) using CO_2 injection into the underground should enable large-scale CO_2 reuse and storage. A miscible gas flood of an oil reservoir is effective because an injected gas like CO_2 becomes miscible, or becomes one liquid phase, with the oil, and helps the oil move through the rock reservoirs and up and out of the wells. This process is called Enhanced Oil Recovery (EOR). For example, about two thousand tons of CO_2 supplied from CO_2 gas fields are handled and rehandled by EOR technology in a west Texas oil reservoir every day.[42] The USA's oil production by EOR technology increased by 87% from 96 thousand barrels per day in 1994 to 179 thousand barrels per day in 1998.[42] An essential problem lies in the fact that CO_2 is currently supplied from CO_2 flood reserves to EOR oilfield customers. This is not CO_2 recovered from flue gas. It is possible to substitute recovered (waste) CO_2 for CO_2 supplied from CO_2 flood reserves.

A methane hydrate is a cage-like lattice of ice, inside of which are trapped molecules of methane. Global estimates of the volume of natural methane hydrate approach $12{,}000 \times 10^{15}m^3$, a staggering figure compared with $0.2 \times 10^{15}m^3$ that make up the world's known gas reserves. These abundant reserves are distributed worldwide (especially in the world's ocean floors), but they have hardly been touched.[43] For the following reason, it may be possible to apply recovered CO_2 to methane hydrate resources. Methane hydrate has a self-preservation effect because of the large enthalpy change of its dissociation.[44] To put it differently, during the dissociation process, methane hydrate causes what is known as an endothermic reaction by absorbing heat from the surroundings. CH_4 hydrate has 54KJ/mol and CO_2 hydrate has 77KJ/mol of variation of overall reaction enthalpy at a temperature of 273K.[44] The heat released by the exothermic reaction of CO_2 hydrate formation exceeds the heat originating from the endothermic reaction of CH_4 hydrate dissociation. Consequently, gas exploitation in CH_4 hydrate becomes possible by using the heat of CO_2 hydration formation.[45] This CO_2 hydrate technique is also applicable to reinforcement of unstable submarine sediment layers in order to facilitate drilling:[46] Before gas hydrate is drilled, CO_2 hydrate is formed by CO_2 injection as an artificial cap rock (block) lying above a CH_4 hydrate zone, in order to guard against well collapse, landslides, falls and sudden gas leakage.

Thus, CO_2 recovered from coal-fired flue gas will contribute to oil production through its use in enhanced oil recovery, and to natural gas production through its application to methane hydrate resources.

8. Conclusions

It is necessary to utilize coal combustion by-products in order to realize the environmentally friendly use of coal. If these by-products are not used, they become waste and cause environmental problems. The utilization rate of FGD by-product (mainly gypsum) is much lower than that of ash. This paper therefore proposed marketable N-fertilizer (mainly ammonium sulfate) as an alternative by-product and discussed the importance of SO_3 treatment of lignite-fired flue gas. Furthermore, a supplementary discussion of a solution to gaseous by-product (CO_2) was also given.

Rapid economic growth has given rise to increasing demand for modern energy in developing countries, but globally nearly 2 billion people still have no access to modern energy.[47] Coal reserves (about 50% low-grade coal) are distributed worldwide and the amounts are vast, making them easily accessible to people everywhere. From this point of view, the realization of the environmentally friendly use of low-rank coal has an important role in the energy sector, global environment and social development.

Acknowledgments. The author is grateful to Mr. S. Aoki and Ms. A. Yokoyama of Ebara Corporation for sending technical information on the electron beam process, Mr. K. Watanabe of Mitsubishi Heavy Industry for sending published data on CO_2 recovery, Dr. Y. Pelovski of the Sofia University of Chemical Technology and Metallurgy for advice with FGD by-product strategy, Prof. C. Costa of Escola Superior Agrária de Coimbra for offering data processing facilities and Mrs. Catherine Lentfer for English review. This paper reports on part of the PhD thesis submitted to Sofia University of Chemical Technology & Metallurgy and to the Bulgarian Ministry of Education, Science and Technology.

References

1. European Commission, *European Energy to 2020, SEC (95) 2283*, European Communities, Luxembourg, 1996.

2. Davis, M., *Institutional Frameworks for Electricity Supply to Rural Communities*, Energy and Development Research Center, Rondebosch, 1995.
3. Nippon Brown Coal Liquefaction Ltd., *Clean Coal Technology—Commercialization towards the 21 Century*, Tokyo, 1999.
4. Edgar, T., *Coal Processing and Pollution Control*, Gulf Publishing, Houston, 1983.
5. Clark, L., *Applications for coal-use residues, IEACR/50*, IEA Coal Research, London, 1992.
6. Western Region Ash Group, <http://*www.wrashg.org/ccps.htm*>, accessed on February 3, 2004.
7. Blomster, D. and Leisio, C., Fly ash—quality recycling material, in *Energy innovation 1996,* Salminen, Y., Laiho, H., Kaikkonen, H., Eds., Imatra Vaimo Group, Helsinki, 1996.
8. Klingspor, J., Tokerud, A. and Ahman, S. Low-cost FGD system for emerging market, *ABB Review* 1, 1998.
9. Sato, T. Fixation of SO_2 with lime. *Inorganic Material*, 3, 6, 1996.
10. Clark, L. Management of FGD residues, *IEACR/62*, IEA Coal Research, London, 1993.
11. Pelovski, Y. Air pollution from energy sector and cleaning system, *paper presented at the Workshop of Pollution and Environment,* 14th IUPAC conference, Osaka, 25–30 August 1996.
12. Ranta, J. Amount, Quality and Utilization of FGD Wastes in Finland, in *Acidification in Finland,* Kauppi, P., and Kenttamies, K., Eds., Springer-Verlag, Berlin, 1990.
13. Tcekov, I. Ecological and Economical Assessments of Industrial Complex for Production of Soda and Phosphorous Fertilizer, *MSc Thesis*, European Masters Degree Course in Sustainable Development, Higher Institute of Chemical Technology and Metallurgy, Sofia, 1995.
14. U.S. Environmental Protection Agency (EPA), Inventory of U.S. Greenhouse Gas Emission and Sinks: 1990–1994, EPA, Washington, D.C., 1995.
15. Ando, J., *Flue gas cleaning technology in the world.* Coal Technology Institute, Tokyo, 1990. [in Japanese].
16. European Topic Center on Waste, Waste generation and management, in *Environment in EU at the turn of the century,* European Environment Agency, Ed., European Environment Agency, Copenhagen, 1999.
17. Haristov, A., Recultivation in the open mines of the energy complex Maritsa Iztok. *Eco* 9, 1996.
18. Elsom, D., *Atmospheric Pollution*, 2nd ed., Blackwell Publishers, Oxford, 1992.
19. Tuv Umwelt Messtecnik GmbH, *Emission Measurements at Marista II Power Station (Bulgaria) Unit 6, Report UM94-0004083*, Tuv Umwelt GmbH, Filderstadt, 1994.
20. Kikuchi, R., Recycling of Air Pollutants from Thermal Power Station, *PhD Thesis*, University of Chemical Technology and Metallurgy, Sofia, 1997.
21. Danielson, J., *Air Pollution Engineering Manual*, National Center for Air Pollution Control, Cincinnati, 1971.
22. Research Triangle Institute, *Toxicological Profile for Sulfur Trioxide and Sulfuric Acid,* Agency for Toxic Substance and Disease Registry, Atlanta, 1997.
23. Krotov,Y., and P. Bespamytnov, *Threshold limit concentrations of harmful substances in air and water.* Chimiy Publishing, Ljeningrad, 1975. [in Russian].
24. Soskolne, C., Pagano, G., Cipollaro, M., Epidemiologic and toxicologic evidence for chronic health effects and the underlying biologic mechanisms involved in sub-lethal exposures to acidic pollutants, *Arch. Environ. Health 44*, 1989.
25. Szalach, P. 1998. New York Electric and Gas Corporation utility experience with SO_3 and H_2SO_4 mist. *Proc., Formation, Distribution, Impact and Fate of SO_3 in Utility Flue Gas Streams Conference*, Pittsburgh, March 30–31, 1998.
26. Incoteco ApS, *Marista TPP—FGD Overview*, European Bank for Reconstruction and Development, London, 1995.
27. Fertilizer Industry Advisory Committee, Sulfur for Crop Production, Food Agriculture Organization, Rome, 1992.
28. Promotion Group of the Ammonium Sulfate Fertilizer Industry, *The Importance of Ammonium Sulfate for Crop Production*, Urea & Ammonium Sulfate Fertilizer Industry Association, Tokyo, 1995.
29. Namba, H., Tokunaga, O., Hashimoto, S., and Tanaka, T., Pilot-scale test for electron beam purification of flue gas from coal-combustion boiler. *Radiation Physics and Chemistry* 46, 4/6, 1995.
30. Mizusawa, K., Ejiri, H., Kashiwagi, M., Taguchi, S., Hamano, M., and Hoshi, Y., The feature of electron beam processing system in past, present and future, *HHV Technical Review* 40, 2, 1995.
31. Asao, S., Fertilizer, in Plant nutrient and application of fertilizer, Chino, M. ed., Asakura Publishing, Co., Ltd., Tokyo, 1995, 217. [in Japanese].
32. Pelovski, Y., Hristov, H., Simeronov, K., Todorova, S., Vasiliev, H., and Kikuchi, R., Reduction of SO_2 and particulate emission, Regional Environmental Center for Central and Eastern Europe, Szentender, 1998.
33. Food Agriculture Organization, *FAO statistical databases*, Rome, 2001.
34. International Fertilizer Development Center, *African Fertilizer Situation*, Alabama, 1998.
35. International Fertilizer Development Center, *World Fertilizer Market Overview* 14, 4, 2001.
36. International Fertilizer Development Center, *Asian Fertilizer Situation*, Alabama, 1998.
37. Cichanowicz, J., Dene, C., and DePriest, W., Engineering Evaluation of Combined NOx/SOx Controls for Utility Application, SO_2 Control Symposium, Washington D.C., December 3–6, 1991.
38. Keeth, R., and Radeliffe, P., Utility Response to Phase 1 and Phase 2 Acid Rain Legislation—An Economic Analysis, SO_2 Control Symposium, Miami Beach, 28–31 March, 1995.
39. Iijima, M., Mitsuoka, S., Mimura, T. and Suda, T., Development of CO_2 recovery technology from combustion flue gas, *JSME International Journal* 41, 4, 1998.
40. Mitsubishi Heavy Industry, <http://*www.mhi.co.jp/power/techno/co2/index.htm* >, accessed on February 12, 2004.
41. Iijima, M., Flue gas CO_2 recovery technology, *Petrotech* 23, 2, 2000.
42. Nippon Steel Corporation, Carbon Dioxide Generator, *technical report PMD 217 89.11*, Tokyo, 1992.
43. Nederlof, H., The scope for natural gas supplies from unconventional sources, *Energy* 13, 1988.
44. Handa, Y., Compositions, enthalpies of dissociation, and heat capacities in the range 85 to 270 K for clathrate hydrates of methane, ethane, and propane, and enthalpy of dissociation of

isobutane hydrate as determined by heat-flow calorimeter. *The Journal of Chemical Thermodynamics* 18, 1986.
45. Ohgaki, K., Takano, K. and Moritoki, M.,. Exploitation of CH_4 Hydrates under the Nankai Trough in Combination with CO_2 Storage, *Journal of Chemical Engineering of Japan* 20, 1, 1994.
46. Kikuchi, R., Vanneste, M. and Andreassen K. Environmentally Friendly Energy System Using Fuel Cell and Methane Hydrate Resources—Conceptual integration of CO_2 reuse with hydrogen production. 7th International Conference on Technologies and Combustion for a Clean Environment, Lisbon, 7–10 July 2003.
47. Davis, M., Institutional Frameworks for Electricity Supply to Rural Communities, Energy and Development Research, Rondebosch, 1995.

4
Fly Ash as a Sealing Material for Spontaneous Combustion and Acid Rock Drainage Prevention and Control

Irena Twardowska and Sebastian Stefaniak

Polish Academy of Sciences, Institute of Environmental Engineering, 34 M. Sklodowska-Curie St., 41-819 Zabrze, Poland

Abstract

Oxidation processes result in a number of environmentally adverse phenomenons such as spontaneous combustion of coal in piles, coal mining waste dumps and coal seams, as well as ARD generation in sulfide-bearing coal and ore mining waste dumps. The most efficient way to prevent, attenuate or intercept both adverse processes is to effectively cut off air penetration into the coal- or/and sulfite-bearing material with use of abundant, cost-effective, efficient and easy-to-handle material of high sealing properties against air penetration. Fly ash (FA) from coal-fired power plants in the form of a dense mixture with water ($\geq$1:1 wt.) appeared to be an excellent sealing/insulation material against air penetration that has no competing natural materials in this respect and fulfills all the conditions of efficiency and availability. A series of experiments carried out to evaluate these properties of different FA: Water mixtures showed that penetration resistance of solidified FA: Water mixtures to air is 1–2 orders of magnitude higher than that of the natural sealing materials, while hydraulic conductivity qualifies this material as permeable to water. Full-scale applications of FA: Water dense mixtures for spontaneous ignition control at coal mining waste dumps confirmed high efficiency of this material. Pending projects and potential application fields of dense mixtures of FA with water for these purposes were also discussed, among them the most promising and not yet sufficiently utilized areas of use such as potential for fire control in burning coal seams (exemplified in Jharia Coalfield, India) and prevention of ARD generation in coal/ore mining waste dumps with use of FA: mine water mixtures.

1. Introduction

A number of the environmentally adverse processes such as endogenous fires of coal piles, dumping sites, and coal seams, which besides of heavy environmental pollution cause severe loss of natural resources, start up from the contact of carbon with air.

Also acid rock drainage (ARD) from mines and sulfidic mining waste disposal sites that generate hazardous high-sulfate, high-metal leachate contaminating ground waters is induced by atmospheric oxygen. Process of ARD generation that results from sulfide oxidation is common for all coalfields and ore mining areas. Its mechanism has been presented in numerous publications and reviews since early sixties[1,2,3]. It is generally recognized as a multi-stage exogenic process involving direct abiotic pyrite oxidation from the reaction with oxygen and water, indirect reaction with ferric iron and biochemical oxidation by bacteria *Thiobacillus ferrooxidans*, which can be summarized by equations below:

$$FeS_2 + 3.5\ O_2 + H_2O \_ Fe^{2+} + 2SO_4{}^{2-} + 2\ H^+$$
$$2FeS_2 + 7.5\ O_2 + 7\ H_2O \_ 2\ Fe(OH)_3 + 4\ SO_4^{2-} + 8\ H^+$$

These processes, besides oxidation of carbon in the presence of moisture, also initiate and support endogenous spontaneous ignition of coal and coal mining waste that has been reported to occur all over the coalfields in the world posing problems of different scale, from local fires in coal piles, coal mining waste disposal sites to vast areas of coalfields set on fire[4–19]. In every case, spontaneous ignition of coal occurs when air and moisture are available.

Coal seams fires appear to be the most problematic. A spectacular example of such environmental disaster is Jharia Coalfield in India that has been operated mostly by Bharat Coking Coal Ltd (in 258 km^2 of total coal bearing area 450 km^2). In this mining area, 17.32 km^2, i.e. 6.7% has been set on fire since 1986 that caused total devastation of the territory, heavy air pollution and huge coal loses (Fig. 1, 2)[17].

Problems at a more limited scale, but also posing serious threat to the local inhabitants and the environment are endogenous fires of coal mining waste dumps. In principle, proper dump construction in compacted layers and with additional insulation layers, if material has weak susceptibility to

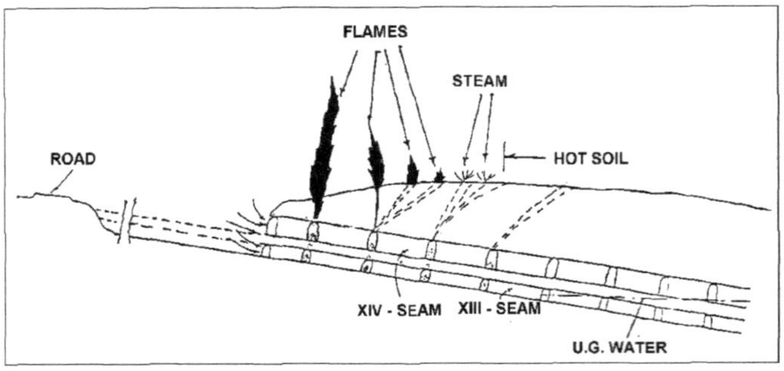

FIGURE 1. Scheme of the seams in Jharia Coalfield (India) set on fire with blazing fire zone, cracks of different size and width, steam issue and the heated surface[17].

compaction, well protects against self-ignition. Though, in old mining areas also historical conic (or reshaped) uncompacted tips may occur that are as a rule thermically active. In the Upper Silesia Coal Mining basin (USCB, Poland) that is the biggest hard coal mining area in Europe, about 20 dumping sites has been identified as showing thermic activity of a different intensity, of these two create serious environmental problems. These sites comprise the biggest in Europe, almost 100 years old and 100 m high Skalny dump, where in 30 hectares 17Mt of coal mining waste has been disposed[15], and a somewhat lesser old reshaped conic dump in Nowa Ruda[18]. Sometimes an initiation of self-ignition happens due to the erroneous realization of contemporary projects. This case can be exemplified in the Siemianowice site of total area 3 hectares (USCB, Poland). In this site, two closed settling ponds of flotation slurry were reclaimed with use of coal mining waste that after levelling had to be covered by an insulation layer to prevent self-ignition. In April 2004, the dump spontaneously ignited; fire proceeded with a speed 2–2.5 m/month. In the burning area, concentrations of CO, NO_x, PAH and VOC exceeded standards up to 15–16-fold (fortunately, no residential area occurred in a close proximity) that caused an urgent need of firefighting action (Fig. 3).

FIGURE 2. Devastation caused by an open fire moving into the residential area (Jharia Coalfield, India)[17].

FIGURE 3. Burning dumping site in Siemianowice, USCB, Poland (state in June 2004)[18]. Self-ignited in April 2004, fire intercepted in November 2004.

Also spontaneous combustion of coal while stored in piles at power plants or coal terminals is a well-known and widely discussed phenomenon[4–12]. High-moisture, highly volatile coal from the Powder River basin (PRB) in Wyoming, USA was reported to smolder and catch fire not only at the piles, but also to be delivered on fire in the rail cars or barges[12]. Wetting is an exothermic process, thus the liberated heat can accelerate the spontaneous heating of coal.

The studies on the relative susceptibility of coal, carbonaceous shales and coal mining waste were also carried out[13]. In general, spontaneous ignition of coal and coal mining waste is mainly abiotic process, though bacteria were reported to be responsible for burning such fossil fuels as peat[14] (material at an early stage of coalification). In all the cases of spontaneous ignition of coal in seams, coal mining waste, or in piles, air (oxygen) availability is a major factor for a process to start and proceed. Oxygen is a major reagent also in ARD formation in coal and ore mining waste dumps, as can be seen from the reactions above.

This leads to the conclusion that the most efficient way to prevent, attenuate or intercept both adverse processes is to effectively cut off air penetration into the coal- or/and sulfite-bearing material with use of abundant, cost-effective, efficient and easy-to-handle material of high sealing properties against air penetration. Fly ash (FA) from coal-fired power plants in the form of a dense mixture with water ($\geq$ 1:1 wt.) has a potential to be an adequate material that fulfills all these conditions. A series of experiments was carried out to evaluate these properties of different FA:water mixtures.

Full-scale applications of FA for spontaneous ignition control were exemplified in case studies. Also pending projects and potential application fields of dense mixtures of FA with water for these purposes were discussed.

2. Studies on Sealing Properties of FA

2.1. Material and Methods

Studies on sealing properties of FA in the form of dense mixture with mine water were conducted as bench experiments, with use of different FA and mine waters.

In the experiments, 1:1 mixtures with mine water of different salinity consisting of (i) "pure" FA; (ii) FA + D-FGDS (FA with flue gas desulfurization products from dry process—an old technology, in the last decade abandoned, in which limestone powder was injected directly to the boiler and thus the end product was a mixture of FA with desulfurization products); (iii) FA + SD-FGDS (FA with flue gas desulfurization products from the contemporary, but less widespread semi-dry process) were used. The studied materials (ii) and (iii) originate from processes that generate end product consisting of FA with admixture of desulfurization products. In particular, the following mixtures that considered FA, "pure" and containing FGD products, and mine water quality specific for the USCB were studied:

A—"pure" FA without FGD products;
I/1—low-alkaline FA (LA-FA); Mine water of Cl-SO_4-Na type, TDS 3.9 g/L, pH 8.15;
II/1—high alkaline FA (HA-FA); Mine water as I/1;
II/2—high-alkaline FA (HA-FA); Mine water: brine of Cl-Na type, TDS 50.0 g/L, pH 7.2;

B—FA + FGDS;
III/1—FA + D-FGDS; Mine water of SO_4-(Cl)-Na-Ca type, TDS 2.9 g/dm^3, pH 8.06;
III/2—FA + D-FGDS; Mine water: brine of Cl-Na type, TDS 50.0 g/L, pH 7.2;
IV/1—FA + SD-FGDS; Mine water as in III/1;
IV/2—FA + SD-FGDS. Mine water: brine as in III/2;
IV/3—FA + SD-FGDS; Mine water: brine of Cl-Na type, TDS 59.2 g/L, pH 7.23.

Mixtures were placed into PVC columns in the layers 0.5 m thick. Solidification process was evaluated as changes of penetration resistance R (in kPa) measured in 2–3 days' intervals, until solidification terminated.

In parallel, hydraulic conductivity of solidified mixtures was measured with use of an equipment that comprised consolidometer with a ring of 100 mm radius and height h = 45 mm, a unit for load application and measurements of a sample deformation, and a unit for input water supply and filtrate measurements. The equipment enables to measure hydraulic conductivity k, volumetric density of a material skeleton, porosity n and effective porosity n_e. The measurements were repeated 5-fold.

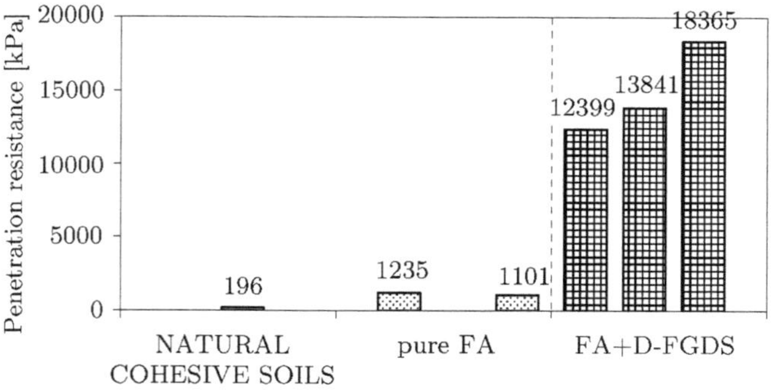

FIGURE 4. Penetration resistance R (kPa) of the solidified mixtures of FA with mine water compared to that of the natural cohesive soils (boulder clay); Penetration resistance of FA-FGDS mixtures with mine water significantly increases with an increase of water salinity.

Both penetration resistance and hydraulic conductivity results for consolidated FA: mine water mixtures were compared with the same parameters for natural materials used for sealing purposes.

2.2. Results and Discussion

The results on hydrogeological parameters of studied FA: Mine water mixtures, including mechanism and kinetics of solidification and changes of penetration resistance, as well as on their hydraulic conductivity have been presented and discussed in detail elsewhere[20]. Here, the data on these parameters are presented in comparison with those of natural sealing materials (Fig. 4, 5, 6)

2.2.1. Air Penetration Resistance

Penetration resistance of the mixtures of FA with Water (1:1 wt.) after solidification (15–24 days) ranged from over 1000 to over 18000 kPa (Fig. 4). These values are 1–2 orders of

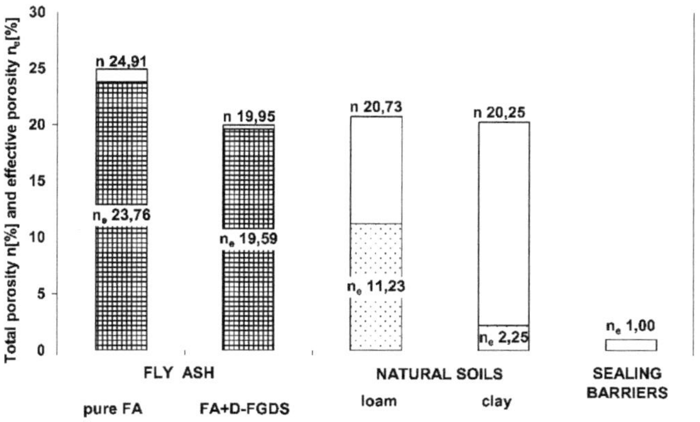

FIGURE 5. Total (n) and effective porosity (n_e) of the solidified mixtures of FA with mine water compared to that of the natural heavy soils; Porosity of FA mixtures with mine water is significantly higher than that of the natural water-tight sealing soils.

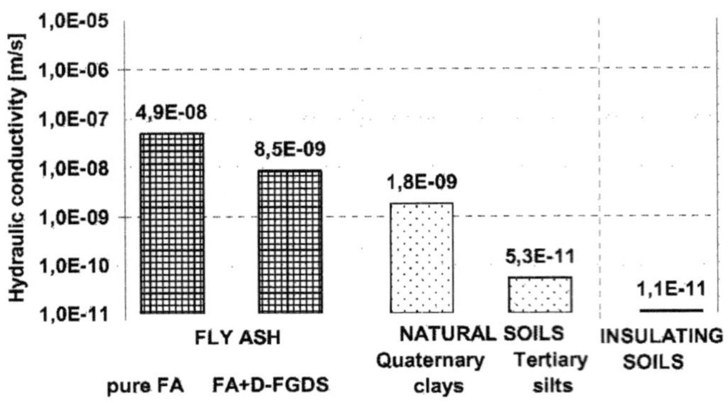

FIGURE 6. Hydraulic conductivity (k) of the solidified mixtures of FA with mine water compared to that of the natural heavy soils; Hydraulic conductivity of FA mixtures with mine water is significantly higher than that of the natural water-tight sealing soils.

magnitude higher than the ones of natural cohesive soils such as boulder clay, for which penetration resistance accounts for 190 kPa. It proves excellent sealing/insulation properties of solidified FA: water mixtures against air penetration that definitely outclass the adequate properties of natural sealing materials.

The R-values of mixtures based on pure FA appeared to be the lowest (1100–1200 kPa) (Fig. 4) and solidification was slow, up to 25 days. FA + FGDS mixtures with mine water displayed R-values roughly an order of magnitude higher. Mixtures consisting of FA + D-FGDS showed the shortest solidification time (15–20 days) and the highest R-values (over 12000–18000 kPa). (Fig 4). This means that lime enrichment in the mixture significantly improves the penetration resistance of the material and reduces solidification time. due to occurrence of complex processes of a chemical binding similar to the hydration and cementation of a lime mortar preceded by a relatively short (about 5 days) period of the physical dewatering/evaporation phase[20].

It should be underlined that also unmodified "pure" FA shows sealing properties against air penetration about one order of magnitude higher than compacted natural cohesive soils commonly used as an air-tight sealing material (Fig. 4). The problematic feature of "pure" FA:water mixtures is a rather slow solidification that occurs mainly due to the gravitational dewatering and evaporation. In the stable climatic conditions with adequately long dry periods, this low speed is a minor deficiency, but in the regions with unstable precipitation a shorter solidification time is preferable. Substantial speeding up solidification period can be achieved through reducing water rate in the mixture. This measure has generally beneficial effect on the environment, as besides faster solidification reduces amount of a leachate. Use of super-dense mixtures requires though mixture preparation at the spot, to avoid hydraulic transportability problems.

The use of high-TDS, low-quality mine water for mixture preparation, besides protection of water resources, resulted in an increase of penetration resistance of the solidified mixture and a shorter solidification process in case of use in a mixture either "pure" FA or FA + FGDS, thus exerts a generally positive effect[20].

The process of solidification of FA+SD-FGDS mixtures appeared to undergo a short (up to 2 days long) phase of a chemical water binding that results in a substantial penetration resistance values gained in this short period (R = 300–2100 kPa) that were 1- up to >2 orders of magnitude higher than the relevant R values for "pure" FA. This phase was followed by a long (up to ≥25 days) period of solidification due to dewatering/evaporation, when penetration resistance increased up to 7000 kPa. Water salinity used for a mixture preparation did not exert significant effect either on solidification time or on the penetration resistance values[20]. A long period of solidification despite of a high penetration resistance values, reduces use of this product for dense mixtures preparation for self-ignition and ARD control in a wet unstable climate of the Northern hemisphere.

2.2.2. *Hydraulic Conductivity*

The major properties that define permeability of the material to water, i.e. effective porosity responsible for transport of solutions n_e and hydraulic conductivity k of the studied FA: mine water dense mixtures, along with pore structure were also characterized in detail in another recent publication[20]. This allows to summarize here these parameters with respect to these materials applicability to ARD and spontaneous ignition prevention and control. It was found that while with respect to air penetration resistance studied mixtures are much more efficient than natural air-tight sealing material, they do not attenuate vertical water flow (Fig. 5, 6)

It was found that the effective porosity of solidified FA: mine water mixtures ranged from 16.9 to 26.4% (mean 23.7%) that places these materials within the range of high and very high porosity. The FA +D-FGDS: mine water mixtures displayed considerably lower porosity than the mixtures based on "pure" FA due to the cementituous properties of calcium carbonate and sulfate, though still much higher than natural heavy soils, in particular those used as a water-tight sealing material.

The relevant values of hydraulic conductivity for solidified mixtures with use of "pure" FA were at the level of $k \geq 10^{-8}$ m/s that means that this material is permeable both for vertical and horizontal water flow. The hydraulic conductivity of solidified mixtures with use of FA + D-FGDS places them within the range of materials permeable to vertical flow, but impermeable to horizontal flow of water. The limit of the vertical impermeability to water is adequate to the hydraulic permeability $k < 3.2 \cdot 10^{-9}$ m/s at a gradient = 1[21]. This precondition has not been fulfilled with respect to the studied FA: mine water mixtures, thus hydraulic regime of the area or the object sealed with use of dense FA:Water mixtures has not been disturbed. This is a rather positive feature of this material that plays the role of a reactive permeable barrier, which can control ARD generation in coal and ore waste dumps, and spontaneous ignition in coal seams, coal mining waste dumping sites and coal piles through sealing against air penetration without changing water flow conditions in the area subject to control measures.

3. Areas of Actual and Potential Application of FA as a Sealing Agent

Taking into consideration the reported sealing properties of dense FA:Water mixtures against air penetration, there are several actual and prospective fields of application of this material, aimed to utilize these properties. In this respect, FA outclasses all available natural sealing materials and should be strongly considered for the control of oxidation in the reactive material in order to control spontaneous ignition processes, as well as to prevent or remediate the generation of ARD and trace metals immobilization.

3.1. Case Studies of Spontaneous Combustion Control with Use of Dense FA: Water Mixtures

There is already growing experience in Poland in using dense FA: water mixtures for spontaneous combustion control at coal mining waste dumping sites. In brief, the practical application of this material consists in covering a dump surface with sealing layers prepared from FA: water super-dense mixture in proportion 0.7: 0.3 (or 2.3:1).

The recent successful firefighting action was carried out at the Siemianowice dump of an area 3 hectares that was set on fire due to errors in the realization of the reclamation project (Fig. 3). The fire control action lasted for 3.5 months and was completed in November, 2004. The fire was intercepted by covering burning borders of the dump with a 5 m wide layer of dense FA: Water mixture in proportion 2.22:1. For this purpose, 533 t of FA and 240 m^3 of water was used. In the next stage, over 330 boreholes were drilled, into which flame retardation mixtures were pumped until thermal processes in the dump thoroughly terminated. The combination of sealing with dense FA:Water mixtures and flame retardants use has been dictated on one hand by the limitation of FA availability that is in the USCB fully utilized, and on the other hand by the requirement of a fast fire control. The methodology of spontaneous ignition control with use of FA: Mine water dense mixtures for interception of thermal activity has been applied successfully also at the Nowa Ruda old reshaped conic dump in 2003–2004. No open fire occurred at this site, though inside the dump the temperatures reached 1000–1500°C For cooling dump bowels, the mixture was pumped into the dump through several tenths of boreholes, each over 40 m deep.

Currently, under realization is another large project, that is a fire interception at the European biggest and oldest thermally active Skalny coal mining dump, where gases and fumes have been emitted from a large area at the surface. The works on

FIGURE 7. Sealing of the biggest in Europe thermally active Skalny dump with FA:mine water dense mixture supplied through the system of absorption and insulation trenches.

spontaneous ignition interception have been carried out since 1999, currently in the area of 9 hectares of total 30 hectares. They comprise dump sealing with dense FA: mine water mixture prepared from FA from the Laziska power plant and mine water from the Boleslaw Smialy coal mine. The sealing mixture has been supplied to the dump through spirals of absorption and insulation trenches (Fig. 7).

The measurements at the dump in the network of 680 points, showed already in 2001 a significant attenuation of the thermal activity of the Skalny dump resulted from the undertaken sealing measures with use of FA: mine water dense mixtures.

In parallel, from cooled old material and waste from current mining, the heavily compacted dump batters has been constructed as supporting measures to prevent further development of thermal activity at the dump.

The presented successful case studies confirm high efficiency of applying dense FA:mine water mixtures for spontaneous combustion of coal control.

3.2. Use of Dense FA: Water Mixtures for Coal Seams Fire Control in Mining Areas

Another very promising but not yet adequately utilized area of FA application in the form of dense mixtures with water is fire control in burning coal seams in mining areas. The scale of this problem is much bigger than the presented above spontaneous combustion interception at coal mining waste sites and is related to the vast areas of coalfields set on fire. The last issue and its environmental and economic consequences is exemplified in presented above case of Jharia Coalfields in India, where 17.32 km^2 was reported to be set on fire since 1986. The current situation is even worse in terms of the total devastation of the territory, air pollution and coal losses.

The currently applied method of fire control is cutting off the sources of air by placing three subsequent layers of repeatedly compacted seals: soil/overburden/soil[17]. This procedure requires moving and digging out enormous amounts of soil and overburden material, and its repeated compaction in

FIGURE 8. Fire control works in progress in the conditions of open flames (Tetulmari coal mine, Jharia coalfield, India)[17].

extremely difficult and dangerous conditions of open flames (Fig. 8).

Application of FA: water mixture as a sealing material of a proven, at least one order of magnitude higher penetration resistance, with no compaction required, would have provided a better effect in fire control and attenuation of environmental damages caused by fire in much simpler, safer, cost-effective and efficient way. FA is an abundant material in the area, thus its utilization as a best seal has no comparable alternative and would allow use of a large amount of FA in the most beneficial way.

3.3. Preventive Sealing of Mining Waste Dumps for ARD and Spontaneous Combustion Control

This area of application sets the task of optimization of FA use in a way that would permit to utilize in full its excellent sealing/insulation properties with respect to air penetration and attenuate adverse impact caused by fairly high leachability and permeability to water infiltration, to achieve the most positive resultant effect.

The developed methods consisted in the placing of horizontal "blanket" layers of dense FA:water mixture alternately with thicker layers of sulfidic mine waste during the dump construction to intercept ARD generation in the mining waste. In parallel, the sealing effect of FA cover plays double role, simultaneously preventing self-ignition of coal mining- and other oxidizable waste. The amount of water used should assure the highest density technically possible to achieve convenient uniform spreading of the FA cover on the top of a mine waste layer and its fast solidification with the minimum of an outflow (Fig. 9)[25]. The proven lowest proportion of FA: low quality water (saline mining water) is 2.33:1, i.e. 70% of FA and 30% of water provided that the mixture is prepared at the spot, directly at the place of application. This proportion of FA:Water in a mixture assures, besides minimization of leachate and leached loads, also the maximum reduction of solidification time. The project of FA: Water dense mixtures application in the redisposed and reconstructed coal mining dump (after coal extraction from the old coal mining waste) primarily for ARD control in parallel with spontaneous combustion prevention is currently pending.

4. Conclusions

1. While FA should compete in the market with other abundant natural materials of comparable properties, its resistance against air penetration outclasses the adequate sealing parameters of natural materials and is 1–2 orders of magnitude

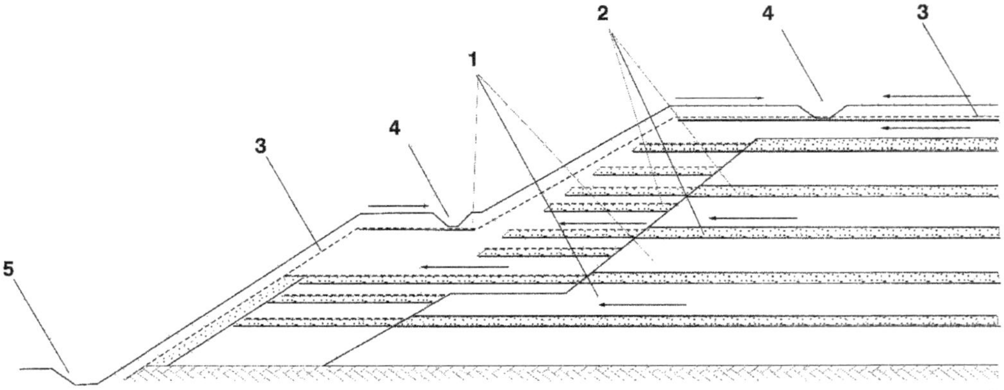

FIGURE 9. Construction of a coal mine waste dump with protective blanket layers of dense water: FA mixtures. 1—coal mining waste; 2—blanket layers of dense FA: water mixtures ($\leq$0.3 m); 3—subsurface drainage system; 4—collecting drainage ditches on the top and terraces; 5—toe drainage ditch.

higher for FA used in the form of solidified dense mixture with water (low-quality mine water use is preferable).

2. Hydraulic conductivity of solidified FA:Water mixture is relatively high, thus this material has no sealing properties against water vertical percolation under the vadose zone conditions and against horizontal water flow in the saturated zone; thus, the hydraulic regime of the engineering construction where FA dense mixtures are used will remain in principle undisturbed

3. FA, both "pure" and containing flue gas desulfurization solids (FGDS from dry and semi-dry processes), can be widely and beneficially used in a form of solidified dense mixture with water for a variety of applications where its sealing properties against air penetration are the most crucial parameters.

5. The best fields of dense FA:Water mixtures application appear to be spontaneous combustion of coal piles, coal mining dumps and coal seams spontaneous combustion control and ARD interception in sulfidic waste dumps from coal and ore mining.

6. Successful practical use of FA:Mine water dense mixtures for spontaneous combustion control at coal mining dumping sites in the USCB (Poland) proves high efficiency of this Material as a sealing/insulation barrier against air penetration.

7. Application of FA: water mixture as a sealing material of a proven, at least one order of magnitude higher penetration resistance, with no compaction required, is a prospective beneficial way of fire control and attenuation of environmental damages caused by fire in Jharia and other burning coalfields in India in much simpler, safer, cost-effective and efficient way than traditional methods being currently in use.

8. Another prospective environmentally beneficial application of FA: water dense mixtures is prevention of ARD generation in sulfidic ore/coal mining waste dumps, in parallel with protection against spontaneous ignition.

References

1. Palmer, M.E., Acidity and nutrient availability in colliery spoil, in *Environmental Management of Mineral wastes.* Goodman, G.T. and Chadwick, M.J., Eds., Sijthoff & Noordhoff, Alphen aan den Rijn—The Netherlands, 1978, 85–126.
2. Hutchinson, J.P.G. and Ellison, R.D., Eds., *Mine Waste Management.* Lewis Publishers, Boca Raton, FL, 1992, 654.
3. Szczepańska J. and Twardowska. I., Mining waste, in *Solid Waste: Assessment, Monitoring and Remediation.* Twardowska, I., Kettrup A.A.F. and Lacy, W.J,. Eds., Elsevier, Amsterdam -Boston-Heidelberg—London—New York—Oxford—Paris—San Diego—San Francisco—Singapore—Sydney—Tokyo, 2004, chap. III.6.
4. Kuchta, J.M., Rowe, V.R. and Burgess, D.S. Spontaneous combustion susceptibility in U.S. coals. *Report on Investigations 8474,* U.S. Bureau of Mines, 1980, 37.
5. Brooks, K., Svanas N., and Glasser, D. Evaluating the risk of spontaneous combustion in coal stockpiles. *Fuel,* 67, 1988, 651–656.
6. Schmal, D. Spontaneous heating of stored coal, in *Chemistry of Coal Weathering.* Nelson, C.R., Ed. Elsevier, New York, *Coal Sci. Technol.,* 14, 1989, 133–213.
7. Chandra, D., and Prasad V.V.S. Effect of coalification on spontaneous combustion of coals. *Internat. J. Coal Geol.,* 16, 1990, 225–229.
8. Misra, B.K. and Singh, B.D. Susceptibility to spontaneoue combustion of Indian coals and lignites: an organic petrographic autopsy. *Internat. J. Coal Geol.,* 25, 1994, 265–286.
9. Clemens, A.H., and Matheson, T.W. Spontaneous heating in New Zealand coals, in *Coal Science.* Pajares, J.A., and Tascon, J.M.D., Eds., Elsevier, New York. *Coal Sci. Technol* 1(24), 1995, 481–484.
10. Misra, B.K., Singh, B.D. and Anand-Prakash. Biopetrographic approach to the spontaneous combustion susceptibility of Indian Gondwana Coals: with reference to some coals of Raniganj and Singrauli coalfields, in *Gondwana nine: Ninth International Gondwana Symposium.* Geological Survey of India, v. 2, 1996, 1257–1264.
11. Beamish, B.B., Barakat M.A., and St. George, J.D. Spontaneous-combustion propensity of New Zealand coals under adiabatic conditions. *Internat. J. Coal Geol.* 45, 2001, 217–224.
12. Lyman, R.M. Phyrophoricity (spontaneous combustion) of Powder River basin coals—considerations for coalbed methane development. Wyoming Geo-Notes, No. 69, 2001, 18–22.
13. Kim, A.G. *Relative* self-heating tendencies of coal, carbonaceous shales, and coal refuse. *Report on Investigations 9537,* U.S. Bureau of Mines, 1995, 21.
14. O'Hanlon, L. Bacteria start underground fires in Mali. *News in Science,* 10.07.2003, 2, http://www.abc.net.au/cgi-bin/common/printfriendly.pl/science/news/stories.htm
15. Dziadul, J. Gates of a hell. *Polityka,* 11/11/2000 (in Polish).
16. Stracher, G.B., Ed. Coal fires burning around the World: a global catastrophe. *Internat. J. Coal Geol., Special Issue,* 59, 2004, 1–151.
17. Malhotra K.K. *Design of a Reclamation Strategy for the Wastelands Created by Coal Mining.* Ph.D. Thesis. Indian School of Mines, 2001.
18. Bień. Devil will not stoke any more. Insulation of the dump, 2003 (in Polish) *http://www.euroglas.ng.pl/archiwim.php?rok=2003&numer=309&count=1*
19. Anonymous. Burning dump in Siemianowice Slaskie (16 June 2004), *http://www.katowice.uw.gov.pl/index.php?trs=&bie=&str=&nr=822&mies=06&rok=2004&licz=&id=&katgr=27*
20. Twardowska I., Szczepańska J. Coal conbustion waste, in *Solid Waste: Assessment, Monitoring and Remediation.* Twardowska, I., Kettrup A.A.F. and Lacy, W.J,. Eds., Elsevier, Amsterdam—Boston—Heidelberg—London—New York—Oxford—Paris—San Diego—San Francisco—Singapore—Sydney—Tokyo, 2004, chap. III.7.
21. Witczak, S. and Adamczyk, A.F. Catalogue of selected physical and chemical parameters of contaminants and methods of their evaluation, in *Library of the Environmental Monitoring,* vol. 1. State Inspectorate of the Environmental Protection (PIO), Warsaw, 1994, 117 (in Polish).
22. Anonymous: Use of fly ash in mining for filling disused workings and for firefighting at burning waste dumps,

http://www.popiol.pl/User/zastosowanie.html?lang=pl&kat=&podkat=&nr=34
23. PAP—Polish Press Agency. Half million zlotys for firefighting at the coal mining waste dump in Siemianowice, 20 July 2004 (in Polish), *http://www.samorzad.pap.pl/?id=12902361&grupa_menu=gaz&klucz=GAZETA-WIADOMO%C5%9AC1*
24. Liczba, H. Coal mining waste dumps—how before and now mines solve the problem of their waste, 2004 (in Polish), *http://www.profesor.pl/mat/n10/n10_h_liczba_040429_2.php*
25. Twardowska, I. Buffering capacity of coal mine spoils and fly ash as a factor in the protection of the aquatic environment. *Sci. Total Environ.* 91, 1990, 177–189.

5
Delineation of Water and Sediment Contamination in River Near a Coal Ash Pond in Orissa, India

S. Tripathy[1] and T. Praharaj[2]

[1]*Department of Geology & Geophysics, Indian Institute of Technology, Kharagpur, West Bengal 721 302, India*
[2]*Department of Earth Sciences, University of Ottawa 140, Louis Pasteur, Ottawa, Ontario KIN 6N5, Canada*

Abstract

High concentrations of trace elements, and in most cases, their enrichment were observed in water in the post-monsoon river water in the vicinity of a coal ash disposal site. Aluminum, Fe, Mn and Pb concentrations were found enriched to varying degrees and exceeded the respective maximum permissible limits prescribed for drinking water. The concentrations of elements increased in the downstream direction irrespective of the sampling periods. Among the elements, Mn followed by Ba, V and Cr showed higher concentrations in the river sediments. The decreases in the clay content, cation exchange capacity and specific surface of the sediments in the downstream as compared to the upstream suggested distinct modifications in their physicochemical characteristics due to ash input from the effluent discharge thereby hindering the sediments' ability to scavenge and fix metals from the river water.

Introduction

Metal contamination in sediments and water of aquatic bodies has received worldwide attention due to its impact on their qualities. The metals are introduced into the environment through anthropogenic activities. While the metals are found in water as dissolved cations, colloidal dispersions and suspended particulates, the sediments become the ultimate sink in an aquatic environment.[1] Estimation of the total heavy metal concentrations in sediments has been extensively used for the purpose of pollution monitoring[2] and water quality management. Toxic metal contamination of river water near mine tailing ponds and other waste disposal sites has been well-known.[3,4]

Aquatic bodies close to a mine settling pond have been reported to contain dissolved Cd and Zn concentrations above toxic levels for aquatic organisms[5] The range of concentrations of As, Cd, Cu, Pb and Zn in streambed sediments around mining sites are generally orders of magnitude higher than the ranges of their concentrations at non-mining sites, particularly in downstream of the source and then gradually decrease with distance[6]. Significant increases in the heavy metal concentrations in the Oka river sediments in northern Spain impacted by anthropogenic wastes[7] and several orders of magnitude higher concentrations of As, Cd, Cr, Cu, Pb, Hg, Ni, Se and Zn in streambed-sediments across the Unites States compared to their average crustal abundances have been reported.[8] The latter study also concluded that anthropogenic activities resulted in higher concentrations of trace elements in the environment than those contributed by natural weathering. Availability of information on the contamination of surface water and sediments around coal ash ponds in particular reference to India is rather meagre though the distribution of Cu in river sediments near a thermal power plant in India was attributed to coal ash disposal.[9]

The current study attempts to understand few other aspects viz. the potential of ash to contaminate the soil-water environment, its pathways of contamination, resultant changes in the physicochemical properties of sediments due to ash, the degree of enrichment of metals in the surface water etc. through a case-study. The study was carried out on the river Nandira which receives effluent through drainage and numerous seepages from the ash pond of a coal based thermal power plant in Angul District of Orissa, India (Figure 1). Previous studies had indicated that the coal ash disposal has the potential to contaminate the surrounding aquatic bodies[10] near to the disposal sites due to the presence of readily leachable elements in ash[11] besides the topsoil.[12]

Materials and Methods

Water samples (NW1 to NW15) in duplicate and bed sediments (0–10 cm) (NS1 to NS6) were collected from the river in both pre and postmonsoon periods along its course (1.5 km upstream to approximately 2 km downstream (Figure 1 and 2). Effluent samples (PL1 to PL10) were also collected during the same periods from the numerous seepages (not shown on map) on

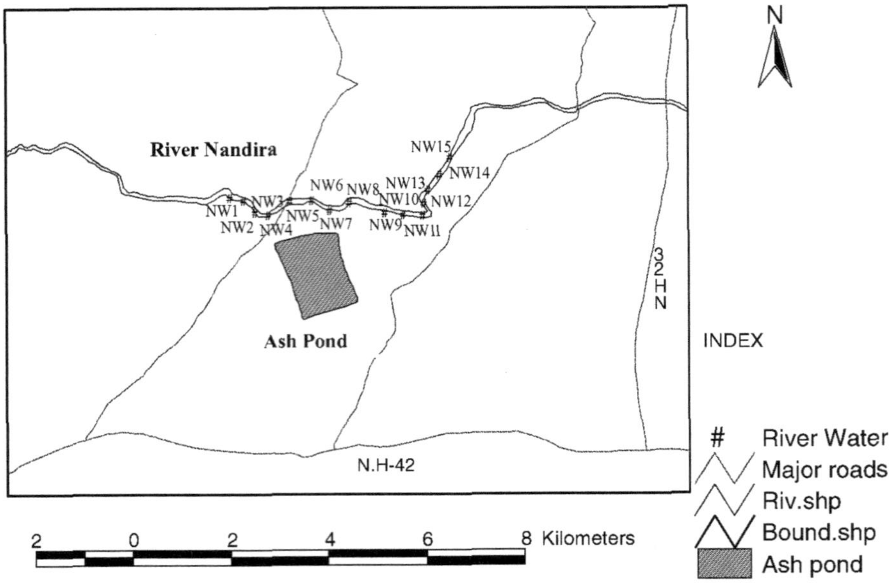

FIGURE 1. Location of the study area and water sampling points in the river.

the northern edge of the ash pond before they discharge into the river.

One part of the water sample was filtered and acidified in the field with 35% HNO_3. The pH of water from the duplicate was determined using ORION 1260 Ion Selective Electrode. The acidified samples were analyzed using an inductively coupled plasma atomic emission spectrometer (ICP OES) Perkin Elmer 3300 for determining the concentrations of As, Ba, Co, Cr, Cu, Fe, Mn, Ni, Pb, S, V, and Zn.

Homogenized bulk sediments (< 2 mm) were ground to –200 mesh (ASTM) before analyzing for element concentrations using wavelength dispersive X-ray Fluorescence Spectrometer (PHILIPS PW 2400). The pH of the sediments was measured at a solid to liquid ratio of 2:5 using Orion 1260 Ion Selective Electrode (ISE).

Diffractorams of the sediments were obtained using Phillips diffractometer (PW 1840) coupled with X-ray generator (PW 1729), using Cu-K_α. Samples were run at a tube voltage of 20 kV and 30 mA. The goniometer speed and chart speed were maintained at 0.05 2θ and 10 mm/2θ, respectively.

Specific surface area of sediments was determined by Quantasorb (Quanta Chrome Corporation, USA). Nitrogen adsorption-desorption experiments (−196 °C) were carried out using BET adsorption-desorption isotherm and the data were analyzed for the specific surface area. The other parameters analyzed for sediments included organic matter,[13] clay fraction[14] and cation exchange capacity (CEC).[15]

Results

Element Concentrations in the River Water

The concentrations of elements determined in river water both in pre and post monsoon samples are given in Table 1. It was observed that while S, Al, Fe, Mn, Ba, Pb, Zn and Ti were present in significant concentrations; Co, Cr, Cu and Ni were below the method detection limits (mdl) in all the samples and As and V were below mdl in most, irrespective of the sampling periods. The concentration profiles of elements from the upstream to the downstream reveal that they increase in their concentrations in the downstream directions irrespective of the sampling seasons (Figure 3). Specifically, very sharp and prominent spikes were observed in the

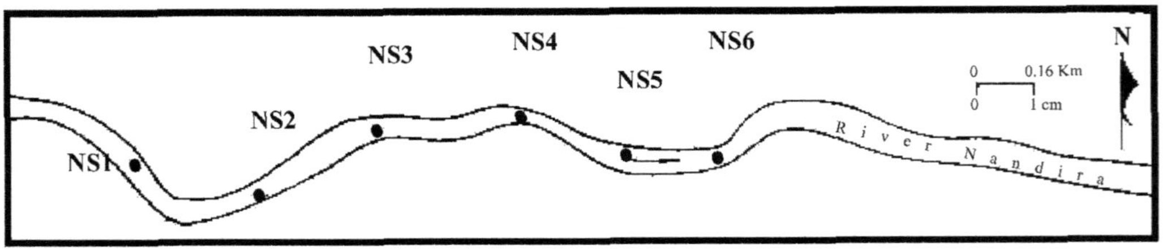

FIGURE 2. Location of sediment sampling points on the river.

TABLE 1. pH and element concentrations (in ppm) in river water (mdl: method detection limit; Pre: premonsoon; Post: postmonsoon).

Element	Period	NW1	NW2	NW3	NW4	NW5	NW6	NW7	NW8	NW9	NW10	NW11	NW12	NW13	NW14	NW15	mdl
Al	Pre	0.551	0.592	0.465	0.450	0.539	1.247	0.687	0.576	0.793	0.510	0.477	0.633	0.650	0.864	0.845	0.005
	Post	0.374	0.379	0.338	0.381	0.290	0.369	0.290	0.315	1.390	0.717	0.735	4.069	1.700	1.854	1.873	
As	Pre	mdl	mdl	mdl	mdl	mdl	mdl	mdl	mdl	mdl	mdl	mdl	mdl	mdl	mdl	mdl	0.029
	Post	mdl	mdl	mdl	mdl	0.085	mdl	mdl	mdl	mdl	mdl	0.032	mdl	mdl	0.073	0.047	
Ba	Pre	0.071	0.065	0.063	0.057	0.060	0.059	0.060	0.066	0.069	0.067	0.041	0.067	0.067	0.071	0.072	0.000
	Post	0.061	0.058	0.058	0.059	0.072	0.068	0.066	0.054	0.086	0.070	0.065	0.080	0.065	0.074	0.077	
Co	Pre	mdl	0.005	mdl	mdl	mdl	mdl	mdl	0.005	0.005	0.005	mdl	mdl	mdl	mdl	mdl	0.005
	Post	mdl	mdl	mdl	mdl	mdl	mdl	mdl	mdl	mdl	mdl	mdl	mdl	mdl	mdl	mdl	
Cr	Pre	mdl	mdl	mdl	mdl	mdl	0.003	mdl	mdl	0.003	mdl	mdl	mdl	mdl	mdl	mdl	0.002
	Post	mdl	mdl	mdl	mdl	mdl	mdl	mdl	mdl	mdl	mdl	mdl	mdl	mdl	mdl	mdl	
Cu	Pre	mdl	mdl	mdl	mdl	mdl	mdl	mdl	mdl	mdl	mdl	mdl	mdl	mdl	mdl	mdl	0.004
	Post	mdl	mdl	mdl	mdl	mdl	mdl	mdl	mdl	mdl	mdl	mdl	mdl	mdl	mdl	mdl	
Fe	Pre	0.148	0.156	0.047	0.053	0.049	0.236	0.309	0.219	0.475	0.177	0.180	0.296	0.311	0.470	0.503	0.003
	Post	0.230	0.246	0.198	0.227	0.198	0.239	0.195	0.275	0.745	0.518	0.550	3.400	1.370	1.873	1.925	
Mn	Pre	0.070	0.062	0.067	0.066	0.063	0.154	0.208	0.138	0.194	0.103	0.103	0.127	0.137	0.133	0.364	0.001
	Post	0.010	0.007	0.007	0.009	0.049	0.040	0.035	0.059	0.056	0.068	0.063	0.281	0.129	0.154	0.139	
Ni	Pre	mdl	mdl	mdl	mdl	mdl	mdl	mdl	mdl	mdl	mdl	mdl	mdl	mdl	mdl	mdl	0.020
	Post	mdl	mdl	mdl	mdl	mdl	mdl	mdl	mdl	mdl	mdl	mdl	mdl	mdl	mdl	mdl	
Pb	Pre	mdl	0.015	0.012	0.013	0.012	0.016	0.013	0.016	0.022	0.025	0.028	0.031	0.029	0.037	0.042	0.010
	Post	mdl	0.014	0.154	0.048	0.052	0.078	0.065	0.050	0.121	0.110	0.122	0.116	0.156	0.177	0.179	
S	Pre	5.296	5.372	5.307	5.328	5.405	10.630	14.220	12.880	16.610	16.190	27.910	17.170	17.280	17.470	30.870	0.024
	Post	3.329	3.246	3.249	3.643	5.760	5.651	6.095	11.054	6.147	6.936	8.272	14.441	14.873	16.237	18.894	
Ti	Pre	0.010	0.006	0.004	0.002	0.003	0.013	0.015	0.010	0.020	0.008	0.006	0.012	0.013	0.019	0.044	0.000
	Post	0.001	0.002	0.001	0.001	0.001	0.002	0.002	0.001	0.022	0.012	0.012	0.105	0.040	0.058	0.056	
V	Pre	mdl	mdl	mdl	mdl	mdl	0.002	0.003	0.003	0.002	0.004	mdl	0.002	0.002	0.003	0.022	0.002
	Post	mdl	mdl	mdl	mdl	mdl	mdl	mdl	mdl	mdl	mdl	0.003	0.010	0.005	0.007	0.007	
Zn	Pre	0.044	0.046	0.047	0.041	0.045	0.028	0.067	0.046	0.079	0.084	0.055	0.121	0.125	0.344	0.815	0.003
	Post	0.022	0.030	0.048	0.060	0.061	0.053	0.066	0.069	0.065	0.053	0.065	0.072	0.076	0.093	0.096	
pH	Pre	8.4	8.3	8.3	8.4	8.2	8.2	8.1	8.3	8.1	8.0	8.0	7.9	7.7	8.0	8.0	
	Post	8.2	8.1	8.1	7.9	8.1	8.0	7.9	8.0	7.8	7.9	7.9	7.7	8.0	8.0	8.0	

concentrations of Al, Fe, Mn, S, Ti and V between the sampling points NW11 and NW12. The ranges of pH variation in the river water were more or less similar in both the sampling periods.

Enrichments in the concentrations of elements to varying degrees were also observed in the downstream direction in both pre and post-monsoon periods. With the exception of Ba, all the elements showed enrichment from 2 to 112 times. The ranges of enrichment of different elements in river water are: Al 2-fold (pre-monsoon), 5-fold (post-monsoon); Fe 3-fold (pre-monsoon), 8-fold (post-monsoon); Mn 5-fold (pre-monsoon), 14-fold (post-monsoon); Pb 3-fold (pre-monsoon), 13-fold (post-monsoon); S 6-fold (both pre and post-monsoon); Ti 4-fold (pre-monsoon), 112-fold (post-monsoon) and Zn 19-fold (pre-monsoon) and 4-fold (post-monsoon).

Concentrations of Elements and Other Properties in the River Sediments

Samples of sediments were collected from both upstream (NS1 to NS4) and downstream (NS5 and NS6) of the confluence of ash pond effluent with the river, respectively. The concentrations of elements in both pre and post monsoon periods in these were determined (Table 2). The cation exchange capacity, specific surface area of the sediments, clay and organic matter contents were also determined (Table 3). The mineral phases of the bulk sediments comprised of quartz, feldspar, illite, kaolinite, hematite and goethite at most of the sampling locations while mullite was observed only in the downstream samples (NS5 and NS6).

Ash Pond Effluent Characteristics

The pH of the ash pond effluent was measured and the concentrations of elements analyzed are given in Table 4. The pH was mildly acidic to slightly alkaline, ranging from 7.1–7.9 in the pre-monsoon and from 6.6–7.6 in the post monsoon periods. In general, it was observed that the concentrations of most of the elements viz. Al, Ba, Fe, Pb, S and Ti were more in the post-monsoon period, compared to the pre-monsoon. The concentrations of As, Co, Cr, Cu, Ni and V were either below mdl or very low in the effluent.

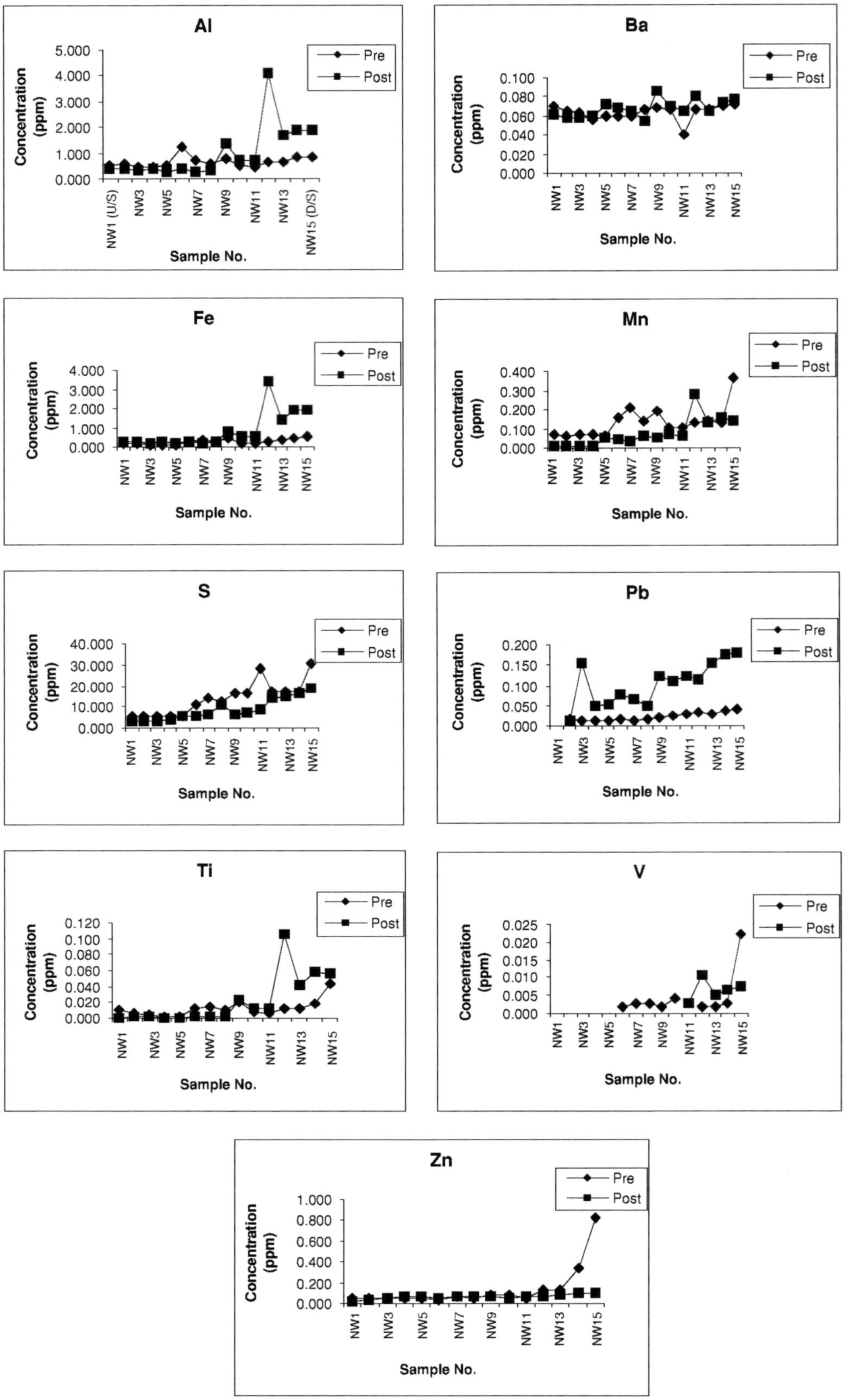

FIGURE 3. Variations in element concentrations from upstream to downstream in river Nandira.

TABLE 2. Concentrations of elements (in ppm) in river sediments

	Premonsoon						Postmonsoon					
	NS1	NS2	NS3	NS4	NS5	NS6	NS1	NS2	NS3	NS4	NS5	NS6
SiO_2(%)	79	88.61	78.6	76.9	50.8	53.2	78.35	80.3	79.92	78.6	61.6	62.7
TiO_2(%)	0.73	0.37	0.48	0.92	0.87	0.91	0.63	0.81	0.61	0.65	0.78	0.89
Al_2O_3(%)	7.18	3.26	7.61	7.72	15.9	16.9	6.34	6.06	6.91	8.57	13.3	12.8
Fe_2O_3(%)	3.81	1.85	3.62	3.46	6.22	5.39	3.55	3.4	3.46	4.27	5.38	5.63
MgO (%)	0.76	0.56	1.24	0.89	1.89	1.64	1.01	0.67	0.72	1.03	1.34	1.3
CaO (%)	1.43	1.19	5.41	2.39	7.85	7.61	1.7	1.43	1.38	3.41	4.34	3.74
K_2O (%)	2.16	1.61	2.06	2.25	2.4	2.26	1.93	1.74	2.02	2.23	2.1	2.76
P_2O_5(%)	0.14	0.06	0.07	0.08	0.11	0.13	0.06	0.06	0.1	0.08	0.07	0.1
Total (%)	95.18	97.51	99.12	94.61	86.06	87.97	93.57	94.48	95.12	98.88	88.93	89.91
LOI (%)	4.82	2.49	0.88	5.39	13.94	12.03	6.43	5.52	4.88	1.12	11.07	10.09
Pb (ppm)	24	17	26	27	31	36	18	21	20	28	29	30
As (ppm)	<2	<2	<2	<2	4	4	3	<2	<2	<2	3	3
Zn (ppm)	24	10	40	33	69	62	31	23	26	47	61	58
Cu (ppm)	11	5	14	7	27	39	10	6	14	15	24	30
Ni (ppm)	13	3	22	17	49	54	15	14	17	30	39	36
Co (ppm)	6	6	16	12	21	20	12	12	17	16	22	26
Mn (ppm)	634	493	917	964	1658	1846	583	750	834	1232	2179	1494
Cr (ppm)	55	23	72	50	107	103	64	48	48	83	106	95
V (ppm)	63	28	67	62	102	97	63	52	57	83	100	108
Ba (ppm)	550	399	558	563	588	632	458	423	568	559	584	698

Discussion

Behaviour of Elements in the River Water

Significant concentrations of S, Al, Fe, Mn, Ba, Pb, Zn and Ti and detectable concentrations of Co, Cr, Cu and Ni were observed in the river water. It has been established that most of these elements are leachable from coal ash, while As and V are non-leachable from the ash in the study area.[10] It was also observed that with the exceptions of Mn, S and Zn, the concentrations of most other elements were higher in the post-monsoon period compared to the pre-monsoon possibly due to higher rates of leaching from the ash pond. The sharp rises in the element concentrations observed between the locations NW11 and NW12 correspond to the mixing of effluent from the ash pond with the river water which observed during sampling. The gradual increase in the concentrations of the most of the elements in the downstream direction could be attributed to their continuous input into the river from numerous seepages from the ash pond as well as from the effluent discharge channel.

Enrichment of Elements

The enrichment of elements calculated with respect to the WHO guidelines (1993) showed that the concentrations of Al, As, Fe, Mn and Pb were enriched and exceeded their maximum permissible limits in drinking water. The range of enrichment values for different elements were: Al 2–6 (pre-monsoon), 1–20 (post-monsoon); Fe 1–2 (pre-monsoon), 1–11 (post-monsoon); Mn 1–4 (pre-monsoon), 1–3 (post-monsoon) and Pb 1–4 (pre-monsoon) and 1–18 (post-monsoon). No enrichment of As in the pre-monsoon period and 3–8 times enrichment in the post-monsoon period was observed. These indicated that the river water is contaminated to varying degrees by all the above-mentioned elements and hence not potable. In general, enrichments in element concentrations were found to

TABLE 3. Physicochemical properties of the river sediments in pre and postmonsoon periods

	NS1		NS2		NS3		NS4		NS5		NS6	
Parameters	Pre	Post	Pre	Post	Pre	Post	Pre	Post	Pre	Post	Pre	Post
Sand (%)	21	26	17	22	18.5	24	25	22	33	32	29	31
Silt (%)	58.5	52	65	59	63	56	66	61	62	62	65	64
Clay (%)	20.5	22	18	19	18.5	20	10	17	5	6	6	5
CEC (meq/100g)	15.1	15.2	17.4	16.3	14.7	16.8	11	13.8	4.2	7.1	5	5.6
Organic Matter (%)	0.95	0.92	1.01	0.96	0.98	0.91	0.81	1.01	0.77	0.87	0.74	0.82
Specific Surface Area (m^2/g)	42.5	45.7	56.4	51.6	52.7	49.8	37.8	44.6	17.4	15.3	19.7	17.6

TABLE 4. pH and concentrations of elements in ash pond effluent (concentrations in ppm; D/S: downstream, U/S: upstream)

Sample ID	Period	Al	As	Ba	Co	Cr	Cu	Fe	Mn	Ni	Pb	S	Ti	V	Zn	pH
PL1	pre	1.464	mdl	0.063	mdl	mdl	0.010	0.794	0.378	mdl	0.010	19.290	0.061	0.014	0.089	7.2
(D/S)	post	1.700	mdl	0.063	mdl	mdl	mdl	1.488	0.488	0.021	mdl	22.441	0.048	0.003	0.044	6.6
PL2	pre	1.238	mdl	0.060	mdl	mdl	0.009	1.304	0.614	mdl	mdl	23.280	0.041	mdl	0.090	7.3
	post	2.269	0.038	0.084	0.010	0.017	0.065	3.606	0.983	0.024	0.046	54.032	0.040	0.024	0.578	7.5
PL3	pre	1.714	mdl	0.072	mdl	mdl	0.009	1.267	0.604	mdl	mdl	19.470	0.029	mdl	0.078	7.4
	post	3.194	0.053	0.088	0.005	mdl	mdl	2.448	0.601	0.025	0.055	26.228	0.075	mdl	0.037	7.4
PL4	pre	1.446	mdl	0.040	mdl	mdl	0.009	1.218	0.210	mdl	mdl	17.390	0.025	mdl	0.215	7.1
	post	1.015	mdl	0.065	mdl	mdl	mdl	0.693	0.036	mdl	0.089	22.287	0.020	mdl	0.028	7.0
PL5	pre	1.536	mdl	0.050	mdl	mdl	0.009	1.960	0.296	mdl	0.010	18.380	0.047	mdl	0.065	7.7
	post	2.911	0.041	0.070	mdl	mdl	mdl	2.774	0.481	mdl	0.056	25.550	0.081	0.002	0.039	7.2
PL6	pre	1.430	mdl	0.039	mdl	mdl	0.027	0.960	0.217	mdl	mdl	20.590	0.049	0.003	0.076	7.3
	post	1.802	mdl	0.051	mdl	mdl	mdl	1.464	0.180	mdl	0.057	25.498	0.063	mdl	0.046	7.4
PL7	pre	4.437	mdl	0.090	mdl	0.005	0.011	3.520	0.536	mdl	0.010	17.100	0.089	0.007	0.058	7.2
	post	2.874	mdl	0.060	mdl	mdl	mdl	1.991	0.272	mdl	0.026	27.018	0.109	0.006	0.033	7.5
PL8	pre	1.387	mdl	0.043	mdl	mdl	0.010	1.231	0.527	mdl	mdl	16.270	0.039	0.003	0.062	7.9
	post	2.280	0.034	0.079	mdl	mdl	mdl	1.854	0.485	mdl	mdl	23.059	0.061	mdl	0.040	7.4
PL9	pre	1.344	mdl	0.035	mdl	mdl	0.010	1.517	0.766	mdl	0.010	33.500	0.053	0.006	0.089	7.3
	post	1.533	mdl	0.041	mdl	mdl	mdl	1.205	0.323	mdl	0.072	22.924	0.033	mdl	0.033	7.2
PL10	pre	0.311	mdl	0.070	mdl	0.002	0.008	0.071	0.194	mdl	0.010	16.060	0.006	0.015	0.044	7.8
(U/S)	post	1.804	mdl	0.055	mdl	mdl	mdl	2.010	0.756	mdl	mdl	25.306	0.068	mdl	0.028	7.6

be higher during the post-monsoon, and may be attributed to higher rate of leaching from the ash pond.

Anthropogenic Impact on Elemental Distribution

It is known that the concentrations of trace elements closely follow lognormal distribution pattern.[16] The background values serve as thresholds between the polluted and the unpolluted and can be derived from these lognormal distributions.[17] The distributions of elements determined in the river water are shown in Figure 4 and it was observed that the poor linearity in the distribution of element concentrations and their low square correlation coefficients (R^2) are likely due to significant anthropogenic impact on the river water through the ash disposal. Very poor linearity and R^2 values of 0.55, 0.50, 0.61, 0.58 and 0.28, respectively suggest maximum impact in the distribution of Al, Fe, Ti, V and Zn in the river water. The inflection points can easily be noted in their distributions. On the other hand, the contrast between the polluted and the unpolluted sediments are not well defined in the case of Ba, Mn, Pb and S which showed slightly better linearity (R^2: 0.87, 0.82, 0.85 and 0.87, respectively). This suggests that their distributions in the river water are affected by the effluent discharge to some extent but not markedly.

River Sediment Characteristics

The upstream sediments were found to have retained their characteristic brownish natural colour typical of clayey weathered ones; while in the downstream they are more greyish in colour due to the predominance of coal ash content in them in the downstream. The ash load was observed to be thinning out towards further downstream.

Element Concentrations

The higher concentration of Mn followed by those of Ba, V and Cr in the river sediments can be attributed to the settling of the suspended particulates[18] from the pond effluent or likely due to the preferential adsorption of dissolved fractions onto their finer particles.[19] Enrichment in Mn concentrations by 2 to 4 times in downstream sediments as compared to the upstream ones confirms its contribution form the ash since the ash contains very high concentrations of Mn.[12] Similar trends have also been observed for all the other elements. The LOI of sediments in downstream were observed to be relatively higher than in the upstream mainly due to the presence of unburnt coal particles in the ash brought in by the pond effluent.

Changes in the Physicochemical Properties of Sediments

Mineralogy

The major mineral phases identified in the diffractograms of river sediments were:- quartz, kaolinite, illite, K-feldspar, hematite and goethite. Interestingly, mullite, one of the major mineral phases in the coal ash was detected only in the diffractograms of downstream sediments, thereby, confirming the presence of ash. The presence of mullite further corroborates

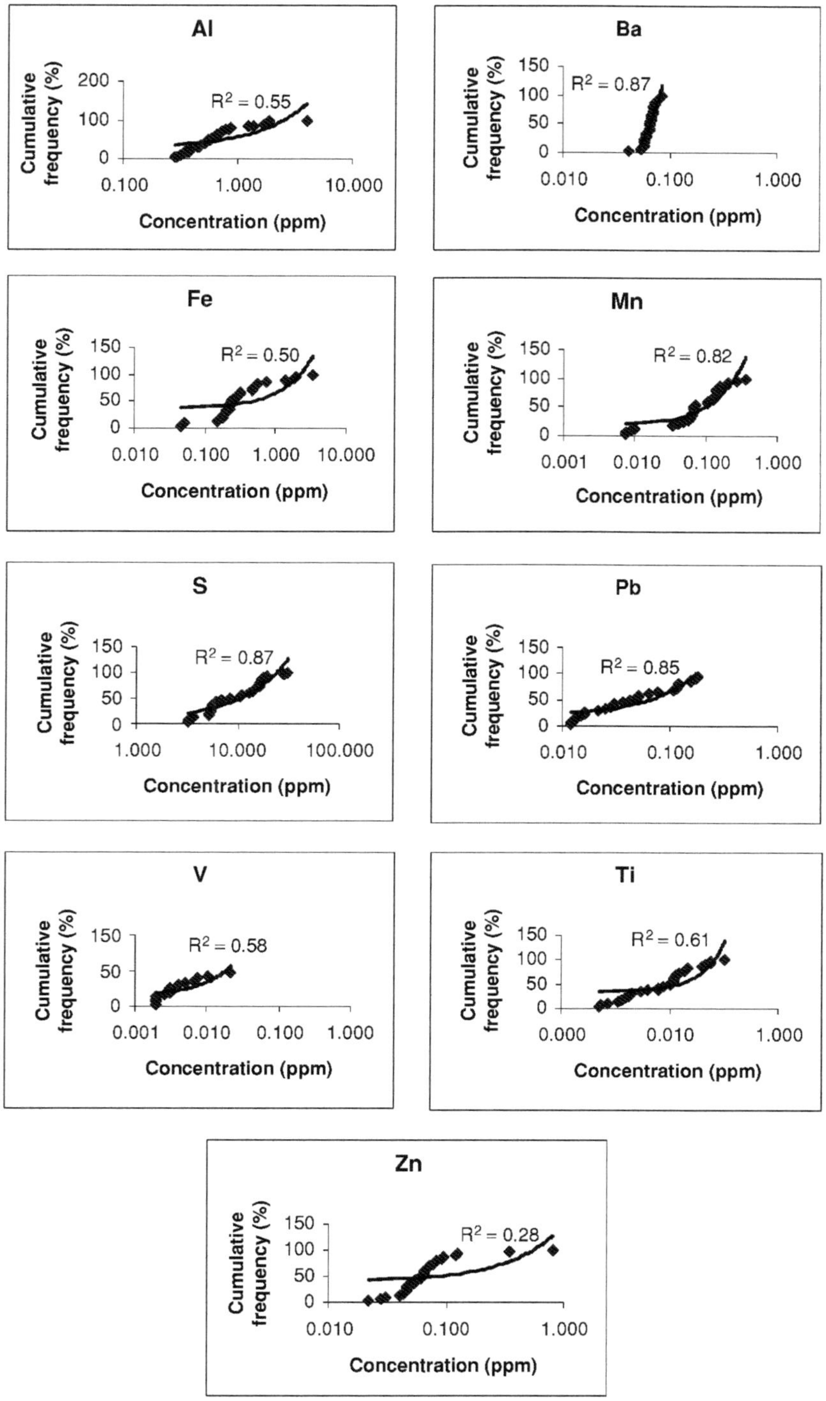

FIGURE 4. Concentration distributions of different elements in river water.

with the field observation of greyish particles in the downstream sediments.

Clay Content

The clay content in pre-monsoon sediments showed distinct variation from the upstream to downstream (Table 3); the upstream sediments being higher ranging from 10 to 20.5% compared to the downstream with lower content of 5 to 6%. The trend was similar in the post-monsoon period. The lowering of clay content of the sediments in the downstream is due to the mixing of coarser pond ash mostly comprising of silt to fine sand sized particles.

Cation Exchange Capacity (CEC)

The CEC of river sediments in the pre-monsoon period were distinctly lower, ranging from 4.2–5.0 meq/100gm in the downstream as compared to the upstream ewhere they ranged from 11.0–17.4 meq/100gm. The trend remained similar in the post-monsoon collected sediments. Both the fly ash and the pond ash from the study area are characteristically devoid of CEC.[20] Therefore, mixing of the ash with the sediments in the downstream resulted in the lowering of CEC which is also confirmed from the lowering of clay content as mentioned earlier.

Specific Surface Area

Distinct lower values (17.4–19.7 m^2/gm) of specific surface area was observed in the downstream sediments as compared to the values of the upstream ones (37.8–55.4 m^2/gm) in pre-monsoon period and the trend was retained in the case of post-monsoon sediments. The lowering of the specific surface area is attributed to the settling of ash predominantly consisting of silt and sand-sized particles in the downstream brought in from by the pond effluent as suspended matter. The changes in the various physicochemical properties clearly confirm that the behaviour of river sediments in the downstream of the effluent confluence have been distinctly modified due to input of ash. These are very much in agreement with previous studies on the physicochemical properties of ash contaminated topsoil around the disposal pond[12] and others conducted on agricultural soils.[21–23]

Ash Pond Effluent Characteristics

pH and Concentrations of Elements

The pH of the ash pond effluent was near neutral as compared to the alkaline pH of the river water. Relatively higher concentrations of S, Al, Fe, Mn, Ba, Zn and Ti known to be readily leachable from coal ash in the study area[11] was observed. The concentrations of all these elements in the effluent were in similar ranges as observed in the case of ash leachates during laboratory leaching.[11] The other elements found in lesser concentration levels in the effluent are mostly non-leachable from the ash. Interestingly, the concentrations of most of the elements, in general, were few orders of magnitude higher in the ash pond effluent compared to the river water. It is possible that the mixing of the ash pond effluent with the river water has resulted in dilution and leading to decreases in their concentrations.

Enrichment of Elements

Different elements were found to have enriched their concentrations with respect to WHO guideline values in varying degrees. The ranges of enrichment were:- Al 2–22, Fe 2–12, Mn 1–2 and Pb 1–9. The concentrations of these elements in the ash pond seepage water exceeded their respective maximum permissible limits prescribed by WHO for drinking water therefore ascertain their potential to contaminate the river water.

Conclusion

The study established that Al, Fe, Mn and Pb are the major contaminants contributed from the ash pond effluent to the river water and their enrichment with respect to the respective prescribed limits confirm that the river water is contaminated to varying degrees and therefore not potable. Most other elements are found to be significantly enriched in both river water and sediments in the downstream from the effluent discharge. Alarmingly, the changes in the physicochemical properties of the sediments in the downstream are critical since their natural ability to scavenge metals from the contaminated river water is altered.

Acknowledgement. The authors are grateful to "Multipurpose Recycling Project" funded by the Canadian International Development Agency (CIDA) and the "Land Restoration through Waste Management" project funded by the India Canada Environment Facility (ICEF), New Delhi, India for providing analytical facilities for this work.

References

1. Sutherland, R. A. (2000). Bed sediment-associated trace metals in an urban stream, Oahu, Hawaii. Environ. Geol. 39(6): 611–627.
2. Tada, F., Nishida, H., Miyai, M. and Suzuki, S. (1983). Classification of Japanese rivers by heavy metals in bottom mud. Environ. Geol. 4: 217–222.
3. Carrillo, A. and Drever, J.I. (1998). Adsorption of arsenic by natural aquifer material in the San Antonio-El Triunfo mining area, Baja California, Mexico. Environ. Geol. 35(4): 251–257.
4. Sprenke, K.F., Rember, W.C., Bender, S.F., Hoffmann, M.L., Rabbi, F. and Chamberlain, V.E. (2000). Toxic metal contamination in the lateral lakes of the Coeur d'Alene River valley, Idaho. Environ. Geol. 39(6): 575–586.
5. Rabe, F.W. and Bauer, S.B. (1977). Heavy metals in the lakes of the Coeur d'Alene river valley, Northwest Sci. 51: 183–197.
6. Deacon, J.R. and Drever, N.E. (1999). Distribution of trace elements in streambed sediment associated with mining activities in the upper Colorado River basin, Colorado, USA, 1995–96. Arch. Environ. Contamination Toxicology. 37: 7–18.
7. Irabien, M.J. and Velasco, F. (1999). Heavy metals in Oka river sediments (Urdaibai National Biosphere Reserve, northern Spain): lithogenic and anthropogenic effects. Environ. Geol. 37(1/2):
8. Rice, C.K. (1999). Trace element concentrations in streambed sediment across the conterminous United States. Environ. Sci. Technol. 33(15): 2499–2504.

9. Ravichander, D.V., Venkobachar, C. and Raymahashay, B.C. (1994). Retention of fly ash-derived copper in sediments of the Pandu river near Kanpur, India. Env. Geol. 24: 133–139.
10. T. Praharaj, S.P. Swain, M.A. Powell, B.R. Hart and S. Tripathy (2002a). Delineation of groundwater contamination around an ash pond: Geochemical and GIS approach. Environment International, 27 (8): 631–638.
11. T. Praharaj, M.A. Powell, B.R. Hart and S. Tripathy (2002b). Leachability of elements from sub-bituminous coal fly ash from India. Environment International, 27 (8): 609–615.
12. T. Praharaj, S. Tripathy, M.A. Powell and B.R. Hart (2003). Geochemical studies to delineate topsoil contamination around an ash pond of a coal based thermal power plant in India. Environmental Geology, 45 (1): 86–97.
13. Walkley A, Black IA (1934) An examination of Degtjareff method for determining soil organic matter and a proposed modification of the chromic acid titration method. Soil Sci 37: 29–38.
14. ASTM (1990) Standard methods for particle size analysis of soils (D422–63; reapproved 1990). In 1996 Annual Book of ASTM Standards, Vol. 4.08. ASTM, Philadelphia, PA., pp. 10–16.
15. Sankaram A (1966) A laboratory manual for Agricultural Chemistry. Asia Publishing House, Calcutta, India, pp 120–122.
16. Förstner U (1983) Applied environmental geochemistry. Academic Press, London, UK.
17. Celo V, Babi D, Baraj B, Cullaj A (1998) An assessment of heavy metal pollution in the sediments along the Albanian coast. Water Air and Soil Pollut 85: 235–249.
18. Hung, J. J. and Chan, C.-L. (1998). Distribution and enrichment of particulate trace metals in the southern East China Sea. Geochemical J. 32: 189–203.
19. Teixeira, E.C., Sanchez, J.C.D., Migliavacca, D., Binotto, R.B. and Fachel, J.M.G. (2000). Environmental assessment: study of metals in fluvial sediments in sites impacted by coal processing and steel industry activities. Fuel. 79: 1539–1546.
20. Sahu KC (1999) Mineralogy of coal and fly ash. Ind J Geol 71: 93–103.
21. Fontana P (1981) Sistema agricolo e produzine di energia termoelettrica. Proc. Conf. on "Centrali termoclettriche, uomo e ambients", Gardone Riviera, 8–9 October, Czechoslovakia, pp 101.
22. Bilski JJ, Alva AK, Sajwan KS (1995) Fly Ash. In: Rechcigl JE (eds) Soil amendments and environmental quality. Chapter 9. Lewis Publishers, New York, pp 327–361.
23. Joseph, L., Fail, J.L. Jr. and Wochok, Z.S. (1997). Soybean growth on fly ash amended strip minespoils. Plant and Soil. 48: 473–484.

6
Prediction of Coal Ash Leaching Behavior in Acid Mine Water: Comparison of Laboratory and Field Studies

Paul, F. Ziemkiewicz[1] and Anna, S. Knox[2]

[1]*West Virginia Water Research Institute, West Virginia University, Morgantown, West Virginia, 26506, USA*
[2]*Savannah River National Laboratory, Westinghouse Savannah River Company, 773-42A, Savannah River Site, Aiken, SC 29808, USA*

Abstract

Strongly alkaline fluidized bed combustion (FBC) ash is commonly used to control acid mine drainage (AMD) in West Virginia coal mines. Objectives include acid neutralization and immobilization of the primary AMD pollutants: iron, aluminum and manganese. The process has been successful in controlling AMD though doubts remain regarding mobilization of other toxic elements present in the ash. In addition, AMD contains many toxic elements in low concentrations. And, each mine produces AMD of widely varying quality. So, predicting the effect of a particular ash on a given coal mine's drainage quality is of particular interest. In this chapter we compare the results of a site-specific ash leaching procedure with two large-scale field applications of FBC ash. The results suggested a high degree of predictability for roughly half of the 25 chemical parameters and poor predictability for the remainder. Of these, seven parameters were successfully predicted on both sites: acidity, Al, B, Ba, Fe, Ni and Zn while electrical conductivity, Ca, Cd, SO4, Pb and Sb were not successfully predicted on either site. Trends for the remaining elements: As, Ag, Be, Cu, Cr, Hg, Mg, Mn, pH, Se Tl and V were successfully predicted on one but not both mine sites.

1. Introduction

Several test procedures have been developed in an attempt to predict the leaching behavior of coal combustion byproducts (CCBs). CCBs include various types of coal ash, flue gas desulfurization solids and boiler slag. The most widely used procedure is the Toxicity Characteristic Leaching Procedure (TCLP), which was designed by the United States Environmental Protection Agency to "determine the mobility of both organic and inorganic analytes present in liquid, solid and multiphasic wastes."[1] However, this test may only account for fast reactions that take place in short term leaching processes.[2] The results obtained with the TCLP on a limited number of fly and bottom ash samples showed that the range of concentrations of specified constituents was well below regulatory limits.[3] These residues are therefore classified as non-hazardous wastes and can be disposed of on land without risk of contaminating groundwaters to the extent of exceeding drinking water standards. However, Bhumbla et al. (2000) concluded that, environmental concerns about toxic element release from fly ash amended soils have been lessened by evaluation of data from short-term studies in highly alkaline environments.[4] They point out that all soils in humid regions ultimately become acidic and that the behavior of ash under conditions pertaining over the long term needs to be addressed. They encouraged a better understanding of the mechanisms and rates of trace element release from CCBs so that behavior over the long term can be predicted.

Regulatory agencies, industry and the public need a tool for quantative prediction of leachate composition to properly assess the risks and benefits of a particular CCB application for mine reclamation. Currently applied laboratory extraction procedures yield imprecise estimates of field leachate composition and field studies alone do not provide the causal relationships for the observed behavior. Also, since TCLP and SPLP use standard leaching solutions, they do not predict interactions between the solid waste and components of a specific mine water. Therefore, our research has focused on developing improved laboratory methods for predicting the leaching behavior of CCBs under field conditions. The geochemical reactions of dissolution/precipitation, adsorption/desorption, and oxidation/reduction are recognized as controlling the mobilization of various constituents from solid residues.[5,6]

To address these uncertainties, a leaching procedure has been developed that, through an empirical procedure, recognizes the interactions among mine water chemistry and specific CCBs. This sheds light on the fundamental geochemical reactions critical to interpretation and prediction of leachate chemistry and interactions with various geological materials. Such laboratory studies can be conducted for a wide range of conditions, such as pH, complexations, and ionic strength of leachates and porewaters that would be encountered at different field sites. It is hoped that the result will yield a laboratory

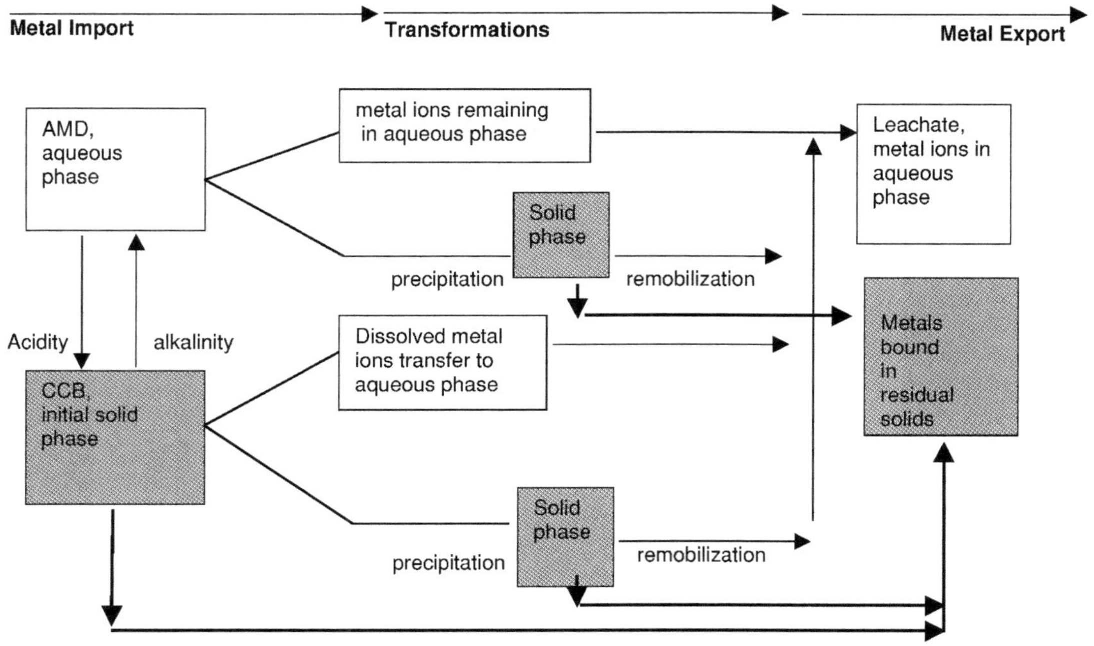

FIGURE 1. **Diagram of the experimental system.** Metals can be imported to the system from either the AMD (aqueous form) or the waste (solid form). Once in the system, metals can either stay in the aqueous phase, precipitate into the solid phase or resolubilize into the aqueous phase.

procedure that will recognize site conditions and predict in a timely and inexpensive manner the risks and benefits of CCB applications in coal mines. We are attempting to first isolate the modeled system first as an interaction between mine water and the CCB. Subsequent study will include interactions between the resulting leachate and the mine spoil.

Figure 1 illustrates the model that we are attempting to simulate. It assumes that all metal ions are imported either via the acidic mine groundwater or the CCB. The shaded squares indicate that the metal ions reside in the solid phase and the white squares indicate aqueous compartments. Initially, acid/base neutralization reactions occur between the CCB and the AMD. This step may persist over many years and it is a function of the alkalinity content of the CCB, its permeability and the acid flux generated by the AMD. Contact between AMD and CCB groundwater can liberate ions through dissolution of the CCB and immobilize others as hydroxides, complexes or through adsorption to precipitated metal hydroxides. These hydroxides may dissolve and adsorbed ions may be released as the pH changes. Heavy lines in figure 1 indicate immobilization pathways leading to storage in residual solids. Thin lines indicate transfer to mobile or aqueous phases as dissolved ions.

The above model does not account for interactions with mine spoil. Rather, this is a first step toward a systematic study of the factors that mobilize and immobilize toxic ions in CCB-minefill settings. Here were only explore the interaction of CCB and the mine's groundwater. Our objective is to test the validity of the laboratory leaching procedure by comparing the results with actual field results using the same CCB and AMD. Throughout we are using the Mine Water Leaching Procedure (MWLP).

It was developed to provide a site-specific risk assessment tool for CCB-minefills. The procedure was developed to determine the long-term leaching potential of toxic elements present in industrial wastes when placed in the groundwater found at a specific mine. Of particular concern is the leaching behavior of these materials when placed in acidic environments associated with reclamation of acidic mine spoils.

It is recognized that correlations between leaching studies and field observations are tenuous at best. We hope that by studying the laboratory leaching behavior of individual parameters and by comparing the results to multiple field sites we can begin to distinguish laboratory artifacts and parameters that have predictive value. Use of CCBs for Coal Mine Reclamation: According to the American Coal Ash Association, 122 million tons of CCBs were produced by the power generating utilities in 2003. Of that total, 2.23 million tons were used in mining applications.[7] CCBs are used as structural fills in mining and are also used to control acid mine drainage, subsidence and for soil reclamation.[8] Class C Fly ash and Class F fly ash mixed with lime exhibit self-cementing properties and can be used to cap surfaces, line pavements and isolate acidic materials in the backfill to prevent AMD formation. In addition, highly alkaline CCBs, such as FGD and FBC residues, are used to directly neutralize acidic materials.

Fly ash generally has a strong influence on spoil pH.[9] Fly ashes are commonly alkaline but may be neutral or acidic as a function of CaO, salts and various amorphous oxides of Fe, which adhere to the exterior of the fly ash spheres.[10] Alkalinity released as CaO is liberated from the dissolving aluminosilicate matrix.[9,11] Alkaline fly ashes normally contain sufficient neutralization capacity to raise the pH of acid soils.[12] While many class F fly ashes have a high paste pH (10 to 12), their neutralization potential is low so extremely high additions (up to 625 tonnes of fly ash per hectare) may be needed to sustain a neutral pH in acidic coal mine soils.[13,14,15] There have been several case studies on the direct neutralization of AMD using highly alkaline CCBs.[16,17]

CCBs are also used during surface reclamation to improve soil physical and chemical properties of acidic minesoils.[18,19] A study conducted by Dhaliwal and associates compared the properties of a mineland area reclaimed with fly ash to an adjacent area that received no fly ash. Even after 22 years the fly ash treated mine soil had a higher pH and thicker organic horizon than the adjacent untreated area.[20] Bhumbla used a fly ash/rock phosphate mixture as a topsoil substitute and found that fly ash addition improved certain physical conditions of the mine soils and reduced Al, Fe and Mn toxicities.[21]

Placement of coal combustion byproducts (CCBs) in coal mines raises the prospect of reactions with geochemically aggressive, acidic groundwater in the mine. The CCPs are generally alkaline and are often seen as valuable amendments for groundwater improvement. However, the weathering behavior of CCPs will change with exhaustion of their alkalinity and regulatory agencies need to know the potential to leach toxic elements over extended time periods.

2. Materials and Methods

Field Sites: Coal ash and acid groundwater (AMD) were collected from two sites where FBC ash had been used to line the acid-generating mine floor rock and also to isolate toxic materials in the backfill of two surface mines in North Central West Virginia. One mine is in Monongalia County (Site M) and the other is in Preston County (Site P). Site M produces a mild AMD with a pH of 4.5, [acidity]: 144 mg/L, [Fe]: 5 mg/L and [Mn]: 0.3 mg/L and [Al]: 23 mg/L. This site is a contour surface mine where a 0.3 m layer of compacted FBC ash was placed on the pit floor prior to backfilling and another 0.3 m layer placed over the pyritic spoil prior to placement of the final soil cover. This method has prevented further AMD formation on the site. We collected our AMD sample from earlier mining works on the same mine that had not received the FBC ash treatment.

The Site P surface mine recovered coal from a more highly pyritic seam. As a result the AMD is more strongly acidic. The pH is 2.9, [acidity]: 766 mg/L, [Fe]: 169 mg/L, [Mn]: 2 mg/L and [Al]: 39 mg/L. FBC ash placement was similar to Site M. The FBC ash used at both mines came from the same FBC unit. FBC ash had been applied to both sites about 10 years prior to sampling.

Field Sampling: The pH and water temperature were taken with a Hanna Waterproof pH Tester. Sample bottles for water samples were acquired from an State certified analytical laboratory. Two sample bottles were used per sample: one non-acidified bottle for analyzing lab pH, acidity, alkalinity, conductivity and SO_4; and one acidified bottle for analyzing metals such as Fe, Al, Mn, Ca, Mg, As, Ag, B, Be, Ba, Sb, Cd, Cr, Hg, Ni, Pb, Se, Tl, V, and Zn. A non-acidified bottle was used to acquire the first sample. The bottle was rinsed with the sampled water three times before it was entirely filled with the sample. The non-acidified sample was refrigerated until taken to the analytical laboratory. The second acidified bottle was filled with a filtered water sample. A cleansed syringe with a 0.45 μm filter was attached and at least 80 mls of water were filtered into the pre-acidified bottle. Samples were taken to the State certified analytical laboratory for analysis within 48 hours.

All samples were collected on 20 September 2004. Estimated flows at Site M were about 20 gpm while those at Site P were about 60 gpm Samples were collected at seeps from both ash treated and untreated locations at each site. Water from the untreated sites was then used as the sites's AMD sample and was used as the leachant for the laboratory leaching tests.

Laboratory leaching procedure: We used the Mine Water Leaching Procedure (MWLP)[22,23] as our laboratory leaching method. It was developed to sequentially leach a particular CCP with the target mine's groundwater to evaluate which elements are likely to be mobilized as alkalinity is exhausted. This provides insight into long-term behavior as well as physical and chemical interactions among the reaction products. This study compared the results of laboratory MWLP leachings with field observations at large-scale CCP applications at acid producing coal mines. Ten grams of coal ash were weighed out and transferred into clean, dry, 2-L plastic reaction bottles. Two liters of AMD were added to each experimental bottle. The bottles were then sealed. The bottles were then agitated for 18 hrs on an end-over-end rotating platform at 30 rpm. After agitation the contents of each bottle were filtered through a 0.7 um glass, borosilicate filter using a stainless steel pressure filtration unit at or below 40 psi. Filtered leachate samples were then analyzed for acidity, pH, electrical conductivity, Al, As, Ag, B, Ba, Be, Ca, Cd, Cr, Cu, Fe, Hg, Mg, Mn, Ni, Pb, Sb, Se, SO_4, Tl, V and Zn. Acidity was calculated from pH and major metal concentrations by the formula:

$$Acidity = 50(1000(10^{-pH}) + \sum_{i=1}^{n}(O_{me}[M_{me}]/A_{me})_i) \quad (1)$$

where:
acidity is given in mg/L $CaCO_3$ *equivalent*
O_{me}= *oxidation state of metal ion i*
[Me] = concentration of metal ion i in mg/L
A_{me}= *atomic weight of metal ion i*

Since pH, Fe, Mn and Al account for virtually all of the acidity in coal mine drainage, the following formula is used:

$$acidity = 50((1000x10^{-pH}) + (3x[Fe]/55.8) + (2x[Mn]/54.9) + (3x[Al]/27)+) \quad (2)$$

TABLE 1. Changes in leachate chemistry through 5 leach cycles using 10g of FBC ash leached with AMD from the Site M surface mine. The far right column is the net difference between AMD and cycle 15 leachant

	units	method detect. limit	US Primary DW Std.	Site M AMD	Cycle 1	Cycle 5	Cycle 10	Cycle 15	AMD vs. last cycle
pH				4.5	9.6	7.6	6.5	6.6	48%
acidity	mg/L			144.0	3.0	133.0	96.7	86.3	−40%
alkalinity	mg/L			0.0	20.6	13.3	37.8	53.8	
Mg	mg/L	0.1		374.1	247.0	420.9	333.5	333.5	−11%
Ca	mg/L	0.1		534.5	834.6	561.4	511.4	543.6	2%
Fe	mg/L	0.1		5.0	0.3	1.3	0.1	0.1	−98%
Al	mg/L	0.1		23.2	0.3	0.2	0.1	0.1	−100%
Mn	mg/L	0.1		0.3	0.2	70.5	52.7	47.0	15344%
SO4	mg/L	10		2590.7	2587.2	2847.4	2660.6	2908.1	12%
Sb	mg/L	0.00087	0.006	0.00087	0.00087	0.00087	0.00087	0.00087	0%
As	mg/L	0.00410	0.05	0.00410	0.00410	0.00410	0.00410	0.00410	0%
B	mg/L	0.01000	2	0.03120	0.11747	0.10200	0.08307	0.11367	264%
Ba	mg/L	0.00116	0.004	0.00116	0.10292	0.02324	0.01528	0.01650	1323%
Be	mg/L	0.00071	0.005	0.01091	0.00710	0.00710	0.00710	0.00710	−35%
Cd	mg/L	0.00191	0.1	0.00191	0.00191	0.00191	0.00191	0.00191	0%
Cr	mg/L	0.00089	0.015	0.00089	0.00089	0.00089	0.00089	0.00089	0%
Pb	mg/L	0.00135	0.002	0.00135	0.00135	0.00135	0.00135	0.00135	0%
Hg	mg/L	0.00050	0.05	0.00050	0.00050	0.00050	0.00050	0.00154	209%
Se	mg/L	0.01350		0.01350	0.04118	0.02197	0.01350	0.01350	0%
Ag	mg/L	0.00115	1.3	0.00115	0.00115	0.00115	0.00115	0.00115	0%
Cu	mg/L	0.00046		0.02139	0.00046	0.00741	0.00046	0.00046	−98%
Ni	mg/L	0.00242	0.002	0.88031	0.02971	1.09037	0.73447	0.50629	−42%
Tl	mg/L	0.00173		0.00173	0.00173	0.00173	0.00173	0.00173	0%
V	mg/L	0.00089		0.00089	0.00089	0.00089	0.00089	0.00089	0%
Zn	mg/L	0.00352		1.52641	0.00894	0.33229	0.25811	0.20112	−87%
Cond	Umhos/cm			1458	2232	1448	1479	1390	−5%

Leaching cycles were repeated until the pH approached that of the mine water. This required only five cycles for the more acidic AMD from Site P while after 15 leaching cycles the Site M samples were still above the pH of the AMD leachant. Method detection limits are given on tables 1 and 2. In order to facilitate calculation, analytical results that were below the detection limit were assigned the detection limit.

3. Results and Discussion

Interpretation of laboratory and field data: The following discussion includes absolute and proportional data. The later may be easily misinterpreted. For example, net gains in elemental concentrations may be almost infinitely large, particular with parameters that start at near detection limits. Apparently large proportional increases may reflect nearly trivial changes in concentration. For example, an increase from 0.001 to 0.010 mg/L represents a 900% increase yet the parameter may still meet primary drinking water standards. Obviously, the converse applies. On the other hand, decreases in a given parameter obviously cannot exceed 100%. For these reasons, both absolute and proportional data as well as U.S. primary drinking water standards are given to help the reader evaluate the significance of a given change in concentration. In summary, proportional changes in concentration are used to identify statistical trends, they do not imply environmental significance.

The laboratory results reflect a pure system with controlled inputs and outputs. The field data are subject to many confounding factors: unknown degree of contact between groundwater and FBC ash, interactions of the leachate with mine spoil, dilution with other groundwater streams, uncertain groundwater flow paths as well as redox zonation with the spoil. Nonetheless, comparison of the laboratory and field data will help identify those elements whose field behavior is predictable from empirical, laboratory testing and those that undergo significant interactions with the mine spoil.

Leaching Studies: Leaching FBC ash with Site M mine drainage caused an immediate rise in pH from 4.5 to 9.6 which then declined very gradually (Table 1). Surprisingly, acidity values remained well above alkalinity even with pH values at or above neutral. This coincides with an increase in Mn values throughout the leaching cycles starting at 0.3 mg/L in the leachant, reaching a maximum of 70.3 mg/L at cycle 5 and falling no lower than 47 mg/L by cycle 15. $Mn(OH)_2$ is soluble at a pH below 9.6 so after the leachate pH fell below this level by cycle 5 Mn began appearing in solution. The spike in Mn concentration probably resulted from solubilization of previously precipitated $Mn(OH)_2$ in the filter cake. Formula (1) illustrates how this soluble Mn would account for this increase in acidity.

The far right column in table 1 shows the net change in each parameter during the leaching procedures. Net change

TABLE 2. Changes in leachate chemistry through 5 leach cycles using 10g of FBC ash leached with AMD from the Site P surface mine. The far right column is the net difference between AMD and cycle 5 leachant

	units	method detect. limit	US Primary DW Std.	Site P AMD	Cycle 1	Cycle 2	Cycle 3	Cycle 4	Cycle 5	AMD vs. last cycle
pH				2.9	9.4	4.4	4.0	3.2	2.9	0%
acidity	mg/L			766.3	12.4	51.8	482.7	504.8	567.0	−26%
alkalinity	mg/L			0.0	49.4	0.0	0.0	0.0	0.0	
Mg	mg/L	0.1		36.6	16.9	46.4	39.2	37.1	33.8	−8%
Ca	mg/L	0.1		48.8	97.4	110.5	39.2	33.7	45.3	−7%
Fe	mg/L	0.1		168.6	0.5	0.6	0.7	1.2	58.6	−65%
Al	mg/L	0.1		39.0	2.0	7.6	84.7	84.4	58.4	50%
Mn	mg/L	0.1		2.0	0.0	3.4	2.0	1.8	1.7	−11%
SO4	mg/L	10		1529.0	1663.7	1409.3	1479.3	4950.3	1442.0	−6%
Sb	mg/L	0.00087	0.006	0.04800	0.05333	0.05567	0.04667	0.05533	0.04867	1%
As	mg/L	0.00200	0.05	0.02500	0.00233	0.00467	0.00133	0.00233	0.00167	−93%
B	mg/L	0.01000	2	0.24700	0.41300	0.35767	0.28833	0.18900	0.11833	−52%
Ba	mg/L	0.01000	0.004	0.01000	0.01000	0.02000	0.01333	0.01333	0.04000	300%
Be	mg/L	0.00010	0.005	0.00850	0.00123	0.00523	0.01050	0.00983	0.00743	−13%
Cd	mg/L	0.00010	0.1	0.00400	0.00403	0.00560	0.00327	0.00203	0.00177	−56%
Cr	mg/L	0.00010	0.015	0.00570	0.00780	0.00343	0.00547	0.01433	0.01733	204%
Pb	mg/L	0.00300	0.002	0.00400	0.02567	0.04067	0.03367	0.05200	0.05567	1292%
Hg	mg/L	0.00010	0.05	0.00070	0.00030	0.00013	0.00040	0.00023	0.00010	−86%
Se	mg/L	0.00300		0.01600	0.00867	0.00867	0.02733	0.01167	0.01133	−29%
Ag	mg/L	0.00115	1.3	0.00200	0.02000	0.00467	0.00567	0.00767	0.00500	150%
Cu	mg/L	0.00010		0.18400	0.00700	0.12367	0.38600	0.36433	0.23567	28%
Ni	mg/L	0.00200	0.002	0.18800	0.02667	0.45867	0.35233	0.31200	0.26067	39%
Tl	mg/L	0.00100		0.02100	0.01400	0.01600	0.05533	0.02833	0.01200	−43%
V	mg/L	0.00089		0.00300	0.00467	0.00900	0.00800	0.00833	0.01100	267%
Zn	mg/L	0.00100		1.24000	0.00833	2.30000	1.72000	1.42667	1.37333	11%
Cond	Umhos/cm			2910	2423	2117	1834	1983	2500	−14%

is expressed as a percent change in concentration between the raw AMD leachant and cycle 15 leachate. The final cycle was chosen since this was expected to most closely approximate long-term conditions in the field. The most dramatic change was a 15,000% increase in Mn concentration. Ba accounted for the second largest increase. Increases in both B and Hg were about 200%. Note that the concentrations of elements such as B, Ba and Hg were well below 1 mg/L and that in the case of Hg, levels were near the method detection limit. Other than Mn, all of the parameters that increased under leaching with the mildly acidic Site M water were either volatile or semi-volatile ions that are known to adhere to the outer surface of the ash particles. Thus leaching, rather than dissolution of the ash's alumino-silicate matrix probably accounts for much of the ionic increase. Major decreases occurred in the concentrations of acidity, Fe, Al, Cu, Be, Ni and Zn. Their proportional reductions ranged from 35 to 100%. Nine of the parameters were unchanged reflecting non-detectable levels in both the AMD leachant and the resulting leachates.

Leaching the same FBC ash with the more acidic Site P water yielded a very different picture (Table 2). For example, even though a pH of 9.4 was reached on the original leach cycle, its strong acidity caused the original pH of 2.9 to be reached in only five leach cycles. Rather than a strong increase in Mn, it decreased by 11%. The greatest increases were in Al, Zn, Ni, Cr, Cu, Pb, Ba, V and Ag. The greatest decreases were in As, Hg, Fe, Cd, B, Tl, Se and acidity. Fourteen of the twenty-five parameters decreased and pH was unchanged. Perhaps the most interesting result was for As. Its concentration in the untreated site's AMD was 0.025 mg/L and it declined to 0.002 mg/L (the method detection limit) after five leach cycles, never exceeding 0.005 mg/L. This may be the result of an interaction with the high iron content of the Site P AMD since the arsenate oxyanion is strongly adsorbed to precipitated ferrihydrite which has an anion exchange capacity at low pH. Behavior As was below the method detection limit in Site M water both in the AMD and through all leach cycles. Sorption to positively charged ferrihydrite surfaces may explain the decreases in the other elements that form oxyanions such as: Tl, B, Se, SO_4, and Be.

Other than Al and Ba, all of the ions mobilized by the strongly acidic Site P water were transition metals. Al increased by 50% suggesting dissolution of the alumino-silicate ash matrix and exposure of the embedded metal ions to mobilization. In contrast, leaching the FBC ash with Site M water resulted in a net loss of Al.

Field Observations: At the Site M surface coal mine, ash application raised the pH from 4.5 to 7.4 changing the groundwater from net acid (144 mg/L) to net alkaline (96 mg/L) (Table 3). On the ash site, ten parameters increased and fifteen decreased.

TABLE 3. Comparison of drainage from sites with and without FBC ash application: Site M surface mine

	units	method detect. limit	US Primary DW Std.	no ash applied	ash applied	net change
pH				4.5	7.4	65%
acidity	mg/L			144.0	3.6	−97%
alkalinity	mg/L			0.0	100.0	0%
Mg	mg/L	0.1		374.1	111.0	−70%
Ca	mg/L	0.1		534.5	238.0	−55%
Fe	mg/L	0.1		5.0	0.4	−93%
Al	mg/L	0.1		23.2	0.1	−100%
Mn	mg/L	0.1		0.3	1.3	311%
SO4	mg/L	10		2590.7	795.0	−69%
Sb	mg/L	0.00087	0.006	0.00087	0.00500	475%
As	mg/L	0.00200	0.05	0.00410	0.00100	−76%
B	mg/L	0.01000	2	0.03120	0.10000	221%
Ba	mg/L	0.01000	0.004	0.00116	0.10000	8521%
Be	mg/L	0.00010	0.005	0.01091	0.00500	−54%
Cd	mg/L	0.00010	0.1	0.00191	0.00500	162%
Cr	mg/L	0.00010	0.015	0.00089	0.05000	5518%
Pb	mg/L	0.00300	0.002	0.00135	0.00100	−26%
Hg	mg/L	0.00010	0.05	0.00050	0.00020	−60%
Se	mg/L	0.00300		0.01350	0.00200	−85%
Ag	mg/L	0.00115	1.3	0.00115	0.01000	770%
Cu	mg/L	0.00010		0.02139	0.02000	−6%
Ni	mg/L	0.00200	0.002	0.88031	0.04000	−95%
Tl	mg/L	0.00100		0.00173	0.00100	−42%
V	mg/L	0.00089		0.00089	0.10000	11136%
Zn	mg/L	0.00100		1.52641	0.05600	−96%
Cond	Umhos/cm			1458	2180	50%

The largest proportional increases were in V, Ba, Cr and Ag. Increases also occurred in Mn, B, Sb and Cd. Nearly 100% decreases occurred in Al, Zn, Ni, Fe and Se. Smaller but substantial decreases occurred in Mg, Be, As, SO4, Hg, Ca, Tl and Pb.

At Site P the pH was 2.9 on the non ash treated location and 4.3 at the ash treated site. Groundwater acidity decreased from 766 mg/L to 308 mg/L (Table 4). Despite ash treatment this remained an acid dominated groundwater. The largest relative increases in elemental concentrations were in V, Ba, Cr, Ni, Ag, Mn, Ca, Mg, Be, Cd and Zn. Of these Mg, Ca and Ba are semi-volatile alkaline earths known to form soluble salts on the exterior of the fly ash particles. The transition metals: Ni, Ag, Mn, Cd and Zn are present in solution as free metal ions are soluble at low pH and form insoluble hydroxides at high pH. All are soluble at pH 4.3. Decreases of between 60 and 100% were observed for Fe, As, Tl, Se, Hg, acidity, B, Cu and Sb. Sorption to precipitated ferrihydrite may partly explain the decreased mobility of the anionic species: As, Tl, Se and B.

Though preliminary it seems that the 22 elemental test parameters fall into three broad, activity classes:

Activity class A: semi-volatile cations: Ba, Ca, Mg, Hg
removal mechanism: Other than Hg, these form the cation in salts coating the ash particles.
Primary removal mechanism: cation exchange with spoil clays

Activity class B: oxyanions: B, Se, As, Tl, Cr, V, SO_4,
While these include semi-volatiles, their behavior is likely dominated by their tendency to form oxyanions in aqueous phase. Redox conditions are important in determining their tendency to sorb to ferrihydrites and clays.
Primary removal mechanism: sorption to ferrihydrite at low pH

Activity class C: metal cations: Al, Be, Zn, Ni, Fe, Cu, Mn, Pb, Cd, Sb, Ag
The soluble phases of these metals are dominated by free cations that form insoluble hydroxides with increasing pH. Some like Fe and Mn have multiple oxidation states, the oxidized forms having lower solubilities.
Primary removal mechanism: precipitation as hydroxides at high pH

While hypothetical, systematic investigation of these removal pathways will enhance our understanding of their behavior in mine groundwater settings.

It is important to note that even after 15 leach cycles the Site M leaching procedure still had not exhausted the alkalinity in the ash sample. Site M water had a pH of 4.5 and after the final leaching the pH was 6.6, while the Site P water had

TABLE 4. Comparison of drainage from sites with and without FBC ash application: Site P surface mine

	units	method detect. limit	US Primary DW Std.	no ash applied	ash applied	net change
pH				2.9	4.3	51%
est. acidity	mg/L			766.3	308.4	−60%
alkalinity	mg/L			0.0	0.0	0%
Mg	mg/L	0.1		36.6	206.0	463%
Ca	mg/L	0.1		48.8	471.0	865%
Fe	mg/L	0.1		168.6	0.2	−100%
Al	mg/L	0.1		39.0	47.6	22%
Mn	mg/L	0.1		2.0	22.4	1049%
SO4	mg/L	10.0		1529.0	1750.0	14%
Sb	mg/L	0.00087	0.006	0.04800	0.00500	−90%
As	mg/L	0.00200	0.05	0.02500	0.00100	−96%
B	mg/L	0.01000	2	0.24700	0.10000	−60%
Ba	mg/L	0.01000	0.004	0.01000	0.10000	900%
Be	mg/L	0.00010	0.005	0.00850	0.03500	312%
Cd	mg/L	0.00010	0.1	0.00400	0.00800	100%
Cr	mg/L	0.00010	0.015	0.00570	0.05000	777%
Pb	mg/L	0.00300	0.002	0.00400	0.00400	0%
Hg	mg/L	0.00010	0.05	0.00070	0.00020	−71%
Se	mg/L	0.00300		0.01600	0.00200	−88%
Ag	mg/L	0.00115	1.3	0.00200	0.01000	400%
Cu	mg/L	0.00010		0.18400	0.07000	−62%
Ni	mg/L	0.00200	0.002	0.18800	1.13000	501%
Tl	mg/L	0.00100		0.02100	0.00100	−95%
V	mg/L	0.00089		0.00300	0.10000	3233%
Zn	mg/L	0.00100		1.24000	3.29000	165%
Cond	Umhos/cm			2910	4310	48%

returned to its original pH of 2.9 after five leach cycles indicating exhaustion of the ash's inherent alkalinity. This may account for some of the differences in results between the two laboratory leaching results.

Comparison of Laboratory and Field Results: The effects of ash on laboratory and field concentrations were compared to identify which field trends (increasing or decreasing) were correctly predicted by the laboratory leaching procedure (Table 5). Trends in twelve of the twenty-five parameters were correctly predicted at Site M. Fourteen of twenty-five parameters were correctly predicted at Site P. Seven parameters were successfully predicted on both sites while six were not successfully predicted on either site. The remaining elements were successfully predicted on at least one of the two field sites. Considering the many other factors that influence field leachate quality, these results indicate significant promise for predicting the effects of CCB minefills on a mine's water quality. Future study will undertake a detailed statistical analysis in the field and the laboratory to identify an improved risk assessment method. Also, sequential extraction of the residual solids would assist in interpreting the sequestration mechanisms. Approached systematically, we will begin to understand the effects of mine water on CCBs and the effects of spoil composition and hydrology on the resulting leachates.

Trends for elements in activity classes A and C: semi-volatile cations and metal cations respectively were evenly divided with respect to predictability. However, trends for all but one of the oxyanions comprising activity class B were not predicted by leaching procedure with Site M AMD while all but one were successfully predicted for Site P. This may be explained by the high iron content of Site P water and the availability of ferrihydrite as a scavenging agent for oxyanions at pH 2.9.

4. Conclusions

This study attempts field verification of the results of a laboratory leaching procedure that uses mine water to sequentially leach coal ash. In addition we attempted to begin the process of identifying which chemical parameters could be predicted in the laboratory.

The Mine Water Leaching Procedure is a sequential leaching procedure that was developed to determine the long-term leaching behavior of industrial wastes in acidic mine environments. The complex hydrology, redox and geochemical conditions found within the backfills of most surface mines are difficult, if not impossible to reproduce in a laboratory setting. However, preliminary comparisons of MWLP leachates with field leachates indicates that this procedure may be a valuable tool for predicting the overall trend of leaching behavior for CCBs placed in unique acidic environments.

TABLE 5. This table illustrates the ability of the MWLP leaching procedure to predict increasing (positive values) or decreasing (negative) trends. Field and laboratory data representing Sites M and P are given. Where the MWLP correctly predicted increasing or decreasing trends in the field data, the row was assigned a 'yes' while those trends that were not correctly predicted were assigned a 'no.' Activity classes represent: A-semi-volatile cations, B-oxyanions, C-metal cations.

	Site M Surface Mine (mildly acidic AMD)					Site P Surface Mine (strongly acidic AMD)			
	Activity class	MWLP results	Field data	predicted trend?		Activity class	MWLP results	Field data	predicted trend?
Al	C	−100%	−100%	yes	Fe	C	−65%	−100%	yes
acidity		−40%	−97%	yes	As	B	−93%	−96%	yes
Zn	C	−87%	−96%	yes	Tl	B	−43%	−95%	yes
Ni	C	−42%	−95%	yes	Se	B	−29%	−88%	yes
Fe	C	−98%	−93%	yes	Hg	A	−86%	−71%	yes
Mg	A	−11%	−70%	yes	acidity		−26%	−60%	yes
Be	C	−35%	−54%	yes	B	B	−52%	−60%	yes
Cu	C	−98%	−6%	yes	Al	C	50%	22%	yes
pH		48%	65%	yes	Zn	C	11%	165%	yes
B	B	264%	221%	yes	Ag	C	150%	400%	yes
Mn	C	15344%	311%	yes	Ni	C	39%	501%	yes
Ba	A	1323%	8521%	yes	Cr	B	204%	777%	yes
Se	B	0%	−85%	no	Ba	A	300%	900%	yes
As	B	0%	−76%	no	V	B	267%	3233%	yes
SO4	B	12%	−69%	no	Sb	C	1%	−90%	no
Hg	A	209%	−60%	no	Cu	C	28%	−62%	no
Ca	A	2%	−55%	no	Pb	C	1292%	0%	no
Tl	B	0%	−42%	no	SO4	B	−6%	14%	no
Pb	C	0%	−26%	no	Cond		−14%	48%	no
Cond		−5%	50%	no	pH		0%	51%	no
Cd	C	0%	162%	no	Cd	C	−56%	100%	no
Sb	C	0%	475%	no	Be	C	−13%	312%	no
Ag	C	0%	770%	no	Mg	A	−8%	463%	no
Cr	B	0%	5518%	no	Ca	A	−7%	865%	no
V	B	0%	11136%	no	Mn	C	−11%	1049%	no

References

1. U.S. Environmental Protection Agency (USEPA). 1992. Toxicity Characteristic Leaching Procedure. SW 846, Method 1311. http://www.epa.gov/epaoswer/hazwaste/test/main.html
2. Yan, Jinying, Luis Moreno, and Ivars Neretnieks. 2000. The Long-Term Acid Neutralizing Capacity of Steel Slag. Department of Chemical Engineering and Technology. Royal Institute of Technology. SE-100 44 Stockholm, Sweden: Elsevier Science Ltd.
3. Muraka, I.P., S.V. Mattigod, and R.F. Keefer. 1993. An Overview of Electric Power Institute (EPRI) Research Related to Effective Management of Coal Combustion Residues. In *Trace Elements in Coal and Combustion Residues*, edited by R.F. Keefer and K.S. Sajwan, Lewis Publishers, Boca Raton, Florida, USA.
4. Bhumbla, D.K., R.N. Singh, and R.F. Keefer. 2000. Coal Combustion By-Product Utilization for Land Reclamation. In *Reclamation of Drastically Disturbed Lands*, edited by R.I. Barnhisel, R.G. Darmody, and W.I. Daniels. Madison, Wisconsin, USA: American Society of Agronomy, Inc., Crop Science Society of America, Inc., and Soil Science Society of America, Inc.
5. Ainsworth, C.C., and D. Rai. 1987. *Chemical Characterization of Fossil Fuel Wastes*, Report EA-5321, Electric Power Research Institute, Palo, California, USA.
6. Roy, W.R., and R.A. Griffin. 1982. A Proposed Classification System for Coal Fly Ash In Mulitdisciplinary Research. *Journal of Environmental Quality*. 11, 563–569.
7. American Coal Ash Association. 2004. 2003 Coal Combustion Product Production and Use Survey. American Coal Ash Association, Inc., Aurora, CO.
8. Butalia, Tarunjit S., and William E. Wolfe. 2000. Market Opportunities for Utilization of Ohio Flue Gas Desulfurization (FGD) and Other Coal Combustion Products (CCPs), Volume 1—Executive Summary. Available at http://ccohio.eng.ohio-state.edu/ccpohio/Marketing/Volume1.PDF
9. Elseewi, A.A., I.R. Straughan, and A.L. Page. 1980. Sequential Cropping of Fly Ash Amended Soils: Effects on Soil Chemical Properties and Yield and Elemental Composition of Plants. *The Science of the Total Environment*. 15:247–259.
10. Martens, D.C. 1971. Availability of Plants Nutrients in Fly Ash. *Compost Science*. 12(6):15–18.
11. Hodgson, D.K., and D.A. Brown. 1982. Nuetralization and Dissolution of High-Calcium Fly Ash. *Journal of Environmental Quality*. 11:93–98.
12. Hodgson, D.K., and R. Holliday. 1966. The Agronomic Properties of Pulverized Fuel Ash. *Chemistry and Industry*. 20:785–790.
13. Jastrow, J.D., C.A. Zimmerman, A.J. Dvorak, and R.R. Hinchman. 1981. Plant Growth and Trace Element Uptake on Acidic Coal Refuse Amended with Lime or Fly Ash. *Journal of Environmental Quality*. 10:154–160.
14. Keefer, R.F., R.N. Singh, O.L. Bennett, and D.J. Horvath. 1983. Chemical Composition of Plants and Soils from Vegetated Mine Soils. *In Proceeding of the 1983 Symposium for Surface*

Mine Hydrology, Sedimentology, and Reclamation. University of Kentucky, Lexington, Kentcky, USA.

15. Phung, H.T., L.J. Lund, and A.L. Page. 1978. Potential Use of Fly Ash as a Liming Material. In *Environmental Chemistry and Cycling Processes*, edited by D.C. Adrinao, and L. Brisbin. U.S. Department of Commerce, Springfield, Virginia, USA: Tech. Inform. Cent. Publ. CONF-760429.
16. Ashby, James C. 2001. Injecting Alkaline Lime Sludge and FGD Material into Underground Mines for Acid Abatement. In *Proceedings of the 22nd West Virginia Surface Mine Drainage Task Force Symposium,* Morgantown, West Virginia, USA. April 3–4.
17. EPRI, Allegheny Energy Supply and US Department of Energy. 2001. *Omega Mine Injection Program: Monongalia County, West Virginia.* #1004032.
18. Capp, J.P. 1978. Power Plant Fly Ash Utilization for Land Reclamation in Eastern United States. In *Reclamation of Drastically Disturbed Lands*, edited by F.W. Schaller and P. Sutton. Madison, Wisconsin, USA: American Society of Agronomy, Inc.
19. Singh, R.N., D.K. Bhumbla, R.F. Keefer, and D.J. Horvath. 1992. Improving Crop Production by Altering Chemical Properties of Mineland With Industrial Wastes. In *Proceedings of the International Symposium on Nutrient Management*, edited by M.S. Bajwa and P.S. Sidhu. Punjab Agricultural University, Ludhiana, India. 366–380.
20. Dhaliwal, S.S., R.N. Singh, D.K. Bhumbla, P. Saini, and R.F. Keefer. 1995. Effects of Weathering on Trace Metal Distribution In Fly Ash Amended Mine Soils. In *Proceedings of the 11th International Symposium On Use and Management Of Coal Combustion By-Prodcts (CCBs), Vol. 2*, Orlando, Florida. American Coal Ash Association, Washington, DC, and Electric Power Research Institute, Palo Alto, California, USA. 57–1 to 57–11.
21. Bhumbla, D.K. 1991. Ameliorative Effect of Fly Ashes. Ph.D. dissertation. West Virginia University, Morgantown, West Virginia, USA.
22. Ziemkiewicz, P.F., J. Simmons and A. S. Knox. 2003. Evaluating the Environment risk of Backfilling Mines with Coal Ash. In: 2003 SME Annual Meeting, Cincinnati, OH, Feb. 24–26, 2003.
23. Ziemkiewicz, P.F., J. Simmons and A. S. Knox. 2003. A Method for Evaluating the Risk of Backfilling Coal Mines with Industrial By-products. In K. S. Sajwan, A. K. Ashok, and R. F. Keefer (eds.) Chemistry of Trace Elements in Fly Ash. Kluwer Academic/Plenum Publishers, New York, USA.

III
Trace Elements in Flyashes

7
Occurrence and Sorption of Radionuclides Onto Coal-Fired Power Plant Combustion Waste

Irena Twardowska[1] and Sebastian Stefaniak
Polish Academy of Sciences, Institute of Environmental Engineering, 34 M. Sklodowska-Curie St., 41-819 Zabrze, Poland

Abstract

Coal combustion results in concentration of many elements, among others also of radionuclides in combustion products. Long-term monitoring of fly ash (FA) from coal combustion in power plant boilers confirms elevated concentrations of isotopes ^{40}K, ^{226}Ra and ^{228}Ra. Though, FA does not belong to the radioactive materials and according to the regulations, can be unrestrictedly used in mine workings and at the land surface. The natural radioactivity level of FA in many cases, however, exceeds parameters f_1 and f_2 that define the maximum acceptable level of γ- and α-radiation, respectively, established by the national Polish Institute of Construction Techniques for assessing the applicability of material for production of construction materials. These parameters define the maximum acceptable level of γ- and α-radiation, respectively. FA that does not fulfil these requirements cannot be used for production of construction materials despite of other beneficial properties. At the same time, it was proved that use of radioactive mine water of A type (low sulfate, high Ba:Ra ratio) for preparing FA: mine water dense mixtures applied in deep mines for filling old mine workings, causes deep reduction of radioactivity in the leachate to the level acceptable for a drinking water. In the radioactive waters of B type (low Ba:Ra ratio) no such effect was observed, though in this case use of these waters for preparing dense mixtures with FA also cause reduction of radioactivity loads discharged to the surface by permanent water binding in a mixture in a proportion of over 60% wt. with respect to FA.

1. Introduction

The process of coal combustion results in the concentration of trace elements in fly ash (FA) and bottom ash (BA) compared to coal and the lithosphere, among others also of radionuclides. It is known since a longer time that not the total concentration, but mobilization potential and leaching behavior of contaminants is of particular importance at evaluation of the environmental impact caused by the material under the specific conditions of the environmental exposure. These properties should be taken into account also for FA safe disposal and use as construction materials for different applications, and in coal mine workings as a dense FA: mine water mixture.[1,2,3,4,5] In particular, it was found that leaching properties of FA could change dramatically in time, resulting in the alteration and increase of the released elements[3,4,5]. Simultaneously, FA displays many beneficial properties, and may be used in a multitude of applications, among them also for attenuation of environmentally adverse processes and binding of contaminant loads.[5]

Some ground waters that occur in deep mines in different parts of the world, display regionally elevated or high level of a natural radioactivity. Such waters, discharged to the surface from the mine dewatering system are highly problematic. Waters of elevated radioactivity, up to 390 kBq/m^3, occur also in different Carboniferous seams of the Upper Silesia coal basin (USCB, Poland). These waters pumped from mines pose a risk to the environment and human health.[6,7]

Up to now, Maximum Concentration Limits (MCL) for radioactivity of drinking water differ in different countries; most of them comes from 70-ties. World Health Organization[8] recommends the assessment of drinking water quality with use of a complex procedure that comprises preliminary measurements of total α and β radioactivity. If total α-radiation is $\leq$0.1 Bq/dm^3, and total β-radiation is $\leq$0.1 Bq/dm^3, water can be used for drinking. If this precondition is not fulfilled, the concentrations of all identified radionuclides should be measured separately. If the total radioactivity dose exceeds 0.1 mSv/a, water does not fit for drinking purposes.[8] In France, total concentration of α, β and γ from all sources should not exceed 0.37 Bq/dm^3.[9] US EPA applies double criterion of radioactivity in drinking water: total concentration of natural radionuclides should be $\leq$0.185 Bq/dm^3, while maximum radioactivity dose from the artificial sources is

[1] Corresponding autor: Ph.: +48 32 2716481; Fax +48 32 2717470; E-mail: irena@ipis.zabrze.pl

limited to 0.04 mSv/a, where Sv is the unit of absorbed dose equivalent.[10]

According to the Directive No 23/70 of the Polish Government Attorney on Use of Nuclear Energy[11], the discharge of radioactive waters to the open surface reservoirs is permitted provided that total Ra concentration does not exceed 1.1 kBq/m^3, i.e. 10-fold MCL for drinking water, which accounts for ^{226}Ra = 0.11 kBq/m^3. Such waters can be discharged to closed collecting sewers if total Ra concentration does not exceed MCL for drinking water by more than 100-fold, i.e. 11.1 kBq/m^3. Both solid materials of elevated radioactivity, and radioactive mine water are met with extremely negative perception of public opinion.

With respect to FA, this sets up two major tasks: (i) to assess of FA radioactivity level for its safe use as a construction material for different applications; (ii) to evaluate possibility of using high-radioactive mine waters for preparation of dense FA mixtures in order to reduce their amount discharged to the surface receiving waters.

Here, the concentrations of radionuclides in FA and binding properties of FA for radioactive elements in water have been discussed.

2. Radioactivity of Coal Combustion Waste

2.1. MCL Criteria for Radionuclide Concentrations in CCW

According to these criteria recommended by the Central Mining Institute in Katowice, Poland[12], waste materials in which the total concentration of radioisotopes does not exceed 10 kBq kg (^{226}Ra<350 Bq/kg, and ^{228}Ra<230 Bq/kg). The natural radioactivity level of FA in many cases, however, exceeds parameters $f_1 = 0.00027\ C_K + 0.0027\ C_{Ra} + 0.0043\ C_{Th} \leq 1$ and $f_2 = C_{Ra} \leq 185$ established by the national Polish Institute of Construction Techniques for assessing the applicability of material for production of construction materials[12]. These parameters (where C_K, C_{Ra} and C_{Th} are the concentrations of isotopes ^{40}K, ^{226}Ra and ^{228}Ra) define the maximum acceptable level of γ- and α-radiation, respectively. FA that does not fulfil these requirements cannot be used for production of construction materials. Variable f_1 and f_2 coefficients for Polish FA imply a requirement of analysis for radioactivity level of every batch of FA, which is considered to be used for this purpose.

As show the results of routine measurements conducted for over 20 years, concentrations of radionuclides in FA are several times higher compared to the lithosphere and coal and also in general considerably higher than other natural and anthropogenic materials considered to be used for construction purposes, with exception of copper smelting slag (Table 1). Nevertheless, this material is not classified as a radioactive waste (total concentration of radionuclides <10 kBq/kg) and fulfils the criteria of unrestricted permit for use in mine workings, as well as for the disposal at the surface and use for civil engineering works.

The study on distribution of a natural radioactivity in size fractions of FA showed distinct increase of radioactivity in finer fractions, in particular of Ra^{226} concentrations, which indicated the level of α-radiation (expressed as f_2 coefficient). It should be, though, pointed out that the coarsest fraction 75–50 μm comprises over 94% wt. of FA and thus determines the overall radioactivity of the material (Table 2). The finest fractions are the most airborne part of the bulk FA, thus its doubled level of α-radiation should be considered for safety reasons, and thus the requirement of hermetic transportation and handling of FA from a filter or precipitator at the power plant to the place of its use is a strictly followed prerequisite of FA management.

3. Material and Methods

Fly ash from three power plants in the USCB (Poland) was used for preparation of dense mixtures 1:1 with mine water of high or elevated radioactivity originated from three coal mines of USCB (100 kg of FA and 100 L of water for each sample). Mixtures were put into the PVC columns of 30.2 cm diameter to collect leachate. FA, mine water and leachate were analyzed for ^{40}K, ^{226}Ra and ^{228}Ra at the certified Laboratory on Technical Acoustics, Laser Technique and Radiometry of the Central Mining Institute in Katowice, Poland.

4. Results and Discussion

4.1. Dense FA:Mine Water Mixtures as a Sink of Radioactive Trace Elements

A considerable part of radioactive mine waters belongs to the Ra-Ba, low-sulfate type A. The rationale of the concept of using FA as a sink of radioactivity loads from high radioactive mine waters has been based on two major assumptions:

(a) FA, in particular with FGD solids (FA+FGDS), contain high concentrations of sulfate, hence has potentially high binding capacity for radium through co-precipitation with $BaSO_4$,
(b) FA has high water retention capacity (50–>70%), hence may bind physically adequate radium loads along with waters also of a type B (low Ba, high–sulfate).

The efficiency of Ra removal from mine water originating from three coal mines of the USCB, Poland, is exemplified in Table 3. For mine water: FA mixture preparation, "pure" FA, and FA + D-FGDS (with FGD solids from dry process) were used, with sulfate content 0.74% and 2.78% SO_4 wt., respectively. The results confirm excellent binding properties of FA for radionuclides that are removed both chemically (co-precipitation with $BaSO_4$) and physically (water retention at the level from 60.0 to 85.75% wt.). For FA + D-FGDS

TABLE 1. Mean concentrations and range (in parentheses) of radionuclides in FA compared to coal, other anthropogenic and natural raw materials, and lithosphere[14,15]

Material	Concentration of radionuclides, Bq/kg			Radiation coefficients*)	
	^{40}K	^{226}Ra	^{228}Ra	f_1	f_2
FA-Rybnik[a)]	**917 (497–1106)**	**116 (<18.5–197)**	**78 (43–122)**	**0.90 (0.47–1.26)**	**116 (<18.5–197)**
FA-all PP[b)]	**701 (33–1386)**	**124 (<1–395)**	**89 (<1–194)**	**0.91 (0.04–1.44)**	**124 (<1–395)**
	Raw materials of anthropogenic origin (1980–2000)[14]				
FA	**712 (20–3303)**	**118 (<1–395)**	**90 (<1–206)**	**0,89 (0,04–1,83)**	**118 (<1–395)**
Boiler slag	570 (1–1260)	97 (2–482)	74 (1–210)	0,73 (0,04–2,08)	97 (2–482)
Copper smelting slag	897 (615–1251)	311 (163–518)	53 (25–183)	1,31 (0,88–2,16)	311 (163–518)
Blast furnace slag	194 (18–1400)	116 (12–351)	35 (2–115)	0,51 (0,06–1,14)	116 (12–351)
		(2001–2002)[15]			
FA	**684 (46–1062)**	**118 (1–276)**	**93 (2–152)**	**0,90 (0,04–1,47)**	**118 (1–276)**
Boiler slag	592 (36–2240)	90 (9–252)	74 (4–132)	0,72 (0,04–1,39)	90 (9–252)
Copper smelting slag	895 (800–970)	310 (292–332)	48 (41–53)	1,28 (1,18–1,38)	310 (292–332)
Blast furnace slag	206 (26–611)	94 (14–163)	40 (4–99)	0,48 (0,06–0,83)	94 (14–163)
	Natural raw materials (1980–2000)[14]				
Lime	44 (<1–331)	24 (1–47)	3 (1–13)	0,09 (0,01–0,26)	24 (1–47)
Gypsum	68 (<1–279)	13 (1–64)	4 (<1–30)	0,07 (0,01–0,35)	13 (<1–64)
Chalk	125 (24–295)	14 (2–43)	5 (<1–14)	0,09 (0,04–0,18)	14 (2–43)
Clinker	200 (<1–534)	37 (9–121)	16 (<1–41)	0,22 (0,07–0,47)	37 (9–121)
Clay material	683 (44–1241)	38 (7–129)	44 (13–144)	0,48 (0,23–1,16[d)])	38 (7–129)
Shale	630 (79–1925)	55 (2–116)	54 (5–219)	0,55 (0,04–1,70[d)])	55 (2–116)
		(2001–2002)[15]			
Lime					
Gypsum	51 (13–140)	10 (2–72)	4 (<1–30)	0,04 (0,01–0,35)	10 (2–72)
Chalk	98 (95–104)	12 (9–14)	3 (2–4)	0,07 (0,06–0,07)	12 (9–14)
Clinker	158 (55–294)	31 (19–53)	16 (12–21)	0,19 (0,13–0,27)	31 (19–53)
Clay material	681 (465–904)	37 (25–49)	41 (35–45)	0,46 (0,38–0,53)	37 (25–49)
Shale	777 (708–846)	67 (64–70)	65 (59–70)	0,66 (0,61–0,71)	67 (64–70)
	Construction materials (1980–2000)[14]				
Cement	227 (<1–608)	50 (4–154)	23 (<1–53)	0,29 (0,06–0,63)	50 (4–154)
Cellular and lightweight concrete	508 (72–1005)	75 (2–200)	54 (3–127)	0,57 (0,07–0,93)	75 (2–200[e)])
Construction ceramics[d)]	685 (25–1410)	53 (<1–190)	48 (<1–130)	0,53 (0,03–1,12[f)])	53 (<1–190[e)])
		(2001–2002)[15]			
Cement	316 (82–812)	53 (19–101)	30 (11–78)	0,35 (0,15–0,70)	53 (19–101)
Cellular and lightweight concrete	514 (150–845)	76 (12–172)	60 (5–98)	0,60 (0,11–0,93)	76 (12–172)
Construction ceramics[d)]	718 (106–1240)	52 (4–102)	46 (2–93)	0,53 (0,04–0,80)	52 (4–102)
Coal[c)]	316 (115–438)	37 (21–81)	24 (14–34)	0.28(0.16–0.39)	37 (21–81)
Lithosphere	380 (100–700)	25 (10–50)	25 (7–50)		

*) Regulatory limits: $f_1 < 1$ (for γ-radiation); $f_2 < 185$ (for α-radiation)
a) Rybnik power plant, Silesia, Poland (1977–1997); b) All Polish power plants (1980–1995) (GUS 1997, after Central Laboratory of Radiological Protection);
c) Coal from the USCB (1977–1997); d) Bricks, structural hollow bricks, tiles, shaped bricks and tiles etc.

TABLE 2. Natural radioactivity of size fractions of FA from the Rybnik power plant

Size fractions (μm)	% wt. rate in FA			Mean %	Radioactivity (Bq/kg)	
					^{226}Ra	$^{228}Ra = Th$
75–50	94.10	93.62	93.81	94.11	80.41	79.55
50–30	2.94	2.78	1.82	2.51	80.41	79.55
30–20	1.48	1.62	1.58	1.56	80.41	79.55
20–10	1.22	1.30	1.28	1.27	88.45	79.94
10–5	0.46	0.11	0.12	0.11	109.35	83.69
<5	0.09	0.11	0.12	0.11	160.82	95.16

radioisotope residuum in outflow was below MCL = 0.11 kBq/m^3 for drinking water, while in leachate from "pure" FA mixture it was somewhat higher, up to 2.5-fold MCL for ^{226}Ra. Due to high water retention capacity of FA (>60% wt.), the overall amount of mine waters to be discharged at the surface was also substantially reduced.

5. Conclusions

The natural radioactivity of FA does not pose a risk to the environment and health, provided that it is considered at FA

TABLE 3. Binding of radionuclides from mine water of A–type onto "pure" FA and FA containing products of flue gas desulfurization by dry method (FA+D-FGDS) from Polish power plants used for preparation of dense mixtures for sealing abandoned mine workings

	Radionuclide concentrations in mine water (kBq/m^3)					
	Input		Leachate		Reduction in leachate (%)	
Fly ash	^{226}Ra	^{228}Ra	^{226}Ra	^{228}Ra	^{226}Ra	^{228}Ra
	Dense mixture FA:mine water (1:1 wt.)					
"Pure" FA, Laziska power plant	103.208 ± 8.26	67.76 ± 13.57	0.234 ± 0.024	0.23 ± 0.08	99.77	99.66
	80.986 ± 6.48	23.10 ± 4.65	0.279 ± 0.027	0.09 ± 0.06	99.65	99.61
	2.171 ± 0.175	1.48 ± 0.30	0.111 ± 0.015	<0.06	94.89	100
	Dense mixture FA + D-FGDS : :mine water (1: 1 wt.)					
"Pure" FA + D-FGDS, Rybnik power plant	103.208 ± 8.26	67.76 ± 13.57	0.058 ± 0.012	<0.05	99.94	100
	80.986 ± 6.48	23.10 ± 4.65	0.095 ± 0.014	<0.06	99.88	100
	2.171 ± 0.175	1.48 ± 0.30	0.051 ± 0.011	<0.06	97.65	100
	Dense mixture FA + D-FGDS: mine water (1: 1 wt.)					
"Pure" FA + D-FGDS, Opole power plant	103.208 ± 8.26	67.76 ± 13.57	0.087 ± 0.013	0.17 ± 0.07	99.92	99.75
	80.986 ± 6.48	23.10 ± 4.65	0.050 ± 0.011	<0.05	99.94	100
	2.171 ± 0.175	1.48 ± 0.30	0.123 ± 0.016	0.17 ± 0.07	94.33	88.51
	Dense mixture FA + D-FGDS: mine water (1: 1 and 1:1.9 wt)					
"Pure" FA + D-FGDS, Rybnik power plant	103.208 ± 8.26	67.76 ± 13.57	0.058 ± 0.012	<0.05	99.94	100
	103.208 ± 8.26	67.76 ± 13.57	0.075 ± 0.013	<0.05	99.93	100
	Dense mixture FA + D-FGDS: mine water (1:1 and 1: 1.1 wt.)					
"Pure" FA + D-FGDS Opole power plant	2.171 ± 0.175	1.48 ± 0.30	0.123 ± 0.016	0.17 ± 0.07	94.33	88.51
	2.171 ± 0.175	1.48 ± 0.30	0.040 ± 0.006	<0.03	98.16	100

use for production of construction materials. At the same time, FA, preferably with FGD solids, but also "pure" material can be used as an efficient agent for radionuclide binding from high-radioactive mining and other waters, in particular of A type (low sulfate, high Ba-Ra content) where radioactive elements are removed through co-precipitation with $BaSO_4$. Radioactivity of waters of B type (high sulfate, low Ba content) may be partially reduced through physical water binding in the FA:water dense mixture in the amounts ≥50% wt. with respect to FA.

References

1. Twardowska, I. Environmental aspects of power plants fly ash utilization in deep coal mine workings, in *Biogeochemistry of Trace Elements in Coal and Coal Combustion Byproducts*, K.S. Sajwan, A.K. Alva, R.F. Keefer, Editors, 1999. Kluwer Academic/Plenum Publishers, New York, Boston, Dordrecht, London, Moscow, 29–58.
2. Twardowska, I. Environmental behavior of power plants fly ash containing FGD solids utilized in deep coal mines, pp. 77–98. in *Biogeochemistry of Trace Elements in Coal and Coal Combustion Byproducts*, K.S. Sajwan, A.K. Alva, R.F. Keefer, Editors, 1999. Kluwer Academic/Plenum Publishers, New York, Boston, Dordrecht, London, Moscow, 77–98.
3. Twardowska I., Szczepanska, J. and Stefaniak S. Occurrence and mobilization potential of trace elements from disposed coal combustion fly ash, in *Chemistry of Trace Elements in Fly Ash*. K.S. Sajwan, A.K. Alva, R.F. Keefer, Editors, 2003. Kluwer Academic/Plenum Publishers, New York, Boston, Dordrecht, London, Moscow, 13–24.
4. Twardowska I., Tripathi P.S.M., Singh G. and Kyziol J., Trace elements and their mobility in coal ash/fly ash from Indian power plants in view of its disposal and bulk use in agriculture, in *Chemistry of Trace Elements in Fly Ash*. K.S. Sajwan, A.K. Alva, R.F. Keefer, Editors, 2003. Kluwer Academic/Plenum Publishers, New York, Boston, Dordrecht, London, Moscow, 25–44.
5. Twardowska I. and Szczepanska J. Coal combustion waste, in *Solid Waste: Assessment, Monitoring and Remediation*, I. Twardowska, H.E. Allen, A.A.F. Kettrup and W.J. Lacy, Editors, 2004, Elsevier B.V., Amsterdam, Boston, Heidelberg, London, New York, Oxford, Paris, San Diego, San Francisco, Singapore, Sydney, Tokyo, 387–452.
6. Twardowska I.. Bulk use of power plant fly ash in deep mines and at the surface for contaminant and fire control, in *Solid Waste: Assessment, Monitoring and remediation*, I. Twardowska, H.E. Allen, A.A.F. Kettrup and W.J. Lacy, Editors, 2004, Elsevier B.V., Amsterdam, Boston, Heidelberg, London, New York, Oxford, Paris, San Diego, San Francisco, Singapore, Sydney, Tokyo, 949–1002.
7. Lebecka, J., and Tomza I., Natural sources of ionizing radiation in coal mines. Protection against natural ionizing radiation in underground mine workings. *Works of Central Mining Institute, Supplementary Series*, 1989. Central Mining Institute, Katowice (in Polish).
8. World Health Organization. *Guidelines for Drinking Water Quality*, 1993, Geneve.
9. Bremond R. and Vuichard R. Parametres de la qualité des eaux, Ministerie de la Protection de la Nature et de l'Environment, 1973, Paris.
10. US Environmental Protection Agency, *Water and Waste Engineering* 14 (8), 15, 1977.

11. *Directive No 23/70 of the Polish Government Attorney regarding Use of Nuclear Energy*, 1970 (in Polish).
12. GIG—Central Institute of Mining, Central Laboratory of Radiological Protection. *Guidelines on Permissible Concentrations of the Natural Radioactive Isotopes in Waste Disposed of or Used for Civil Engineering Works att the Surface or Deposited in the Deep Mines*, 1994. GIG, Katowice (in Polish).
13. ITB—Institute of Construction Techniques. Guidelines for testing natural radioactivity of raw materials and construction materials. *Instruction No. 234/95,* 1995, Warsaw (in Polish).
14. Central Statistical Office. *Environment Protection 2002. Information and Statistical Papers*, 2002. GUS, Warsaw (in Polish).
15. Central Statistical Office. *Environment Protection 2003. Information and Statistical Papers*, 2003. GUS, Warsaw (in Polish).

8

Heavy Metals Adsorption and Their Distribution in Three Soil Types of India: Effect of Coal Fly Ash and Sewage Sludge Amendment

S. Tripathy[1], H. Veeresh[2], D. Chaudhuri[3], M.A. Powell[4], and B.R. Hart[4]

[1]*Department of Geology & Geophysics, Indian Institute of Technology, Kharagpur, West Bengal, 721 302, India*
[2]*Department of Soil Science and Agricultural Chemistry, U.A.S., G.K.V.K., Bangalore, Karnataka, 560 065, India*
[3]*Department of Applied Science, College of Engineering, University of California, Davis, Livermore, CA 94550, USA*
[4]*Department of Earth Sciences, University of Western Ontario, London, N6A 5B7, Canada*

Abstract

Even though both coal fly ash and sewage sludge are rich sources of bio-essential nutrients, one of the major limiting factors in their use either individually or in mixture proportions as for amendment into agricultural soil is the presence of various metals likely to be potentially toxic at their elevated concentrations. This study evaluated the adsorption and distribution behaviours of selected metals in three different soils from India amended with mixtures of ash and sludge in various proportions at a maximum application rate of 52t ha^{-1} and incubated up to 90 days at near field capacity moisture level. The properties of amended soils were affected significantly by sludge than the ash and the Freundlich distribution coefficients (K_{D}s) of metals were marginally higher compared to their respective controls. The degree of impact of amendment on soil properties, distribution coefficients of metals and their adsorption affinity sequences, based on K_{D}s, within and across soil types were determined. Sequential extraction indicated that, the concentrations of native metals in each soil type tends to be less in highly mobile and moderately mobile fractions than the resistant; and changed marginally following amendment while the distribution patterns remained more or less undisturbed. The sequential extraction of adsorbed metals at various equilibrating concentrations of their addition indicated shifts in the distribution patterns in each soil type. However, the changes in the partitioning of adsorbed metals were wide and clearly noticeable only at higher loadings and the metals showed propensity to accumulate preferentially in more mobile fractions, depending on the type of soil surface and substrate. The results indicated that at low rate of application, ash and sludge, though capable of changing the soil properties and its metal adsorption capacities to certain extent; they had no major influence on metal distribution patterns in the amended soils which mainly depended on the soil properties, type of metal and its concentration.

1. Introduction

Rapid industrialization and ever growing population have resulted in generation of huge quantity of waste and attempts are being made world over to develop new avenues for their recycling with minimum environmental hazard. Among the solid waste, coal ash from thermal power plants and sludge from sewage treatment plants are significant in consequences due to their high volume and potential to contaminate the environment.

In India, coal based thermal power plants mainly use bituminous to sub-bituminous coal generating about 100 million tones of ash annually manly due to high ash content (30–50%) in coals. Besides the cost of transportation, use of vast quantities of water and land for disposal, ash is also known to cause ecological problems. Nearly 65,000 acres of land is occupied by ash ponds in India and are possibly rendered unfit at varying degrees for further use. India's dependence on coal as an energy source shall continue into the next millennium; and, therefore ash management would remain an important area of national concern.

Sewage sludge, on the other hand, is a residue consisting of mixtures of organic and inorganic solids. In India, the capacity of municipalities to receive and treat human waste is extremely limited with only a handful of cities and town possessing and operating sewage treatments facilities. It is believed that approximately 73 million tones per year of sewage sludge being produced from the 217 waste treatment facilities. With the concern over disposal of sludge in ocean and the high cost of sludge incineration, land application and land filling is becoming increasingly popular. The main constraint in the land application of sewage sludge is its potential for contaminating the food chain from heavy metals present whose concentrations largely dependent on the types and amounts of discharges into the waste water treatment systems.

Sewage sludge benefits crop production since it is a rich source of organic matter, N, P and many minor elements[1,2] and acts much like a slow-releasing organic fertilizer and needs to be seen as a valuable resource but with care. Coal ash due to its alkaline pH, nutrient contents and silt-sized particles helps in increasing pH, concentrations of most macro and micronutrients in soil besides improving the structure and moisture holding capacity.[3–6] Mixture of coal ash and sewage sludge produces a material having more balanced properties that can improve physical and nutrient properties of soils[7,8] depending on the proportion and quantity applied with less potential for environmental contamination than either of them applied individually.[9–11] Nevertheless, damage to the soil quality, crops grown and groundwater can occur, if used indiscriminately and without proper scientific management.

Crop responsesin agricultural soils amended with coal ash and organic by-product have been found to be positive in different soils[12–14] though cannot be related with assurance to only a single factor. Besides, improving plant growth such as nutrient supply, factors may mask the negative effects of the heavy metals present.

Heavy metals[15] (relatively higher atomic density i.e. >5.0 mg/m^3) viz. Cd, Cu, Co, Fe, Pb, Zn, Ni etc. when applied to soils in excess, may depress plant yield, degrade the quality of food or fiber produced besides being toxic at elevated concentrations to the living organisms. The specific metal concentration of a particular soil reflects the chemical composition of the parent material albeit modified to a degree over time by weathering. Concentrations of heavy metals, greatly in excess of those expected in a particular soil type are signs of soil contamination from anthropogenic activities or simply by accumulation from natural biogeochemical processes[16]. The primary concerns are therefore the transfer of metals through plants into the food chain besides the threat of leaching into the surface and groundwater, both of which depend on the metal concentrations in soil solution. Therefore, the potential risk from metal contamination associated with disposal and utilization of ash and sludge on soil depends in part on the metal sorption characteristics of the soil vis-à-vis its other properties temporarily altered either positively or negatively depending on the quality and quantity of amendment application. Therefore, understanding the adsorption behaviour of metals in amended soils is of great importance since it varies among soil types and depends on one or multitude of soil properties.[17–19] The retention of heavy metals added is often well correlated to the organic matter content in soils[20] since the presence of ligands or carboxyl, phenolic, alcoholic and carbonyl groups can form chelates with these metals.[21] Mineralogy of soils, particularly the presence of 2:1 permanent charge clays is more important for Cd and Pb sorption than the quantity of organic matter and the external surface area in oxisol, mollisol and ultisol.[22] In sewage sludge amended soils, low adsorption affinities and greater phyto-availability of Cd and Zn than Pb and Cu has been observed[23] warranting continuous assessment of sludge as well as the amended soils.

Heavy metal retention and release by soil matrix are described by both equilibrium-type and kinetic or time-dependent models while the sorption isotherms provide useful information about the retention capacity and the strength by which metals are held on to the soil. However, the ability of a particular model does not reveal the actual nature of the retention process. Since the interaction of the metals with solid phases cannot be determined from adsorption isotherms, therefore their actual partitioning in various chemical phases cannot be identified. Owing to the complexity of the retention processes of metals in soil and the limitations associated with their respective isotherms in understanding the sorption mechanism, it has been found worthwhile to combine metal adsorption with sequential extraction.

Sequential extraction procedures are based on the principle that, sorbed metals can be displaced from their geochemical phases by using appropriate extracting reagents.[24–26] Even though several sequential extraction schemes have been proposed, no single fractionation method is totally effective in dissolving each distinct form of a metal. Despite uncertainties in the selectivity of various extractants and possible problems due to re-adsorption, sequential extraction procedures have been commonly used to study metal partitioning among various solid phases of the soil and hence provide qualitative evidences regarding their various associations in soils and indirectly their bioavailability.[24,25] Thus, combining the information obtained from both sequential extraction and sorption isotherms of metals, a deeper insight into their fate and affinity in the constituent phases of soil can be ascertained.

This study attempted to understand the level of impact on soil properties, heavy metal adsorption and their distribution in three soils from India amended with coal ash and sewage sludge in various mixture proportions.

2. Materials and Methods

Representative surface samples of acidic alfisol (Oxyaquic haplustalf—SL1), neutral red alfisol (Typic haplustalf—SL2), alkaline black vertisol (Typic haplustert—SL3), and coal fly ash from Kolaghat Thermal Power Station (KTPS) and sewage sludge from Howrah Sewage Treatment Plant (HSTP) of India were used. Samples of soil, ash (FA) and sludge (SS) were air-dried and ground to pass through 2mm sieve before use. The major physical and chemical characteristics soils, ash and sludge (Table 1) were analyzed by using standard procedures and techniques.

Mixtures of Soil-FA-SS were prepared at different proportions on dry weight basis (w/w) as per treatment details (Table 2). Two kilograms of each mixture were incubated in pre-water soaked earthen pots and the moisture level was maintained approximately at near field capacity by watering and stirring at regular intervals. Representative samples of these mixtures were collected after 90 days and analyzed for

TABLE 1. Selective properties of soil, coal fly ash and sewage sludge samples

Parameters	Soil 1 (SL1)	Soil 2 (SL2)	Soil 3 (SL3)	Coal fly ash (FA)	Sewage sludge (SS)
Soil Classification[1]					
Order	Alfisol	Alfisol	Vertisol		
Sub order	Ustalf	Ustalf	Usterts		
Great group	Haplustalf	Haplustalf	Haplustert		
Sub group	Oxyaquic	Typic	Typic		
Physical					
Size fractions (%)[a]					
Sand	62.7	52.6	9.8	19.8	
Silt	22.5	19.8	27.0	50.2	
Clay	14.8	27.6	63.2	30.0	
Texture	SSiL	SCL	C	SiCL	
Main clay minerals[b]	K	K, I	M		
Chemical					
pH (1:2.5)[c]	5.20	7.10	8.60	7.80	4.30
EC ($\mu S\ cm^{-1}$)[c]	43.0	65.3	133	127	2560
CEC (Cmol (+) kg^{-1})[c]	5.48	12.5	32.7	4.50	19.7
OC ($g\ kg^{-1}$)[c]	5.30	5.10	6.20	8.60	144
Total N (%)	0.05	0.05	0.06	0.02	1.01
Available N ($mg\ kg^{-1}$)[d]	76.9	84.3	94.6		2625
Carbonates ($g\ kg^{-1}$)[e]	6.70	8.10	84.3	88.8	3.53
Total Metal Concentration in ppm (unless mentioned)[f]					
Al (%)	2.98[g]	5.34[g]	6.49[g]	0.68	1.84
Ca (%)	0.12[g]	0.44[g]	1.62[g]	0.44	2.40
Cd	0.75	0.76	0.93	0.58	2.65
Cu	10.67	16.01	25.41	38.2	303
Fe (%)	1.27[g]	1.60[g]	3.64[g]	2.14	8.53
K	476	1394	1605	685	2148
Mg	446	1069	6089	1145	3305
Ni	8.08	6.45	29.0	13.0	41.2
P (%)	0.05	0.06	0.05	0.15	1.24
Pb	14.8	11.5	18.8	50.0	303
S	21.4	21.0	76.3	178	27750
Zn	18.9	33.3	54.9	98.0	3085
Mehlich III Extractable Concentrations (ppm)					
Cd	0.24	0.21	0.19	0.10	0.62
Cu	4.56	5.63	8.88	12.6	97.6
Ni	2.32	3.06	4.93	3.92	10.2
Pb	6.74	5.13	7.22	28.2	86.3
Zn	6.05	13.7	16.9	23.6	714

SSiL—Sandy silt loam; SCL—Sandy clay loam; C—Clay; SiCL—Silt clay loam; K—Kaolinite; I—Illite; M—Montmorillonite EC—Electrical conductivity; CEC—Cation exchange capacity; OC—Organic carbon

[1] USDA, Soil survey staff (1999)

[a] Piper, 1966; [b] Analysis by XRD (Rigaku,Miniflex); [c] Jackson (1967); [d] Subbiaha and Asija (1956); [e] Rowell (1994)

[f] Analysis by ICP-OES (Perkin-Elmer PE-3300 DV), after aqua regia digestion

[g] Analysis by WDXRF (Philips, PW2404)

various physical and chemical parameters and also used for batch equilibration study.

2.1. Heavy Metal Adsorption

Adsorption of individual heavy metal in each soil type was determined using batch equilibrium technique. Soil (2.0 g) was weighed into each of six 50 ml capacity polypropylene centrifuge tubes and individually equilibrated with 30 ml of 0.005 M $Ca(NO_3)_2.2H_2O$ containing graded levels of Cd, Cu, Ni, Pb and Zn ranging from 10 to 200 $\mu M\ L^{-1}$ in form of $Cd(NO_3)_2.4H_2O$, $Cu(NO_3)_2$, $Ni(NO_3)_2.6H_2O$, $Pb(NO_3)_2$ and $Zn(NO_3)_2$, respectively. Calcium nitrate was used as the background electrolyte to eliminate non-specific adsorption of metals, so as to saturate the low affinity sorption sites by calcium ions[27]. Soil suspensions were shaken for 24 h at $25 \pm 1°C$ in an incubator shaker then centrifuged at 7000 rpm for 15 min followed by filtration of the supernatant solutions through Whatman No. 42 filter paper to remove particulates. The solutions were acidified (pH $\sim$ 2) using 0.1 ml concentrated

TABLE 2. Treatments used in the study

		Quantity of waste used (t ha^{-1})	
Soil type	Treatments	Sewage sludge	Coal fly ash
	M1—Control	0	0
	M2—FA 100	0	52
SL1, SL2 or SL3	M3—SS 25 + FA 75	13	39
	M4—SS 50 + FA 50	26	26
	M5—SS 75 + FA 25	39	13
	M6—SS 100	52	0

Note: Weight of one hectare plough (upper 15 cm) depth of soil = 2,240,000 kg
SS—Sewage sludge; FA—Coal fly ash

HNO_3 and the concentrations of the metals in the filtrate were determined using FAAS (Aanalyst 300, Perkin Elmer), GFAAS (HGA 850, Perkin Elmer). Metal adsorbed by soil was calculated from the difference between the amounts added and that recovered. All measurements were made in duplicate and the average is reported. The sorption data for all metals were interpreted in terms of linearized Freundlich adsorption equation:

$$\log \frac{X}{m} = \log K_D + n \log C_e$$

where,
X/m = Amount of metal adsorbed in $\mu M\ kg^{-1}$
C_e = Equilibrium solution concentration of metal in μML^{-1}
K_D = Distribution Co-efficient or Adsorption affinity of metal
n = Freundlich's equation constant

In general, the concentrations of metals used during adsorption study represent their respective concentrations in natural soils to contaminated ones except for Cd, which only represents for contaminated conditions.

Metal standards of required concentrations were prepared from the 1000-ppm stock solutions (Merck, Germany) and the standards for all trace metals analyzed were ensured to be in the same matrix as the extractant to minimize matrix effects.

2.2. Sequential Extraction of Heavy Metals

After adsorption, a four-step sequential extraction[28,29] scheme was used determine the distribution of each of the adsorbed metal at various equilibrating concentrations (Figure 1). The distributions of metals in native and in the mixture amended soil were also studied by following the same method.

In this study, the exchangeable fraction (EXC) includes nonspecific adsorption of free and complex metal ions; the organic fraction (ORG) represents metals held by complexation, adsorption and chelation; inorganic or carbonate fraction (INORG) includes surface precipitated or co-precipitated metal compounds (e.g., carbonates or hydroxides) while the residual fraction (RES) refers to metals bound to sulfides or held within the lattices of crystalline minerals in soils[30,31].

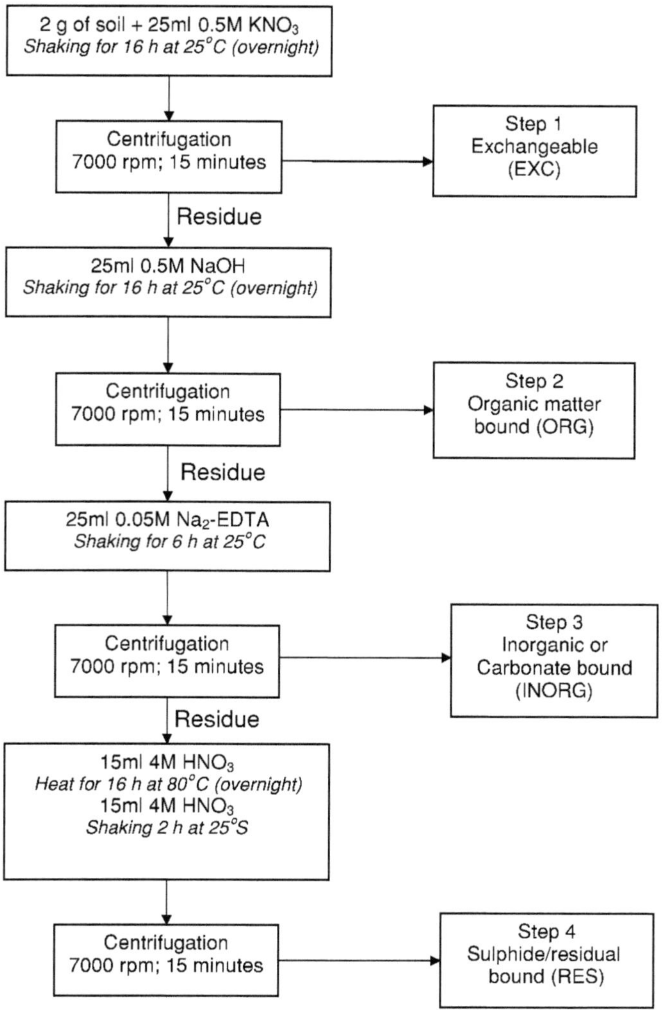

FIGURE 1. Schematic diagram of the sequential extraction procedure.

3. Results and Discussion

3.1. Physicochemical Properties of Soil, Coal Ash and Sewage Sludge

Baseline soil samples were found to be sandy silt loam in texture with acidic pH for alfisol (SL1), sandy loam texture with neutral pH for red alfisol (SL2) and clayey texture with alkaline pH for the black vertisol (SL3). Electrical conductivity (EC), available N (N) and organic carbon (OC) contents were low in all the soils except for cation exchange capacity (CEC) which was low, medium and high in SL1, SL2 and SL3, respectively. The dominant clay mineral phases identified were kaolinite in SL1, montmorillonite in SL3 and kaolinite with an admixture of illite in SL2 (Table 1).

The differences in properties among the soil types are mainly attributed to their parent material and to the degree of weathering besides presence of organisms and topography etc. Though the influence of parent materials on the formation of clay minerals in soil is most conspicuous at early stages of formation; the effect is progressively erased under the impact of more active soil forming factors with the advancement of weathering. The likely general conditions that governed the formation of clay minerals in these soils have been summarized[32]:

1. Intensive leaching at low pH conditions and in absence of high concentrations of magnesium and other bases leading to the formation of kaolinite in SL1.
2. Presence of mica-type minerals leading to the formation of illite, the extent of formation depending on environment with magnitude of low pH but containing higher concentration of potassium in SL2.
3. Impeded drainage conditions and the presence of high silica and magnesium concentrations giving rise to smectite-type of clay in SL3.

The properties of coal ash (FA) mostly depend on the type and source of coal used, combustion conditions and ash handling process while that of the sewage sludge (SS) vary depending on the source and composition of wastewater and the type treatment process. The coal ash used is slightly alkaline (pH 7.8), low in salinity (EC = 127 μScm^{-1}) with high proportions of silt-sized particles and high water holding capacity whereas the sewage sludge is highly acidic (pH 4.3), high in salinity (EC = 2,560 μScm^{-1}) with OC (144 g kg^{-1}), sulphur (270 g kg^{-1}) and N (2,625 ppm available N and 1.01% total N) contents (Table 1).

3.2. Heavy Metals Content in Soil, Coal Ash and Sewage Sludge

The "total" concentrations of heavy metals in soil, ash and sludge were determined through aqua regia extraction,[33–36] a standard protocol used by international environmental protection agencies for heavy metal analysis[37]. Mehlich III (M III) extraction, instead of DTPA extraction, was used to determine the plant available fractions of the metals since it is a universal extractant over broad range of soil pH.[38–40]

The "total" concentrations of Cd, Cu, Ni, Pb and Zn were lowest in soils compared to both ash and sludge. In general, the sequence of metal concentrations followed the order of SS > AH > SL3 > SL2 > SL1. The concentrations of metals in soils were within their ranges of world averages[41] while in sludge they were much below the USEPA limits for its agricultural use (Table 1).

The plant available concentrations of metals of interest did not vary much among soil types (Table 1). The plant availabilities of metals in amended soils were comparatively higher in case of sludge than coal ash. Determinations of chemical forms of metals in addition to their "total" and "bioavailable" concentrations in native soil and the amendments are very important in order to have a more meaningful understanding of their mobility, plant availability and related ecotoxicity. The concentrations of metals in sequentially extracted fractions and the percent recovery in three soil types are given in Table 3 and in coal ash and sludge are reported in Table 4. The comparison of the sum of the metal concentrations in each fraction with their respective "total" concentration (aqua regia digest) show recoveries are within 20%, indicating satisfactory efficiency of the extraction procedure adopted.

In general, the heavy metals were found to be distributed in low percentages in the mobile fraction (EXC) and in higher percentages in relatively less mobile fractions in all soil types. The distributions of selected metals were primarily as Cd_{RES}, Ni_{RES}, Zn_{RES}, Cu_{ORG} and as both Pb_{RES} and Pb_{INORG}. The distribution patterns of metals observed are similar to the earlier findings; e.g. Cd[42], Cu and Ni[29,43–45], Pb[43], and Zn.[46,47] However, different soil types, change in sequential extraction

TABLE 3. Chemical fractions of heavy metals in three soil types

Sample	Element	Total concentration[a] ($\mu M\ kg^{-1}$)	Exchangeable ($\mu M\ kg^{-1}$)	Organic ($\mu M\ kg^{-1}$)	Inorganic ($\mu M\ kg^{-1}$)	Residual ($\mu M\ kg^{-1}$)	Recovery[b] (%)
SL1 (Acidic alfisol)	Cd	6.7	0.31	1.61	0.56	3.34	87
	Cu	168	14.2	84.3	14.6	24.8	82
	Ni	138	26.5	16.2	6.60	78.1	92
	Pb	72	9.29	11.8	19.7	23.6	90
	Zn	289	58.8	25.6	44.5	137	92
SL2 (Neutral alfisol)	Cd	6.8	0.22	1.18	0.71	3.61	85
	Cu	252	11.2	115	25.8	57.9	83
	Ni	129	22.9	8.90	11.7	69.1	88
	Pb	56	5.07	9.02	16.6	20.4	92
	Zn	509	93.0	68.7	113	232	99
SL3 (Alkaline vertisol)	Cd	8.3	0.18	1.09	0.82	4.78	83
	Cu	400	12.0	193	61.2	104	92
	Ni	324	23.6	13.0	28.7	193	80
	Pb	91	7.12	8.14	38.1	30.1	92
	Zn	839	92.1	73.7	192	366	86

[a]—Total concentration by aqua regia digestion
[b]—Calculated as % of sum of fractions with respect to the total concentration (aqua regia)

TABLE 4. Chemical fractions of heavy metals in coal fly ash and sewage sludge

Sample	Element	Total concentration[a] (μM kg^{-1})	Exchangeable (μM kg^{-1})	Organic (μM kg^{-1})	Inorganic (μM kg^{-1})	Residual (μM kg^{-1})	Recovery[b] (%)
Coal fly ash (FA)	Cd	5.16	ND	0.49	1.02	2.80	84
	Cu	602	1.57	187	55.0	268	85
	Ni	221	20.0	10.0	55.0	121	93
	Pb	463	12.1	25.3	205	139	82
	Zn	1498	63.0	188	394	612	84
Sewage sludge (SS)	Cd	23.6	ND	3.23	6.54	10.2	86
	Cu	4772	11.0	2164	367	1863	88
	Ni	702	133	10.0	103	318	80
	Pb	1462	33.0	17.0	403	687	78
	Zn	47171	4847	8517	10061	16131	84

[a]—Total concentration by aqua regia digestion
[b]—Calculated as % of sum of fractions with respect to the total concentration (aqua regia)
ND—Not detected

procedures and other experimental conditions would possibly produce different results.

The heavy metals were predominantly found to be present in the immobile RES fraction in the coal ash with the exception of Pb recovered mainly from Pb_{INORG}. On the other hand, the sludge had 24% Ni and 12% Zn in the mobile EXC fractions even though the metals were mainly present in RES fraction. Copper was primarily associated with the ORG fraction while majority of Pb was recovered from as Pb_{RES}. The distribution of metals in sludge is almost consistent with earlier findings in the cases of Cd[48], Cu[48–50] and Ni and Pb[51]; but not for Zn. However, comparison of metal distributions in this study, particularly in amendments, with earlier such findings is not warranted. Because, the major forms of metals in ash depend on the source of coal, combustion temperature and also type of ash collector[52]. Similarly, metal associations in sludge depend upon the source and composition of wastewater and the treatment processes.[53] Therefore, distribution of metals must be considered for each amendment on an individual basis in order to predict their behavior in amended soils. In this study, since the metals in both ash and sludge are mainly associated with less mobile fractions, it is less likely that their application will cause major changes in the metal distribution in amended soils.

3.3. Effect of Amendment on Physicochemical Parameters of Soils

The change in the selected properties of each soil on amendment with various mixture proportions of ash and sludge is given in Table 5.

It is observed that, application of amendments to various soils types influenced pH, EC, OC, N, bulk density (BD) and maximum water holding capacity (MWHC) etc. more than they did to CEC and $CaCO_3$ content.

Initial increases in pH of both SL1 and SL2 with 100% ash addition were noticed but decreased with increase in sludge proportions in the mixture; whereas, in SL3 it reduced with ash addition and further with increasing sludge proportion in the mixtures. The overall trend of pH was reducing with increase in sludge proportion in the mixture irrespective of the soil types; maximum in the case of SL1. In the present study, sludge applied at 52t ha^{-1} reduced pH in all cases, 5.2 to 4.7 in SL1, 7.1 to 6.9 in SL2 and 8.6 to 8.2 in SL3 reflecting its nature. The decrease in pH, in proportion to the sludge application rates[54], in sludge-amended soils has been attributed to acids present or produced during the microbial decomposition of organic matter in the sludge[55]. Besides the amendment application rates, the change in pH on ash addition is attributed to the moderate alkalinity (pH = 7.8) of ash[4], which is higher than that of SL1 and SL2 but less in SL3 (vertisol).

The three soil used had low salinity (Table 1) which increased moderately on ash addition and sharply further with increase in sludge proportions in the amendments; the sludge-amended ones showing comparatively higher values. The sharp increase in the salinity of sludge amended soils is due to the high salinity of the sludge compared to either ash or the soils; as observed earlier in sludge amended gypsisols[56]. Gradual increases in OC and N (available nitrogen) with increase in sludge proportion in the amendment occurred in all the three soils mainly due to higher organic carbon and nitrogen content in the sludge (Table 1). The increase was highest at 52t ha^{-1} amendment with 100% of sludge compared to the control.

The increase in MWHC was associated with decrease in the bulk density[56–58] (BD) of all amended soils, irrespective of their types; the highest of MWHC coupled with lowest of BD was observed at M6 (100% sludge application @ 52 t ha^{-1}) and the impact was more in SL2 and SL3 than in SL1.

3.4. Heavy Metal Concentration in Amended Soils

Concerns about the presence of heavy metals in coal ash and sewage sludge and their consequent transfer to plants and eventual into the food chain are well established.[59,60] Both major

TABLE 5. Changes in selective properties of different mixtures after 90 days of incubation at near field capacity moisture level

Soil type	Mixtures	pH (1:2.5)	EC (μScm^{-1})	OC (g kg^{-1})	Available N (ppm)	CEC (Cmol kg^{-1})	$CaCO_3$ (g kg^{-1})	BD (g cc^{-1})	MWHC (%)
SL1 (Acidic alfisol)	M1	5.2	43	5.3	76.9	5.48	6.70	1.64	30
	M2	5.4	49	6.0	75.4	5.53	7.80	1.63	31
	M3	5.1	48	6.3	77.4	5.65	7.30	1.63	31
	M4	5.1	58	6.8	80.3	5.65	7.10	1.62	32
	M5	4.9	67	7.2	81.7	5.72	7.03	1.61	32
	M6	4.7	65	7.5	82.1	5.86	6.57	1.58	32
SL2 (Neutral alfisol)	M1	7.1	65	5.1	84.3	12.5	8.10	1.62	37
	M2	7.2	79	5.6	85.1	12.4	9.63	1.56	38
	M3	7.1	82	5.8	86.3	12.7	8.83	1.56	39
	M4	7.0	83	6.2	86.4	12.8	8.23	1.54	40
	M5	6.9	81	6.7	88.8	13.5	8.47	1.52	40
	M6	6.8	84	7.6	91.4	14.3	8.10	1.49	42
SL3 (Alkaline vertisol)	M1	8.6	133	6.2	94.6	32.7	84.3	1.48	53
	M2	8.5	145	6.5	95.1	32.9	84.4	1.44	54
	M3	8.5	150	6.8	96.2	33.7	81.1	1.44	53
	M4	8.4	147	7.2	96.2	34.6	81.0	1.45	55
	M5	8.3	163	7.6	97.3	34.9	78.9	1.41	56
	M6	8.2	160	7.8	99.8	36.1	77.7	1.41	58

Note: M1—CONTROL; M2—FA100; M3—SS25 + FA75; M4—SS50 + FA50; M5—SS75 + FA25; M6—SS100 (FA—Coal fly ash; SS—Sewage sludge) OC—Organic carbon; HA—Humic acid; FA—Fulvic acid; CEC—Cation Exchange Capacity; BD—Bulk Density; MWHC—Maximum Water Holding Capacity

and trace element concentrations determined in control soil and after mixture amendment are given in Tables 6 and 7, respectively. In general, "total" concentrations of all the major elements barring a few, increased linearly with mixture amendment irrespective of the soil type and were related with sludge proportions in the mixture. Concentrations of Al, Fe, P, Mg, S and K were higher in the amended soils compared to the respective controls but Ca remained more or less unchanged (Table 6). In the case of trace metals, slightly higher values were observed in mixture applied soils over respective controls (Table 7). Particularly, higher concentration of Zn was noticed with 100% sludge added at 52 t ha^{-1}; its concentration increased from 18.89 to 40.14 ppm (112% increase) in SL1; 33.28 to 68.25 ppm (105%) in SL2 and 54.86 to 88.18 ppm

TABLE 6. Total (Aqua regia) and Mehlich III extractable concentrations of selective elements in different mixtures after 90 days of incubation at near field capacity moisture level

		Al		Ca		Fe		K		Mg		Mn		P		S	
Soil type	Mixtures	AQR	M III	AQR	M III	AQR	M III	AQR	M III	AQR	M III	AQR	M III	AQR	M III	AQR	M III
SL1 (Acidic Alfisol)	M1	7428	823	598	379	9358	427	476	92	446	64	58	20	235	33	21	13
	M2	7409	832	621	400	9354	453	491	95	461	76	60	23	261	37	25	14
	M3	7324	831	580	381	9490	467	492	96	464	79	58	22	285	44	28	15
	M4	7873	844	650	425	9964	577	506	95	466	78	57	25	340	56	31	17
	M5	7725	847	582	406	9537	591	519	98	446	82	49	28	350	66	34	18
	M6	7919	852	648	408	10521	584	600	104	465	93	53	30	395	71	37	20
SL2 (Neutral Alfisol)	M1	8154	566	1956	1189	14347	218	1394	272	1069	195	256	117	327	99	21	15
	M2	9073	567	1965	1231	14363	286	1439	290	1195	214	304	136	353	102	23	16
	M3	9808	631	1956	1245	15653	250	1504	313	1257	225	258	136	406	107	26	15
	M4	9750	645	1553	1284	14809	282	1580	310	1258	251	261	145	331	114	26	15
	M5	9552	618	1949	1276	16834	330	1553	311	1263	238	265	148	444	122	24	17
	M6	9620	580	1905	1280	15903	330	1598	322	1272	241	253	151	446	122	29	16
SL3 Alkaline Vertisol)	M1	28767	794	13138	7648	27248	127	1605	280	6089	834	527	214	246	25	76	55
	M2	28901	819	13135	7694	27259	135	1674	291	6109	854	546	218	249	24	79	58
	M3	29805	842	16841	7708	27866	145	1639	287	6131	871	583	220	280	30	81	60
	M4	31027	836	16336	7718	29104	159	1690	296	6330	874	644	222	346	36	87	64
	M5	29603	824	17481	7715	27833	164	1687	299	6393	870	648	226	329	40	93	65
	M6	29941	826	16785	7724	22758	163	1695	302	6433	887	571	233	336	43	95	64

Note: M1—CONTROL; M2—FA100; M3—SS25 + FA75; M4—SS50 + FA50; M5—SS75 + FA25; M6—SS100 (FA—Coal fly ash; SS—Sewage sludge) Aqua regia (AQR) and Mehlich III (M III) values are in ppm

TABLE 7. Aqua regia and Mehlich III extractable concentrations (ppm) of trace elements in different mixtures of three soil types after 90 days of incubation at near field capacity moisture level

		Cd		Cu		Ni		Pb		Zn	
Soil type	Mixtures	AQR	M III	AQR	M III	AQR	M III	AQR	M III	AQR	M III
SL1 (Acidic Alfisol)	M1	0.75	0.24	10.7	4.56	8.08	2.32	14.8	6.74	18.9	6.05
	M2	0.78	0.26	11.2	4.97	8.63	1.81	21.0	7.21	37.0	7.35
	M3	0.81	0.23	12.2	5.53	8.35	1.92	19.2	7.44	44.2	9.64
	M4	0.79	0.27	14.6	6.39	8.24	2.01	19.8	7.50	55.3	12.0
	M5	0.83	0.30	14.7	6.94	8.14	1.98	20.7	8.73	55.6	13.1
	M6	0.86	0.35	16.4	7.30	8.15	2.05	23.1	8.48	60.1	14.3
SL2 (Neutral Alfisol)	M1	0.76	0.21	16.0	5.63	7.54	3.06	11.5	5.13	33.3	13.7
	M2	0.78	0.22	16.8	5.84	8.55	3.08	14.3	6.44	45.5	13.0
	M3	0.80	0.22	18.5	6.19	9.31	2.91	12.1	6.04	50.9	15.9
	M4	0.82	0.25	19.5	5.96	9.48	2.86	11.8	5.63	54.1	16.0
	M5	0.86	0.22	22.5	7.50	9.82	3.21	15.0	6.58	64.9	18.0
	M6	0.88	0.25	24.2	7.58	10.1	3.20	16.0	6.52	68.3	17.7
SL3 (Alkaline Vertisol)	M1	0.93	0.19	25.4	8.88	19.0	4.93	18.8	7.22	54.9	16.9
	M2	0.98	0.16	27.5	8.96	18.7	4.99	19.6	7.40	73.1	17.0
	M3	1.06	0.17	28.5	9.63	18.8	4.99	19.3	7.95	71.0	18.5
	M4	1.03	0.15	28.7	10.3	19.7	5.09	19.0	8.24	77.0	21.4
	M5	1.01	0.15	30.5	10.7	20.8	5.05	20.9	8.69	87.7	23.4
	M6	1.05	0.15	31.3	12.2	21.3	5.05	21.2	8.85	88.2	23.1

Note: M1—CONTROL; M2—FA100; M3—SS25 + FA75; M4—SS50 + FA50; M5—SS75 + FA25; M6—SS100 (FA—Coal fly ash; SS—Sewage sludge)
Aqua regia (AQR) and Mehlich III (M III) values are in ppm

(61%) in SL3. Greater increases were registered in soils containing higher proportions of sludge than ash in the amendments principally due to higher Zn presence in the sludge (3,085 ppm). The mode and amount of accumulation of metals following application of either sludge or ash depends on the nature of the material added and the rates of application[56,61]. The metal concentrations though increased due to amendment, still were below the respective maximum permissible limits.

On the other hand, no distinctive changes were noticed in the bioavailable concentrations of both major and trace elements in soils following ash and sludge applications except for Zn. The concentration of Zn increased to some extent linearly with increasing sludge proportions in the amendment in each soil type and is primarily attributed to its presence in relatively mobile fractions (59%, equivalent to 1532 ppm) in sewage sludge.

3.5. Heavy Metal Distribution in Amended Soils

Negligible changes in the distribution of metals except for Zn were observed due to low application rates of amendments i.e. low input of metals contributed by ash or sludge or both since the metals are mainly associated with less mobile fractions in them. Therefore, only the distribution pattern of Zn is discussed further.

The fractions of zinc present in different mixtures of three soil types are given in Table 8 and their distribution is portrayed in Figure 2. Amendment application increased Zn_{EXC} with associated decreases in Zn_{INORG} and Zn_{RES} in all the soil types. The neutral alfisol showed the highest increase (7%) followed by acidic alfisol (5%) and alkaline vertisol (2%) in the 100% sludge application over respective controls (Figure 2) owing to very high concentration of Zn_{Total} and presence in appreciable amounts as Zn_{EXC} in sludge. On the other hand, ash had negligible amounts of Zn_{EXC} but higher "Zn_{Total}" than native soils. Therefore, within the amendments, higher increases were noted in 100% sludge amendments irrespective of soil types.

3.6. Adsorption Behavior of Heavy Metals in Amended Soils

The adsorption behavior of selected metals in three types of soil after amendment was studied to understand the possible changes in their retention capacity and adsorption affinities. In all the cases, characteristic increases in the amount of metals adsorbed with increase in equilibrating concentrations were observed. In some cases, the percentage sorption decreased with increase in the equilibrating concentration of metals depending on metal and the type of soil indicating likely gradual saturation of sorption sites. Application of amendments to vertisol (SL3) resulted in highest sorption for all the metals.

Simple isotherms obtained from equilibrium solution concentration of metal (C_e) with respect to the amount sorbed (X/m). For reference, isotherms of Cd and Ni are shown in Figures 3 and 4, respectively. Normally, in the case where the steep initial slopes leveled off with subsequent increase in the equilibrium concentration of metal giving rise to a plateau or a linear section with positive slope is classified as L-type

TABLE 8. Chemical fractions of zinc in different mixtures of three soil types

Soil type	Mixture	Total concentration[a] (μM kg^{-1})	Exchangeable (μM kg^{-1})	Organic (μM kg^{-1})	Inorganic (μM kg^{-1})	Residual (μM kg^{-1})	Recovery[b] (%)
SL1 (Acidic Alfisol)	M1	289	58.75	25.61	44.53	136.7	92
	M2	566	109.9	51.54	66.06	252.2	85
	M3	676	134.6	77.66	68.68	276.8	83
	M4	846	187.8	108.0	85.02	344.3	86
	M5	850	187.4	108.0	78.64	351.3	85
	M6	920	203.7	128.4	79.59	350.9	83
SL2 (Neutral Alfisol)	M1	509	93.01	68.67	113.0	231.6	99
	M2	696	104.0	85.55	124.9	245.0	80
	M3	778	123.6	91.09	126.8	263.9	78
	M4	828	126.0	103.2	147.1	283.2	80
	M5	993	206.0	119.3	112.7	351.4	80
	M6	1044	207.8	147.4	128.4	362.3	81
SL3 (Alkaline Vertisol)	M1	839	92.06	73.70	192.31	366.1	86
	M2	1118	106.1	110.2	271.5	409.9	80
	M3	1086	93.29	121.8	263.5	383.3	79
	M4	1177	102.7	129.2	294.4	397.7	78
	M5	1342	143.5	131.2	357.6	455.9	81
	M6	1348	166.9	148.6	342.8	461.7	83

Note: M1—CONTROL; M2—FA100; M3—SS25 + FA75; M4—SS50 + FA50; M5—SS75 + FA25; M6—SS100 (FA—Coal fly ash; SS—Sewage sludge)

[a]—Total concentration by aqua regia digestion

[b]—Calculated as % of sum of fractions with respect to the total concentration

isotherms and reflect relatively high affinity between the metal and soil surface, and usually indicate chemisorption. But, in the present study, the plateau was not clearly established for most of the metals, mainly because of their low to moderately high concentration (10 to 200 μM L^{-1}) ranges. On the other hand, in few cases, the amount of metal adsorbed increased linearly with insignificant increase in its equilibrium concentration and showed nearly a vertical line and been classified as H-type isotherms suggesting very strong metal-soil surface interaction, an actually the extreme case of L-types[16]. The isotherms showing both moderate L-type and H-type characteristics have been classified as L-2 type in this study.

The classification of isotherms deduced from the plots (note: different scales on X-axis in the Figure 4 for SL3) is given in Table 9. Different scales have been used to steer clear of adsorption curves overlapping with Y-axis due to the presence of very low metal concentrations in the equilibrium solution, more particularly in case of SL2 and SL3. The differences among isotherms of various mixtures (M1 to M6) for each soil type cannot be identified owing to small difference among amendment loading rates. Therefore, it may be generalized that, in acidic alfisol all the metals show L-type isotherms except Pb, which showed H-type isotherm. In neutral alfisol both Cd and Ni showed L-2 isotherm while Cu, Pb and Zn showed H-type isotherms. All the metals showed H-type isotherms except for Cd with L-2 type in the case of alkaline vertisol. The order of adsorption affinities for metals relating to isotherm is found to be L > L-2 > H indicating vertisol has greater adsorption affinity for added metals than other soil types; and, among metals, Pb with the highest adsorption affinity. Similar observations[62] on simple isotherms of L or H-type for different metals have also been made in most of the earlier cases. In particular, more strongly adsorbed Pb and Cu tend towards high affinity (H-type) isotherms and a shift toward high affinity adsorption with increased pH is expected.

From simple adsorption isotherms it was possible only to distinguish the adsorption affinities of metals in different soil types. Therefore, further efforts were made to quantify the adsorption affinity of metals in various treatments using Freundlich isotherms. The distribution coefficients (K_D) for respective metals were obtained (Table 10) from the regression of log C_e with log X/m. The K_D is analogous to the equilibrium constant of a chemical reaction and has been used to describe the partitioning of metals between solid and liquid phases and therefore higher the values greater is the metal adsorption affinities or capacities[63,64]. In other words, low values of K_D indicate most of the metals present in the system will remain in solution and hence are available for various chemical processes and plant uptake. It was further confirmed that, among the soil types, SL3 had the highest K_D for all metals than others. Among the metals, Pb showed very high affinity irrespective of soil/or mixture type. Lead adsorption was found to be quite high even in the acidic alfisol. In case of alkaline vertisol (SL3), Freundlich isotherms of Pb could not be constructed; because, at lower equilibrations (10 and 25 μM L^{-1}), Pb concentrations were below the analytical detection limits in the equilibrium solution indicating very high affinity i.e. all the added Pb adsorbed on the soil surface. Therefore, it is clear that Pb adsorbs very strongly in alkaline vertisol followed by neutral and acidic alfisols. It has been confirmed that soils

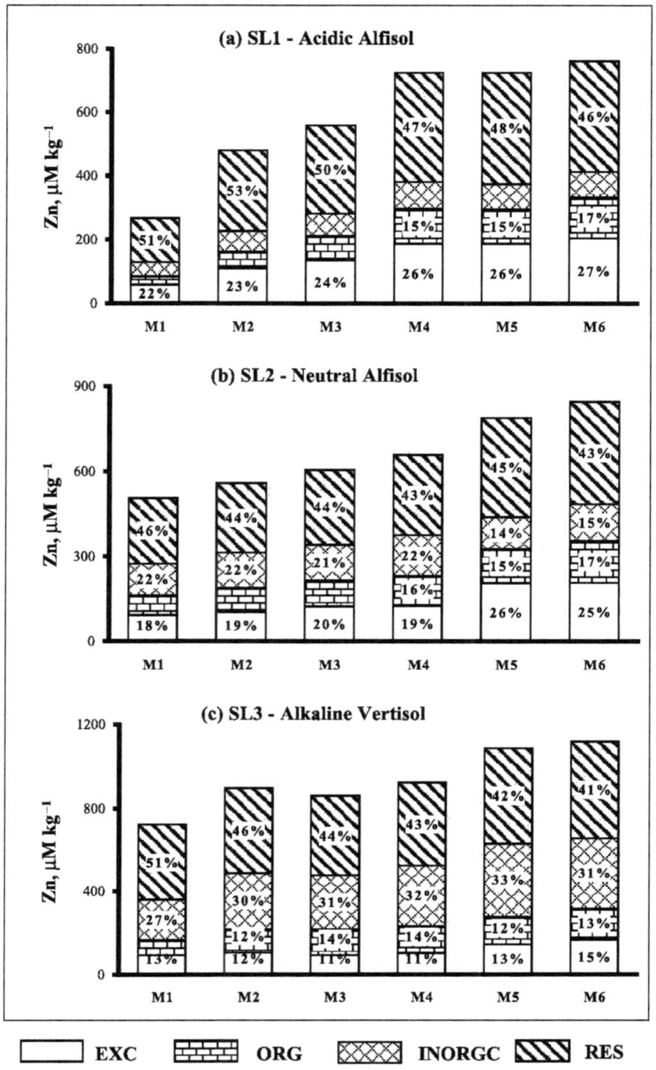

M1 - CONTROL; M2 - FA100; M3 - SS25+FA75; M4 - SS50+FA50; M5 - SS75+FA25; M6 - SS100 (FA - Fly ash; SS - Sewage sludge)
Note: Scales on Y - axis are different

FIGURE 2. Chemical distribution of Zn in different mixtures of three soil types.

generally show higher adsorption affinity for Pb over a broad range of pH[65,66].

The K_D of metals in amended treatments were comparatively higher (Table 10) within a given soil type over the respective controls, the difference being smallest in the case of vertisol. However, it is summarized that, the impact of amendments on K_D was neither systematic across metals nor soil types.

By considering all the metals together, it is clear that, in each soil type, Pb had higher K_D than others indicating its reduced mobility, and, therefore, the bioavailability compared to others. Similarly, among all the soil types, the mobility of metals was highest in acidic alfisol, moderate in neutral alfisol and the least in alkaline vertisol indicating the individual characteristic of soil being the major controlling factors for the difference in sorption behavior of each individual metal. Therefore, it is understood that the degree of influence on metal sorption due to amendments in a particular soil type is governed primarily by its native properties while the quality and quantity of amendment affect it to a lesser extent. The ash, sludge and mixture application increased the K_D for some of the metals similar to observed elsewhere[67–69]. The results also indicate that amendment application at 52t ha^{-1} had less impact on sorption of metals in fine textured vertisol compared to relatively coarse textured alfisols; which is, similar to earlier findings[69].

3.7. Distribution of Adsorbed Metals

From the Freundlich isotherm parameters (K_D) obtained for different metals it was possible to differentiate their affinities in different soil types. It should be noted that, K_D represents the adsorption behavior of metals over a range of concentrations (in this study it is low to moderately high). Owing to the complexity of metal adsorption processes, it is not possible to decipher from the sorption isotherms, neither the extent of chemisorption nor precipitation involved at different metal additions. In addition, the soil components providing the sorption sites at higher metal concentrations are not necessarily those responsible for sorption of metals at lower concentrations[64]. Therefore, the sequential extraction of the adsorbed metals at various equilibrating concentrations was more helpful in understanding the partitioning of metals in their various geochemical phases and hence facilitated to distinguish the predominant mechanisms of adsorption of each metal in each soil type.

The concentrations of metals extracted, in each fraction, at different equilibrating concentrations indicated that the changes in the distribution of adsorbed metals were contingent on the soil type and metal loadings irrespective of the amendments. The differences observed in the amount of metals extracted under each fraction of various amendments relative to the respective controls in each soil type were lower at lower metal additions; and, became more distinct only at higher metal loadings. Among different mixtures, amended into a particular soil type, unsystematic trends were observed in the concentrations of metals extracted in different fractions; and, the changes in the distribution pattern of adsorbed metals were clearly observed only at higher metal loadings.

The changes in the distribution pattern of adsorbed metals subsequent to metal additions in varying concentrations to the control soils are therefore discussed first. Blank treatments (0 $\mu M\ L^{-1}$) indicate the distribution of native metals in different soil types: Cd_{EXC} was predominant at all metal additions in all the soil types and the degree of predominance linearly increased with metal additions (Figure 5). In the case of SL2 and SL3, sufficient amount of Cd_{EXC} was also extracted in addition to the Cd_{INORG} (Figure 5 B and C) and contributions from Cd_{ORG} and Cd_{RES} became insignificant at higher metal loadings. At highest Cd addition, 78% and 53% respectively were adsorbed as Cd_{EXC} in SL1 and SL2 whereas distributed both as Cd_{EXC} (45%) and Cd_{INORG} (43%) in the case of SL3.

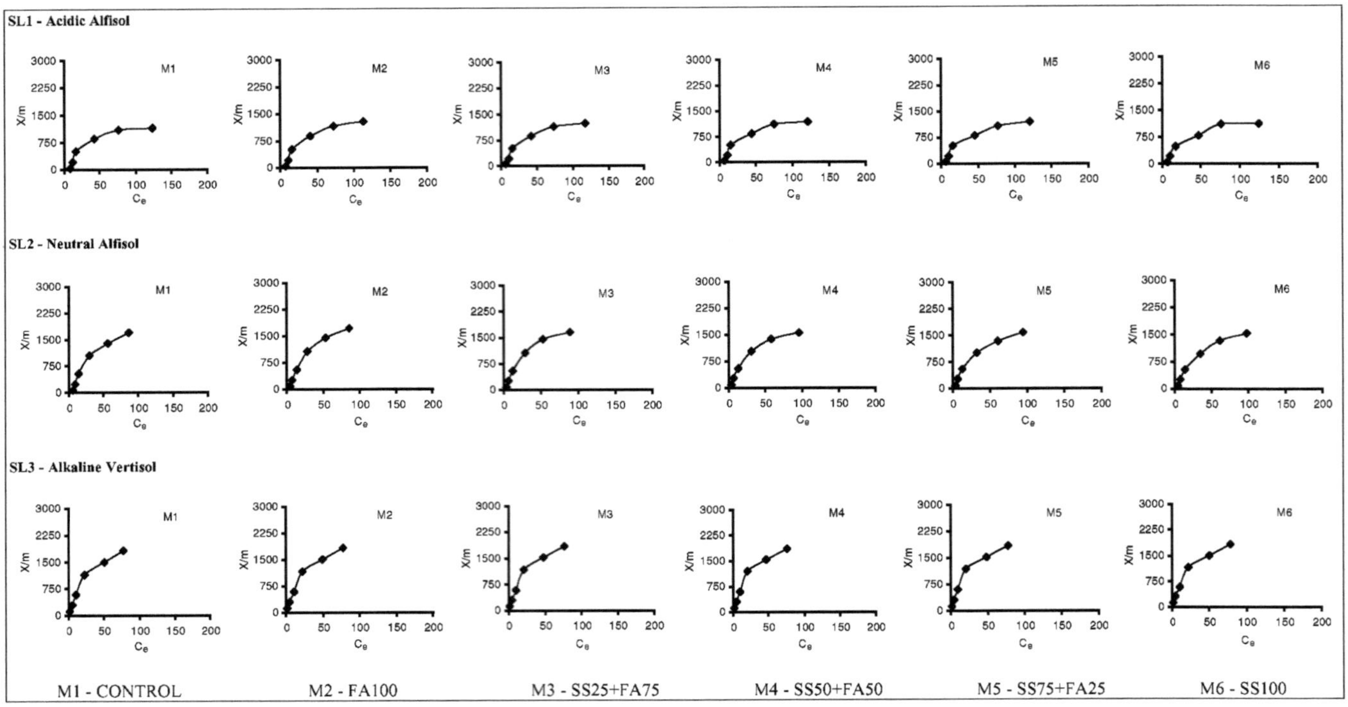

X/m = Amount of metal adsorbed in μM kg^{-1} ; Ce = Amount of metal remained in solution at equilibrium in μM L^{-1}; FA - Coal fly ash; SS - Sewage Sludge

FIGURE 3. Adsorption isotherms for cadmium in three soil types and their respective mixtures.

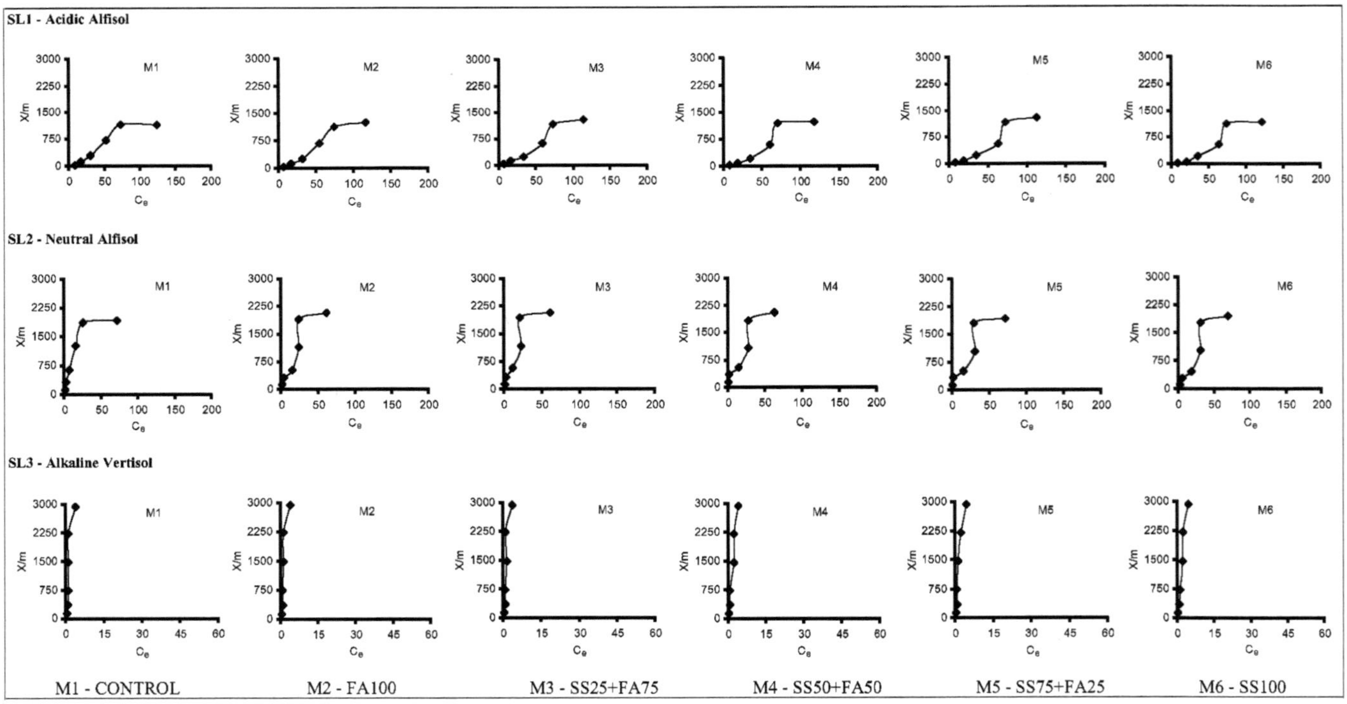

X/m = Amount of metal adsorbed in μM kg^{-1} ; Cc = Amount of metal remained in solution at equilibrium in μM L^{-1}; FA - Coal fly ash; SS - Sewage Sludge

FIGURE 4. Adsorption isotherms for nickel in three soil types and their respective mixtures.

TABLE 9. Classification of isotherms deduced from plots of X/m against C_e

Metal	Isotherm type		
	SL1 Mixtures	SL2 Mixtures	SL3 Mixtures
Cd	L	L-2	L-2
Cu	L	H	H
Ni	L	L-2	H
Pb	H	H	H
Zn	L	H	H

C_e—Concentration of metal remained in equilibrium solution, μM L^{-1}
X/m—Amount of metal adsorbed per unit weight of soil, μM kg^{-1}

Copper predominantly existed in the organic fraction at all metal loadings in the cases of both SL1 and SL2. However, at higher Cu additions, significant amount was adsorbed as Cu_{EXC} in SL1 and Cu_{INORG} in SL2 (Figure 6 A and B). At lower loadings (up to 50 μM L^{-1}), most of the adsorbed Cu was found in all the fractions in significant amounts except as Cd_{EXC} (<3.0 %) in SL3. Further increase in the loading showed downward trends of Cu_{ORG} and Cu_{RES} but upward trend of Cu_{INORG} (Figure 6 C). At the highest addition, 40% of the adsorbed Cu was extracted as Cu_{EXC} and 48% as Cu_{ORG} in SL1 and as Cu_{ORG} (48 and 26%, respectively) and Cu_{INORG} (36 and 56%, respectively) in the cases of SL2 and SL3.

Ni_{EXC} was more predominant at higher metal addition in SL1 and SL2 while at lower additions (up to 25 μM L^{-1} in SL1 and up to 50 μM L^{-1} in SL2), sufficient amount was also extracted from other fractions (Figure 7 A and B). Though the distribution of Ni in SL3 was unsystematic over the entire concentration range, Ni_{EXC} prevailed distinctively over others at the highest addition (Figure 7 C). In general, the accumulation of Ni_{EXC} showed upward trend with 74%, 68% and 49% in SL1, SL2 and SL3, respectively on increased additions while the reverse was true for other fractions.

Redistribution of adsorbed Pb into Pb_{EXC} and Pb_{INORG} with simultaneous decrease in Pb_{ORG} occurred with increased additions in SL1 (Figure 7 D). On the other hand, Pb_{INORG} was found predominantly and increased with further loading in both SL2 and SL3 (Figure 7 E and F). This preferential movement of Pb towards Pb_{INORG} was more pronounced in alkaline vertisol with the highest $CaCO_3$ content among the soil types. At highest addition, the adsorbed Pb was extracted 51% as Pb_{EXC} and 41% as Pb_{INORG} in SL1; while in case of SL2 and SL3, it was mainly extracted as Pb_{INORG} (>80%) fraction.

Most of the Zn adsorbed was found as Zn_{EXC} in SL1 and SL2, but higher in SL1 (Figure 7 G and H) while in SL3 Zn_{INORG} was predominant at higher loadings (Figure 7 I). Zinc decreased in other fractions on further addition and varied among fractions and soil types viz. Zn_{ORG}, Zn_{INORG}, Zn_{RES} in SL1; Zn_{ORG}, Zn_{RES} in SL2; and Zn_{EXC}, Zn_{ORG}, Zn_{RES} in SL3. At highest addition, 64% of the adsorbed Zn was extracted as Zn_{EXC} in SL1, Zn_{INORG} in SL3, while in SL2 as both Zn_{EXC} (45%) and Zn_{INORG} (33%).

From the above, it is confirmed that, on addition at low concentrations, metals are distributed significantly in all fractions but at higher concentrations they tend to show preferential distribution varying across soil types. The observed changes in the partitioning of adsorbed metals among the untreated soil in all types were wide and distinctly noticeable only at higher metal

TABLE 10. Freundlich's distribution coefficient values of metals in different mixtures of three soil types

Soil type	Mixture	Distribution coefficient (K_D)				
		K_D-Cd	K_D-Cu	K_D-Ni	K_D-Pb	K_D-Zn
SL1 (Acidic Alfisol)	M1	11.1	266	1.43	288	24.5
	M2	15.9	328	4.88	578	26.7
	M3	24.1	374	5.65	546	26.9
	M4	21.8	352	1.92	312	28.3
	M5	23.7	372	1.59	376	29.0
	M6	20.2	334	1.60	338	29.4
SL2 (Neutral Alfisol)	M1	23.4	678	104	2455	166
	M2	38.9	822	108	2018	206
	M3	42.2	902	135	1726	206
	M4	56.6	1472	233	2904	215
	M5	48.0	1734	187	1837	216
	M6	39.0	2080	157	2773	222
SL3 (Alkaline Vertisol)	M1	106	6902	685	NC	410
	M2	114	7499	791	NC	445
	M3	116	6577	849	NC	456
	M4	114	7499	675	NC	480
	M5	116	8492	680	NC	490
	M6	114	9612	491	NC	520

Note: M1—CONTROL; M2—FA100; M3—SS25 + FA75; M4—SS50 + FA50; M5—SS75 + FA25; M6—SS100 (FA—Coal fly ash; SS—Sewage sludge)
NC—Isotherms not constructed

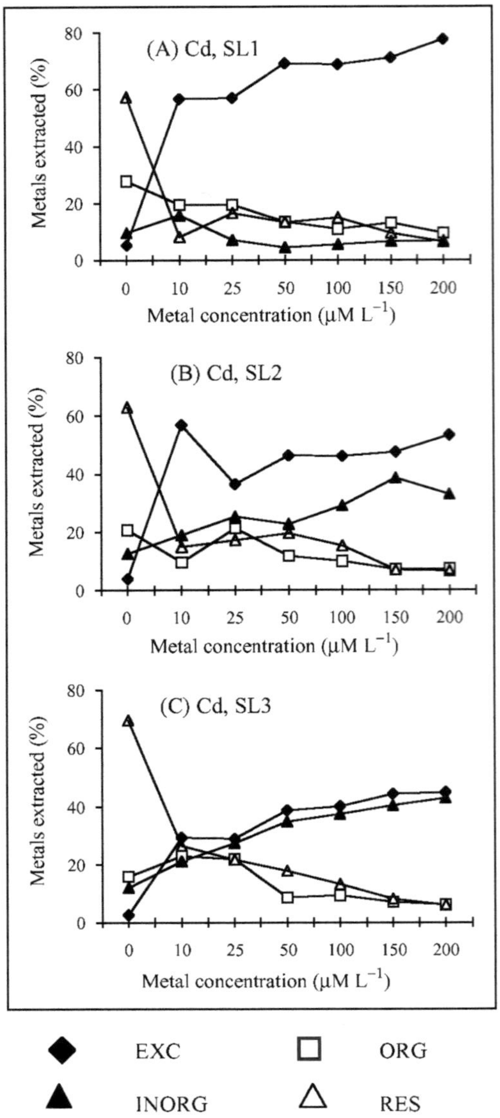

FIGURE 5. Extraction of Cd in different fractions in three soil types at different metal additions.

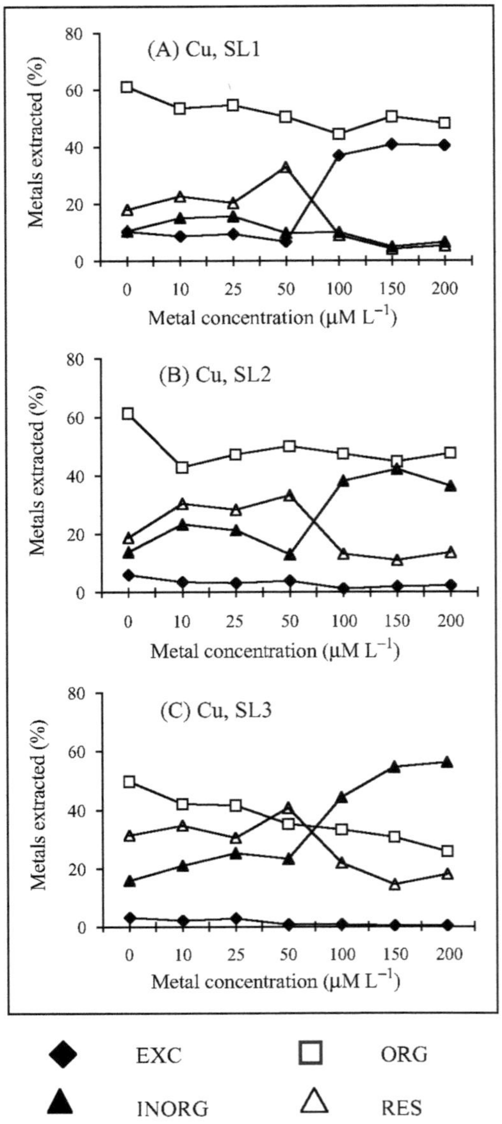

FIGURE 6. Extraction of Cu in different fractions in three soil types at different metal additions.

loadings mainly due to this preferential distribution. Therefore, in order to understand the difference in impact among treatments in each soil type and between the soil types, most parts of the discussion have been limited to metal partitioning at highest addition (200 $\mu M\ L^{-1}$) only. Figure 8 depicts the distribution of adsorbed metals in sludge, ash and their mixture amended soils at the highest metal addition in the soil types. No appreciable changes in metal distribution patterns between control and amended soils at the highest metal loading were noted within each soil type except for increase in the Cu_{ORG} on sludge addition, more prominently in SL1 and SL3 than in SL2. In the amended SL1, having acidic pH, most of the adsorbed Cd, Ni and Zn were extracted in the EXC fraction, while Cu and Pb were additionally found associated with ORG and INORG fractions, respectively. In the amended SL2, having neutral pH, the adsorbed metals were recovered mainly as Cd_{EXC}, Ni_{EXC}, Pb_{INORG}, both Cu_{ORG} and Cu_{INORG} and Zn_{EXC} and Zn_{INORG}, respectively. Amended SL3 soil with alkaline pH had most of the sorbed Cu, Ni and Pb distributed as in the case of alfisol while significant amounts of Cd_{INORG} was also extracted along with Cd_{EXC} and Zn predominantly as Zn_{INORG}.

It was observed that, added Cd had higher tendency to locate predominantly as Cd_{EXC} similar to earlier reported[70]. The finding is in agreement to the presence of significant amounts of Cd_{EXC} in calcareous soils[71] and on addition in form of soluble salts to highly acidic German soils (pH 3.3–4.5) majority being distributed as Cd_{EXC} decreasing with increase in soil pH[72].

The distribution pattern of adsorbed Cu varied with soil types and was found mainly associated as Cu_{ORG} in amended al-

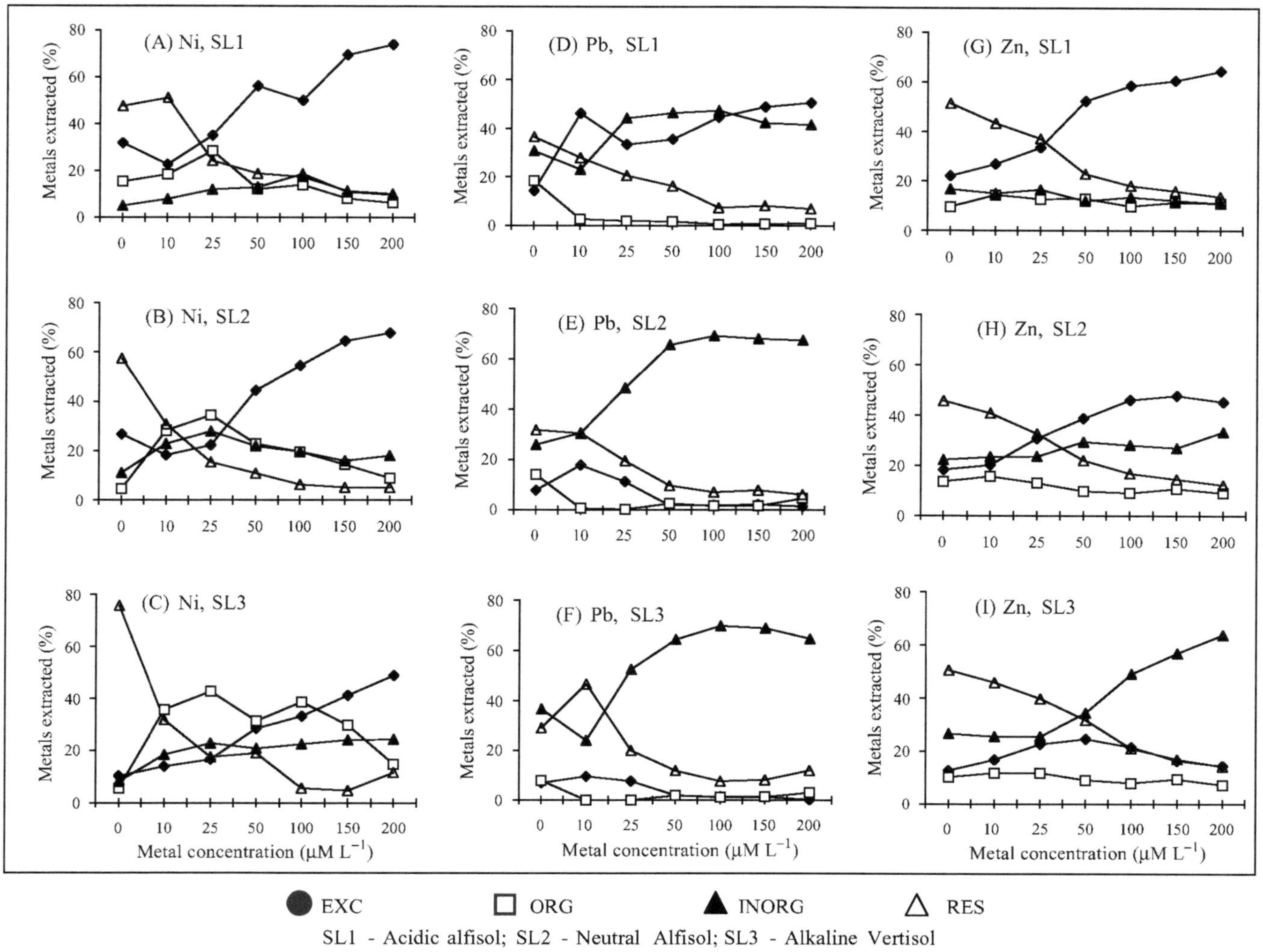

FIGURE 7. Extraction of Ni, Pb and Zn in different fractions in three soil types at different metal additions.

fisols, due to the likely formation of organic complexes, more so in sludge amended ones. Others have also reported Cu predominantly existing as Cu_{ORG} in Cu-contaminated sites.[73,74] In soils amended with soluble salts and in highly acidic soils, Cu is present mainly as Cu_{EXC}[72] while in arid-zone soils as Cu_{INORG}, Cu_{ORG} and Cu_{EXC}[75]. The Cu_{RES} gradually decreases with increased metal loadings because most of the added Cu adsorb onto the more reactive forms[76]. The distribution of Cu predominantly as Cu_{INORG} is similar to earlier findings[70] about increased precipitation of Cu in form of carbonates owing to alkaline pH and high $CaCO_3$ (8.43%) content in vertisol.

Similar to Cd, the added Ni showed higher tendencies to locate predominantly as Ni_{EXC} fraction in all the soil types agreeing more or less to earlier findings regarding the adsorbed Ni to be predominantly in the mobile fractions involving water-soluble and exchangeable as well as easily soluble metal-organic complexes[77].

The extent of adsorbed Pb distribution as Pb_{INORG} varied with soil type and the highest was in vertisol followed by neutral alfisol while in acidic alfisol Pb_{EXC} and Pb_{INORG} were predominant. Added Pb has been mainly found as Pb_{INORG} in arid-zone soils[78]; and as Pb_{EXC} in strongly acidic (pH 3.7) soil[72].

The adsorbed Zn in amended soil was present as Zn_{EXC} at higher levels in acid alfisol, as both Zn_{EXC} and Zn_{INORG} in neutral alfisol and mainly as Zn_{INORG} in alkaline vertisol and confirm to earlier findings[70,75,76]. In another study[72], Zn concentration has been found inversely related to soil pH with most added in the form of soluble salts existing mainly as Zn_{EXC} in acid soil.

Based on the above results, it is summarized that metal sorption is mainly ascribed to ion exchange in acid soil while in neutral and alkaline soils it was aided by carbonate precipitation except for Cu. The sorption of Cu, on the other hand, is mainly due to both ion exchange and organic complexation in acidic alfisol and both complexation and carbonate precipitation in

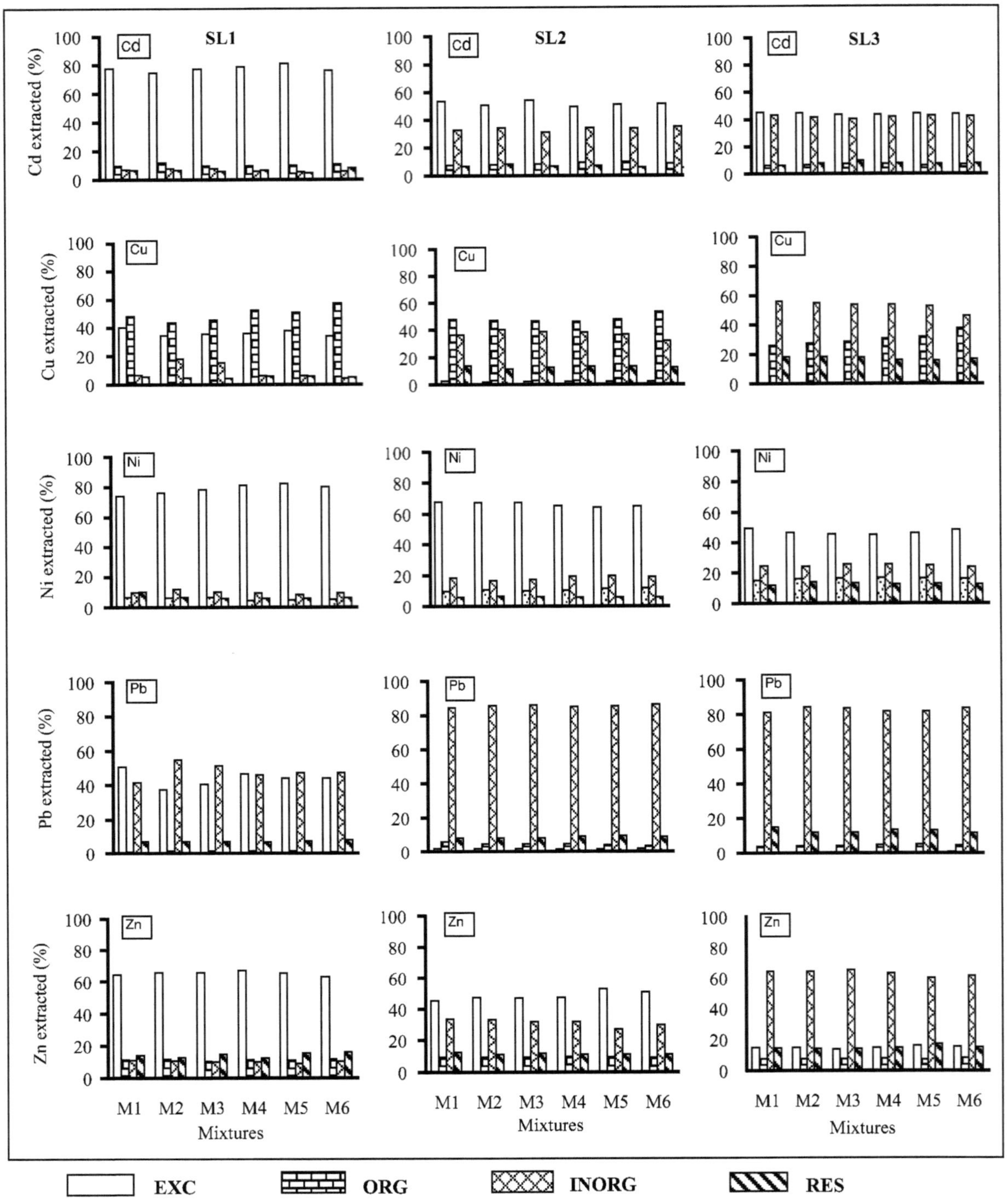

M1 - CONTROL; M2 - FA100; M3 - SS25+FA75; M4 - SS50+FA50; M5 - SS75+FA25; M6 - SS100; SL1 - Acidic alfisol; SL2 - Neutral alfisol; SL3 - Alkaline vertisol

FIGURE 8. Distribution of adsorbed metals at 200 μM L^{-1} concentrations in various mixtures of three soil types.

neutral alfisol and alkaline vertisols. Therefore, it is more likely that at elevated concentrations, metals will be potentially more bioavailable in acidic soils as compared to both neutral and alkaline soils wherein they are present in relatively lesser mobile fractions.

Taking into account that distribution of metals varies according to the type and loading of substrate, type of soil, and the gradient of association stability within the solid phases; the predominance of exchangeable fraction indicated higher metal mobility in acidic soil than the others.

4. Conclusion

It therefore can be concluded that amendment with sludge had greater impact on the soil properties than the coal fly ash. Heavy metals being predisposed in less mobile fractions in both ash and sludge did not alter the distribution pattern of metals in post amended soils. Amendments provided additional sorption sites for the metals; and, therefore, post amendment, the soils irrespective of the type showed marginally higher adsorption affinities. Addition of metals in the form of soluble salts changed metal distribution pattern irrespective of the amendments; at lower metal additions they showed a tendency to distribute among all the geochemical phases; but as the loading increased they accumulated more on the surface-active mobile fractions than in the resistant ones. Based on the distribution coefficients and sequential extraction behaviour of adsorbed metals it is understood that, all metals were highly mobile in acidic alfisol followed by neutral alfisol and alkaline vertisol. By considering the consequences associated with application of ash and sludge in different soil types, it is inferred that due to higher metal retention capacity, the application of amendment in alkaline vertisol and neutral alfisols is of lesser concern in the short term but with possibilities of metal build up on repeated amendment and in long term not only affecting the cultivated crops but also the soil quality itself. On the other hand, due to low metal retention capacity of acidic alfisol, the application of these amendments necessitates close monitoring of both groundwater and crops grown mainly due to the associated risk by increased metal mobility. However, the amendments, more particularly the sewage sludge, though adds heavy metals to soils, it also supplies organic and inorganic components that bind the metals strong enough to restrict their mobility during the initial period of application. Therefore, sewage sludge with lower metal concentrations and near neutral pH is more appropriate for agriculture use on a wide variety of soils. The sequential extractions pattern of metals from the metal-added soils represents freshly contaminated soil environments. Therefore, it is pertinent to note that the metals added as soluble salts or through amendment will mostly persist in their mobile forms; but the possibilities of their redistribution into lesser mobile/residual forms over a period of time also cannot be ruled out and requires further insight.

Acknowledgements. This work was financed by India Canada Environment Facility, New Delhi, India, Canadian International Development Agency, Ottawa and International Development & Research Center, Ottawa, Canada through the project "Land Restoration through Waste Management" and Fly Ash management in India.

References

1. Chang AC, Page AL, Lund LJ, Pratt PR, Bradford GR (1978) Land application of sewage sludge—a field demonstration. Final report for Regional Waste Water Solids Program. Dept of Soil and Environ Sci, Univ of California, Riverside
2. Smith SR (1994) Effect of soil pH on availability to crops of metals in sewage sludge-treated soils. II. Cd uptake by crops and implications for human dietery intake. Environ Poll. 86: 5–13
3. Adriano DC, Page AL, Elseewi AA, Chang AC, Straughan I (1980) Utilization and disposal of fly ash and other coal residues in terrestrial ecosystems: A review. J. Environ. Qual. 9: 333–343
4. Warren CJ (1992) Some limitations of sluiced fly ash as a liming agent for acidic soil. Waste Manage. Res. 10: 317–327
5. Ghodrati M, Sims JT, Vasilas BL (1994) Evaluation of fly ash as a soil amendment for the Atlantic coastal plain: I. Soil hydraulic properties and elemental leaching. Water Air Soil Poll. 81: 349–361
6. Sims JT, Vasilas BL, Ghodrati M (1995) Evaluation of fly ash as a soil amendment for the Atlantic coastal plain: II. Soil chemical properties and crop growth. Water Air Soil Poll. 81: 363–372
7. Sims JT, Vasilas BL, Ghodrati M (1993) Effect of coal fly ash and composted sewage sludge on emergence and early growth of cover crops. Comm. Soil Sci. Plant Anal. 24: 503–508
8. Wong JWC (1995) The production of an artificial soil mix from coal fly ash and sewage sludge. Environ. Technol. 16: 741–751
9. Alva AK, Paramasivam S, Prakash O, Sajwan KS, Ornes WH, Van Clief D (1999) Effects of fly ash and sewage sludge amendments on transport of metals in different soils. In: Sajwan KS, Alva AK, Keefer RF (eds) Biogeochemistry of Trace Elements in Coal and Coal combustion Byproducts, Kluwer Academic/Plenum Publishers, New York, pp 207–222
10. Chaney RL (1983) Potential effects of waste constituents on the food chain. In: Parr JF (ed) Land treatment of hazardous wastes, Noyes Data Corp., Park Ridge, NJ, pp 50–76
11. Wong JWC, Su DC (1997) Reutilization of coal fly ash and sewage sludge as an artificial soil-mix: Effects of pre-incubation on soil physico-chemical properties. Bioresource Technol 59: 97–102
12. Adriano DC, Page AL, Elseewi AA, Chang AC (1982) Cadmium availability to Sudan grass grown on soil amended with sewage sludge and fly ash. J. Environ. Qual. 11: 197–203
13. Sajwan KS, Ornes WH, Youngblood T (1995) The effect of fly ash/sewage sludge mixtures and application rates on biomass production. J. Environ. Sci. Health 30: 1327–1333
14. Sajwan KS, Ornes WH, Youngblood T (1996) Growth and elemental composition of sorghum sudan grass grown on fly ash/organic waste amended soils. J. Environ. Sci. Health 31: 1729–1736
15. Pias I, Jones Benton Jr J (1997) The hand book of trace elements, St. Lucie Press, Florida.
16. McBride MB (1994) Environmental chemistry of soils. Oxford Univ. Press, New York
17. Gerritse RG, van Driel W (1984) The relationship between adsorption of trace metals, organic matter and pH in temperate soils. J. Environ. Qual. 13: 197–204
18. Adriano DC (1986) Trace elements in the terrestrial environment. Springer-Verlag, New York
19. McBride MB (1989) Reactions controlling heavy metal solubility in soils. Adv. Soil Sci. 10: 1–55
20. Alloway BJ (1990) Cadmium. In: Alloway BJ (ed) Heavy metals in soils, John Wiley and Sons, New York, pp 100–124

21. Yong RN, Mohammed AMO, Warkentin BP (1992) Principles of contaminant transport in soils, Elsevier Publishers, Amsterdam
22. Appel C, Ma L (2002) Concentration, pH, and surface charge effects on cadmium and lead sorption in three tropical soils. J. Environ. Qual. 31: 581–589
23. Chaney RL (1990) Twenty years of land application research. Biocycle 31: 54–59
24. Lake DI, Kirk PWW, Lester JN (1984) Fractionation, characterization and speciation of heavy metals in sewage sludge and sludge-amended soils: A review. J. Environ. Qual. 13: 175–183
25. Tessier A, Campbell PGC (1988) Comments on the testing of the accuracy of an extraction procedure for determining the partitioning of trace metals in sediments. Anal. Chem. 60: 1475–1476
26. Sims JT (1986) Soil pH effects on the distribution and plant availability of manganese, copper and zinc. Soil Sci. Soc. Am. J. 50: 367–373
27. Gray CW, McLaren RG, Roberts AHC, Condron LM (1998) Sorption and desorption of cadmium from some New Zealand soils: effect of pH and contact time. Austr. J. Soil Res. 36 (2): 199–216
28. Walter I, Cuevas G (1999) Chemical fractionation of heavy metals in a soil amended with repeated sewage sludge application. The Sci. Total Environ. 226: 113–119
29. Emmerich WE, Lund LJ, Page AL, Chang AC (1982) Solid phase forms of heavy metals in sewage sludge treated soil. J. Environ. Qual. 11: 178–181
30. Tessier A, Campbell PGC, Bisson M (1979) Sequential extraction procedure for the speciation of particulate trace metals. Anal. Chem. 51 (7): 845–851
31. McBride MB (1980) Chemisorption of Cd on calcite surfaces. Soil Sci. Soc. Am. J. 44: 26–33
32. Biswas TD, Mukherjee SK (1999) Text book of soil science, Tata McGraw Hill Publishers, New Delhi, pp 286–339
33. Ščančar J, Milačič R, Stražar M, Burica O (2000) Total metal concentrations and partitioning of Cd, Cr, Cu, Fe, Ni and Zn in sewage sludge. The Sc. Total Environ. 250: 9–19
34. Andrewes P, Town RM, Hedley MJ, Loganathan P (1996) Measurement of plant available cadmium in Newzealand soils. Aust. J. Soil Res. 34: 441–452
35. Canet R, Pomares F, Tarazona F (1997) Chemical extractability and availability of heavy metals after seven years application of organic wastes to a citrus soil. Soil Use Manage. 13: 117–121
36. Forster JC (1995) Heavy metals In: Alef K, Nannipieri P (eds.) Methods in applied soil microbiology and biochemistry, Academic Press, London, pp 101–104
37. ISO 11466 (1995) International standard soil quality-extraction of trace elements soluble in aqua regia. 03–01
38. Mehlich A (1984) Mehlich 3 soil test extractant: A modification of Mehlich 2. Comm. Soil Sci. Plant Anal. 15: 1409–1416
39. Jones JB (1990) Universal soil extractants, their composition and use. Comm Soil Sci. Plant Anal. 21: 1091–1101
40. Chilimba ADC, Mughogho SK, Wendt J (1999) Mehlich 3 or modified olsen for soil testing in Malawi. Comm. Soil Sci. Plant Anal. 30: 1231–1250
41. Davies BE (1980) Trace element pollution. In: Davies BE (ed) Applied soil trace elements, Wiley Publ., New York, pp 287–351
42. Dudka S, Chlopecka L (1990) Effect of solid phase speciation on metal mobility and phyto availability in sludge-amended soil. Water Air Soil Poll. 51: 153–160
43. Sims JT, Kline JS (1991) Chemical fractionation and plant uptake of heavy metals in soil amended with co-composted sewage sludge. J. Environ. Qual. 20: 387–395
44. Tu C (1996) Distribution and transformation of native and added Ni fractions in purple soils from Sichuan Province. Pedosphere 6: 183–192
45. Wang P, Qu E, Li Z, Shuman LM (1997) Fractions and availability of nickel in Loessial soil amended with sewage or sewage sludge. J. Environ. Qual. 26: 795–801
46. Singh JP, Karwasra SPS, Singh M (1988) Distribution and forms of copper, iron, manganese, and zinc in calcareous soils of India. Soil Sci. 146: 359–366
47. Zhang M, Alva AK, Li YC, Calvert DV (1997) Chemical associations of Cu, Zn, Mn, and Pb in selected sandy citrus soils. Soil Sci. 162 (3): 181–188
48. Taylor R, Xiu H, Mehadi A, Shuford J, Tadesse W (1995) Fractionation of residual cadmium, copper, nickel, lead, and zinc in previously sludge-amended soil. Comm. Soil Sci. Plant Anal. 26: 2193–2204
49. Planquart P, Bonin G, Prone A, Massiani C (1999) Distribution, movement and plant availability of trace metals in soils amended with sewage sludge composts: application to low metal loadings. The Sci. Total Environ. 24: 164–179
50. Sposito G, Levesque CS, Le Claire JP, Chang AC (1983) Trace metal chemistry in arid-zone field soils amended with sewage sludge: III. Effect of time of the extraction of trace metals. Soil Sci. Soc. Am. J. 47: 898–902
51. McGrath SP, Cegarra J (1992) Chemical extractability of heavy metals during and after long-term application of sewage sludge to soil. J. Soil Sci. 43: 313–321
52. Finkelman RB (1995) Modes of occurrence of environmentally sensitive trace elements in coal. In: Swaine DJ, Goodarji F (eds) Environmental aspects of trace elements in coal, Kluwer Academic Publishers, Dordrecht, pp 24–50
53. Ravishankar B, Auclair JC, Tyagi R (1994) Partitioning of heavy metals in some Quebec municipal sludges. Water Poll. Res. J. Can. 29: 457–470
54. Brofas G, Michopoulos P, Alifragis D (2000) Sewage sludge as an amendment for calcareous bauxite mine spoil reclamation. J. Environ. Qual. 29: 811–816
55. Fresquez PR, Francis RE, Dennis GL (1990) Influence of sewage sludge on soil and plant quality in a degraded semi arid grass land. J. Environ. Qual. 19: 324–330
56. Navas A, Bermudez F, Machin J (1998) Influence of sewage sludge application on physical and chemical properties of Gypsisols. Geoderma 87: 123–135
57. Pagliai M, Guidi G, La Marca M, Giachetti M, Lucamante G (1981) Effects of sewage sludges and compost on soil porosity and aggregation. J. Environ. Qual. 10: 556–561
58. Cripps RW, Wimfree SK, Reagan JL (1992) Effects of sewage sludge application method on corn production. Commun. Soil Sci. Plant Anal. 23: 1705–1715
59. Logan TJ, Chaney RL (1983) Metals In: Page AL, Gleason TL, Smith IE, Iskander IK, Sommers LE (eds.) Utilization of municipal waste water and sludge on land. University of California, Riverside, pp 235–326

60. Vanni, A., Gennaro, M., Petruzzelli, G., Liberatori. A. (1994) Leachability of heavy metals in municipal sludge particulate. Environmental Technology, 15, pp. 71–78.
61. Chang AC, Page, AL, Warneke JE, Grgurevic E (1984) Sequential extraction of soil heavy metals following a sludge application. J. Environ. Qual. 13: 33–38
62. Leckie JO, James RO (1974) Control mechanisms for trace metals in natural waters. In: Alan JR (ed.) Aqueous-environmental chemistry of metals. Ann Arbor Publ., Michigan, pp 1–76
63. Anderson PR, Christensen TH (1988) Distribution coefficients of Cd, Co, Ni, and Zn in soils. J. Soil Sc. 39: 15–22
64. Christensen TH (1989) Cadmium soil sorption at low concentrations: VIII. Correlation with soil parameters. Water Air Soil Poll. 44: 71–82
65. Lumsdon DG, Evans LJ (1987) Predicting chemical speciation and computer simulation. In: Ure AM, Davidson CM (eds.) Chemical speciation in the environment, Blackie Acad. Publ., New York, pp 408–419
66. Perez-Sirvent C, Martinez-Sanchez J, Garcia-Rizo C (1999) Lead mobilization in calcareous agricultural soils. In: Selim HM, Iskander IK (eds.) Fate and Transport of heavy metals in the vadose zone, Lewis Publishers, Washington DC, pp 177–199
67. Petruzzelli G, Lubrano L, Petronio BM, Gennaro MC, Vanni A, Liberatori A (1994) Soil sorption of heavy metals as influenced by sewage sludge addition. J. Environ. Sci. Health Part A, Environ. Sci. Eng. A29: 31–50
68. Stewart DPC, Cameron KC, Cornforth IS (1998) Effects of spent mushroom substrate on soil chemical conditions and plant growth in an intensive horticultural systems: A comparison with inorganic fertilizers. Aust. J. Soil Res. 36: 185–198
69. Shuman LM (1999) Effect of organic waste amendments on zinc adsorption by two soils. Soil Sci. 164 (3): 197–205
70. Morera MT, Echeverría JC, Mazkiarán C, Garrido JJ (2001) Isotherms and sequential extraction procedures for evaluating sorption and distribution of heavy metals in soils. Environ. Poll. 113: 135–144
71. Han FX, Hu AT, Qin HY (1990) Fractionation and availability of added cadmium in soil environment. Environ. Chem. 9: 49–53
72. Banin A, Nir S, Brummer GW, Han FX, Serban C, Krumnohler J (1995) Cadmium pollution in soils: long-term processes in the solid phase, their characterization and models for their prediction, Report ISC-8911-ISR (Env), Joint Israel-Commission of the European Committees Research Projects
73. Ma LQ, Rao GN (1997) Chemical fractionation of cadmium, copper, nickel, and zinc in contaminated soils. J. Environ. Qual. 26: 259–264
74. Levy DB, Barbarrick KA, Siemer EG, Sommers LE (1992) Distribution and partitioning of trace metals in contaminated soils near Leadville, Colorado. J. Environ. Qual. 21: 185–195
75. Han FX, Kingery WL, Selim HM (2001) Accumulation, redistribution, transport and bioavailability of heavy metals in waste-amended soils. In: Iskander IK, Khirkham MB (eds.) Trace elements in soil-bioavailability, flux and transfer, Lewis Publishers, pp 145–173
76. Nyamangara J (1998) Use of sequential extraction to evaluate zinc and copper in a soil amended with sewage sludge and inorganic metal salts. Agri. Ecosystem. Environ. 69 (2): 135–141
77. Zehetner F, Wenzel WW (1999) Adsorption isotherms of nickel in acid forest soils. In: Selim HM, Iskander IK (Eds) Fate and Transport of Heavy metals in the vadose zone. Lewis Publishers, London, pp 29–55
78. Han FX, Banin A (1999) Long-term transformation and redistribution of potentially toxic heavy metals in arid-zone soils. I: Incubation under saturated condition. Water Air and Soil Poll. 114: 221–250

IV
Transport and Leachability of Metals from Coal and Ash Piles

9

Impact of Grassed Swales on the Fate of Metals Leached from Roads Built with Municipal Solid Waste Incineration Bottom Ashes

Jurate Kumpiene, Holger Ecke, and Christian Maurice
Division of Waste Science & Technology, Luleå University of Technology, SE-97187 Luleå, Sweden

Abstract

Roads built with municipal solid waste incineration (MSWI) bottom ashes may possess a risk to the environment if contaminants (e.g. heavy metals) are leached out. In this study, roadside swales were constructed and studied as a trap for pollutants leached from an experimental road. Soil from the slopes and bottoms of the swales was sampled twice yearly (in spring and autumn) and analysed for metal content. Common plant species in Northern Sweden were planted and a grass mixture was sowed in the swales. The aerial plant parts were sampled in the autumn and analysed for Cu, Cr, Pb, and Zn. One season of observations did not reveal any significant effect of the road on the soil and plants. The metal concentrations in the soil of the swales varied insignificantly between the samples collected in the spring and autumn. Metal concentration in the soil did not exceed the generic guideline values for Swedish contaminated soils. The dry summer of 2002 could have hindered the formation of ash leachate from the road. As the soil properties (pH, metal content) of all studied swale sections remained unaltered, a lack of humidity might be the reason for the observed poor plant establishment in the experimental and reference swales. The impact of the road material on the soil quality started to show up during the second season (2003). Slopes of the experimental swale became contaminated with Cu. Yet the metal translocation from grassed swales through the trophic chain is unlikely; in soil containing higher concentrations of metals grew vegetation with lower metal uptake.

Keywords: *ash leachate, soil, vegetation, phytomonitoring.*

1. Introduction

Increasing amounts of bottom ashes generated by municipal solid waste incineration (MSWI) have prompted a search for using the ash. In alternative to landfiling the ash is to use it in urban constructions such as roads.

The BARC (bottom ash in road constructions) project was initiated in Sweden to evaluate the appropriateness of MSWI bottom ashes as a road construction material and assess the environmental impact of an experimental road built on bottom ashes. Such roads may possess a risk to the environment if contaminants (e.g. heavy metals) are leached out. Heavy metals leached from the road and traffic might be adsorbed by soil particles at the roadsides and, if not stabilised, might be carried to surface waters or blown from the area by wind. Managing soil properties and maximizing the vegetative cover of roadside swales could reduce soil loss and the entry of pollutants into surface and groundwater.[1] In this study, roadside grassed swales were constructed and studied as a trap for pollutants (Figure 1). Thus, phytostabilisation, or *in situ* immobilization of metals, was employed to diminish the pollution spread from the road. The role of plants in phytostabilisation is physical stabilisation of the soil using a dense root system to prevent erosion and to reduce water runoff and percolation through the soil.[2] Only plants capable of immobilising metals within the soil, but unable to translocate them to their aerial parts, are of interest in phytostabilisation.

The main aim of the work was to study the potential of the grassed swales to serve as a contaminant trap along the road build on bottom ashes. Objectives of this study were: (i) to evaluate the impact of the road leachate on soil quality and plant establishment, (ii) to study the correlation between the metal concentrations in the soil and aerial parts of the plants to identify the spread of the contaminants from one medium to another, and (iii) to study the establishment and sustainability of the vegetative cover regarding different substrates present in the roadside swales.

2. Method and Materials

2.1. Experimental Road

During the summer of 2001, a 500 m long test road was constructed with bottom ash (Table 1) from the MSW incinerator *Dåva kraftvärmeverk* (Figure 2). The road is located near the

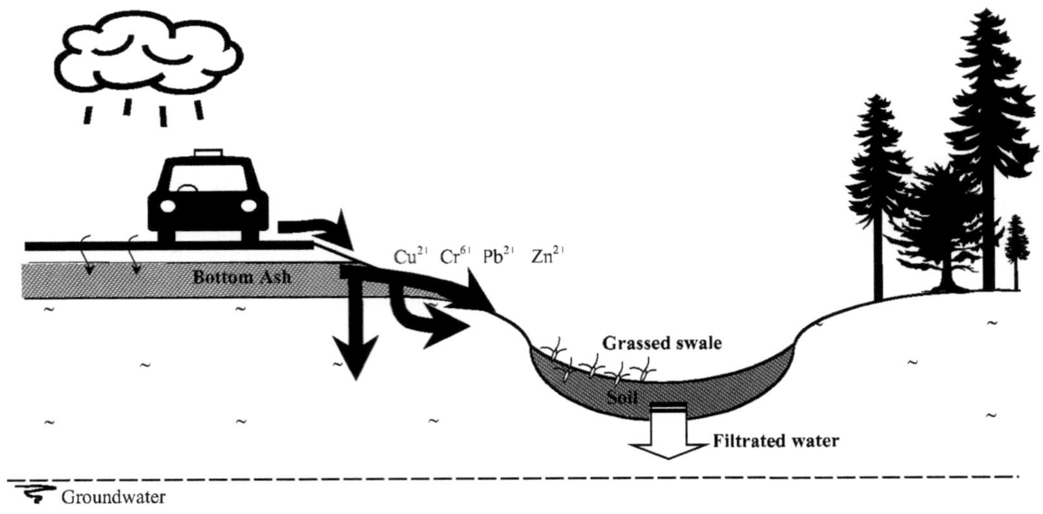

FIGURE 1. Basic principle of the pollutant trap in the BARC-project.

incinerator, outside the city of Umeå, in Northern Sweden and predominately used by trucks for transport of incineration residues. In this region, the soil surface is frozen for about 6 to 8 months and up to half a meter snow covers the ground during the winter. The area is a natural peat wetland, where the water table varies between 0 and 1 meter. The bedrock is covered by about 5 m till and 0.5 to 1.5 m of peat.

TABLE 1. Total content of oxides and elements (±SD, n = 3), total solids TS (±SD, n = 12), and loss of ignition LOI (±SD, n = 12) in bottom ash[3]

Element	Unit	Amount
SiO_2	wt. %	37 ± 0.9
Al_2O_3	"	13 ± 0.3
CaO	"	15 ± 0.2
Fe_2O_3	"	15 ± 0.5
K_2O	"	1.4 ± 0.02
MgO	"	2.5 ± 0.1
MnO_2	"	0.3 ± 0.02
Na_2O	"	2.8 ± 0.1
P_2O_5	"	1.0 ± 0.1
TiO_2	"	1.6 ± 0.1
As	mg (kg dw)$^{-1}$	33 ± 3.2
Cd	"	6 ± 1.1
Co	"	34 ± 6.0
Cr	"	568 ± 435
Cu	"	11570 ± 3261
Mo	"	24 ± 2.6
Ni	"	567 ± 446
Pb	"	2260 ± 983
S	"	5103 ± 38
Zn	"	9117 ± 1198
TS	wt. %	83.0 ± 0.3
LOI	"	1.2 ± 0.08

The test road is divided in four sections. Two sections are covered with asphalt while no asphalt is used in the two remaining sections (Figure 3). The two first sections are reference sections, built with conventional construction material i.e. gravel. In the following two sections, 50-cm bottom ash is used for construction. The two middle sections are covered with asphalt.

2.2. Road Leachate

The leachate from each section is collected separately in lysimeters built with a high density polyethylene (HDPE) geomembrane (Figure 4). The lysimeters are 3 m wide and 8 m long, filled with 0.42 m bottom ash. The water flow from the lysimeters is measured with flow-meters. Samples are taken monthly, during the ice-free period of the year. The leachate generated in the road has been sampled since the autumn 2001.

The pH of the ash leachate from the road was relatively constant and varied between 10 and 11.5. The salinity (measured as electrical conductivity) of the leachate was decreasing together with the concentration of Cl (Figure 5).

Concentrations of Cu and Cr were constantly decreasing with a slight secondary peak at the beginning of the second sampling season (after approximately 500 sampling days) (Figure 6), while the concentrations of Pb and Zn in the leachate were mostly below instrument detection limits, specifically $<7.5\ \mu g\ l^{-1}$ Pb and $<5.8\ \mu g\ l^{-1}$ Zn.

2.3. Roadside Swales

A 10 m long, 1.5 m wide, and 0.75 m deep swale was constructed along the road. Swales were designed as constructed

FIGURE 2. The 500 m test road used as a research object for the BARC project. The section in the foreground is asphalted while the one in the background is not. Wells collecting water from the lysimeters are visible on the side of the road. The chimney of the incineration plant is visible on the background of the view.

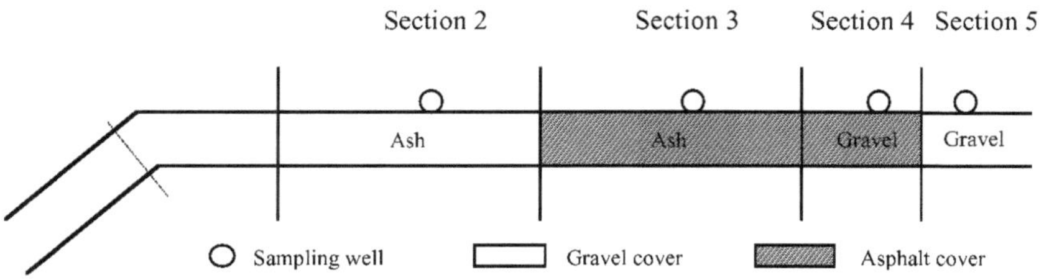

FIGURE 3. Schematic description of the test road seen from above.

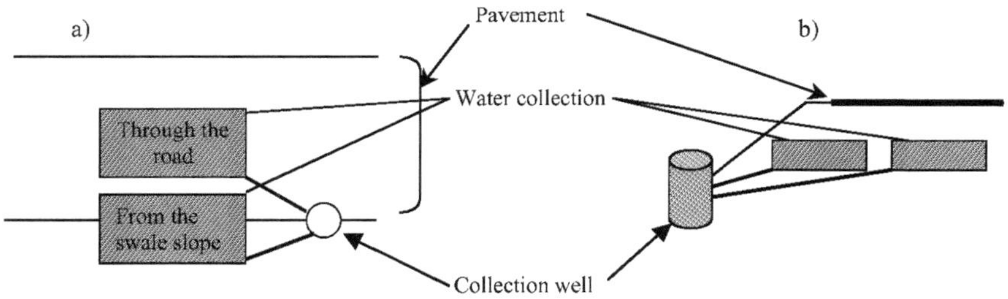

FIGURE 4. Schematic description of the water sampling equipment at the site. The road section is shown from a) above and from b) the side.

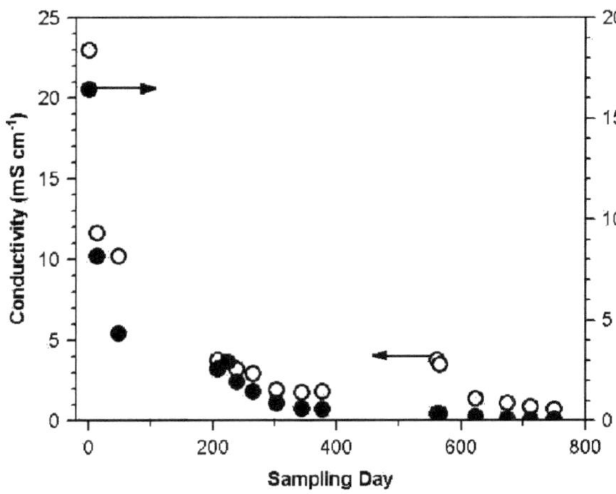

FIGURE 5. Changes of electrical conductivity and Cl concentration in ash leachate from the road section 3 versus time.

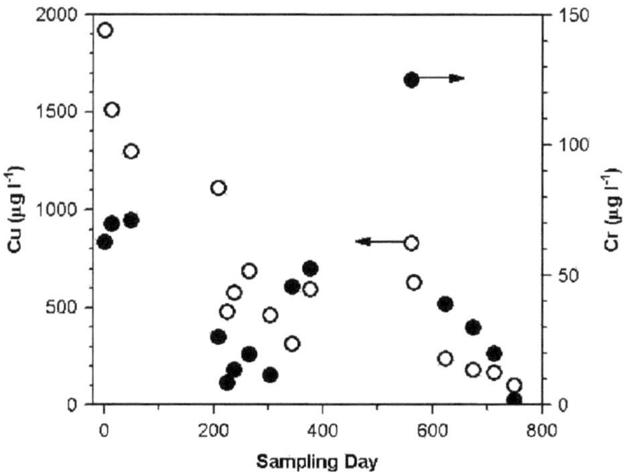

FIGURE 6. Changes of Cu and Cr concentrations in the ash leachate from the road section 3 versus time.

wetlands to treat pollutants transported in the water phase. The reference area with three swales corresponding to the above dimensions was designed and as well as the experimental swale was covered with mould (topsoil-peat mixture). A HDPE-geomembrane was placed under the soil cover to retain surface runoff water from the road within the swale soil. The remaining roadside swales were covered with peat taken from the surroundings of the road. Hence, five swale sections were studied: one experimental, two reference (all covered with mould and having a geomembrane), and two sections covered with peat (thereafter called swales 1 and 2).

2.4. Soil Samples

Soil from the slopes and bottoms of the swales was sampled at the beginning of May 2002 and the end of September 2002 and 2003 from depth to 5 cm. Three sub-samples taken 1 m from each other along the swale were pooled into one sample.

The acid extractable concentration of the metals was determined in 1 M HNO_3 extraction from the air dried samples (<2 mm). The total metal content of the soil samples was determined by digesting the air dried samples (<2 mm) for 2 h in HNO_3(65%) at 120°C. Extracts were filtrated through a 0.45 µm nitrocellulose membrane filter and analysed for Cr, Cu, Pb, and Zn by inductively coupled plasma optical emission spectroscopy (ICP-OES). The mould substrate (thereafter called initial sample) was analysed before covering the experimental and reference swales (Table 2).

The organic matter content of the substrates was determined by dry combustion at 550°C for 2 h. Soil texture characterization was done by the hydrometer sedimentation method.[4]

TABLE 2. Initial characteristics of the mould used for the experimental and reference swales cover

	Unit	Amount	
pH	—	6.3	
EC	mS m^{-1}	3.0	
SOM	%	6.4	
Texture	"		
Sand	"	38.5	
Silt	"	40.9	
Clay	"	20.6	
Metals	mg $(kg\ dw)^{-1}$ ± SD, n=3	Total	1 M HNO_3 extractable
Cr	"	16.5 ± 1.8	1.0 ± 0.0
Cu	"	10.6 ± 0.7	4.8 ± 0.1
Pb	"	9.7 ± 1.0	5.2 ± 0.1
Zn	"	18.0 ± 4.7	12.0 ± 0.5

EC—electrical conductivity
SOM—soil organic matter

TABLE 3. The composition of the mixture of the plant species used for the vegetation of the roadside swales.

Herbs	Grasses
Proportion of the total amount of seeds, %	
10	90
Achillea millefolium	*Agrostis capillaris*
Anthyllis vulneraria	*Anthoxanthum odoratum*
Campanula rotundifolia	*Bromus hordeaceus*
Hieracium pilosella	*Deschampsia flexuosa*
Hieracium umbellatum	*Festuca rubra*
Hypericum maculatum	*Festuca ovina*
Leucanthemum vulgare	
Lotus corniculatus	
Linaria vulgaris	
Rumex acetosa	
Silene dioica	
Thlaspi cearulescens	
Vicia cracca	

2.5. Vegetation Samples

At the beginning of June 2002, six plant species were planted (*Phalaris arundinacea* (reed canary grass), *Potentilla (Argentina)anserina* (silverweed), *Typha latifolia* (broadleaf cattail), *Plantago maritima* (sea plantain/goose tongue greens), *Linaria vulgaris* (yellow toadflax/butter and eggs), and *Leymus arenarius* (Lyme-grass)) and a grass mixture (Table 3) was sowed in the experimental and two reference swales, as well as several roadside sections with peat cover. Plants were planted so that the salt marsh and wetland species (*P. arundinacea, P. anserina, T. latifolia, and P. maritima*) were located lowest on the slope and in the swale bed, while plants growing on well-drained soil (*L. vulgaris and L . arenarius*) were planted closest to the road and highest on the swale slope.

The Braun-Blanquet scale[5] was used to visually evaluate the relative surface cover by the sawed grass mixture.

The harvestable (aerial) parts of the plants sampled at the end of September 2002 were rinsed with double distilled water, dried at 65°C for 72 h, ground to <0.85 mm, and wet digested for 4 h in HNO_3(65%) at 125°C. Samples were analysed for heavy metals by inductively coupled plasma mass spectroscopy (ICP-MS). Plant roots were not sampled to avoid disturbing the established vegetation.

Bioaccumulation factor (BAF) was calculated as a ratio between the 1 M HNO_3 extractable metal concentrations in the harvested plant parts and the soil. The metal concentrations in the compartments were calculated on a dry weight basis.

2.6. Statistical Analysis

Statistical analyses were performed using the software *Statgraphics Plus 5.0.* A multiple sample comparison, Fisher's least significant difference (LSD) procedure ($p < 0.05$),[6] was used to discriminate among the sample means.

3. Results

3.1. Soil

After one season none of the analysed metals in the soil of the experimental swale exceeded the generic guideline values for contaminated soils in Sweden[7] (Table 4). While at the end of the second year (2003), the total Cu concentration in the slope of the experimental swale significantly increased and exceeded the guideline value of 100 mg (kg dw)$^{-1}$. Data variance from the year 2003 was very high. Standard deviations of Cu, Pb, and Zn concentrations even exceeded the average values.

The 1 M HNO_3 extractable metal content of the soil samples taken from the experimental swale did not change during one season, nor did they significantly differ from the samples collected from the reference swales (Figure 7).

Correspondingly to the increase of the total metal content at the end of the second season, an increase in 1 M HNO_3 extractable metal concentrations on the slope of the experimental swale was also observed. Likewise, Cu, Pb, and especially Zn concentrations in the slope soil had larger variance than in the bed of the experimental swale (Figure 8).

The box-and-whisker plots (Figure 9 a and b) display a comparison of the EC and pH of the soil from the experimental swale. In both figures, each box encompasses the middle 50 percent of the data, where the median is drawn as a horizontal line inside the box and the cross represents the average value.

The average EC of the soil samples from the experimental swale did not significantly differ during two seasons (Figure 9 a). A large variability of data from autumn 2002 was due to the high EC determined in the soil from the swale slopes. The pH of the soil in autumn 2003 significantly increased since the autumn 2002 (Figure 9 b).

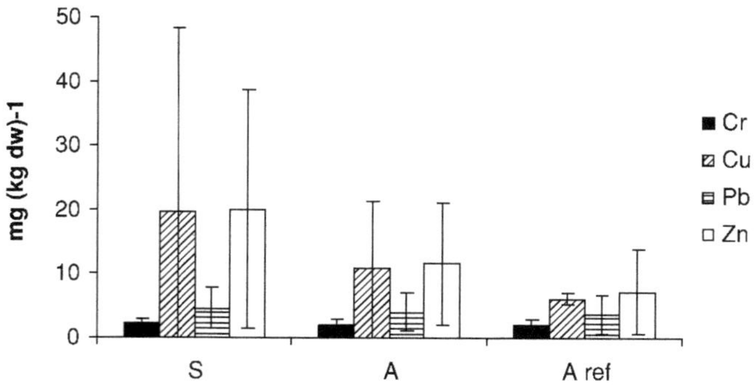

FIGURE 7. Comparison of the 1 M HNO_3 extractable metal concentrations determined in the soil of the experimental swale in the spring (S) and autumn (A) of 2002, n = 12, and the reference (A ref, n = 8) swales in the autumn of 2002.

3.2. Vegetation in the Roadside Swales

After the first season, insufficient plant material was available to analyse metal concentrations in aerial plant parts in duplicates. No samples of *T. latifolia* were collected at the end of the second sampling season. The plants did not recover after they were harvested during the first sampling season in 2002.

The establishment of vegetation in the roadside swales was very uneven. Plants in the experimental and reference swales covered with mould were not as established as compared with those growing on peat. According to the Braun-Blanquet scale[5] swale sections with mould can be placed into the class "+", corresponding to the less than 1% surface cover by grass. Swale sections with peat can be placed into the class 4, i.e. the surface covering between 50 and 75%.

The concentrations of metals in the aerial parts of the plants from the swales were related to the metal concentrations of the soil on which they grew and were ranked as follows:

TABLE 4. Average total and 1 M HNO_3 extractable metal concentrations of the soil samples collected in autumn 2002 and 2003 , mg (kg dw)$^{-1}$, ±SD

	Cr		Cu		Pb		Zn		
Swale	2002	2003	2002	2003	2002	2003	2002	2003	n
				Total					
1	6.1 ± 1.4	54.1 ± 2.1	41.4 ± 4.0	1092 ± 892	9.0 ± 0.2	745 ± 817	18.51 ± 5.83	2413 ± 2055	2
2	6.1 ± 0.5	79.6 ± 4.0	75.3 ± 6.8	32147 ± 42557	26.4 ± 3.7	979 ± 469	59.4 ± 6.7	14210 ± 16829	2
exp sl	15.7 ± 0.8	23.9 ± 5.5	11.0 ± 0.9	280 ± 444	2.1 ± 1.3	42.7 ± 54.1	24.8 ± 7.7	173 ± 180	6
exp b	14.6 ± 2.3	16.9 ± 4.2	11.5 ± 2.3	48.6 ± 71.7	2.6 ± 0.7	13.5 ± 16.6	20.0 ± 12.8	63.4 ± 49.0	6
ref	15.5 ± 3.0	14.2 ± 1.7	10.3 ± 1.2	10.8 ± 2.4	2.2 ± 1.4	2.9 ± 0.6	18.2 ± 5.5	10.9 ± 2.8	8
GV	120		100		80		350		
				1 M HNO_3 extractable					
1	1.2 ± 1.7	10.5 ± 4.8	19.2 ± 22.7	380 ± 14	8.6 ± 0.1	455 ± 111	9.8*	1201 ± 156	2
2	1.4 ± 1.4	14.0 ± 0.5	33.6 ± 19.7	975 ± 45	13.6 ± 5.5	403 ± 50	36.6 ± 11.5	3207 ± 923	2
exp sl	2.09 ± 0.6	1.7 ± 0.8	5.9 ± 1.0	67.1 ± 77.6	1.2 ± 0.4	36.4 ± 37.8	13.1 ± 11.8	163 ± 176	6
exp b	2.0 ± 0.8	2.7 ± 0.3	6.1 ± 0.7	6.1 ± 3.2	0.8 ± 0.3	0.6 ± 0.5	7.2 ± 6.7	2.3 ± 2.4	6
ref	2.0 ± 0.6	2.0 ± 0.9	8.3 ± 3.9	18.7 ± 16.7	1.5 ± 1.0	8.7 ± 11.5	10.0 ± 7.5	54.7 ± 79.6	8

exp—the experimental swale, sl—slope, b—bed, ref—average of two reference swales, n—number of replicates,
* —one replicate, GV—guideline values for contaminated soil in Sweden.[7]

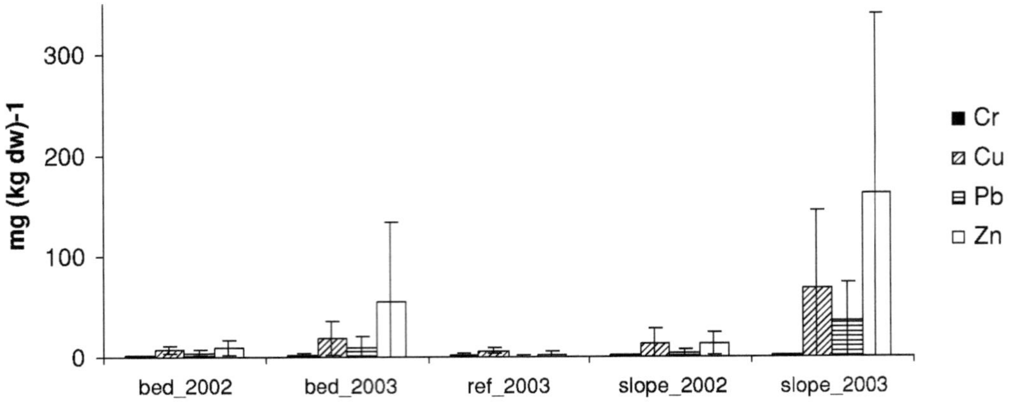

FIGURE 8. Comparison of the 1 M HNO_3 extractable metal concentrations of the soil sampled from the bed of the experimental and reference swales in the autumn of 2002 and 2003, and from the slope of the experimental swale in the autumn of 2002 and 2003, n = 6.

Zn>>Cu>Pb>Cr (Table 5). Zn concentration in plants was significantly higher than of any other metal.

Figure 10 summarises the metal accumulation in the plants, expressed as a bioaccumulation factor (BAF), i.e. the relative concentration of a component in the live tissue of the plants as compared to the soil concentration. For all analysed plants and metals the BAF was very low, ranging from 0.0 to 8.1 in year 2002 and to 49 in year 2003 for Zn and to 4.7 for the other metals. A toxic chemical is considered to be bioaccumulative if its bioaccumulation factor is 1,000 or greater.[8]

BAF plotted versus metal concentration in the soil on which plants grew show a decreasing trend (Figure 11), i.e. the metal accumulation in the aerial plants parts decreased with increasing the metal concentration in the soil. The pattern was characteristic for all the metals and the samples from both sampling seasons.

4. Discussion

4.1. Soil in the Swales

Metals extracted with 1 M HNO_3 include those fractions complexed by organic matter and sorbed by or co-precipitated with hydrous oxides, carbonates, and sulphides,[9] and were assumed to represent a potentially bioavailable element fraction. Which metal fractions should be measured to assess contamination and need for clean up is still an open question,[10] though the bioavailability and potential toxicity of the element is observed to correlate with the weak acid-soluble metal fraction.[11] In this study, 1 M HNO_3 extractable metal fraction was used for the calculations of the bioaccumulation factor. As well, the total metal content of the soil samples was determined to estimate the soil contamination level, i.e. if metal concentrations exceed the generic guideline values for contaminated soils in Sweden based on the total metal concentrations in the soil.

Collecting the representative soil samples from the roadside sections where the swale slopes were covered with only ashes was difficult due to the dusting of the ashes from the slopes over the peat in the swale bottom. This could be the reason of high metal concentrations and variance observed in the soil of swales 1 and 2, where some of the soil sub-samples were contaminated with the ash particles. Hence, sections of slopes with the exposed bottom ashes may bear the risks for slope erosion, deflation, particle transportation, and consequently the airborne soil contamination.

Relatively dry summer 2002, with cumulative precipitation from May to October 2002 being 144.5 mm, could have reduced if not hindered the formation of the leachate runoff from the road during the first sampling season. It is possible that polluted water (ash leachate) did not reach the swales, possibly

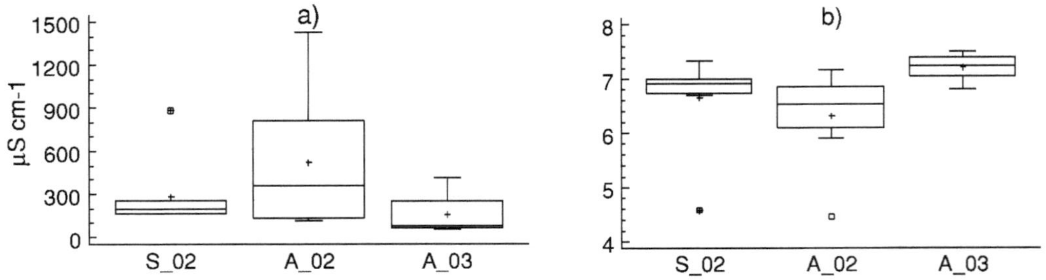

FIGURE 9. The comparison of (a) the electrical conductivity (EC) and (b) pH of the soil samples from the spring (S) and autumn (A) 2002 and autumn 2003.

TABLE 5. Metal concentrations determined in the aerial plant parts

		Cr		Cu		Pb		Zn	
		mg (kg dw)$^{-1}$							
Swale	Plant	2002^{I}	2003^{II}	2002^{I}	2003^{II}	2002^{I}	2003^{II}	2002^{I}	2003^{II}
1	grass mixture	0.1	1.9	7.0	6.6	1.1	1.5	37.4	45.5
2	grass mixture	0.1	—	10.4	—	1.0	—	81.3	—
exp	grass mixture	1.2	6.7	4.6	7.9	0.4	1.7	31.9	24.5
ref ^{III}	grass mixture	2.6	8.6	7.1	6.0	0.7	0.5	60.8	25.7
1	*Leymus arenarius*	0.9	—	7.6	—	2.0	—	41.9	—
2	*L. arenarius*	1.1	1.8	8.6	7.0	1.3	0.8	62.8	31.5
exp	*L. arenarius*	0.4	0.6	8.8	6.4	1.7	0.1	67.0	13.1
ref ^{III}	*L. arenarius*	0.4	1.2	4.1	4.6	0.6	—	15.0	—
1	*Linaria vulgaris*	0.2	8.3	2.8	35.0	0.0	10.1	126.3	208.3
2	*L. vulgaris*	0.9	5.2	8.0	21.4	1.3	17.7	152.9	207.1
exp	*L. vulgaris*	0.8	6.5	6.6	18.2	1.3	17.7	129.2	106.8
ref ^{III}	*L. vulgaris*	0.0	1.1	3.0	9.20	0.0	10.1	108.5	88.2
1	*Phalaris arundinacea*	1.3	—	12.1	—	2.4	—	110.8	—
2	*P. arundinacea*	0.6	4.1	12.2	8.6	1.7	0.8	98.7	95.2
exp	*P. arundinacea*	1.1	3.8	8.3	6.4	1.0	0.2	49.9	42.7
ref ^{III}	*P. arundinacea*	0.7	6.3	8.1	7.1	1.2	0.5	46.5	45.0
1	*Plantago maritima*	0.5	2.0	12.9	11.0	1.2	1.6	32.8	20.5
2	*P. maritima*	0.9	3.6	13.5	19.1	1.4	1.9	86.0	31.5
exp	*P. maritima*	0.3	3.5	11.9	15.8	1.1	1.7	23.6	26.3
ref ^{III}	*P. maritima*	0.3	3.5	5.4	15.8	1.6	1.7	25.3	26.3
1	*Potentilla anserina*	1.4	2.5	8.8	13.4	2.5	3.6	103.5	74.8
2	*P. anserina*	2.7	1.4	13.7	14.2	3.1	2.6	108.1	66.1
exp	*P. anserina*	0.4	3.5	7.3	23.8	1.0	5.6	26.0	68.3
ref ^{III}	*P. anserina*	1.3	0.9	7.6	9.9	1.9	1.3	63.1	44.2
1	*Typha latifolia*	1.2	—	4.5	—	1.5	—	30.7	—
2	*T. latifolia*	1.4	—	8.2	—	95.1^{IV}	—	78.3	—
exp	*T. latifolia*	0.2	—	5.5	—	0.1	—	29.4	—
ref ^{III}	*T. latifolia*	0.8	—	3.9	—	1.1	—	22.8	—

ref—reference swale, exp—experimental swale, 1,2—the first and the second swales, respectively. ^{I}—one replicate; ^{II}—average of two replicates; ^{III}—average of values from two reference swales;^{IV}—excluded from the statistical analyses as an outlier;—no sample analysed.

explaining the unaltered element concentration in the experimental swale soil in 2002 (Figure 7).

The impact of the road material on the soil quality started to show up during the second season. The slopes of the experimental swale were contaminated with Cu. Even the concentrations of Pb and Zn in the individual samples of the slope soil were over the guideline values, causing the high variance of the average metal concentrations (Table 4). As the Zn and Pb concentrations in the leachate from the road were below detection limits, the elevated metal concentrations is believed to be caused by the dusting of ashes, probably from the trucks transporting ashes from the incinerator to the landfill. The concentrations of Zn and Pb in bottom ashes are in a range of several g (kg dw)$^{-1}$ (Table 1).

4.2. Vegetation in the Swales

Four plant species were selected for their capability to survive in saline wet environments, as was expected in the swales because of the ash leachate run-off from the road. *P. maritima* is a halophyte species that has adapted to alkaline and saline conditions,[12] and together with *P. arundinacea, P. anserina,* and *T. latifolia,* are common species of salt marshes and wetlands. These species were used to create a system that acts as a physical barrier to pollutants, retarding their translocation from the swales. *L. vulgaris* and *L. arenarius* are rather invasive species easily growing along roadsides on well-drained soil, thus suitable for the vegetation and stabilisation of the swale slopes.

It is generally assumed that Cu, Cr, and Pb have little mobility within plants and tend to accumulate in the roots, while Zn is easily transported up to the shoots;[10,13]hence, the accumulation of Zn by all analysed plants was higher than any other metal. Other authors[14,15] also observed such metal partitioning. In our study, the sequence of the average metal bioaccumulation according to the calculated BAF was as follows: Zn>Cu>Cr>Pb in 2002 and Zn>Cr>Cu>Pb in 2003. The BAF of Pb placed the metal last as compared with the sequence of the metal accumulation in plants expressed in mg (kg dw)$^{-1}$ (Zn>Cu>Pb>Cr). It indicates that Pb is the least bioavailable metal for the analysed plant species. Root tissue is reported to be the main barrier for Pb uptake and trans-location.[14]

Seeing as the BAF is calculated as a ratio between metal concentration in plant and in soil, higher BAF of Cr than that

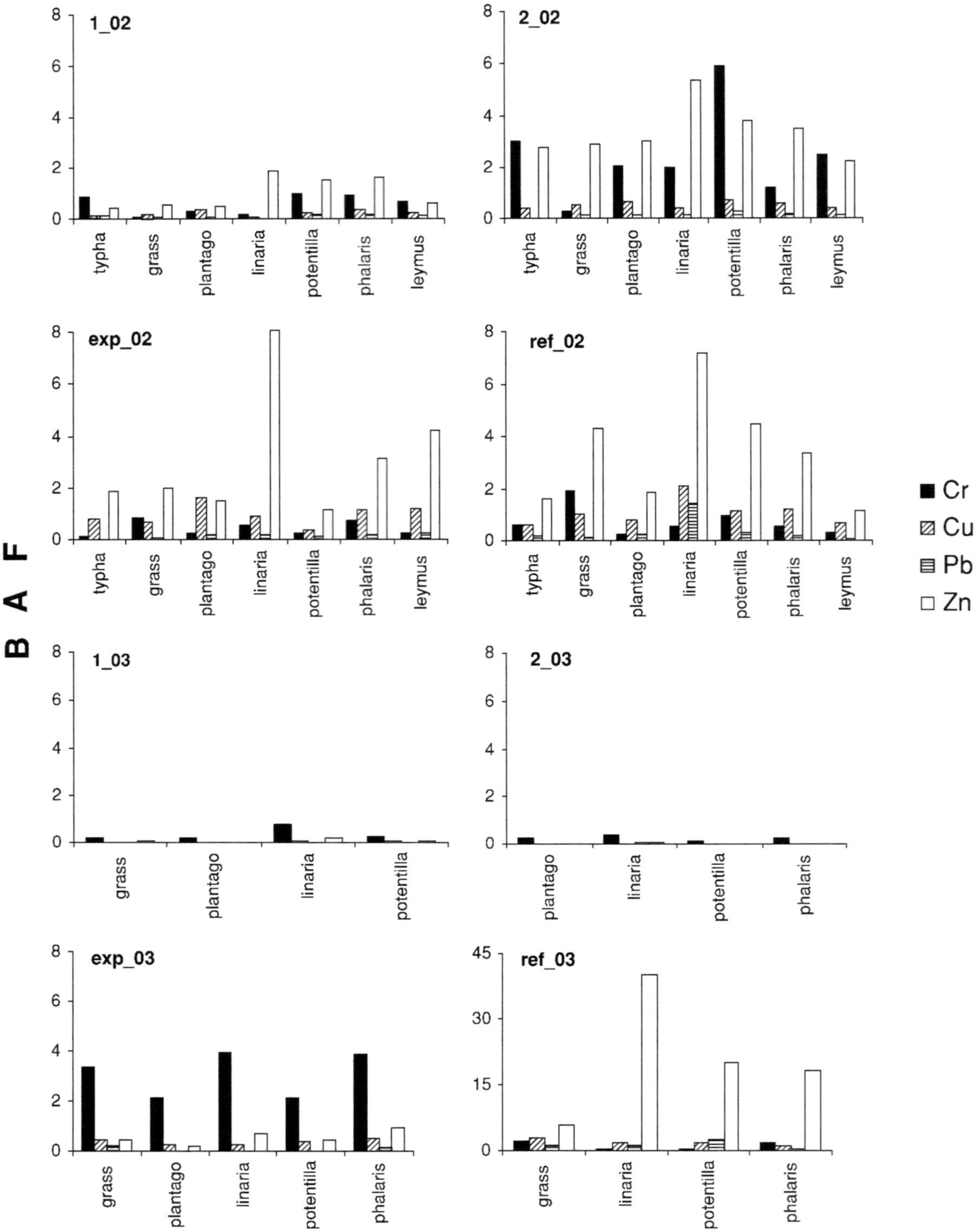

FIGURE 10. Bioaccumulation factors for vegetation in the road swales harvested in 2002 and 2003. Ref—average of two reference swales (2002 n = 2, 2003 n = 4); exp—experimental swale (2002 n = 1, 2003 n = 2), and 1and 2 represent the first and the second swales (2002 n = 1, 2003 n = 2).

of the other metals in experimental swale 2003 is due to the low Cr concentration in soil (Figure 10). While low BAF in plants from swales 1 and 2 in 2003 were due to the high metal concentrations in the soil. Hence, plants that grew in soil containing higher metal concentrations had lower metal uptake into aerial plant parts and none of the analysed plants had a tendency to accumulate elevated concentrations of the metals. This agrees with the findings of other authors who observed a minor heavy

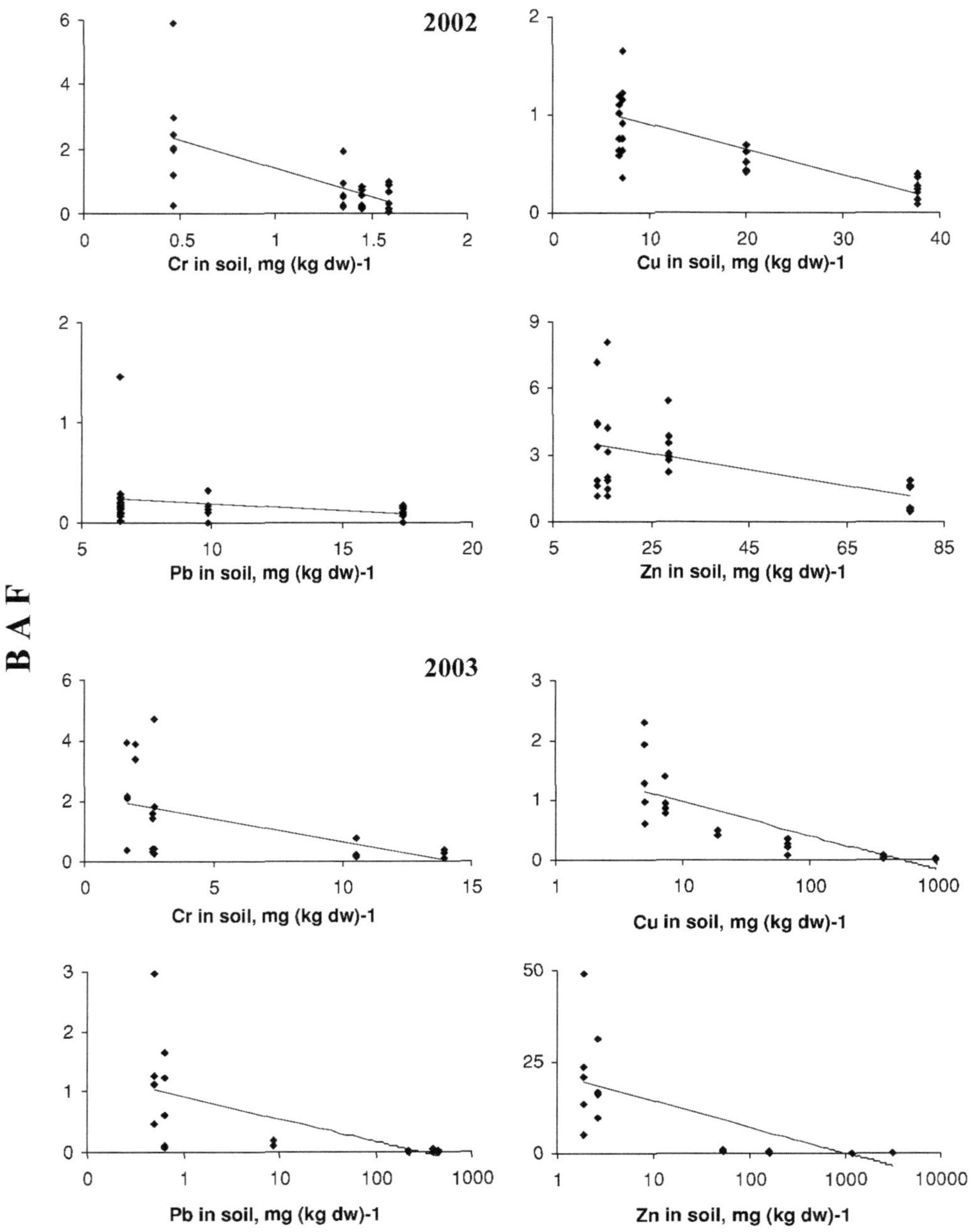

FIGURE 11. The BAF for all analysed plant species in respect to the 1 M HNO_3 extractable metal concentrations determined in soil on which plants grew.

metal fraction (1–2%) taken up by *P. arundinacea*,[16,17] and for *P. maritima*.[15] Based on our observations, increasing concentrations of metals in the soil does not necessarily cause a higher metal uptake into aerial plant parts. BAF had a negative correlation with the 1 M HNO_3 metal concentration of the soils (Figure 11). However, such a pattern might be characteristic for metal uptake under micronutrient deficiency, primarily from Fe and Mn, when plant roots increase metal bioavailability by additionally extruding protons, phytosiderophores, and organic acids to acidify the soil and mobilize the metals and nutrients.[18,19,20]Also, the higher biomass of plants growing in the peat can have a dilution effect on the metal concentration of the aerial plant parts. The range of metal concentrations during the observations was not large; hence, the higher metal concentrations in soil may cause a different metal uptake by plants.

Soltan and Rashed[21] reported decreasing metal uptake by plants with an increasing concentration of Pb in growth media. Williams et al.,[22] Leendertse et al.,[23]and Fitzgerald et al.[15] observed that an increase in Pb concentrations in soil and sediments did not lead to any proportional increase in the metal concentrations in plant shoots. Similar results were also obtained for Cu.[15,23] However, Szymanowska et al.[24] reported that increasing metal concentrations in sediments increased Cr uptake by T. *latifolia*. This illustrates that metal uptake can vary within the cultivars of the same species and some other factors (e.g. amount of nutrients, salinity, irrigation, growing media) besides the compartment metal concentration account for the metal uptake by plants. This study however showed that higher metal concentrations in the soil did not increase the biotranslocation of contaminants via analysed plant species.

As the concentrations of the toxic metals in all swales in 2002 were low, pH and electric conductivity (EC) unaltered (except the slope of the experimental swale) lack of humidity could have affected plant establishment in the experimental and reference swales during the first year. Peat has good humidity retention capacity and provides water to vegetation longer during dry periods, hence facilitating the establishment of vegetation in the swales.

4.3. Sustainability of the Metal Retention

Variations in pH and drying out can influence soil retention capacity for heavy metals, although parts of bound metals may stay strongly incorporated into organic matter by the formation of coordinate linkages and ring structures. These are predominant forms of complexed metals when humic substances are present in abundance.[25] Since only a finite amount of metals can be leached from the road, sufficient peat layering can be applied to the swales to ensure undersaturation of the reactive sites, thus complexation of the metals through highly stable coordinate linkages. However, there is a risk that alkaline ash leachate could dissolve soluble organics from soil and increase the mobility of metals as metal-OM complexes. Even if this occurs the risk of contamination spread via analysed vegetation is unlikely; the bioavailability and toxicity of such complexes are several orders of magnitude less than that of free metal ions.[26,27] Besides, an additional layer of acid soil (e.g. peat) can lead to acidification of the environment and was observed to reduce the amount of metals leached trough ash-soil profiles.[28]

Samecka-Cymerman and Kempers[17] observed that an accumulation of metals in *P. arundinacea* was negatively correlated to the nutritional element content (Na, K, Ca etc.) in all compartments, suggesting that a high concentration of major cations in ash leachate may prevent metal uptake by plants. Conversely, leachate salinity may result in a more even distribution of essential elements, such as Cu, between plant parts and was reported to increase the translocation of Pb from plant roots to shoots.[15] Based on the relationship of metal release from leaf tissues onto the leaf surface with salt excretion, Weis & Weis[29] suggested that greater metal release at higher salinities can be expected due to more active salt excretion by leaves. The authors refer that the translocation of metals from soil to aboveground parts and their release from leaf tissue due to the saline conditions may cause metal flux to adjacent ecosystems. Even if the salinity of the ash leachate rapidly decreases with the time (Figure 5), the flux of salts to the soil might still be expected due to ash dusting over the swales.

Degradation of organic matter may lead to the release of bound metals, therefore additional stabilising material could be used in swale substrate, e.g. iron and manganese (hydro)oxides are effective in metal binding.[30] Plants themselves are also capable of immobilising contaminants by absorption and compartmentalisation into cell vacuoles, adsorption, or precipitation of contaminants within a root zone. For example, formation of Fe and Mn plaques (coatings of Fe and Mn (hydro)oxides deposited on roots of plants) would immobilize metal ions in the rhizosphere.[14] Under micronutrient deficiency, plant roots increase metal bioavailability by additionally extruding ion transporters, most of which have a broad-range allowing nonessential elements to enter root cells.[31] The presence of high concentrations of major cations in ash leachate might outdo the toxic effect of metals employing different mechanisms at plant cellular and genetic levels.

5. Conclusions

One season of observations did not reveal any significant effect on the soil and plants from the road built on MSWI bottom ashes. Cu, Cr, Pb, and Zn concentrations in the soil of the swales varied insignificantly between samples collected in the spring and autumn 2002. Metal concentration in the soil in 2002 did not exceed the generic guideline values for contaminated soils in Sweden. The dry season could have hindered the ash leachate from the road to reach the roadside swales. As the soil properties (pH, metal content) of the studied roadside swales remained unaltered, the lack of humidity could have affected poor plant establishment in the experimental and reference swales. Peat, which has good water holding capacity, had 50 to 75 times better surface cover with the sowed grass compared with mould, hence being a better environment for the establishment of vegetation.

The effect of the road material on the soil quality appeared the second season through significantly elevated Cu concentrations of the experimental swale slopes. Even Pb and Zn, having concentrations in the road leachate below detection limits, had elevated concentrations in the slope soil. Consequently, road leachate is not the only pathway of the contaminant spread in the area. Most likely, soil contamination is due to the ash dusting from the bare slopes and trucks transporting ashes from the incinerator to the landfill.

Although a direct contaminant leakage to the groundwater via infiltration through the slope of the swales with the exposed bottom ashes may still be possible, the metal translocation from grassed swales through the trophic chain is not likely. The trend of lower metal uptake by vegetation with higher metal concentrations in the soil where the plants grew was observed. However, the ash leachate can modify the metal translocation from plant roots to shoots by e.g. increasing soil solution salinity, while high nutrient content in the ash leachate may help the plants to overcome the uptake of toxic metals. Moreover, dusting of ashes over the aboveground plant tissues might be a source of the trophic chain contamination.

The sustainability of the metal fixation in the soil depends greatly on the vegetation establishment. Plants are responsible for the revenue of organic matter in the soil, which in turn has a large metal binding capacity. However, the long-term effect has to be evaluated by continuous monitoring the leaching of the elements from the road, changes in soil salinity, acidity, and vegetation.

Acknowledgements. This work was financially supported by The Swedish Research Council for Environment, Agricultural Sciences and Spatial Planning (project no. 20022148), and European Union Structural Funds and New Objective 1, North Sweden Soil Remediation Center (contract no. 113-12534-00).

References

1. U.S. EPA (United States Environmental Protection Agency), *Erosion, sediment and Runoff control for Roads and Highways* EPA-841-F-95-008d, Office of Water (4503F), 1995.
2. Berti, R. and Cunningham, S.D., Phytostabilisation of metals, in *Phytoremediation of toxic metals: using plants to clean up the environment*, Raskin, I., Ensley, B.D., Eds., John Wiley & Sons, New York, 2000, 71–88.
3. Åberg, A. and Ecke, H., manuscript, Quantification of the effects of different leaching factors on long-term emissions from bottom ash in road construction.
4. SIS (Swedish Standards Institute), Geotechnical tests—Particle size distribution—Sedimentation, hydrometer method SS 02 71 24, 1992.
5. Causon, D.R., An introduction to vegetation analysis: principles, practice and interpretation, Unwin Hyman, London, 1988.
6. Montgomery, D.C., Design and analysis of experiment, John Wiley & Sons, New York, 2001.
7. SNV (Swedish Environmental Protection Agency), *Guidance in environmental technical investigations of soil, part II: Field investigations* (Vägledning för miljötekniske markundersökningar, del II: Fältarbete), Report 4311, Solna, Sweden, 1994.
8. Mayer, A. and Pieters, J., Environment PBTS: How low can you go? *Electric Perspectives.* 28, 2001.
9. Carter, M.B., Soil sampling and methods of analysis, Lewis publishers, Boca Raton, Florida, 1993.
10. Pulford, I.D. and Watson, C., Phytoremediation of heavy metal-contaminated land by trees—a review. *Environ. Intern.* 29, 529–540, 2003.
11. 11. Luo, Y., Jiang, X., Wu, L., Song, J., Wu, S., Lu, R., and Christie, P., Accumulation and chemical fractionation of Cu in a paddy soil irrigated with Cu-rich wastewater. *Geoderma* 115, 113–120, 2003.
12. Forbes, B.C. and Jefferies, R.L., Revegetation of disturbed arctic sites: constraints and applications. *Biological Conservation* 88, 15–24, 1999.
13. Kabata-Pendias, and A.H., Pendias, Trace Elements in Soils and Plants. London, Library of Congress, 1992.
14. Ye, Z., Baker, A.J.M., Wong, M-H, and Willis, A.J., Zinc, lead and cadmium accumulation and tolerance in *Typha latifolia* as affected by iron plaque on the root surface. *Aquatic Botany.* 61, 55–67, 1998.
15. Fitzgerald, E.J., Caffrey, J.M., Nesaratnam, S.T., and McLoughlin, P., Copper and lead concentrations in salt marsh plants on the Suir Estuary, Ireland. *Environ. Pollution.* 123, 67–74, 2003.
16. Olsen, F.J. and Chong, S.K., Reclamation of acid coal refuse. *Landscape and Urban Planning.* 20, 309–313, 1991.
17. Samecka-Cymerman, A. and Kempers, A.J., Concentrations of heavy metals and plant nutrients in water, sediments and aquatic macrophytes of anthropogenic laces (former open cut brown coal mines) differing in stage of acidification. *The Science of the Total Environ,* 281, 87–98, 2001.
18. Marschener, H., Role of root growth, arbuscular mycorrhiza, and root exudates for the efficiency in nutrient acquisition. *Field Crops Research.* 56, 203–207, 1998.
19. Rengel, Z., Heavy metals as essential nutrients, in *Heavy metals stress in plants: from molecules to ecosystems* Prasad, M.N.V., Hagemeyer J., Eds., Springer-Verlag, Berlin, 231–251, 1999.
20. Garbisu, C. and Alkorta, I., Phytoextraction: a cost-effective plant-based technology for the removal of metals from the environment. *Biores. Technol.* 77, 229–236, 2001.
21. Soltan, M.E. and Rashed, M.N., Laboratory study on the survival of water hyacinth under several conditions of heavy metal concentrations. *Adv. Environ. Research.* 7, 321–334, 2003.
22. Williams, T.P., Bubb, J.M., and Lester, J.N., The occurrence and distribution of trace metals in halophytes. *Chemosphere.* 28, 1189–1199, 1994.
23. Leendertse, P.C., Scholten, M.C.Th., and van der Wal, J.T., Fate and effects of nutrients and heavy metals in experimental salt marsh ecosystems. *Environ. Pollution.* 94, 19–29, 1996.
24. Szymanowska, A., Samecka-Cymerman, A., and Kempers, A.J., Heavy Metals in Three Lakes in West Poland. *Ecotox. Environ. Safety.* 43, 21–29, 1999.
25. Stevenson, F.J., Humus chemistry: genesis, composition, reactions, John Wiley & Sons, New York, 1994.
26. Lorenzo, J.I., Nieto, O., and Beiras, R., Effect of humic acids on speciation and toxicity of copper to Paracentrotus lividus larvae in seawater. *Aquatic Toxic.* 58, 27–41, 2002
27. Tipping, E., Smith, E.J., Lawlor, A.J., Hughes, S., and Stevens, P.A., Predicting the release of metals from ombrotrophic peat due to drought-induced acidification. *Environ. Pollution.* 123, 239–253, 2003.
28. Chirenje, T.L. and Ma, Q., Effects of acidification on metal mobility in a papermill-ash amended soil. *J. Environ. Qual.* 28, 760–767, 1999.

29. Weis, J.S. and Weis, P., Metal uptake, transport and release by wetland plants: implications for phytoremediation and restoration. Review. *Environ. International.* In press.

30. Mench, M., Vangronsveld, J., Clijsters, H., Lepp, N.W., and Edwards, R., *In situ* metal immobilisation and phytostabilisation of contaminated soils, in *Phytoremediation of contaminated soil and water*, Norman, T., Banuelos, G., Eds., Lewis Publishers, Boca Raton, Florida, 323–358, 2000.

31. Korshunova, Y.O., Eide, D., Clark, W.G., Guerinot, M.L., and Pakrasi, H.B., The IRT1 protein from *Arabidopsis thaliana* is a transporter with a broad substrate range. *Plant Molecular Biology.* 40, 37–44, 1999.

10

Removal of Mercury from Aqueous Streams of Fossil Fuel Power Plants Using Novel Functionalized Nanoporous Sorbents

S. V. Mattigod[1], G. E. Fryxell[1], X. Feng[2], K. E. Parker[1], and E. M. Piers[1]

[1]*Pacific Northwest National Laboratory, Richland, WA 99352, USA*

[2]*Ferro Corporation, Cleveland, OH 44114, USA*

Abstract

A new class of hybrid nanoporous materials has been developed at the Pacific Northwest National Laboratory for removing toxic heavy metals such as mercury from aqueous and nonaqueous waste streams. These novel materials consist of functional molecules capable of selectively binding mercury (thiol groups) covalently bound (as densely populated monolayers) to the synthetic nanoporous substrates. Tests indicated that this sorbent (Self-Assembled Monolayers on Mesoporous Silica—SAMMS) can achieve mercury loading as high as ~635 mg/g. The high affinity for Hg adsorption by this material was reflected by K_d values as high as 3.5×10^8 ml/g. Data indicated that SAMMS can adsorb both inorganic and organic forms of mercury. Experimental data indicated that mercury adsorption performance of SAMMS was not significantly affected by pH, ionic strength, presence of other cations (Na, Ca, Cd, Cu, Fe, Ni, Pb, and Zn), and complexing anions (Cl, CN, CO_3, SO_4, and PO_4) in solution. Adsorption kinetics studies indicated that SAMMS adsorbed mercury very rapidly (about 99.9% adsorption occurring within first five minutes). Tests conducted using samples of different aqueous and nonaqueous waste streams have confirmed the exemplary performance characteristics of SAMMS sorbents. Preliminary cost estimates indicated that using SAMMS would result in significant savings in mercury remediation costs as compared to the use of conventional adsorbents such as ion exchange resin and activated carbon.

Introduction

There is a significant need for a technology to remove toxic constituents such as mercury, arsenic, and selenium from aqueous effluents of coal-fired power plants such as, blow down water, wet scrubber effluents, and ash pond waters. These effluents typically contain high concentration of dissolved major and dissolved toxic constituents, and exhibit a wide range of pH values (~4 – 12 SU)[1–3]. There are a number of existing technologies for mercury removal from water and wastewater. These include, sulfide precipitation, coagulation/coprecipitation, adsorption, ion exchange, and membrane separation. Recent reviews have included detailed discussion of the performance characteristics, advantages and disadvantages of these treatment methods[4,5]. These reviews indicated that there is a need for a new mercury removal technology that overcomes the significant process, performance and cost-related shortcomings of these conventional technologies.

To meet this need we have designed and developed a new class of high-performance nanoporous sorbent materials for heavy metal removal that overcomes the deficiencies of existing technologies. These novel materials are created from a combination of synthetic mesoporous ceramic substrates that have specifically tailored pore sizes (2 to 10 nm) and very high surface areas (~1000 m^2/g) with self-assembled monolayers of well-ordered functional groups that have high affinity and specificity for specific types of free or complex cations or anions (Figure 1).

Typically, the nanoporous supporting materials are synthesized through a co-assembly process using oxide precursors and surfactant molecules. The synthesis is accomplished by mixing surfactants and oxide precursors in a solvent and exposing the solution to mild hydrothermal conditions. The surfactant molecules form ordered liquid crystalline structures, such as hexagonally ordered rod-like micelles, and the oxide materials precipitate on the micellar surfaces to replicate the organic templates formed by the rod-like micelles. Subsequent calcination to 500 °C removes the surfactant templates and leaves a high surface area nanoporous ceramic substrate. Using surfactants of different chain length produces nanoporous materials with different pore sizes.

These nanoporous materials can be used as substrates for self-assembled monolayers of adsorptive functional groups that are selected to specifically adsorb heavy metals. Molecular self-assembly is a unique phenomenon in which functional molecules aggregate on an active surface, resulting in an organized assembly having both order and orientation. In this approach, bifunctional molecules containing a hydrophilic head

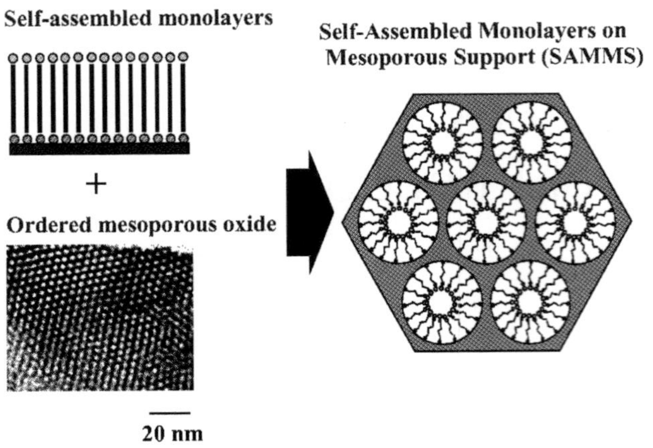

FIGURE 1. Technological basic of novel nanoporous sorbents.

group and a hydrophobic tail group adsorb onto a substrate or an interface as closely packed monolayers. The driving forces for the self-assembly are the intermolecular interactions between the functional molecules (such as van der Waals forces). The tail group and the head group can be chemically modified to contain certain functional groups to promote covalent bonding between the functional organic molecules and the substrate on one end, and the molecular bonding between the organic molecules and the metals on the other. Populating the head group with alkylthiols (which are well-known to have a high affinity for various soft heavy metals, including mercury) results in a functional monolayer that specifically adsorbs heavy metals. Using this technology, we have successfully synthesized a novel sorbent (thiol-SAMMS—thiol -self-assembled monolayers on mesoporous silica) for efficiently scavenging heavy metals such as mercury from waste streams. Detailed descriptions of the synthesis, fabrication, and adsorptive properties of these novel materials have been published previously[6–12].

Materials and Methods

A series of experiments were conducted to establish the mercury adsorption performance as function of pH, ionic strength, competing cations, and complexing ligands. Comparative tests were conducted to measure the kinetics of mercury adsorption. The results of these experiments are described in the following sections.

Adsorption experiments were conducted by contacting known quantities of thiol-SAMMS material with 0.1M $NaNO_3$ solutions containing Hg (II) added as a nitrate salt $[Hg(NO_3)_2 \cdot H_2O]$. The initial concentrations mercury ranged from 17–800 mg/L, and the solution to sorbent ratio was fixed at 5000 ml/g. A similar set of experiment was conducted to examine the adsorption of the most toxic form of mercury namely, the methylated form (CH_3Hg^+). In this experiment, 10 mg quantities of thiol-SAMMS was contacted with 50 ml aliquots of 0.1M $NaNO_3$ solution containing methyl mercury (CH_3HgOH) in concentrations ranging from ~12 – 360 mg/L The suspensions in both sets of experiments were continually shaken and allowed to react under ambient temperature conditions (~25°C) for approximately 8 hours. Next, the sorbent and the contact solutions were separated by filtration and the residual mercury concentrations in aliquots of contacting solutions were measured by using inductively-coupled plasma mass spectrometry (ICP-MS). The adsorption parameters were derived by fitting the data with the linearized form of Langmuir isotherm,

$$\frac{C}{x/m} = \frac{1}{Kb} + \frac{C}{b}$$

Where, C is the residual metal concentration in mg/L, x/m is the adsorption density in mg/g, b is the adsorption maximum (mg/g), and K is the Langmuir Affinity constant (g/L).

The selectivity of a sorbent for a contaminant is typically expressed as a distribution coefficient (K_d ml/g) which defines the partitioning of the contaminant between sorbent and solution phase at equilibrium. Distribution coefficient is the measure of an exchange substrate's selectivity or specificity for adsorbing a specific contaminant or a group of contaminants from matrix solutions, such as waste streams. The distribution coefficient (sometimes referred to as the partition coefficient at equilibrium) is defined as a ratio of the adsorption density to the final contaminant concentration in solution at equilibrium. This measure of selectivity is defined as

$$K_d = \frac{(x/m)_{eq}}{c_{eq}}$$

Where, K_d is the distribution coefficient (ml/g), $(x/m)_{eq}$ is the equilibrium adsorption density (mg of contaminant per gram of adsorbing substrate), and c_{eq} is the contaminant concentration (mg/ml) in contacting solution at equilibrium.

Adsorption kinetic experiments were conducted by contacting 200 mg of thiol-SAMMS with a 0.1 M $NaNO_3$ solution spiked with 10 mg/L concentration of mercury. The mixture was stirred constantly, and periodically aliquots of solution were drawn to monitor the residual mercury concentration. For comparison, mercury adsorption kinetic performance of a resin (GT-73) was also studied.

The effect of pH on mercury adsorption was tested by adjusting the pH (2–10) of a series of aliquots of 0.1M $NaNO_3$ solution each spiked with 0.09 mM of mercury. The effect of ionic strength on mercury adsorption was assessed by using a series of mercury-containing (~0.09mM) $NaNO_3$ solutions with ionic strength ranging from 0.1–4 M. For control, deionized distilled water containing only mercury was used with the resulting ionic strength of 0.0003M.

The experiment to assess the effect of competing cations on mercury adsorption consisted of 100 mM sodium nitrate solution containing ~0.09 mM of mercury and equimolar (~0.09 mM each) concentrations of four [Ca, Cu(II), Fe(II), Pb] and seven [Ca, Cd, Cu(II), Fe(II), Ni, Pb, Zn] cation mixtures equilibrated with thiol-SAMMS sorbent. The adsorption of mercury by thiol-SAMMS in competition with these cation mixtures was evaluated at two different pH values of 4 and 7. After equilibrating the solution-sorbent (ratio of 5000 ml/g) mixture for 4 hours at ~25°C, the residual mercury concentrations in filtered aliquots were measured.

Tests were also conducted to evaluate the effects of selected complexing ligands on mercury adsorption by thiol-SAMMS. These experiments were conducted using pH 4 and 7 solutions each containing ~0.1 mM of mercury and concentrations of fives times the molar excess of each ligand separately (~0.5 mM, Cl, CN, CO_3, SO_4, and PO_4), in a constant ionic strength matrix of 100 mM sodium nitrate solution. A solution-sorbent ratio of 5000 ml/g was used in all experiments. Following equilibration for 4 hours the residual concentrations of mercury were measured.

To assess the leachability of thiol- SAMMS material loaded with different amounts of mercury (~500–610 mg/g), Toxicity Characteristics Leaching Procedure (TCLP–USEPA) was used, and concentration of mercury released into the solution was measured.

A bench-scale treatability test was conducted to evaluate the mercury adsorption performance of thiol-SAMMS from a condensate waste stream originating from a pilot-scale waste glass melter operation. The principal dissolved components in this alkaline waste stream (pH: 8.5) consisted of mainly sodium borate (~30 mM), and sodium fluoride (~9 mM) with minor concentrations of sodium chloride(~3 mM), sodium nitrite (~0.9mM), sodium sulfate (~0.8mM), sodium nitrate (~0.6mM) and sodium iodide (~0.2 mM). The mercury concentration in solution was measured to be 4.64 mg/L. Other dissolved components such as, Al, Ba, Ca, Cd, Co, Cr, Cu, Fe, Mo, Ni, PO_4, Pb and Zn were present in trace concentrations (<2 mg/L). To test the effectiveness of thiol-SAMMS in adsorbing mercury from this complexing matrix, a 50 mL aliquot of the filtered solution was treated with 40 mg of thiol-SAMMS (Solution to solid ratio 1250 ml/g) and after 8 hours of equilibration, the residual mercury concentration was determined.

Results and Discussion

The data from the adsorption experiments indicated that mercury [Hg (II)] adsorption by thiol-SAMMS fit the Langmuir isotherm (Figure 2) with a predicted adsorption maximum of 625 mg/g, (measured maximum of 635 mg/g). The affinity parameter was 0.1074 g/L respectively. The calculated distribution coefficients were $9.5 \times 10^2 - 3.5 \times 10^8$ ml/g. Adsorption of methyl mercury was also Langmuirian in nature (Figure 3). The predicted adsorption maximum for methyl mercury was 357 mg/g with the affinity parameter being 0.1677 g/L. The distribution coefficients for adsorption of this toxic form of mercury ranged from $1.3 \times 10^3 - 8.4 \times 10^5$ ml/g by weight) of both inorganic and organic forms of mercury with high specificity.

The kinetics data indicated that thiol-SAMMS adsorbed ~99% of the dissolved mercury within first 5 minutes of

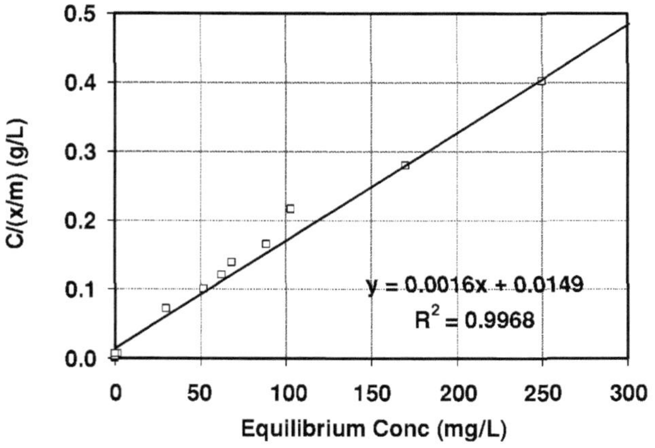

FIGURE 2. Langmuir isotherm for Hg (II) adsorption by thiol-SAMMS.

reaction (Figure 4). Comparatively, the resin (GT-73) adsorbed only ~18% of the dissolved mercury during the initial 5 minutes. These data showed that thiol-SAMMS substrate adsorbs mercury about 2 to 3 orders of magnitude faster than the commercial GT-73 ion-exchange resin. After 8 hours of reaction, thiol-SAMMS reduced the residual concentration of mercury to ~0.04 mg/L, whereas, the resin material was not capable of reducing the residual dissolved concentration of mercury any lower than 1 mg/L. Calculated distribution coefficients (K_d) indicated that thiol-SAMMS adsorbed mercury at about one to three orders of magnitude higher selectivity (7×10^3–3.6×10^5 ml/g) than the resin material (4.5×10^2–1.9×10^3 ml/g).

The results the pH effects experiment (Figure 5) indicated that K_d values for mercury adsorption ranged from ~1×10^6 to ~4.5×10^7 ml/g. These data indicated that the pH does not significantly affect mercury adsorption by thiol-SAMMS even under highly acidic (pH 2) or alkaline conditions (pH 10).

The data (Figure 6) showed that ionic strength had no measurable effects on adsorption of mercury by thiol-SAMMS (K_d ~3×10^7 to ~6×10^7 ml/g). Typically the ionic strength

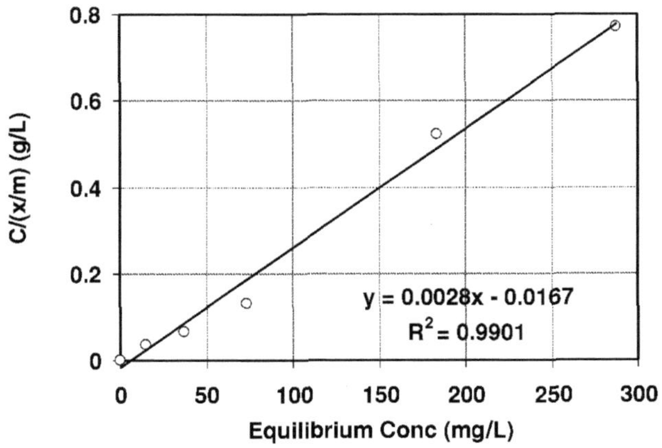

FIGURE 3. Langmuir isotherm for methyl mercury adsorption by thiol-SAMMS.

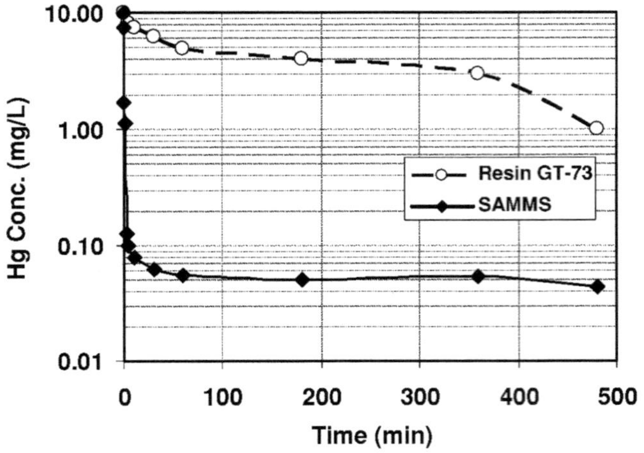

FIGURE 4. Kinetics of Hg (II) adsorption by thiol-SAMMS and an ion exchange resin.

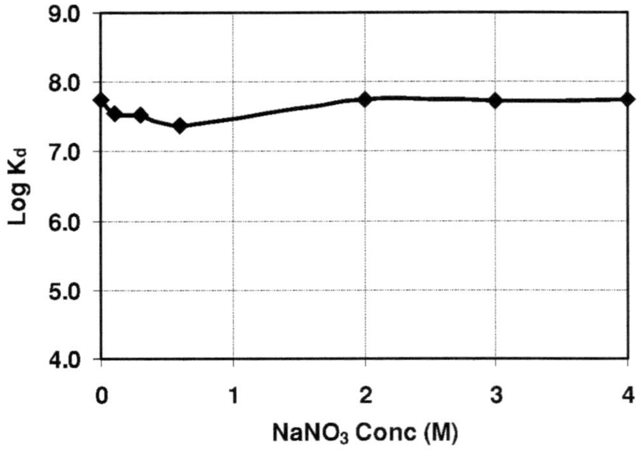

FIGURE 6. Effect of ionic strength on Hg (II) adsorption by thiol-SAMMS.

effects on adsorption These data confirmed that although the concentration of sodium exceeded the concentration of mercury in solution by 3 to >4 orders of magnitude, it had no significant effect on the adsorption of mercury.

The data from the cation effects experiment (Table 1) showed that equimolar concentrations of competing cation combinations at either pH did not significantly affect the high affinity of thiol-SAMMS for mercury in solution (K_d: 1–4.5 × 10^7ml/g). Such selectivity and affinity in binding mercury by thiol-SAMMS can be explained on the basis of the hard and soft acid base principle (HSAB)[13–15] which predicts that the degree of cation softness directly correlates with the observed strength of interaction with soft base functionalities such as thiols (-SH groups). According to the HSAB principle, soft cations and anions possess relatively large ionic size, low electronegativity, and high polarizability (highly deformable bonding electron orbitals) therefore, mutually form strong covalent bonds. In competing cation combinations, all cations except lead and cadmium, were either hard or borderline cations therefore, were less likely to interact significantly with soft base moiety (-SH). Because the relative degree of softness[16] is in the order Hg≫Pb≫Cd, both lead and cadmium would bind less preferably to the thiol functionality. These tests showed that thiol-SAMMS can very selectively bind dissolved mercury from solutions containing a mixture of other heavy metals, alkali and alkaline earth cations.

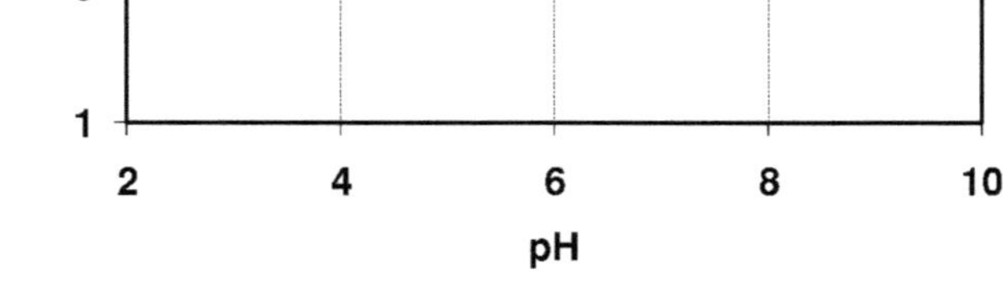

FIGURE 5. Effect of pH on Hg (II) adsorption by thiol-SAMMS.

Tests conducted with various ligands (Cl, CN, CO_3, SO_4, and PO_4) indicated that complexation did not affect the binding affinity of thiol-SAMMS for mercury (Table 2). In these experiments the molar concentrations of each ligand was five times higher than mercury concentration, therefore, the mercury complexes in these solutions were expected to be in the form of $HgL_n^{(nz-2)}$, where n is the ligand stoichiometry, and z is the ligand charge. The calculated speciation of mercury in these solutions showed (Table 3) that as expected, a major fraction of dissolved mercury, existed mainly in the form of mainly hydrolytic species [e.g $Hg(OH)_2^0$] in CO_3, SO_4, and PO_4 media, as $Hg(CN)_2^0$ in CN medium, and as $HgCl_2^0$ in Cl medium at pH 4.

Because, the affinity for thiol-SAMMS to bind mercury (log K: ~53 for $=S_2Hg$) is orders of magnitude greater than the stability of major mercury complexes (log K = 14.03, 21.8, 32.75 for $HgCl_2^0$, $Hg(OH)_2^0$, and $Hg(CN)_2^0$ respectively[17]) present in these experiments, mercury adsorption by thiol-SAMMS was not significantly affected by different forms of mercury complexes in solution.

The data from the bench-scale test of the condensate waste stream indicated that thiol-SAMMS had adsorbed ~98.9% of the mercury initially present in this solution resulting in

TABLE 1. Effect of cation mixtures on mercury adsorption by thiol-SAMMS

pH	Cation mixtures	K_d(ml/g)
4	Na, Ca, Cu(II), Fe(II), Pb	1.0 × 10^7
4	Na, Ca, Cd, Cu(II), Fe(II), Ni, Pb, Zn	1.0 × 10^7
7	Na, Ca, Cu(II), Fe(II), Pb	1.1 × 10^7
7	Na, Ca, Cd, Cu(II), Fe(II), Ni, Pb, Zn	4.5 × 10^7

TABLE 2. Effect of complexing ligends on mercury adsorption by thiol-SAMMS

Complexing Ligand	pH: 4 K_d (ml/g)	Complexing Ligand	pH: 7 K_d (ml/g)
Cl	3.3×10^6	Cl	3.3×10^6
CN	3.7×10^7	CN	5.9×10^7
CO_3	3.7×10^6	CO_3	6.1×10^6
SO_4	3.2×10^6	SO_4	5.5×10^6
PO_4	3.7×10^6	PO_4	7.6×10^6

a residual mercury concentration of 0.052 mg/L. This solution contained ~0.2 mM of iodide, a strong mercury complexing ligand and the speciation calculations indicated that all the soluble mercury existed in the form of iodide complexes (HgI_2^0: ~52%; HgI_3^-: ~47%, and HgI_4^{2-}: ~1%). These results confirmed previous observations[11] that thiol-SAMMS can effectively remove iodide-complexed mercury from solutions.

Mercury adsorption by thiol-SAMMS from a non-aqueous medium (meteorology oil from a U.S Department of Energy site) has been previously tested at the Oak Ridge National Laboratory[18]. These batch tests consisted of equilibrating for 24 hours, 50mL aliquots of oil containing about 650 ppb mercury with thiol-SAMMS. (w/v 0.1 – 1%). Following equilibration, the residual mercury concentration in the treated oil was measured. The data showed that residual concentration of mercury decreased logarithmically as a function of thiol-SAMMS added to the solution (w/v). This test demonstrated that thiol-SAMMS could also effectively remove mercury from nonaqueous medium such as oil.

The results of the TCLP test (Figure 7) indicated that concentrations of mobilized mercury were very low (~0.0002–0.001 mg/L). These concentrations are two to three orders of magnitude lower than the EPA regulatory requirement of 0.2 mg/L. The TCLP data confirmed the strong covalent nature of mercury bonding with thiol-SAMMS material and because of its very high resistance to leaching mercury loaded thiol-SAMMS

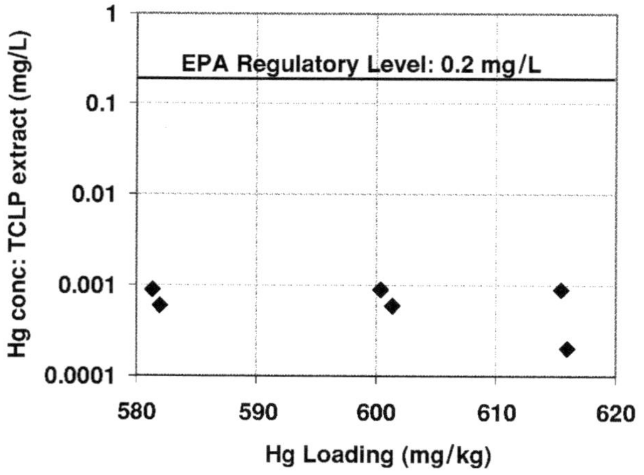

FIGURE 7. Mercury extractability from spent thiol-SAMMS.

be a very stable waste form thus could be disposed of with ease.

These tests showed that the novel synthetic nanoporous thiol-SAMMS sorbent has superior mercury removal characteristics (loading capacity and kinetics). However, for deployment of this novel material for mercury removal we need to assess the cost effectiveness of this technology. Therefore, we conducted a preliminary material lifetime cost comparison for mercury removal using thiol-SAMMS, a resin and a conventional adsorbent such as granulated activated carbon (GAC).

Based on costs for synthesis in 200 kg batches we estimated the cost of thiol-SAMMS material to be \$300/kg. Scaled up manufacture is expected to bring the cost down to ~\$200/kg. The cost of commercial resin shipped in 225L quantities was estimated to be ~\$55/kg, and the GAC was estimated to cost ~\$2/kg. We based our calculations on achieving 99.9% reduction in concentration of a waste stream containing an influent mercury concentration of 10 mg/L. The quantity of sorbent needed were based on the loading data and for removing a total of 1 kg of mercury from the waste stream. The disposal costs were estimated at ~\$60/ft^3 based on the assumption that all the sorbents would pass the EPA-TCLP test.

Preliminary lifetime cost comparison (Figure 8) showed that the higher material cost of thiol-SAMMS (which is about 6 times higher than resin and more than an order of magnitude higher than GAC), is significantly offset by its superior loading capacity (more than 1–3 orders magnitude) as compared to the other two sorbents. Consequently, the total cost of thiol-SAMMS for removing 1 kg mercury is ~60% and ~95% less than the cost of using resin, and GAC respectively. Similarly, due to large solid waste volume generated from resin and GAC use, the disposal cost of these spent solids are significantly higher (~1 to 3 order of magnitude) than the cost of disposing of spent thiol-SAMMS.

These preliminary cost estimates showed that using a novel nanomaterial such as thiol-SAMMS would result in significant savings in mercury remediation costs as compared to the use of conventional adsorbents such as resins and GAC.

TABLE 3. Calculated aqueous speciation of mercury in the presence of complexing ligands at two pH values

Complexing ligand	Aqueous Hg species	%Total pH 4	%Total pH 7
CN	$Hg(CN)_2^0$	100	100
Cl	$Hg(OH)_2^0$	0	47
	$HgCl_2^0$	100	9
	$HgOHCl^0$	0	44
CO_3	Hg^{2+}	3	0
	$HgOH^+$	7	0
	$Hg(OH)_2^0$	84	100
	$HgHCO_3^+$	4	0
	$HgCO_3^0$	2	0
SO_4	Hg^{2+}	4	0
	$HgOH^+$	7	0
	$Hg(OH)_2^0$	89	100
PO_4	Hg^{2+}	4	0
	$HgOH^+$	7	0
	$Hg(OH)_2^0$	89	100

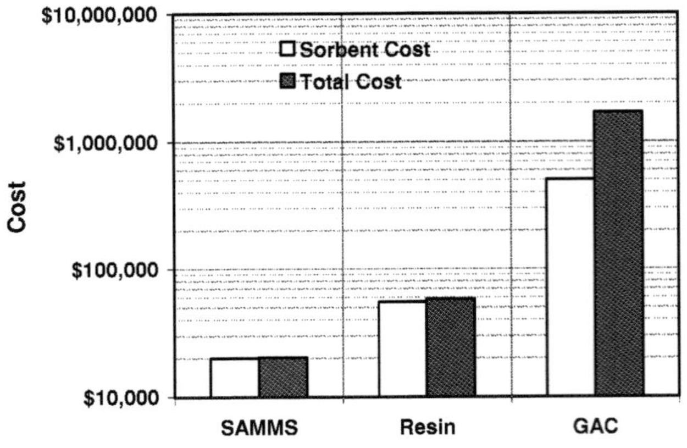

FIGURE 8. Preliminary life time cost comparison for mercury removal.

Summary and Conclusions

Synthetic thiol-SAMMS, a nanoporous sorbent adsorbed significant quantities of mercury (adsorption maximum 635 mg/g with relatively high specificity (K_d: 9.5×10^2–3.5×10^8 ml/g). SAMMS substrate adsorbed mercury about 5 times faster than a heavy metal specific commercial ion-exchange resin. Varying ionic strength and pH did not significantly affect mercury adsorption characteristics of SAMMS material. SAMMS can very selectively bind dissolved mercury from solutions containing a mixture of other competing heavy metal cations. Data also showed that mercury adsoption was not significantly affected by different forms of mercury complexes in solution. Tests conducted using real aqueous and nonaqueous waste stream samples confirmed the excellent mercury adsorption performance of thiol-SAMMS. The results of the TCLP test indicated that concentrations of mercury released from loaded thiol-SAMMS were very low (~0.0002 – 0.001 mg/L) confirming the highly stable nature of the waste form.

Preliminary cost estimates indicated that using a novel nanomaterial such as thiol-SAMMS would result in significant savings in mercury remediation costs as compared to the use of conventional adsorbents such as ion exchange resin and activated carbon.

Acknowledgements. This study was supported by the Office of Science and Technology of the U.S. Department of Energy and the IR&D funds from Battelle. Pacific Northwest National Laboratory is operated for the U. S. Department of Energy by Battelle under contract DE-AC06-76RLO 1830.

References

1. Rubin, E. S. Toxic Releases from Power Plants. *Environ. Sci. Technol.* 33, 3062, 1999.
2. U. S. Environmental Protection Agency. http://www.epa.gov/tri/ 2004.
3. Export-Import Bank of U. S. Environmental Guidelines. Thermal, Gas Turbine, and Diesel Driven Power Plants. http://www.exim.gov/products/policies/environment/envtbl6.html 2004.
4. Ebadian, M.A, M. Allen, Y. Cai, and J. F. McGahan. Mercury Contaminated Material Decontamination Methods: Investigation and Assessment. Final Report -Hemispheric Center for Environmental Toxicology, Florida International University, Miami, FL 33174. 2001
5. U. S. Environmental Protection Agency. Aqueous Mercury Treatment—Capsule Report. EPA/625/R-97/004. Office of Research and Development, Washington DC 20460. 1997.
6. Feng, X.; Fryxell, G. E.; Wang, L. Q.; Kim, A. Y.; Liu Functionalized Monolayers on Ordered Mesoporous Supports., J.*Science*, 1997, 276, 923–926.
7. Jun L., X. Feng, G. E. Fryxell, L.Q. Wang, A. Y. Kim, M. Gong, Hybrid Mesoporous Materials with Functionalized Monolayers. *Adv. Materials,* 10, 161–165. 1998
8. Fryxell G. E, J Liu, AA Hauser, Z Nie, KF Ferris, SV Mattigod, M Gong, RT Hallen. Design and Synthesis of Selective Mesoporous Anion Traps, *Chemistry of Materials.* 11, 2148–2154. 1999.
9. Fryxell G. E, J Liu, SV Mattigod. Self-Assembled Monolayers on Mesoporous Supports (SAMMS)—an Innovative Environmental Sorbent, *Materials Technology & Advanced Performance Materials.* 14, 188–191, 1999.
10. Fryxell G. E, J Liu, SV Mattigod, L. Q. Wang, M. Gong, T. A. Hauser, Y. Lin and K. F. Ferris, X. Feng. Environmental Applications of Interfacially-Modified Mesoporous Ceramics. Proceedings of the 101st National Meetings of the American Ceramic Society. 1999.
11. Mattigod, S. V, X Feng, G. E Fryxell, J. Liu, and M. Gong. 1999. Separation of Complexed Mercury from Aqueous Wastes using Self-Assembled Mercaptan on Mesoporous Silica. *Separation Science and Technology*, 34, 2329–2345. 1999.
12. Liu, J., G. E. Fryxell, S.V. Mattigod, T. S. Zemanian, Y. Shin and L. Q. Wang. 2000. Synthesis and Applications of Functionalized Nanoporous Materials for Specific Adsorption, *Studies in Surf. Sci. and Catalysis.* 129, 729–738, 2000.
13. Pearson R.G. Hard and Soft Acids and Bases Part 1, *Journal of Chemical Education*, 45, 581–587, 1968.
14. Pearson R. G. Hard and Soft Acids and Bases Part 2, *Journal of Chemical Education*, 45, 643–648, 1968.
15. Hancock R. D. and A. E. Martell. Hard and Soft Acid and Base Behavior in aqueous solution: Steric effects make some metal ions Hard: A Quantitative Scale of Hardness-Softness for Acids and Bases. *Journal of Chemical Education*, 74, 644, 1996.
16. Misono, M., E. Ochiai, Y. Saito, and Y. Yoneda. A Dual Parameter Scale for the Strength of Lewis Acids and Bases with the Evaluation of their Softness. *Journal of Inorganic and Nuclear Chemistry*, 29, 2685–2691, 1967.
17. Smith R.M. and A. E. Martell. *Critical Stability Constants. V 4. Inorganic complexes*. Plenum Press, New York, 1976.
18. Klasson, K.T. Treatment of Mercury Contaminated Oil from the Mound site. Topical Report. Oak Ridge National Laboratory, Oak Ridge, TN 37831, 2000.

11
Leachability of Trace Metals from Sandy or Rocky Soils Amended with Coal Fly Ash

Yuncong Li[1] and Jianjun Chen[2]
[1]*Tropical Research and Education Center, IFAS, University of Florida, Homestead, FL 33031, USA*
[2]*Mid-Florida Research and Education Center, IFAS, University of Florida, Apopka, FL 32703, USA*

Abstract

Application of coal fly ash as a soil amendment for Florida sandy or rocky soils is an alternative to disposal of coal fly ash and also will improve soil physical and chemical properties and increase crop production. However, environmental fate of trace metals in coal fly ash is a major concern. The objective of this experiment was to evaluate movement of trace metals in coal fly ash amended soil columns. Three soils were collected from south Florida to represent Alfisols, Entisols, and Spodosols. Each of three representative soil was packed into separate columns (32.5 cm long and 7.5 cm inner diameter), amended with coal fly ash at rates of 0, 40, 80 Mg ha^{-1} and leached with 10 pore volumes of water, which is equivalent to 113.8 cm rainfall. Leachates were analyzed for Zn, Cu, Mn, Fe, Cd, Pb, and Ni. There was no detectable Cd in leachate from any of the soil columns. The maximum concentrations of Zn from all three soils with or without coal fly ash amendment were below the Maximum Contamination Level (MCL) for drinking water. High concentrations of Fe, Pb and Mn in leachate were more closely related to the background concentrations of these metals in three soils than to the coal fly ash application. Application of 80 Mg ha^{-1} coal fly ash to Riviera soil elevated Cu concentration in leachate above MCL. After completion of leaching, soil columns were divided into five sections (5 cm each) and analyzed for total concentrations of selected metals. The results showed that application of coal fly ash increased all of the trace metals measured in soils at top 5 cm or 5–10 cm depth after leaching. Concentrations of Zn, Cu, Mn, and Cd were also elevated in soil depths from 5 to 25 cm of Riviera soils mainly because the soil has very sandy texture and low organic carbon.

Introduction

There are a number of benefits that result from the application of coal ash to agricultural soils, especially sandy or rock soils, which are common in south Florida. The benefits include improving soil texture, modifying soil pH, and providing essential plant nutrients for crop production.[1,2] Townsand and Hodgson[3] reported that the particle fractions of coal ash samples ranged from 45–70% silt and 1–4% clay and the fine-sized ash particles should increase the total porosity of soil and improve soil water and nutrient holding capacities. Ghodrati et al.[4] reported that moisture holding capacity in a soil amended with 30% fly ash was increased from 12% to 25%, and that the rate of water flow through the soil was reduced three-fold. Coal ash consists of over 40 elements, and at least 9 of trace elements are essential nutrients for plant growth. Application of coal ash increases bioavailability to plants of nutrients in soils. Chen and Li[5] used coal fly ash to replace dolomite and showed that coal fly ash could be used as alternative to commercial dolomite in soilless substrates for ornamental plant production. Adriano et al.[6] studied the effects of coal ash on soil chemical properties, growth and mineral nutrition of corn (*Zea mays* L.) and bush bean (*Phaseolus vulgaris* L.), and reported that the application of fly ash increased concentrations of extractable K, Ca, Mg, Cu, Fe, Mn, and Zn in the soil, and concentrations of K and Fe in corn shoots, and K, Ca, Mg, S, and Cu in bush bean shoots. Similar results have been reported for bermudagrass (Cynodon dactylon L.),[7] barley (*Hordeum vulgare* L. var. 'Leduce'),[8] rice (*Oryza sativa* cv. 'PR 106'), wheat (*Triticum aestivum* L.),[9] tomato (*Lycopersicon esculentum* Mill. cv. 'Pusa Ruby')[10] and other vegetable crops.[11] Application of coal ash also increased boron contents of alfalfa (*Medicago sativa* L.)[12] and corn.[13] Therefore, it is possible that coal fly ash can be used as an amendment for Florida's sandy or coarse soils to improve crop production.

Environmental concern is always an issue in the land application of coal fly ash products. Of particular concern is the possibility that trace metals released from coal fly ash may be leached into groundwater.[14,15] In the U.S., coal combustion by-products are generally exempt from hazardous waste regulations and some states have elected to regulate these materials as solid, specific or industrial wastes. Minnesota Pollution

Control Agency[16]has set maximum limits for trace metals in coal ash to be used as a fertilizer or liming agent. Under Florida regulations, coal ash and other wastes generated from the combustion of coal and other fossil fuels are exempt from regulation as hazardous waste. However, the state environmental agency is opposed strongly to the unregulated use of coal waste products. Study on the leachability of trace metals from coal ash amended soils may help regulatory agencies in evaluating their practices.

Laboratory batch-leaching tests have been commonly used to determine the leachability of trace metals from coal ash. The most common method is the toxicity characteristic leaching procedure (TCLP), EPA method 1311.[17] This method is designed for the evaluation of leaching of wastes when co-disposed in a sanitary landfill and is not a suitable leaching test for evaluating field behavior of coal ash.[18] Acetic acid solution at low pH is used for the procedure causing dissolution of soil metals and leads to overestimating leachability of metals from coal ash especially for calcareous soils in south Florida. In this study, column leaching was used to simulate rainfall events to more closely represent field conditions. The objective of this study was to determine leachability of selected metals (Zn, Cu, Mn, Fe, Cd, Pb, Ni, and Mo) from coal fly ash amended soils using rainfall simulation.

Materials and Methods

Coal fly ash was obtained from a power plant in Florida. Soil samples were collected from citrus or tropical fruit groves in St. Lucie, Martin and Miami-Dade Counties, Florida. These samples represented three soil series namely Myakka (sandy, siliceous, hyperthermic Aeric Haplaquods), Riviera (sandy, siliceous, hyperthermic Arenic Glossaqualf), and Krome (Loamy-skeletal, carbonatic, hyperthermic Lithic Udorthents). Coal fly ash and soil samples were air dried, ground, and sieved to pass through a 2-mm sieve. Particle size distribution was analyzed following a micropipette method.[19] Soil and fly ash pHs were measured in a 1:1 (w/v) ratio of solid and water suspension. Organic carbon content was analyzed by the Walkley-Black method.[20] Total concentrations of metals were determined using the USEPA 3050 method.[17] Selected properties of the coal fly ash and soil samples are presented in Table 1. The concentrations of listed trace metals in the coal fly ash used in this study are well below the maximum limits set by the Minnesota Pollution Control Agency for land use of coal ash.[16]

Leaching columns (32.5 cm long and 7.5 cm inner diameter) were constructed using clear plexiglass columns. The plexiglass column was cut longitudinally into halves. The two halves were joined with silicon glue. A filter paper was placed on the bottom of the column, which had a nylon mesh over a Plexiglass plate with several 5 mm diameter holes. The soil (1.6 kg per column) was packed to a mean bulk density of 1.5 g cm^{-3}. Soil columns were saturated using deionized water and excess water was allowed to drain for 24 hr. Coal fly ash was applied at rates of 0, 17.7, and 35.4 g per column and mixed with top 3 cm soils. On a surface area basis, the amounts of coal fly ash applied were equivalent to 0, 40, 80 Mg ha^{-1}. A filter paper was placed on the soil surface and deionized water was applied using a peristaltic pump at a rate of 1.5 ml min^{-1}. The leachate was collected at each half pore volume (240 mL) for a total of 10 pore volumes. The quantity of water applied to the soil column for each pore volume of leachate was equivalent to 11.38 cm of rainfall on the surface of the soil column. The leachate fractions were filtered through 0.45 μm filter paper and concentrations of Zn, Cu, Mn, Fe, Cd, Pb, and Ni were determined using inductively coupled plasma emission spectroscopy (ICPES; Plasma 40 unit; Perkin-Elmer, Norwalk, CT).

After completion of leaching of ten pore volumes of leachate, the plexiglass columns were split into two halves. The soil column was divided into five sections with each of 5 cm height. The soil in each section was collected separately, air-dried, extracted and analyzed following EPA3050 method.

Results and Discussion

The leaching patterns of trace metals through soil columns are presented in Figs. 1–3 and the maximum (peak) concentrations of trace metals in 10 pore volumes of leachate are shown in Table 2. There was no detectable Cd in leachate from all soil columns which is different from the results of Papamasivam et al.[21] They reported the leaching of Cd was detected from leachate of soils amended with 74.1 Mg ha^{-1} coal ash, but did

TABLE 1. Chemical composition and physical properties of soils and a coal ash used in this study

Soil or coal fly ash	pH	Organic C	Zn	Cu	Mn	Fe	Cd	Pb	Ni	clay	silt	sand	gravel
		g kg^{-1}				mg kg^{-1}						%	
Krome	7.7	21	65	63	244	11950	1.3	34	11	6.2	11.2	16.5	66.1
Myakka	5.5	16	32	21	16	183	0.2	3	1	2.2	3.0	94.8	0
Riviera	6.2	9	92	152	63	980	0.5	12	2	1.4	2.8	95.8	0
Coal fly ash	9.4	61	198	285	96	45348	3.0	113	79	11.3	82.5	6.2	0

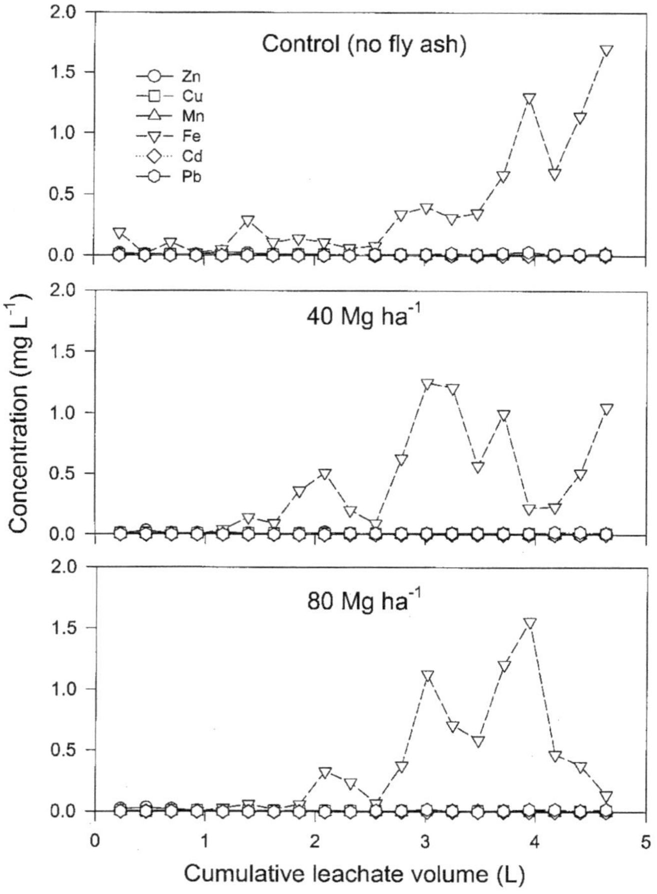

FIGURE 1. Concentrations of trace metals (Zn, Cu, Mn, Fe, Cd, and Pb) in leachate from a Krome very gravelly loam soil amended with 0, 40, and 80 Mg ha^{-1}coal fly ash and leached continuously with ten pore volumes of water.

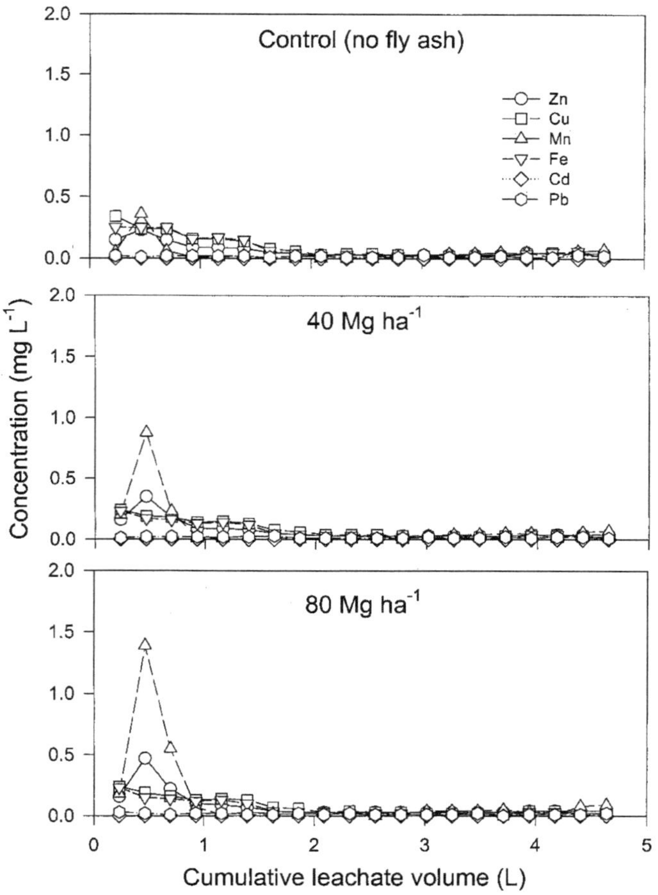

FIGURE 2. Concentrations of trace metals (Zn, Cu, Mn, Fe, Cd, and Pb) in leachate from a Myakka sand soil amended with 0, 40, and 80 Mg ha^{-1} coal fly ash and leached continuously with ten pore volumes of water.

not exceed the Maximum Contamination Level (MCL), which is the highest level of a contaminant that is allowed in drinking water in the US (Table 2). The maximum concentrations of Zn from all three soils with or without coal fly ash amendment were below MCL.

Iron leaching started at 3–4 pore volumes and through most of the leaching process for both Krome and Riviera soils. The maximum concentrations of Fe in leachate were 1.25–1.7 mg L^{-1}, while MCL for Fe is only 0.3 mg L^{-1}. Similar Fe concentrations in groundwater from Krome soil area were reported previously.[22] The high levels of Fe in the soil and the bedrock probably are the cause of the high concentrations of Fe in groundwater. Concentrations of Fe in unfarmed natural calcareous soils in the area are about 16 times higher than the mean concentration of Fe in 448 Florida surface soils. 2[4] Iron is in secondary drinking water standards, which are not enforceable.

There is no clear leaching pattern for Pb for treated or untreated soils (Figs. 1–3). However, the maximum concentrations of Pb in leachate from all soils were higher than MCL and ranged from 0.02–0.06 mg L^{-1} (Table 2). Our previous field experiment also showed similar results.[23] The high concentrations of Pb in leachate may be related to relative high concentration of Pb in soils used in this experiment. The mean concentration of Pb in Florida soils is 4.1 mg kg^{-1}while concentrations of Pb in soils used for this study were 3, 12 and 34 mg kg^{-1} for Myakka, Riviera and Krome, respectively.[24] Howie[22] reported that lead concentrations in the Biscayne aquifer beneath the south Miami-Dade area ranged from <0.001 to 0.016 mg L^{-1}. Therefore, coal fly ash may not be the source for the high concentration of Pb in leachate.

Most of the Mn leaching occurred in the first two pore volume of leaching for both Myakka and Riviera soils (Figs. 1–3). The highest leaching peak (4.6 mg L^{-1}) of Mn was at 1 pore volume of Riviera soil amended with 80 Mg ha^{-1} of coal fly ash. The maximum concentrations of Mn in leachates from Myakka and Riviera soils were above the Secondary Standard for drinking water with or without coal fly ash amendment (Table 2). Both soils used in this study have the higher Mn

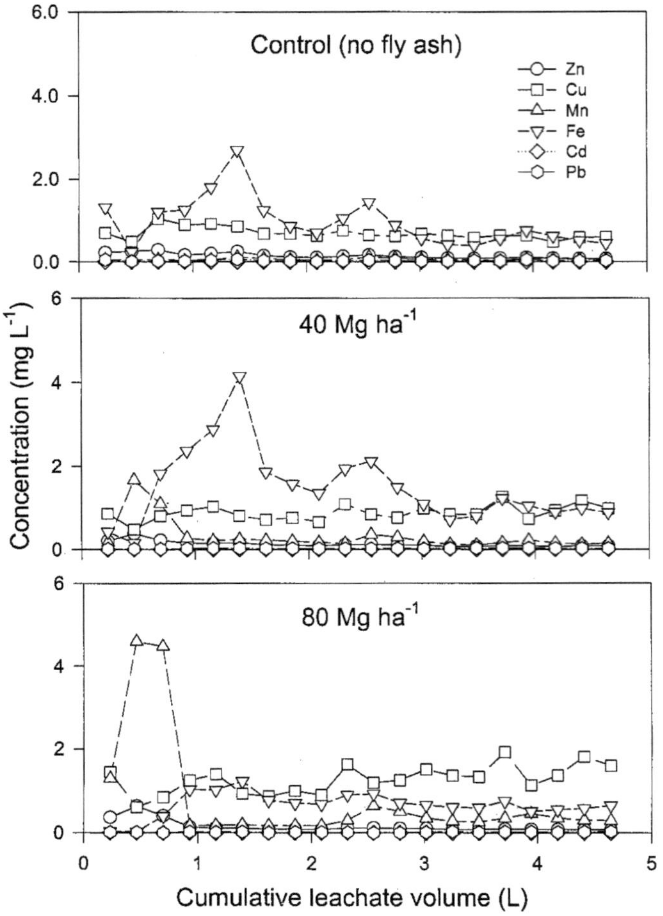

FIGURE 3. Concentrations of trace metals (Zn, Cu, Mn, Fe, Cd, and Pb) in leachate from a Riviera loam soil amended with 0, 40, and 80 Mg ha^{-1} coal fly ash and leached continuously with ten pore volumes of water.

TABLE 2. Maximum concentrations of trace metals in leachate

Trace metal (mg L^{-1})	Application rate of coal ash mixture (Mg ha^{-1}) 0	40	80
		Krome	
Zn	0.03	0.03	0.03
Cu	0.01	0.01	0.01
Mn	0.02	0.01	0.02
Fe	1.70*	1.25*	1.56*
Cd	0.00	0.00	0.00
Pb	0.03*	0.02*	0.02*
		Myakka	
Zn	0.23	0.35	0.47
Cu	0.34	0.24	0.24
Mn	0.36*	0.88*	1.39*
Fe	0.25	0.23	0.23
Cd	0.00	0.00	0.00
Pb	0.03*	0.03*	0.03*
		Riviera	
Zn	0.29	0.38	0.64
Cu	1.04	1.26	1.81*
Mn	0.12*	1.67*	4.60*
Fe	2.70*	4.15*	1.23*
Cd	0.00	0.00	0.00
Pb	0.06*	0.03*	0.03*

* The concentration of element in leachate is higher then MCL (Cu = 1.3 mg L^{-1}; Pb = 0.015 mg L^{-1}; and Cd = 0.005 mg L^{-1}) or secondary standard (Zn = 5 mg L^{-1}; Fe = 0.3 mg L^{-1}; Cu = 1.3 mg L^{-1}; and Mn = 0.05 mg L^{-1}) for drinking water.[29] MCL (Maximum Contaminant Level) is the highest level of a contaminant that is allowed in drinking water and is an enforceable standard.

concentrations than the mean concentration of Mn of Florida surface soils. However, the maximum concentration of Mn in leachate from soils treated with either 40 or 80 Mg ha^{-1} of coal fly ash were up to 38 times higher than those from untreated soil columns. This indicates that using coal fly ash could increase leaching of Mn from south Florida sandy soils.

Application of 80 Mg ha^{-1} coal fly ash to Riviera soil elevated Cu concentration in leachate above MCL (1.3 mg L^{-1}) (Table 2). South Florida Water Management District also reported the elevated concentrations of Cu in the Indian River Lagoon sediments and cannel water flowing from agricultural area.[25] The copper is suspected in the appearance of fish lesions in the Indian River Lagoon.

Amendment of coal fly ash at rates of 40 and 80 Mg ha^{-1} increased Zn concentrations for Krome and Myakka at the 5–10 cm depth (Fig. 4). Very small increase of soil Zn at lower depths indicated that movement of Zn from coal fly ash was very limited after leaching with 10 pore volume water. However, concentrations of Zn in Riviera soil were elevated through soil columns which were amended with coal fly ash. Sandy texture (95.8% sand) and low organic carbon content (9 g kg^{-1}) are probably main factors for high mobility of Zn in this soil. Untreated Riviera soil also had the highest concentration of Zn (152 mg kg^{-1}) because zinc is applied regularly as a micronutrient fertilizer for crop production in south Florida.

Concentrations of Cu and Mn in the two top depths (0–5 and 5–10 cm) increased with application of coal fly ash in these three soils (Figs. 5–6). There is no difference of Cu and Mn concentrations at lower depth for Krome and Myakka soils. Many researchers have reported low mobility of Cu in soils. Merry et al.[26] reported that Cu primarily accumulated in the surface soil (0.25 cm) of orchard (apple and pear tree) when Cu was applied as a pesticide for many years and Cu concentrations was up to 320 mg kg^{-1} but with very little evidence of Cu transport and accumulation below 25 cm in the soil profile. Labeled Cu applied to the soil surface was leached to a maximum depth of 5 cm in a coarse textured soil, which received 46 cm water.[27] Mobility of Cu and Mn in Riviera soil was similar to Zn in this soil. High Cu concentrations in Riviera soil in south Florida came from regular application of fungicides.[28]

The concentrations of soil Fe, Pb, and Ni in the top 5 cm increased by as much as 19, 7, and 6 times, respectively with

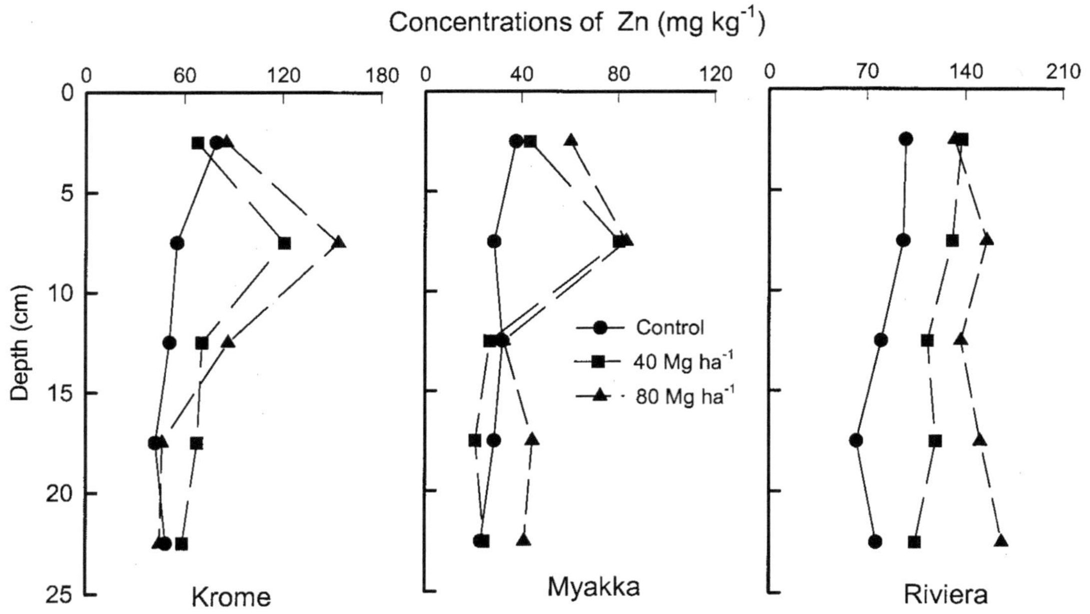

FIGURE 4. Distribution of soil Zn in soil columns following the addition of various amounts of coal fly ash and leaching with ten pore volumes of water.

application of 40-80 Mg ha^{-1} coal fly ash even after leaching with 10 pore volumes of water (Figs. 7–9). Total Fe contents were very high in both untreated soils (183–11950 g kg^{-1}) and coal fly ash (43348 mg kg^{-1}). The concentrations of Pb and Ni in coal fly ash were about 3–38, and 7–79 -fold higher than in those soils (Table 1). Concentration changes of these metals in the lower soil depth were almost negligible. These results indicate that high concentrations of Fe and Pb detected in leachate were not due to the application of coal fly ash.

Application of coal fly ash increased soil Cd in 0–15 cm depth and also showed slight elevation of soil Cd in the 20–25 cm depth (Fig. 10). However, there was no Cd detected in the leachate, probably because concentrations of

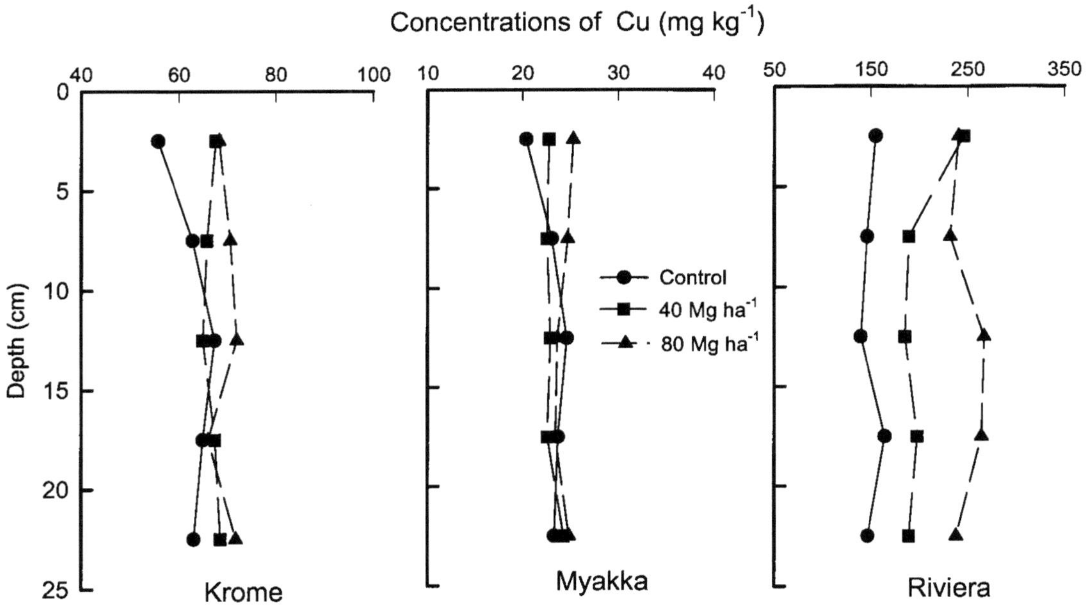

FIGURE 5. Distribution of soil Cu in soil columns following the addition of various amounts of coal fly ash and leaching with ten pore volumes of water.

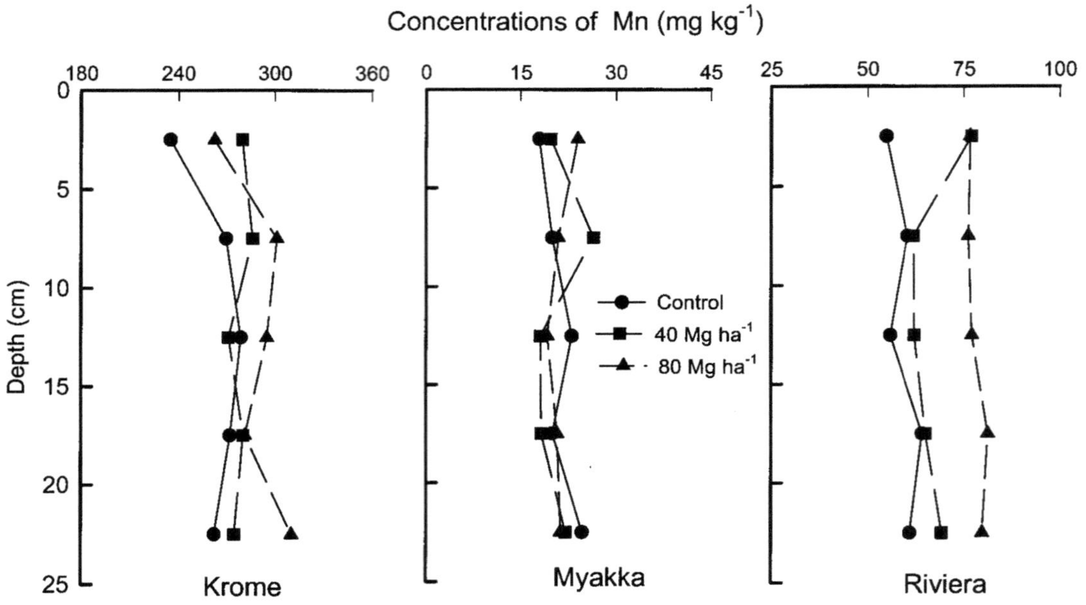

FIGURE 6. Distribution of soil Mn in soil columns following the addition of various amounts of coal fly ash and leaching with ten pore volumes of water.

Cd in these soils and coal fly ash used in this study are very low.

Conclusions

Concentrations of trace metals in the coal fly ash used in this study were relatively low, and such coal fly ash should be considered for using as a soil amendment in south Florida's sandy or rocky soils. High application rates (40 or 80 Mg ha^{-1}) of the coal fly ash increased concentrations of trace metals in surface soil. These metals may have potential to be leached through sandy or rocky soils into the groundwater. The maximum concentrations of some metals in leachate were higher than the maximum contaminations level for drinking water. However, high concentrations of most metals in leachate are probably not due to the application of coal fly ash since soils in south Florida

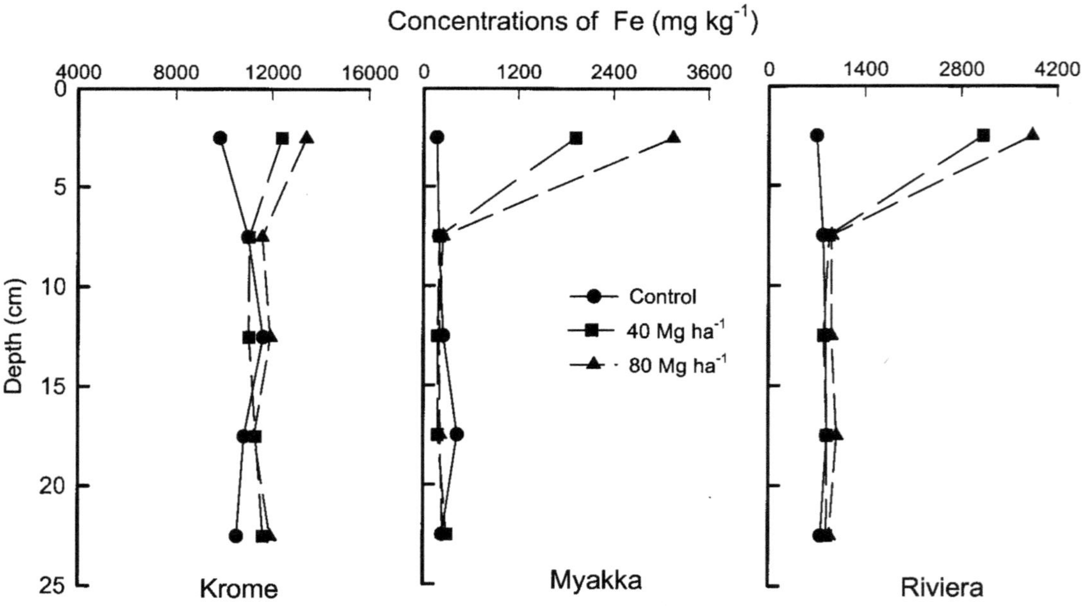

FIGURE 7. Distribution of soil Fe in soil columns following the addition of various amounts of coal fly ash and leaching with ten pore volumes of water.

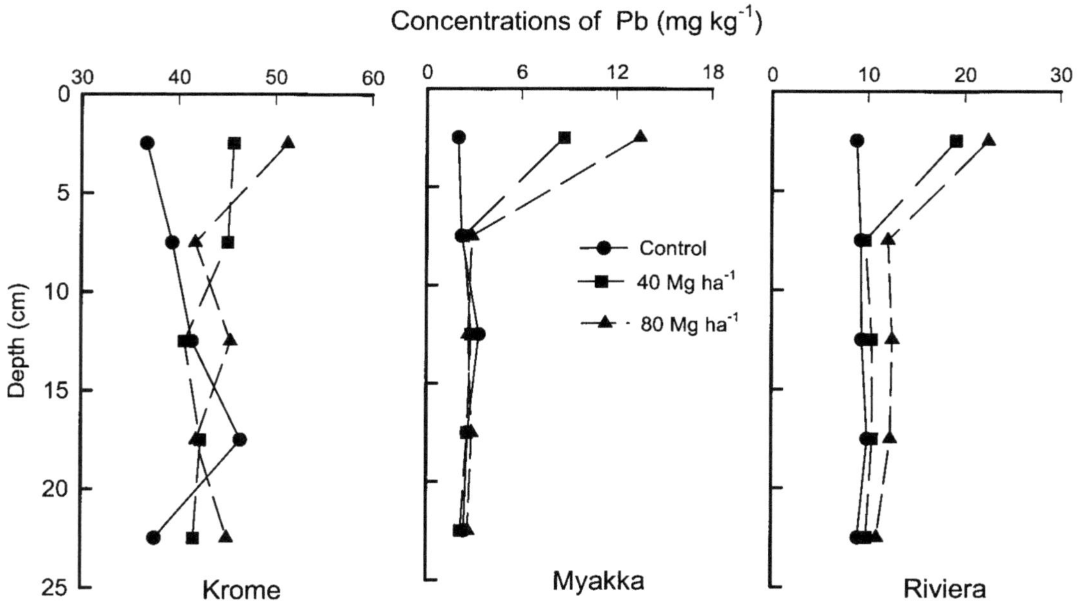

FIGURE 8. Distribution of soil Pb in soil columns following the addition of various amounts of coal fly ash and leaching with ten pore volumes of water.

have relatively high concentrations of Fe, Cu, Zn, Cd, and other trace metals. In order to be environmentally sound use of coal fly ash as a soil amendment, it suggested that the maximum limits of trace metals in coal ash should be established.

Acknowledgments. This research was supported by the Florida Agricultural Experiment Station and a grant from the U.S. Department of Energy, Federal Energy Technology Center, through it's Cooperative Agreement No. DE-FC26-98FT40028 with West Virginia University Research Corporation. However, any opinions, findings, conclusions, or recommendations expressed herein are those of the authors and do not necessarily reflect the views of WVU or DOE.

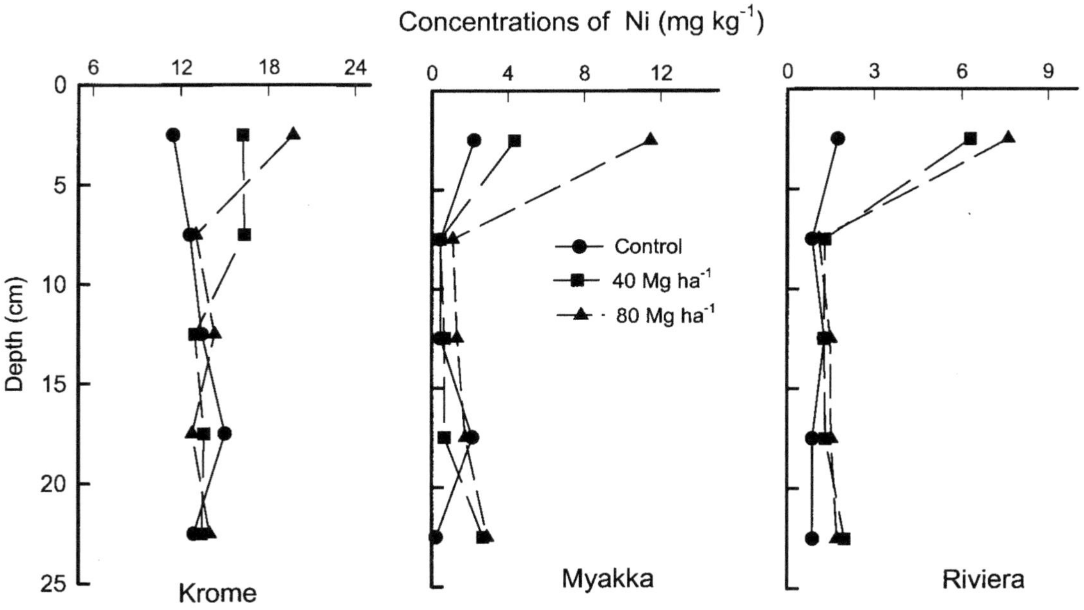

FIGURE 9. Distribution of soil Ni in soil columns following the addition of various amounts of coal fly ash and leaching with ten pore volumes of water.

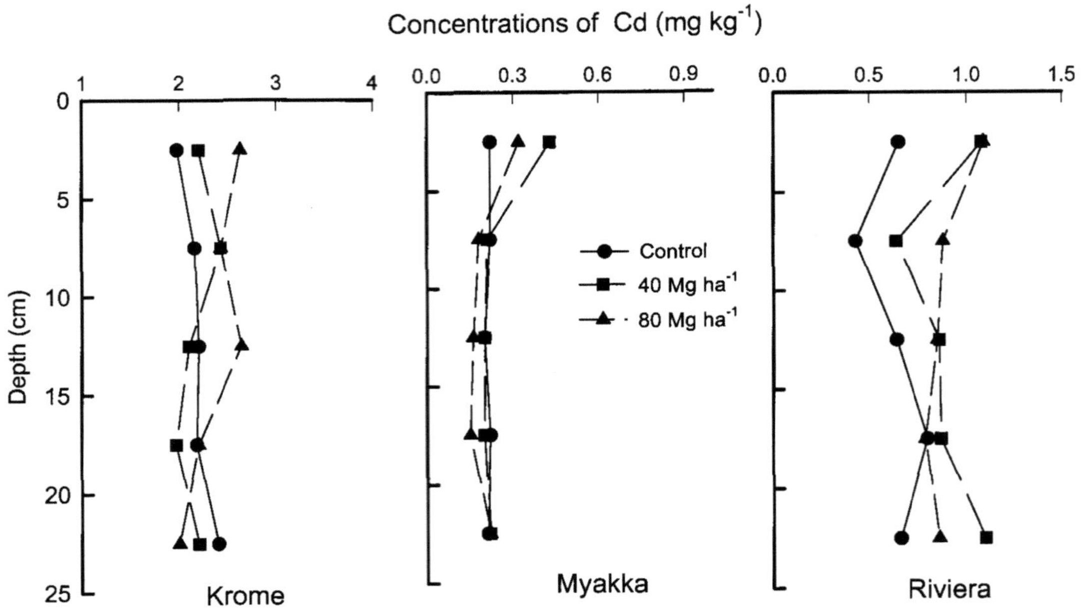

FIGURE 10. Distribution of soil Cd in soil columns following the addition of various amounts of coal fly ash and leaching with ten pore volumes of water.

References

1. Bilski, J.J., Alva, A.K., and Sajwan, K.S., Fly ash, in *Soil Amendments and Environmental Quality*, Reckcigl, J.E., Ed., CRC Press, Inc., Boca Raton, Fla., 1995, 327.
2. Korcak, R. F., Utilization of coal combustion by-products in agriculture and horticulture, in *Agricultural Utilization of Urban and industrial By-Products*, Karlen, D.L., Wright, R.J., Kemper, W.D., Eds., American Society of America (Special Publication No. 58), Madison, WI, 1995, 107.
3. Townsand, W.N. and Hodgson, D. R., Edaphological problems associated with deposits of pulverized fly ash, in *Ecology and Reclamation of Devastated Land, Vol. I.*, R.J. Hutuik and Davis, G., Eds., Gorden and Breach, London, 1973.
4. Ghodrati, M., Sims, J. T., and Vasilas, B. L., Evaluation of fly ash as a soil amendment for the Atlantic coastal plain: II Soil hydraulic properties and elemental leaching. *Water Air Soil Pollu.* 81, 349, 1995.
5. Chen, J, and Y.C. Li. 2004. Amendment of fly ash to container substrates for ornamental plant production. pp. xxx–xxx. In K.S. Sajwan, I. Twardowska, T. Punshon, and A.K. Alva (Eds.) Coal and Coal Combustion Byproducts. Kluwer Academic/Plenum Publishers, New York. (In press)
6. Adriano, D. C., Woodford, T. A., and Ciravolo, T.G., Growth and elemental composition of corn and bean seedlings as influenced by soil application of coal ash. *J. Environ. Qual.* 7, 416, 1978.
7. Schlossberg, M.J., W.P. Miller, and S. Dudka. 2003. Use of coal combustion by-products (CCBP) in horticultural and turfgrass industries. pp. 275–288. In K.S. Sajwan, A.K. Alva, and P.F. Keefer (Eds.) Chemistry of Trace Elements in Fly Ash. Kluwer Academic/Plenum Publishers, New York.
8. Sale, L. Y., Naeth, M. A., and Chanasyk, D. S., Growth response of barley on unweathered fly ash-amended soil, *J. Environ. Qual.*, 25, 684, 1996.
9. Sikka, R. and Kansal, B.D., Effect of fly-ash application on yield and nutrient composition of rice, wheat and on pH and available nutrient status of soils. *Bioresource Technology*, 51, 199, 1995.
10. Khan, M. R., Khan, M. W., The effect of fly ash on plant growth and yield of tomato. *Environ. Pollution.*, 92,105, 1996.
11. Cary, E.E., Gilbert, M., Bache, C.A., Gutenmann, W.H., and Lisk, D.J., Elemental composition of potted vegetables and millet grown on hard coal bottom ash-amended soil, *Bull. Environ. Contam. Toxicol.*, 31, 418, 1983.
12. Mulford, F.R. and Martens, D.C., Response of alfalfa to boron in fly ash. *Soil Sci. Soc. Am. Proc.* 35, 296, 1991.
13. Kukier, U. and Sumner, M. E., Boron availability to plants from coal combustion by-products. *Water Air Soil Pollu.* 87, 93, 1995.
14. Alva, A.K., Paramasivam, S., Prakash, O., Sajwan, K.S., Ornes, W.H., and Van Clief, D., Effects of fly ash and swage sludge amendments on transport of metals in different soils, in *Biogeochemistry of Trace Elements in Coal and Coal Combustion Byproducts*, Sajwan K.S., Alva, A.K., and Keefer, R.F., Eds., Kluwer Academic/Plenum Publishers, New York, 1999a, 193.
15. Alva, A. K., Bilski, J. J., Sajwan, K. S., and Van Clief, D., Leaching of metals from soils amended with fly ash and organic byproducts, in *Biogeochemistry of Trace Elements in Coal and Coal Combustion Byproducts*, Sajwan K.S., Alva, A.K., and Keefer, R.F., Eds., Kluwer Academic/Plenum Publishers, New York, 1999b, 193.
16. MPCA (Minnesota Pollution Control Agency). 1999. Applying coal ash to fields as fertilizer or liming agent. Water/Land Application/#2-02. Minnesota Pollution Control Agency, St. Paul, Mn.
17. USEPA. 2004. Test Methods for Evaluating Solid Waste, Physical/Chemical Methods (SW-846). Government Printing Office (GPO), Superintendent of Documents, Washington, DC.
18. Heebink, L.V. and D.J. Hassett. 2001. Coal fly ash trace element mobility in soil stabilization. International Ash Unitization

symposium, Center for Applied Energy Research, University of Kentucky, Lexington, KY.
19. Miller, W.P., and D.M. Miller. 1987. A micropipette method for soil mechanical analysis. Commun. Soil Sci. Plant Anal. 18:61–73.
20. Walkley, A. and I.A. Black. 1934. An examination of the Degtjareff method for determining soil organic matter and a proposed modification of the chromic acid titration method. Soil Sci. 37:29–38.
21. Paramasivan, S. K.S. Sajwan, A.K. Alva, D.C. Adriano, T. Punshon, D. van Clief, and K.H. Hostler. 2003. Comparative study of elemental transport and distribution in soils amended with fly ash and sewage sludge ash. pp. 189–202. In K.S. Sajwan, A.K. Alva, and P.F. Keefer (Eds.) Chemistry of Trace Elements in Fly Ash. Kluwer Academic/Plenum Publishers, New York.
22. Howie, B., Effects of dried wastewater-treatment sludge application on ground-water quality in south Dade County, Florida, *Water-Resources Investigations Report* 91–4135, U.S. Geological Survey, Denver, CO., 1992.
23. Li, Y.C., M. Zhang, P.J., Stoffella, Z. He, and H.H. Bryan. 2003. Influence of coal ash/organic waste on distribution of trace metals in soil, plant, and water. pp. 251–263. In K.S. Sajwan, A.K. Alva, and P.F. Keefer (Eds.) Chemistry of Trace Elements in Fly Ash. Kluwer Academic/Plenum Publishers, New York.
24. Chen, M., Ma, L. Q., and Li, Y.C., Concentrations of P, K, Al, Fe, Mn, Cu, Zn, and As in Soils from South Everglades. *Soil Crop Sci. Soc. Florida Proc.* 59, 124, 2000.
25. SFWMD (South Florida Water Management District). 1998. Ecosummary, Southeast District Assessment and Monitoring Program, West Palm Beach, Fl.
26. Merry, R.H., K.G. Tiller, and A.M. Alston. 1981. The accumulation of copper in Australian orchard soils. p. 364. *In* J.F. Loneragem, A.D. Robson and R.D. Graham (eds.) Copper in Soils and Plants. Academic Press, New York, NY.
27. Jones, G.B. and G.B. Belling. 1967. The movement of copper, molybdenum, and selenium in soils as indicated by radioactive isotopes. Aust. J. Agric. Res. 18:733–40
28. Alva, A.K. and J.H. Graham. 1991. The role of copper in citriculture. Adv. Agron. 1:145170
29. USEPA. 2002. Current drinking water standard, Office of Water, U.S. Environmental Protection Agency, Gov, Print Office, Washington, D.C., 2002.

12
Arsenic and Selenium Speciation in Aged Flue Gas Desulfurization Amended Soil

Tracy Punshon[1], Brian P. Jackson[2], John C. Seaman[2], Domy C. Adriano[2], and Joanna Burger[1]

[1] *Consortium for Risk Evaluation with Stakeholder Participation, Environmental and Occupational Health Sciences Institute, Division of Life Sciences, Rutgers University, 604 Allison Road, Piscataway, NJ 08854, USA*

[2] *The University of Georgia, Savannah River Ecology Laboratory, Aiken, SC 29802, USA*

1. Abstract

The application of flue gas desulfurization (FGD) residue to soil was thought to have potential as a viable re-use option for coal combustion waste products, as well as being a solution for nutrient deficient soil. Previous studies found that application of certain FGD materials increased crop biomass and calcium (Ca) concentration of the soil, but also added boron (B), selenium (Se) and arsenic (As). These findings are a cause for concern and could hinder research into how various by-products can be recycled. Studies of long term metalloid bioavailability arising from FGD application are lacking but are needed to assess the safety of coal by-products. We investigated the concentration and speciation of As and Se in Southeastern coastal plain soils that has been amended with a weathered, stabilized FGD material (0, 2.5, 5, 7.5 and 10%) five years previously, using ion chromatography coupled to an inductively coupled plasma mass spectrometer (IC-ICP-MS). We also studied the speciation of As and Se in naturally invading plant tissue (*Vicia cracca* L., Leguminosae), as a simulation of a site of previous FGD application that had been allowed to vegetate naturally. Total As and Se concentrations in mesocosm soils were elevated in proportion to the FGD application rate, but water soluble As and Se concentrations were low five years after application (0.04 and 0.08 mg kg^{-1} respectively at 10%). In all treatments arsenate [As (V)] and selenite [Se (IV)] were the predominant As and Se species. Ligand-exchangeable metalloids were significantly elevated to a depth of 0–0.15 m only; again As (V) and Se (IV) predominated, generally considered potentially available for plant uptake. While a large proportion of the Se had leached from the soil profile or been taken up by plants, the majority of the As from applied FGD remained in the soil profile.

2. Introduction

2.1. Coal Combustion By-Product Recycling

The beneficial re-use or recycling of coal combustion waste products (hereafter, CCPs) is by far the most desirable end-point for the vast quantities of CCPs produced worldwide. The cost of landfilling waste material is increasing alongside the demand for space, and the environmental impacts of stockpiling and aquatic disposal are considerable[1]. A recycling solution that fulfils demand, is profitable, and yet poses little or no threat to the environment remains elusive, even though the need for such a solution increases annually; both for developed countries whose demand for energy continues to grow, and for developing countries, whose reliance on coal for energy production is rapidly increasing. Coal is our most significant source of energy, as well as our most abundant source of waste, but enhancing the sustainability of coal combustion via various recycling options currently poses significant environmental issues.

The application of CCP materials to soil was thought to hold a great deal of potential, because it involved both recycling of the CCP and could restore nutrients to leached or eroded soils. Certain CCPs such as fly ash (FA) have been used at high application rates to increase the water-holding capacity of coarse textured soils[2–4], and can be applied to both sandy and clayey soils. When the Clean Air Act was revised in 1995, flue-gas scrubbing technologies for the removal of sulfur dioxide produced a new waste product known as flue-gas desulfurization residue (FGD). This material contained variable concentrations of $CaSO_4$ (gypsum) and was thought to have potential for use in place of agricultural gypsum. This prompted a variety of studies into the conditions, application rates and effects of FGD application on soils, groundwater and crop plants. CCP materials, although widely variable, tend to contain potentially toxic metals and metalloids, such as boron (B)[5–8], molybdenum (Mo), arsenic (As)[6,9] and selenium (Se)[10]. FGD residue does not usually contain appreciable concentration of metals or metalloids, but when FA impurities picked up during the scrubbing process are not separated from the FGD material the result is a final material that can contain up to 20% FA[11].

Fly ash typically contains metals and metalloids from the parent coal, often at very high concentrations. It is made up from small, glassy hollow spheres, onto which metals and metalloids condense during the combustion process[12]. When these materials are applied to the soil, the elements leach out, and can reside in the soil, enter groundwater and/or be taken up by plants. From this point, it is assumed—lthough not rigorously tested in long term mesocosm or field experiments—that the potentially toxic elements released from CCPs pose a risk to the wider environment, and eventually to humans.

In addition, FGD and FA mixtures behave differently in the environment than either material applied alone. The readily available Ca found in FGD material can enhance the mobility of metal cations and plant-essential elements by competing with them for electrostatic charge sites[13]. Similarly, sulfate can compete with oxyanions (e.g. PO_4^{3-} and MoO_4^{2-} etc.) for adsorption sites which influences their mobility in the soil. In addition, studies have shown that FGD application enhanced the leaching of potassium (K) and magnesium (Mg) through the soil profile, and raised concerns about plant deficiency[11].

Research into the use of CCPs as soil amendments has seriously contracted in recent years, perhaps as a result of discouraging findings of metal and metalloid enrichment. New reports on the environmental safety of CCP materials, which is central to their re-use, have also almost completely ceased. Despite this, there is a pressing need for recycling solutions in developing countries such as India, which is in a phase of exponential CCP production. Without an effective recycling solution, developing countries are set to suffer from the same environmental problems as the US and Europe; stockpiling CCPs in areas where toxic metals and metalloids leach into the soil and groundwater, chronic toxicity effects on aquatic flora and fauna[1], the occupation of vast quantities of landfill space, and inflated disposal costs.

Determining the long term safety of CCP amendments to soil should be a priority, and should employ appropriate analytical tools and advances made in the study of metal and metalloid bioavailability. There is as yet no reliable knowledge base from which we can assess the long term safety of CCP application to land, because the study of metal and metal bioavailability is still a relatively new science, and one which has only made significant advances in aquatic ecosystems. As such the study of CCPs and their effects may provide the ideal opportunity to advance our understanding of metal and metalloid bioavailability.

Our focus should be on sites of previous CCP application to understand whether these perceived health risks can actually be substantiated, and if so, whether metalloid enrichment in particular is an issue which can be prevented during CCP production or handling. Our aim should ultimately be to use this information to develop CCP production processes that are complimentary to CCP recycling, as well as developing new recycling avenues.

2.2. Bioavailability of As and Se

The study of bioavailability in recent years concludes that the total concentration of a contaminant in the environment is not sufficient information to assess its effect on biota.[14]; in particular whether it will move or magnify through successive trophic levels and cause unanticipated damage to wildlife or humans. It is thought that the operational form (e.g. soluble, exchangeable, tightly bound) and the chemical form (i.e. the speciation) of a metal or metalloid is of primary importance in determining whether it will move across biological membranes to exert an deleterious effect. The dominant forms of As in natural waters are As (III), As (V), dimethylarsenate (DMA) and monomethylarsenate (MMA)[15]. The toxicity of arsenical compounds in the environmental are (from highest to lowest): arsines (e.g. AsH_3 and derived compounds by substituting hydrogen atoms with hydrocarbyl groups) > inorganic arsenites > organic trivalent compounds (arsenoxides) > inorganic arsenates > organic pentavalent compounds > arsonium compounds > elemental arsenic[16]. Arsenate [As (V)] is the most prevalent form of the element in the environment, and although arsenite [As (III)] is thought to be the most toxic to biota[17], its increased toxicity may be due solely to the fact that it is less strongly sorbed to soil minerals, and is therefore generally more available for uptake by biota.

The dominant forms of Se in the soil are the selenates (VI) (e.g. SeO_4^{2-}), selenites (IV) (e.g. SeO_3^{2-}) and selenides (-II) (e.g. Se^{2-}). In highly oxidizing conditions, SeO_4^{2-} predominates. Selenium can be absorbed by plants in both the inorganic form (i.e. selenate and selenite) and in organic forms, and can be converted from inorganic to organic forms by green plants[18]. Plants without specialized uptake systems tend to contain a protein-bound Se, such as selenomethionine, as well as inorganic Se. Forage plants contain compounds such as selenocysteine, selenomethionine, and methylselenomethionine. Selenomethionine is the predominant form of Se ingested by animals and becomes associated with proteinaceous body tissues, whereas inorganic Se can bind to various other tissues within the body. Additionally, selenomethionine is less readily absorbed by the body, but is retained for longer periods of time. Selenium in particular is not generally considered to be toxic to plants, although As toxicity has been reported[19]. Furthermore, although Se has been shown to be essential in animal nutrition, it is not essential to plants[20].

Studies of metal and metalloid speciation within soil frequently use chemical extraction schemes (macroscopic speciation) such as that put forward by Tessier et al.,[21]. Extractants of increasing strength, intended to represent a specific soil fraction are used on samples of air dried soil. In this study the fractions are readily available, ligand exchangeable, sorbed to amorphous Fe oxyhydroxides and sorbed to crystalline Fe oxides. It is assumed that the fractions to which plants and the majority of subterranean biota will have access to are the first two fractions, with the latter describing the more strongly bound fractions[22].

A number of techniques have been used for the determination of As and Se species in environmental samples, typically involving hydride generation atomic absorption spectrometry (HG-AAS), or inductively coupled plasma optical emission spectroscopy (ICP-OES). However these techniques have had limited success in separating As (III) and As(V) compounds. Ion chromatography (IC) coupled with inductively coupled plasma mass spectrometry (ICP-MS) is a particularly successful technique[23], and has been applied previously to FA extracts, and found that the main metalloid species were As (V) and Se (IV) with As (III) detected in low pH FA.

2.3. Objectives

In this study our focus was to build upon a body of work initiated in March 1998, where an extensive, replicated mesocosm experiment was established to investigate the impact—in terms of metal and metalloid accumulation and toxicity—of the application of a locally-obtained FGD material to an agricultural cropping mesocosm. Results of the short term experiments[12,24], include characterization of the FGD material used, which was stabilized with FA and allowed to weather before use. These early studies showed that adding stabilized FGD residue to local soils at treatment levels of 2.5–10% resulted in a short term biomass increase, a longer term increase in leachate salinity, and the persistence of As and Se in the soil. In particular, the increase in B, which is a frequent finding of CCP amendment studies, was less severe than in many other studies[25], probably because the B concentration in the original material was not as high as in undiluted FA, and the material underwent a period of weathering before it was applied. Further, over time, the salinity of the mesocosm leachate began to return to normal[26]. Only As and Se remained, constituting an enduring environmental issue. Herein we examine the concentration and speciation of As and Se five year old FGD-amended mesocosms, and invading vegetation, using IC-ICP-MS.

3. Materials and Methods

3.1. Mesocosm Establishment

A mesocosm study, consisting of galvanized steel cattle tanks (4.67 m^2) filled with soil and an upper layer (0.15 m depth) of FGD-amended soil was established in March, 1998[24]. The FGD residue was obtained from a local coal-fired power plant (Cope, SC), and had been stabilized with FA and allowed to weather. Details of the As and Se concentrations in the material

TABLE 1. Standard operating conditions for ICP-MS and IC instruments

ICP-MS	
Rf forward power/W	1200
Plasma gas flowrate/l min^{-1}	20
Nebulizer gas flow rate/l min^{-1}	0.75–0.85
Lens voltage/V	7.9
Monitoring masses (m/z)	75,82
Dwell time per u/ms	15
Sweeps per reading	10
Estimated reading time/s	0.36
Readings per replicate	1168
Replicates	1
Total sampling time/min	7.00
Read delay/s	30
IC	
Anion-exchange column	
SOIL	Dionex AS16 [As(III), As(V), DMA, MMA, Se(IV), Se(VI)]
PLANTS	Dionex AS7 [Se(IV), Se(VI), Selenomethionine]
Mobile phase	50mM Tetramethylammonium hydroxide, 50 mM HNO_3
Flow rate	2 ml min^{-1}
Injection volume	100μl

used are given in Table 1, and a complete elemental composition analysis is given in Punshon et al.,[12]. The coal from which it was derived was an eastern anthracite coal, with a 2% sulfur content. The FGD material had a pH of 9.3 and an EC of 3.56 dS m^{-1}. Details of the set-up and monitoring of the FGD-amended soil mesocosm are provided in full in Punshon et al.,[24,26]. Flue-gas desulfurization material was air-dried and mixed with locally obtained Orangeburg soil (fine-loamy, siliceous, thermic, Typic Paleudult) in a large volume soil homogenizer to obtain treatment mixes which were equivalent to 0, 2.5, 5, 7.5 and 10% FGD (0, 55.5, 111.1, 166.6 and 222.2 Mg FGD ha^{-1}).

3.2. Determination of As and Se Concentration in Soil and Plant Tissues

The concentration of HNO_3-extractable As and Se in the soil and above ground biomass of volunteer plants (seed pods and leaves) was determined using EPA protocol 6020 + 3051, using microwave digestion techniques and inductively coupled plasma mass spectrometry (ICP-MS) under standard operating conditions. Plant species that had invaded the mesocosms naturally over a period of several years were sampled, removing the above ground biomass only. The species common to all of the mesocosms was tufted vetch (*Vicia cracca* L., Leguminosae), although the species composition varied between the control and 10% FGD treatment considerably. Plant material was dried in an air-assisted oven at 60°C until no further weight loss, and ground in a sample mill to pass through a 1 mm stainless steel mesh screen before digestion.

3.3. Sequential Extraction of FGD-Amended Soil

Samples of soil were collected from the upper 0.3 m of the mesocosm soil profile, and divided into 0–0.15 and 0.15–0.3 meter sections. This is equivalent to the FGD amended layer (upper 15 cm) and the unamended Orangeburg soil underlying this. Sequential extraction of mesocosm soil was performed on 1 g of air-dried, homogenized soil using 10 ml of the following extractants: 18 MΩ H_2O (readily available), 0.1 *M* PO_4(pH 3.2) (ligand exchangeable), 0.25*M* $NH_2OH \cdot HCl$, 0.25 *M* HCl, 0.1 *M* PO_4 (sorbed to amorphous oxyhydroxides) and 0.1 *M* oxalic acid, 0.175*M* Na oxalate, 0.1*M* PO_4 (sorbed to crystalline oxides)[27]. Each fraction was collected, filtered (0.22 μm polycarbonate syringe filter) and analyzed using ICP-MS. Duplicate samples of the first two fractions were analyzed for As and Se speciation using Ion Chromatography ICP-MS.

3.4. Determination of As and Se Species by Ion Chromatography ICP-MS

Arsenic and Se speciation in plant tissues was determined using 1g of dried, ground plant material, and 2 ml of a 1:1 water: methanol extractant, and utilized methodology successfully tested in previous studies[28]. A Dionex (Sunnydale, CA) GP50 gradient pump was connected to a Thermo Separations autosampler, so that timed events on the autosampler initiated both the method clock on the pump and data collection by the ICP-MS software upon sample injection. This also allowed the clock to be re-set, so that the method could be re-started under identical conditions for the next sample. Arsenic separations in the soil extracts employed an AS16 column, which has ultra-low hydrophobicity, 55% cross-linking and an alkanol quaternary ammonium functional group. For the plant extracts an AS7 column was used; this column has a medium-high hydrophobicity, 2% cross-linking and an alkyl quaternary ammonium functional group. The ion exchange columns were both supplied by Dionex (Sunnydale, CA), and were 4 × 250 mm and used with appropriate guard columns. Eluents used in the separations were 50 m*M* tetramethyl ammonium hydroxide (AS16) and 2.5 and 50 m*M* HNO_3 (AS7), prepared from reagent grade salts of solutions and 18 MΩ-cm deionized water. The end of the analytical column was connected to the nebulizer of the ICP-MS (Perkin-Elmer, Elan 6000) by 93 mm polyetheretherketone tubing (0.25 mm i.d.). Before analysis, ICP-MS conditions (i.e. static lens voltage and nebulizer gas flow) were optimized for maximum signal intensity at mass 75 by aspirating a 50 μg L^{-1} As standard under normal sample introduction conditions. A pneumatic nebulizer and cyclonic spray chamber were used. For IC-ICP-MS, the injection volume was 100μL for all standards and samples. The calibration curves were generated for each separation by running a three-point calibration with mixed standards solutions in the range of 1–100 μg L^{-1} (as As) for each As species. All calibrations were linear with $R^2 > 0.995$. The standards used were for As: As(III), As(V), DMA, MMA and for Se: Se(IV), Se(VI), selenomethionine. The standard operating conditions for the IC-ICP-MS analysis are given in Table 1.

4. Results

4.1. Total Metalloid Concentration in FGD and Amended Soils After 5 Years

The unamended Orangeburg soil contains 0.9 mg kg^{-1}As, and 0.45 mg kg^{-1}Se. The concentration of As and Se within FGD residue is given in Table 2, indicating that every 1% of FGD applied contributed approximately 1 mg kg^{-1} As and 0.4 mg kg^{-1} Se to the surface soil. The total initial (March 1998) and final (March 2003) As and Se concentrations of mesocosm soils (0–0.3 meters depth) are shown in Figure 1. A comparison of the two metalloids indicates there has been a significant drop in soil Se concentration, but no loss of As over the five years since the amendment was applied. The percentage decrease in soil Se concentration over five years is annotated on Figure 1, showing that the greater the rate of FGD applied, the greater the Se concentration, and consequently the longer the time period needed for Se to leach out of the soil profile.

The total metalloid concentration of FGD-amended layer and the underlying Orangeburg soil were also examined separately. Arsenic in the FGD-amended layer was within the range 0–15 mg kg^{-1} (Figure 2), with a proportional relationship between FGD application rate and As concentration in the amended layer, and significant ($F_4 = 14.81$; $P > 0.001$) differences between As concentration over the FGD treatment range. In the underlying Orangeburg soil, As was within the range 0–3 mg kg^{-1} (Figure 2); lower than that of the amended layer. The As concentrations did not differ significantly over the FGD treatment range ($F_4 = 2.16$; $P = 0.113$), although the 10% treatment level contained the highest As concentration of 2.5 mg kg^{-1} (Figure 2).

Selenium concentration in the FGD-amended layer was between 0–2.5 mg kg^{-1}, and there were significant differences ($F_4 = 38.34$; $P = 0.001$) between concentrations across the FGD treatment range (Figure 3). In common with As, the highest concentration of Se resulted from the application of 10% FGD, and this treatment also gave the most variable response. The unamended soil beneath the incorporation layer contained between 0–1 mg kg^{-1} Se five years after FGD application; also differing significantly across the treatment range ($F_4 = 5.16$; $P = 0.005$) (Figure 3).

TABLE 2. Metalloid concentration (μg g^{-1}) within FGD and FGD amended soils. DI = deionized. Values are arithmetic means

	Total	Extractable	
(μg g^{-1})	HF + Aqua Regia	Double-Acid[a]	DI Water
Arsenic	68.9	0.032	0.030
Selenium	29.2	0.306	0.269

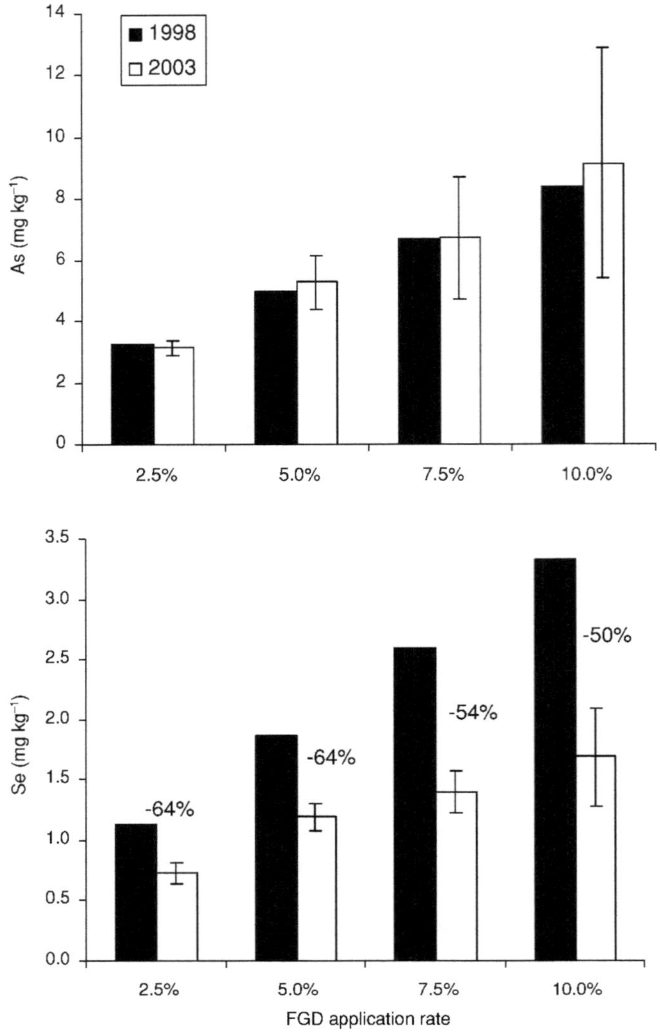

FIGURE 1. Total concentration of As and Se (expressed as mg kg^{-1}) within FGD-amended soil (0–0.3 m sampling depth) at initial set up (1998) and five years later (2003). Data are means (±SD) where $n = 4$.

In both cases, metalloid concentrations were higher in the incorporation layer than in the layer immediately beneath. Significant differences in Se concentration of the soil underlying the amended layer, however, indicates that Se has moved from the FGD material into the soil beneath, but this was not the case for As.

4.2. Sequential Extraction of FGD Amended Soil After 5 Years

The concentration of As and Se in the readily available, ligand exchangeable and amorphous Fe oxide bound soil fractions are shown in Figure 4. There were no significant differences in the concentration of As or Se in the unamended Orangeburg soil beneath the FGD-amended layer for these fractions. As stated earlier, it is the former two fractions in this series that are

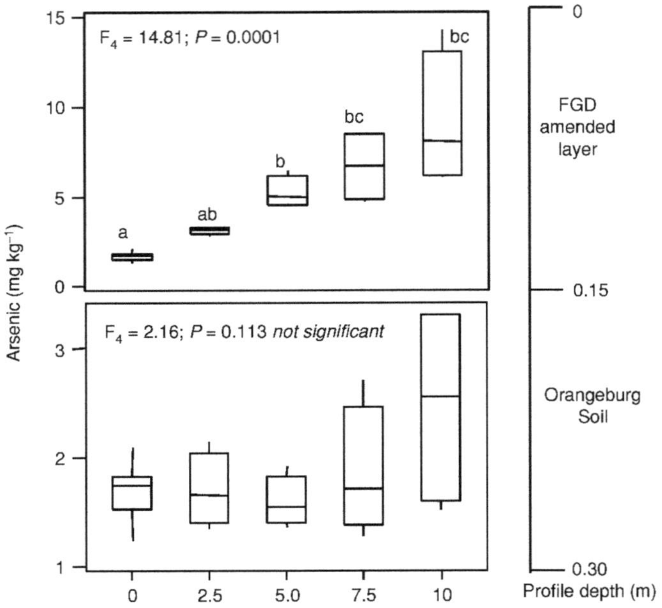

FIGURE 2. Total As concentration (expressed as mg kg^{-1}) in the FGD-amended layer of mesocosm soil (0–0.15m) and the unamended Orangeburg soil underneath (0.15–0.3 m), from a range of FGD treatments (0–10%). Boxes represent the 25th and 75th percentile, bisected by the median, and whiskers indicate the data range. Annotated letters indicate the results of a Tukey's means separation analysis (where significance is established when $P < 0.05$), and the results of an ANOVA test are shown in the upper left corner.

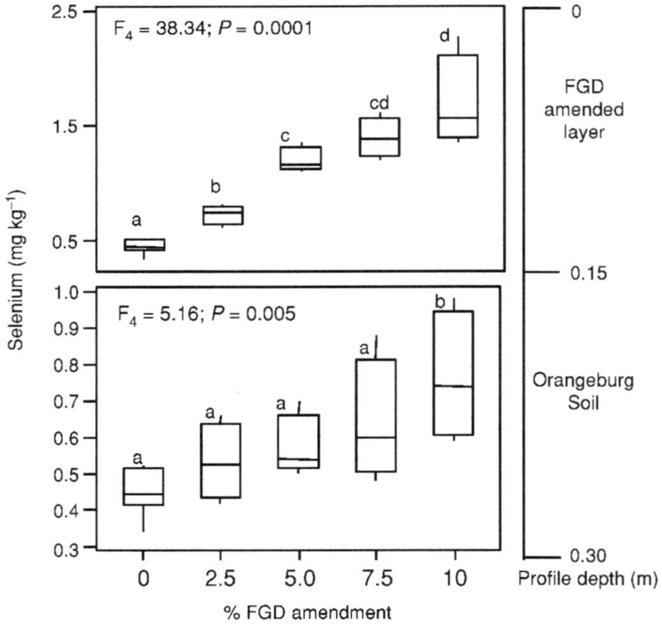

FIGURE 3. Total Se concentration (expressed as mg kg^{-1}) in the FGD-amended layer of mesocosm soil (0–0.15m) and the unamended Orangeburg soil underneath (0.15–0.3 m), from a range of FGD treatments (0–10%).

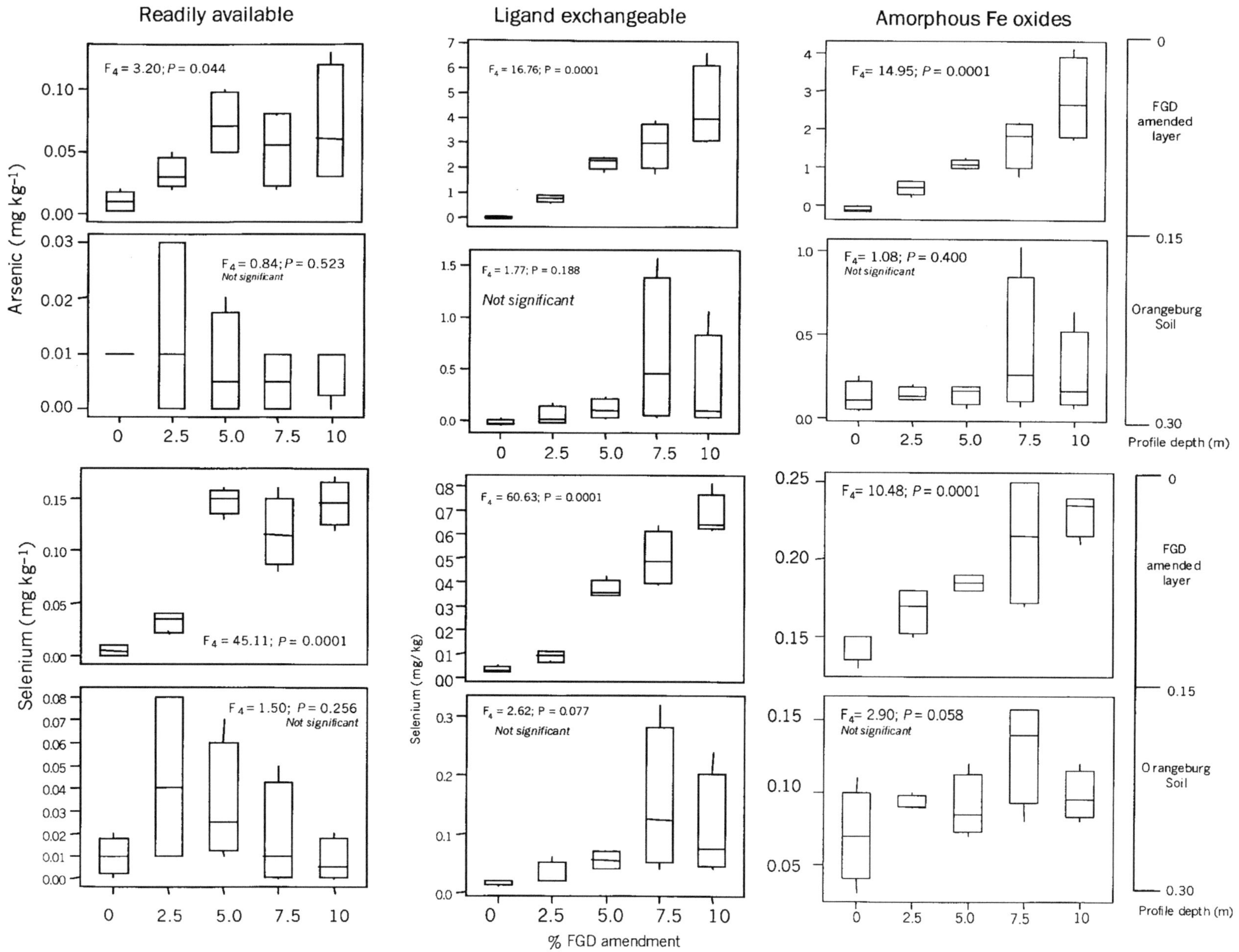

FIGURE 4. As and Se concentration (expressed as mg kg^{-1}) within readily available, ligand exchangeable and amorphous Fe oxide bound fractions of the FGD-amended soil layer (0–0.15 m) and the unamended Orangeburg soil underneath (0.15–0.3 m). Results of an ANOVA statistical test (where significance is established when $P < 0.05$) are shown in the upper left corner.

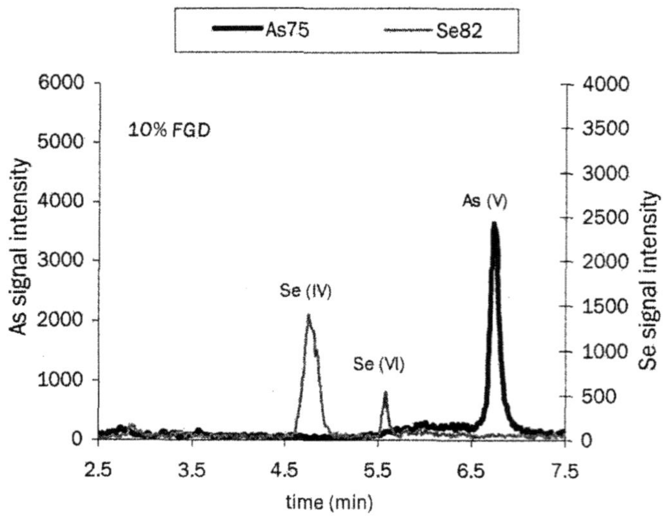

FIGURE 5. Chromatograms of As and Se speciation in the water soluble (readily available) extracts of 10% FGD-amended soils.

considered to be most available to biota. The highest concentration of both As and Se were in the ligand exchangeable fraction of the soil with 0–7 mg kg^{-1} As and 0–0.8 mg kg^{-1} Se. In both the ligand exchangeable and amorphous Fe oxide fractions of the soil, As concentrations exceeded that of Se by an order of magnitude.

4.3. Metalloid Speciation in Soils

Analysis of FGD treated soil by IC-ICP-MS included standards for four As species [dimethylarsenic, As(III), monomethylarsenic and As (V)], and two for Se species [Se (IV) and Se (VI)], with detection limits <0.5 μg kg^{-1} for all detected species. The readily available fraction of the soil contained only one form of As; As (V), and two Se species; Se (IV) and Se (VI) (Figure 5). Only Se (IV) and As (V) were detected in ligand exchangeable extracts of FGD amended soil (Figure 6), and a comparison between 10% and 5% FGD indicates that the signal intensity of both metalloids is influenced by FGD amendment level.

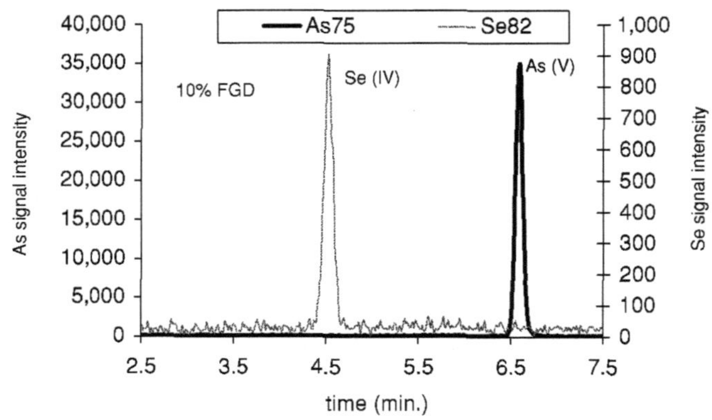

FIGURE 6. Chromatograms of As and Se speciation in the ligand exchangeable extracts of 10% FGD-amended soils.

4.4. Metalloid Concentration in Plant Tissues

The concentrations of As and Se within the above ground biomass of tufted vetch (*Vicia cracca*) are shown in Figure 7. Concentrations of Se within above ground tissues is an order of magnitude higher than As; in the range 0–3.5 mg kg^{-1} as opposed to 0–0.3 mg kg^{-1}. Both As and Se were predominantly found within the leaf tissue of tufted vetch, and the concentrations of As within the seed pods were very similar to concentrations detected in plants collected from the untreated mesocosms. Selenium concentrations in *V. cracca* seed pods were in the range of 0–1 mg kg^{-1}.

When these concentrations are compared to those reported in Punshon et al.[24] it is evident that the concentration of As detected within plant tissues is almost an order of magnitude less than five years previously. Although there is bound to

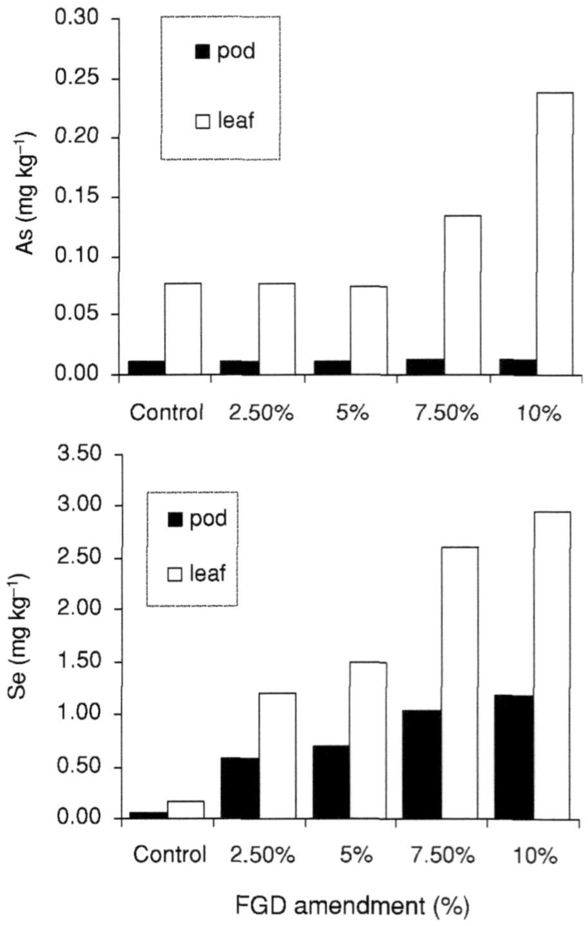

FIGURE 7. Concentration of As and Se (expressed as mg kg^{-1}) within the leaf and seed pods of the legume *Vicia cracca* (tufted vetch) collected from soils amended with 0–10% FGD material.

be considerable species variation in As uptake, the lowest concentrations in 1998 were detected in cotton (*Gossypius hirsutus* L.) plants, and were in the range of 0–1 mg kg^{-1}. When corn plants (*Zea mays* L.) were grown in a second season, they contained between 41–51% less As in their tissues than the first year of growth in amended soil. Selenium concentrations have remained higher than that of As throughout the five year study period. Leaf Se concentrations were in the range of 0–10 mg kg^{-1} in corn plants in the first season, and fell by between 50–61% the following year. In contrast to As, Se was also found within the seed pods at between 0.1–1.2 mg kg^{-1}.

4.5. Metalloid Speciation in Plant Tissues

The seed pods and leaf tissues of *V. cracca* growing in mesocosms amended with 10% FGD material were analyzed via IC-ICP-MS. The analysis clearly identified Se (VI) in both plant organs, with almost an order of magnitude more Se (VI) within leaf tissues (Figure 8). For As, however, peaks arising from analysis of seed pods did not coincide with the standard peaks of As (III), As (V), DMA or MMA, and no clear identification could be made. The As detected within leaf tissue differed from that of the seed pod, and consisted of numerous, low intensity As peaks which were thought to be MMA, As (III) and As (V), although further work is necessary to clarify this. In addition, the peak intensity for the unknown As species detected within seed pods was considerably higher than those detected within leaf tissue.

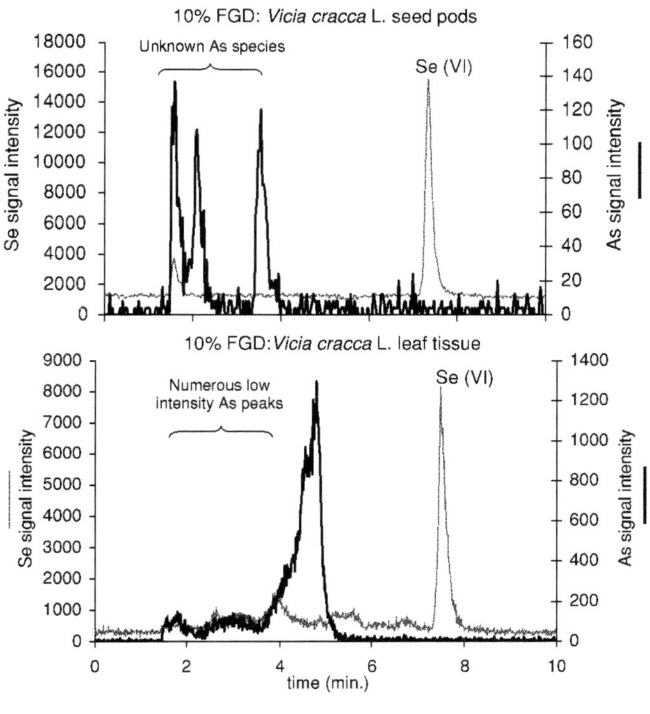

FIGURE 8. Chromatograms of As and Se speciation in the seed pods and leaves of *Vicia cracca* (tufted vetch) collected from soils amended with 10% FGD material.

5. Discussion

Metalloid enrichment of the soil and plants (tufted vetch) resulting from application of a stabilized FGD material is evident five years after initial application. The total concentration of As and Se in amended soil was low, however, and was in the same range as uncontaminated agricultural soils. The mobility and bioavailability of As and Se differed markedly, however. Concentrations of As in the FGD-amended layer remained consistent over a period of five years, and As was generally absent from the layer underlying the amended soil. There was no significant downward movement of As through the soil profile, consistent with relatively strong binding of As in the FGD amended layer. Further, the very low concentrations of As detected within naturally invading plants collected from the mesocosms (0.01–0.25 mg kg^{-1}) confirmed that both mobility and bioavailability are very low.

Selenium mobility and bioavailability exceeded that of As; with 50–64% lower total Se within the soil five years after initial application. Plant tissue and leachate analysis suggests that Se removal may be largely due to plant uptake, although soil analysis also found significantly higher total Se concentrations in the unamended Orangeburg soil underlying the FGD-amended layer, which suggests that Se may also be leaching through the soil. Although Se was detected within mesocosm leachate at the outset of the study[24], the concentration was low (0.02 mg kg^{-1}) and rapidly diminished; falling 50% in the four months following application, whereas there are considerably greater concentrations found within plant tissue—pproximately 3 mg kg^{-1}. It is well known that the 'safe range' of Se within plant tissues, in terms of its safety to grazing animals, is very narrow; a minimum Se concentration (needed to prevent animal deficiencies such as white muscle disease) is 0.1 mg kg^{-1}, whereas concentrations of 3–4 mg kg^{-1} are considered toxic[16]. The concentration of Se within tufted vetch growing on 10% FGD may be considered too close to this upper limit to allow animal grazing.

The speciation of As and Se within the soil using the sequential extraction scheme found that the majority of As was located in the ligand exchangeable fraction, which is considered moderately available. Although our analysis has shown that invading plants have not removed a high concentration As from the system, its presence in this fraction may be a cause for concern if amended soil were to be subsequently treated with phosphatic fertilizers, or poultry litter, both of which are enriched with phosphate and can potentially mobilize As. Poultry litter in particular is often co-applied with CCPs as a source of organic matter and plant nutrients[29], and has proven a successful mix for land restoration in the past[30], however this study indicates it is particularly important that the As content and mobility of a CCP is known before proceeding with co-application with a P-enriched material. Speciation analysis using IC-ICP-MS found As present as As (V) within the soil. This species is less toxic As (III), and more strongly sorbed to the soil mineral phase. Speciation data obtained for As in plants suggested DMA and MMA, but were inconclusive. Selenium was present within the soil predominantly as Se (IV) with lesser amounts of Se (VI), but existed in plant tissue only as Se (VI). The predominance of Se (VI) within plant tissues may arise from the biochemical oxidation of Se (VI) in the S systems in plant metabolism, due to the similarity of Se (IV) with the SO_4^{2-} ion.

Clearly the metalloids detected in this study originated from the FA with which the FGD residue was deliberately mixed or contaminated during the handling process: the species of both As and Se are consistent with those detected in a survey of 23 different FA materials conducted by Jackson and Miller[23]. The concentrations of As and Se detected here are relatively low, however, and many agricultural soils have been found to contain similar if not greater concentrations of As and Se than those reported for soil amended with 10% FGD. Surveys have shown that uncontaminated orchard soils contain As up to approximately 80 mg kg^{-1}, and Se concentrations between 0.0–2.0 mg kg^{-1} are considered normal[16]. The total soil concentrations reported here for the highest rate of FGD amendment fall within these normal ranges.

It is particularly important to point out that the environmental risks posed by As and Se derived from the application of stabilized FGD material could not have been inferred from their total concentrations in amended soil, or the layers beneath it. The total Se concentrations were an order of magnitude lower than that of As, and yet the concentration in plant tissues was an order of magnitude higher. This is a direct example of the pertinence of bioavailability to the application of CCPs to land.

The persistence of As in the soil profile in a moderately bioavailable form is an important limiting factor to the repeated application of CCPs, and restricts the subsequent use of phosphatic fertilizers and poultry litter, the latter of which can also contain appreciable concentrations of As from poultry feedstuffs. It is suggested here that the risks posed to the environment from the application of stabilized FGD are minimal, and are currently less than that associated with the routine application of arsenical pesticides. This risk could be effectively eliminated, however, if the FGD and FA waste streams were kept separate.

Acknowledgements. This research was supported by Financial Assistance Award number DEFG2600NT-40938 from the US Department of Energy (DOE) to the Consortium for Risk Evaluation and Stakeholder Participation (CRESP). This research was partially supported by the Environmental Remediation Science Division of the Office of Biological and Environmental Research, US DOE through the financial assistance award n. DE-FC09-96SR18546 to the University of Georgia Research Foundation and NIEHS (ESO 5022).

Literature Cited

1. Rowe, C.L., Hopkins, W. and Congdon, J.D. Ecotoxicological implications of aquatic disposal of coal combustion residues in the United States: A review *Environmental Monitoring and Assessment* 80, 207–276. 2002.
2. Chang, A.C., Lund, L.J., Page, A.L. and Warneke, J.E. Physical properties of fly-ash amended soils. *Journal of Environmental Quality* 6, 267–270. 1977.
3. Ghodrati, M., Sims, J.T. and Vasilas, B.L. Evaluation of fly ash as a soil amendment for the Atlantic Coastal Plain: I. Soil hydraulic properties and elemental leaching. *Water Air and Soil Pollution* 81, 249–261. 1995.
4. Jacobs, L.W., Erickson, A.E., Berti, W.R. and MacKellar, B.M. Improving crop yield potentials of coarse textured soils with coal fly ash amendments *Proceedings of the Ninth International Ash Use Symposium* 3, 59–1 to 59–16. 1991.
5. Aitken, R.L. and Bell, L.C. Plant uptake and phytotoxicity of boron in Australian fly ashes *Plant and Soil* 84, 245–257. 1985.
6. Furr, A.K., Parkinson, T.F., Gutenmann, W.H., Pakkala, I.S. and Lisk, D.I. Elemental content of Vegetables, Grains and Forages Field-Grown on Fly-Ash Amended Soil. *J. Agric. Food Chem.* 26(2), 357–359. 1978.
7. Kukier, U. and Sumner, M.E. Boron availability to plants from coal combustion by-products. *Water, Air & Soil Pollution* 87(1–4), 93–110. 1996.
8. Clark, R.B., Zeto, S.K., Ritchey, K.D. and Baligar, V.C. Boron accumulation by maize grown in acidic soil amended with coal combustion by-products. *USDA Publication.* 1995.
9. Labus, K. Heavy-metal emissions from coal combustion in southwestern Poland *Energy* 20(11), 1115–1119. 1995.
10. Arthur, M.A., Rubin, G. and Woodbury, P.B. Uptake and accumulation of selenium by terrestrial plants growing on a coal fly ash landfill. 2. Forage and root caps. *Environ. Toxicol. Chem.* 11(9), 1289–1299. 1992.
11. Ishak, C.F., Seaman, J.C., Miller, W.P. and Sumner, M. Contaminant mobility in soils amended with fly ash and flue-gas gypsum: Intact soil cores and repacked columns *Water Air and Soil Pollution* 134(1–4), 287–305. 2002.
12. Punshon, T., Knox, A.S., Adriano, D.C., Seaman, J.C. and Weber, T.J. Flue Gas Desulfurization residue (FGD): Potential Applications and Environmental Issues., in *Biochemistry of Trace Elements in Coal and Coal Combustion Byproducts.*, K.S. Sajwan and R.F. Keefer, Editors. 1999, Lewis Publishers: Boca Raton, FL. 7–28.
13. Ishak, C.F., Seaman, J.C., Sumner, M.E. and Miller, W.P. Contaminant mobility in soil columns amended with fly ash and flue gas desulfurization Gypsum., in *Biogeochemistry of Trace Elements in Coal and Coal Combustion Byproducts.*, K. Sajwan, Editor. 1999.
14. NRC *Bioavailability of contaminants in soils and sediments: procedures, tools and applications.* 2003, Washington, DC: Water Science Technology Board, National Research Council. 432.

15. Yalcin, S. and Chris, L. Speciation of arsenic using solid phase extraction cartridges *Journal of Environmental Monitoring* 3, 81–85. 2000.
16. Adriano, D.C. *Trace Elements in Terrestrial Environments*. 2nd ed. 2001, New York: Springer Verlag. 867.
17. McLaughlin, M.J., Parker, D.R. and Clarke, J.M. Metals and micronutrients—food safety issues *Field Crops Research* 60(1–2), 143–163. 1999.
18. Ohlendorf, H.M. Bioaccumulation and Effects of Selenium in Wildlife., in *Selenium in Agriculture and the Environment.*, L.W. Jacobs, Editor. 1989, SSSA Special Publication No. 23: Madison, WI. 133–177.
19. Jlang, Q.Q. and Singh, B.R. Effect of different forms and sources of arsenic on crop yeild and arsenic concentration. *Water, Air and Soil Pollution* 74, 321–343. 1994.
20. Ohlendorf, H.M., Kilness, A.W., Simmons, J.L., Stroud, R.D., Hoffman, D.J. and Moore, J.F. Selenium toxicosis in wild aquatic birds. *J. Toxicol. Environ. Health* 24, 61–37. 1988.
21. Tessier, A., Campbell, P.G.C. and Bisson, M. Sequential extraction procedure for the speciation of particulate trace metals *Analytical Chem.* 51(7), 844–851. 1979.
22. Jackson, B.P., Miller, W.P., Schumann, A.W. and Sumner, M.E. Trace Element Solubility from Land Application of Fly Ash/Organic Waste Mixtures. *Journal of Environmental Quality* 28(2), 639–647. 1999.
23. Jackson, B.P. and Miller, W.P. Arsenic and selenium speciation in coal fly ash extracts by ion chromatography-inductively coupled plasma mass spectrometry *Journal of Analytical Atomic Spectrometry* 13(10), 1107–1112. 1998.
24. Punshon, T., Adriano, D.C. and Weber, J.T. Effect of Flue Gas Desulfurization Residue on Plant Establishment and Soil and Leachate Quality. *Journal of Environmental Quality* 30(3), 1071–1080. 2001
25. Clark, R.B., Zeto, S.K., Ritchey, K.D. and Baligar, V.C. Boron accumulation by maize grown in acidic soil amended with coal combustion products. *Fuel* 78, 179–185. 1999.
26. Punshon, T., Seaman, J.C. and Adriano, D.C. The effect of flue gas desulfurization residue on corn (*Zea mays* L.) growth and leachate salinity: Multiple season data from amended mesocosms, in *Chemistry of Trace Elements in Fly Ash*, K.S. Sajwan, A.K. Alva, and R.F. Keefer, Editors. 2002, Kluwer Academic/Plenum Press: New York, NY. In press.
27. Jackson, B.P. and Miller, W.P. Effectiveness of phosphate and hydroxide for desorption of arsenic and selenium species from iron oxides *Soil Science Society of America Journal* 64(5), 1616–1622. 2000.
28. Jackson, B.P. and Bertsch, P.M. Determination of arsenic speciation in poultry wastes by IC-ICP-MS *Environmental Science & Technology* 35(24), 4868–4873. 2001.
29. Schumann, A.W. and Sumner, M.E. Formulation of environmentally sound waste mixtures for land application *Water Air and Soil Pollution* 152(1–4), 195–217. 2004.
30. Punshon, T., Adriano, D.C. and Weber, J.T., 1999. Restoration of Eroded Land Using Coal Fly Ash and Biosolids. Electrical Power Research Institute, Pal Alto, CA., TR-113940.

13
Trace Element Transport in Putting Green Root Mixes Amended by Coal Combustion Products (CCP)

Maxim J. Schlossberg and William P. Miller
Dept. of Crop and Soil Sciences, University of Georgia, Athens, GA 30602-7272, U.S.A.

Abstract

Golf course putting green (PG) construction methods rely on homogenous mixtures of coarse and medium–sized sands as primary components of the root mix. In some locations, cost of quartz sand delivery is prohibitive. Coal combustion products (CCP) have been used to amend soil in agricultural, horticultural, and land reclamation endeavors. A greenhouse study evaluated the hydraulic properties of PG root mixes constructed with sand–sized bottom ash (BA) and silt–sized fly ash (FA) as substitutes for mineral sand and soil components. Columns (5 cm ID × 42 cm) were packed with one of three CCP–containing root mixes or a non-amended, control root mix; all of soil textural class commonly used in putting greens. Pore volumes of leachate were collected 0, 11, 24, and 100 days after [turfgrass] sodding (DAS), and analyzed for elemental content. Compactive forces were applied to root mix surfaces throughout the 6, 12, or 18 month experimental periods, following which hydraulic parameters and particle size distribution (PSD) were measured. Root mixes and compaction treatments interactively affected K_s. The control PG mix generally demonstrated the greatest saturated hydraulic conductivity (K_s), regardless of compaction treatment. Root mixes containing FA and sand (SFA) exhibited severe reductions in K_s. Similarly, the PSD of the surface 7 cm of the CCP–root mixes became more finely-textured with cumulative compaction events. This soil textural alteration was likely responsible for impaired entry and flow of water in the CCP–amended root mixes, yet conductivity rates observed after 18 months in all but the sand–FA (SFA) root mixes mix fell within accepted ranges. Trace element concentrations in root mix leachate varied with CCP inclusion rate and cumulative pore volumes, and were highly correlated to the presence of fly ash (FA). Bottom ash may be better suited for putting green root mixes amendment than FA, yet further research is necessary to determine optimal rates of inclusion.

Introduction

Dependable recreational use of a turfgrass site requires a well–drained root mix. Fine–textured soils often have limited macroporosity and do not support heavy traffic and use. The oldest golf course putting greens (PG), originally constructed of native, fine–textured soils, were topdressed with sands for physical improvement. The attributes of sand in putting green construction have been recognized for years. When comprising a majority (>0.7 m^3 m^{-3}) of a root– mix, sand–sized particles bridge and increase porosity appreciably.[1,2] Desirable physical characteristics of a putting green root mixes include 10–15% non–capillary porosity, rapid infiltration rate, and a volumetric water holding capacity >10%.[3] A root mix in the sand textural class meets most of these properties, with the common exception of >10% volumetric water–holding. Although sandy soils are resistant to compaction, they often have low moisture retention and nutrient adsorption capacity.[4]

In order to increase moisture and nutrient retention, researchers have amended sand with inorganic materials such as zeolites, calcined clay, and diatomaceous earth.[5,6,7] These sand–sized materials possess internal porosity that retains moisture and nutrients more effectively than mineral sands.[4] The two greatest concerns regarding the value of these materials are (1) their ability to significantly increase drainage and CEC of root mixes and (2) their resistance to the forces of weathering, impact, and abrasion.

Wasura and Petrovic found that all of the inorganic amendments they tested were physically less stable and more susceptible to weathering than quartz sand.[8] Their results showed ceramic clay and diatomaceous earth products possessed significantly greater resistance to weathering than the zeolite products tested as determined by the sulfate soundness test (ASTM 88–90). The ceramic clay also possessed significantly higher stability during abrasion and impact testing (ASTM C 131–89) than the diatomaceous earth and zeolite.

Researchers have also amended sand with coal combustion products (CCP) in attempts to improve soil physical properties. Studies investigating attributes of coal ash as soil amendments date back over 40 years.[9] More recently, CCP have increased capillary porosity in sandy soils and total porosity in fine–textured soils.[10,11,12,13] Gangloff et al. mixed fly ash into an acidic, excessively–drained sand at rates of 316, 632, and 950 Mg per ha[14]. Fly ash raised the water–holding capacity in the plow depth of the treated soil, but decreased infiltration rates by 80% compared to the unamended control plots. Similarly, Campbell et al. reported sizable increases in capillary porosity of fly ash amended fine and coarse sands, coupled with reduced air–filled porosity and saturated hydraulic conductivity.[15]

Golf course builders and superintendents have inherent interest in substitutes possessing equivalent stability and size of quartz sand, but improved physicochemical properties. Bottom ash (BA), a relatively inert and stable aggregate, is commonly produced in the sand–sized fraction (0.05–2 mm). Fly ash (FA) is a finer CCP (<0.05 mm) with a more highly–variable chemical makeup.[16] Both BA and FA possess lesser bulk densities than sand, with values ranging from 0.7–1.2 g cm^{-3}, and when used as a supplemental root mix aggregate could lower construction costs by minimizing transport requirements (J. Moore, pers. comm., 2002).

Thus, considering the positive reports of low bulk density inorganic amendment usage in putting green mixes, and the potential reduction in cost associated with use of ubiquitous and inexpensive CCP, an investigation of CCP bottom and fly ash as inorganic amendments/sand–substitutes in golf course putting green root mixess was initiated. The objectives of this study were to determine: (1) soil hydraulic properties and physical stability of CCP–amended root mixes receiving typical compaction forces and maintenance regimes applied to golf course putting greens, and (2) trace element concentrations in leachate as affected by cumulative pore volumes and CCP inclusion rate.

Materials and Methods

Root Mix and Column Preparation

Bottom ash (BA) and FA were collected from electric–generating plants in Georgia. Quartz sand and sphagnum peat moss were acquired at local commercial outlets. Chemical properties of these root mix components are shown in Table 1. Sphagnum peat moss was granulated by passage through a 2–mm sieve. Sand and BA were air–dried, and separated by particle diameter with hand sieves. The sized CCP and sand were mixed by volume in varying proportions into four root mixes (Table 2). Organic fertilizer (6-2-0; Milorganite®, Milwaukee, WI) was included in each root mix at 0.016 m^3 m^{-3}, providing slow-release N and P fertilization rates of 1600 and 232 kg ha^{-1}, respectively. Each root mix formulation was

TABLE 1. Physicochemical properties of CCB used in putting green root mixes

	Bottom ash (BA)	Fly ash (FA)
	mg (dry kg)$^{-1}$	
Trace Elements†		
As	2	104
Cd	<1	1
Cr	25	13
Cu	274	200
Mo	2	17
Ni	179	94
Pb	6	39
Se	2	24
Water Chemistry		
1:1 pH	6.4	8.7
EC	0.1 dS m^{-1}	1.2 dS m^{-1}

†USEPA 3050A method.[29]

thoroughly homogenized in a plastic drum. The control rooting mix (CON), containing 8 parts quartz sand to 2 parts peat moss (v:v), was comprised of coarse, medium, and fine sands in proportions commonly used in putting green construction (Table 2). Two of three root mix treatments contained bottom ash and fly ash for replacement of sand and silt/very fine sand, respectively (BAFA), or bottom ash as a replacement for half of the sand aggregate (v:v) and fly ash as a replacement for all the silt/very fine sand (BASFA) (Table 2). The third root mix treatment contained the control quartz sand with fly ash as a replacement for the silt/very fine sand sized particles (SFA).

Once homogenized, root mixes were transferred to 42–cm long sections of schedule 40, polyvinyl–chloride (PVC) pipe (5.1 cm ID). The columns were closed on one end by a PVC end- cap with a 1.2–cm diameter center hole, and filled with 210 cm^3 (10 cm height) of fine pea-gravel. Each PVC section was then loaded with 630 cm^3 (30–cm height) of the root mix in 5–cm high (106 cm^3) lifts. Lifts were firmly tamped in an effort to maintain consistent bulk density throughout each column. Surplus root mixes (of each type) were stored for subsequent analysis.

Creeping Bentgrass Establishment and Putting Green Leachate Collection

Crenshaw creeping bentgrass [*Agrostis stolonifera* L. var. *palustris* (Huds.). Farw., (syn. *Agrostis palustris* Huds.)] was seeded in plastic trays (20 kg pure certified bentgrass seed ha^{-1}) for transplant onto the columns. Each tray was lined with ~2 cm of either bottom ash or sand (in equal parts of 1–2 mm, 0.5–1 mm, and 0.25–0.5 mm), and covered with a thin layer (~4 mm) of sphagnum peat moss/bentgrass seed mixture.

On 7 Oct. 2000, 10-cm PVC extensions were attached to the top of the 24 columns, and leached with ~1 pore volume (300–420 mL) of de-ionized (DI) H_2O. A leaching fraction of ~0.25 (75–105 mL) was collected, filtered (Whatman 42),

TABLE 2. Component proportion and particle size distribution of three experimental putting green root mixes compared with an unamended mix (CON).

	Sand-sized components						
	1–2 mm	0.5–1 mm	0.25–0.5 mm	0.15–0.25 mm	Peat moss	Fly ash	Coal combustion by-products Bottom ash and/or Fly ash
				% by volume			
Mix Type							
CON	7 S[†]	42 S	21 S	10 S	20	0	0
BAFA	7 BA	42 BA	21 BA	0	20	10	80
BASFA	3.5 S + 3.5 BA	21 S + 21 BA	10.5 S + 10.5 BA	0	20	10	45
SFA	7 S	42 S	21 S	0	20	10	10

	1–2 mm	0.5–1 mm	0.25–0.5 mm	<0.25 mm	LOI[‡]	BD[§]
			% by mass			$g\ cm^{-3}$
Mix Type						
CON	[¶]5.3 ± 0.4	39.6 ± 2.2	32.1 ± 1.3	23.1 ± 2.0	5.5 ± 0.3	1.35 ± 0.0
BAFA	4.0 ± 0.2	33.1 ± 0.1	24.3 ± 0.2	38.5 ± 0.6	13.7 ± 0.9	0.74 ± 0.0
BASFA	5.7 ± 0.2	37.7 ± 0.7	30.8 ± 0.7	25.8 ± 0.8	7.5 ± 0.9	1.03 ± 0.0
SFA	5.7 ± 0.1	41.9 ± 0.9	34.5 ± 0.5	17.9 ± 1.1	7.9 ± 0.2	1.34 ± 0.0
Fly Ash	0	0	1.0 ± 0.1	99.0 ± 0.9	3.0 ± 0.5	1.20 ± 0.1

† S, sand; BA, bottom ash; FA, fly ash.
‡ LOI, Loss on ignition (ASTM D 2974-87, Method C).
§ BD, Bulk density.
¶ $\mu \pm$ SE; means followed by sample standard error (SE).

spiked to pH 2 with trace-metal nitric acid, and stored at 4°C. Following this first leach event, the columns were 'sodded' with pre-established discs of the creeping bentgrass sod. The BAFA and BASFA root mix columns received plugs propagated in the bottom ash trays, whereas the SFA and CON root mix columns received plugs propagated in the sand trays. Columns were irrigated with de-ionized water as needed, but never in excess of 50-mL over a 48-h period. This restriction prevented drainage outside of scheduled leaching events during the first 100 days of the experimental period. Column leachings were repeated 11, 24, and 100 days after sodding (DAS) in the manner described above. The remaining 48 cylinders were divided into 2 blocks and initially leached on January 22 or March 4 (2001) and sequentially thereafter (11, 24, and 100 DAS) as described above. These columns were sodded with bentgrass and irrigated lightly and frequently during the 100-d leaching period.

All putting green columns were identically maintained in the greenhouse. Supplemental lighting was not implemented in this study, thus daily photoperiod did not extend beyond that naturally observed at 32° N lat. Maximum photosynthetically-active photon flux density levels recorded in the greenhouse reached 1665 mmol m^2 sec^{-1} at solar noon on 16 July 2001, approximately 70% of levels measured near-simultaneously outdoors. Seasonal fluctuation of environmental conditions in the greenhouse was unavoidable, but temperatures were maintained in a favorable range for creeping bentgrass growth (10–24 °C).

Putting green leachate was analyzed for trace element content by ICP-MS (Elan-6000, PerkinElmer, Wellesley, MA). When necessary, samples were diluted with a 5% trace metal grade nitric acid solution. All leachate samples were analyzed within 8 days of collection.

Compaction treatments were initiated ~130 days after sodding (DAS), by dropping a 1.24–kg cylindrical stainless steel weight from a height of 80 cm onto each cylinder turfgrass surface 3 consecutive times (the weight possessing kinetic energy equivalent to 9.5 J immediately prior to impact). A second application of 3 impacts was made ~20–d later. These six impacts were reapplied as compaction treatments every 6 months. Thus, the 6 month-old columns received 6 total impacts, the 12 month-old columns received 12, and the 18 month-old columns received 18 total impacts.

At 184 DAS, 24 columns (comprising 3 replications of 8 PG mix–compaction combinations) were treated with a 7.5 kg ha^{-1} application of glyphosate (Roundup Pro™) herbicide by a pressurized spray volume equivalent to 200 L ha^{-1}. This identical termination procedure was performed on 24 experimental units 367 DAS, and the remaining columns at 543 DAS.

Root Mix Hydraulic Conductivity and Physical Properties

Following termination of each PG, the surface 7–cm of the root zone mixture (with the intact 10–cm extension), was sectioned by band saw and removed. The base of the 7–cm segment was secured with cheesecloth and baling wire, then saturated from the bottom-up with degassed 0.01 M $CaCl_2$ solution. Hydraulic conductivity was measured by the steady state constant–head method, using a 4–L mariotte bottle and degassed 0.01 M $CaCl_2$ solution. Average saturated conductivity (K_s) through the upper 7–cm of each column was calculated using Darcy's Law for vertical flow after correcting for the solution specific gravity.[17]

Following conductivity measurements, the upper 7–cm of the root mix was carefully extruded from each column intact. The turfgrass leaves and shoots were discarded, and the root mass was separated from the root mix of the 0–7 cm depth segment by an elution procedure. The de–rooted mixes were dried at 70°C in a forced–air oven prior to particle size analysis. Duplicate ten gram samples of root mix from the 0–7 cm depth segment were transferred to a tarred glass bottle. Soil organic matter was oxidized in 10% H_2O_2 at 60°C until the root mix dilution appeared colorless (24–72 h).[18] Soil colloids were dispersed in dilute sodium hexametaphosphate solution while horizontally shaken for 15 h. Clay–sized particles were quantified using the pipette method.[19] Sand–sized particles were collected and dry fractioned in a nest of sieves.[20] Silt content in the root mixes was inferred by subtraction, after correcting for soil organic matter using loss-on-ignition values.

Experimental Design

The greenhouse columns were prepared in a completely randomized design, comprising 72 experimental units. The root zone mixes and compaction treatments were fixed effects randomly assigned to each empty column, configuring a two–way effects model with nine replications. At 184, 367, or 543 days after sodding (DAS); 3 replications of each of the 8 mix-compaction treatment combinations were destructively analyzed as mentioned above. Hydraulic conductivity rates (cm hr^{-1}) were distributed neither normally nor with equal variance among treatments. Thus, this dependent variable was successfully log–transformed. Once requisite assumptions for parametric analysis were validated, PROC GLM subroutines of SAS (Version 8.2, SAS Inst., Inc.) facilitated analysis of data. Repeated measures of leachate quality were analyzed (without inclusion of compaction effect) by the linear mixed model (PROC MIXED). Predetermined multiple comparisons of treatments within significant ($P \leq 0.05$) main or interactive effects were conducted using Tukey's Studentized range procedure ($\alpha = 0.05$).

Results and Discussion

Root Mix Hydrology and Physical Properties

A rapid saturated conductivity rate (K_s) is an important attribute of a putting green root mix. Saturated conductivity measuredin the upper 7–cm of root mix was significantly affected by root zone mix and compaction (Table 3). However, at any given age, compaction treatment and root mix did not significantly interact to affect K_s (Table 4). Thus, compaction treatments significantly lowered all mix types proportionally. Comparison of root mixes at 184 DAS, indicated the CON possessed the significantly highest K_s rate of all root mixes, except

TABLE 3. Analysis of variance and mean comparisons of saturated conductivity rate (K_s) in the top 7.0 cm of putting green root zone mix treatments, all experimental periods combined (data shown are means of 9 observations)

	Saturated hydraulic conductivity	
	Compaction	No compaction
Mix Type	cm hr^{-1}	
CON†	57 B a‡	137 A§ a
BAFA	39 B ab	145 A a
BASFA	25 B bc	119 A a
SFA	22 B c	65 A b

ANOVA¶

Source	df	main effects
Age (Experimental period length)	2	***
Mix	3	***
Compaction	1	***
Mix X Compaction	3	*
Mix X Age	6	**
Compaction X Age	2	***
Mix X Compaction X Age	6	NS

*,**,*** Significant at the 0.05, 0.01, and <0.001 probability levels, respectively.

† By volume: CON, 80% sand, 20% organic matter (OM); BAFA, 70% bottom ash (BA), 10% fly ash (FA), 20% OM; BASFA, 35% BA, 35% sand, 10% FA, 20% OM; SFA, 70% sand, 10% FA, 20% OM.

‡ In columns, means followed by different lowercase letters signify differences by root mix, according to Tukey's Studentized Range Test ($\alpha = 0.05$).

§ In rows, means followed by different uppercase letters signify differences by compaction, according to Tukey's Studentized Range Test ($\alpha = 0.05$).

¶ Saturated conductivity data (cm hr^{-1}) were log transformed for ANOVA and multiple comparison analysis.

TABLE 4. Saturated conductivity (K_s) of top 7.0 cm of putting green columns by compaction treatment and root mix, sorted by column age

		184 Days after sodding (DAS)			Saturated hydraulic conductivity 367 DAS			543 DAS		
		Compaction	No compaction	Treatments averaged	Compaction	No compaction	Treatments averaged	Compaction	No compaction	Treatments averaged
			cm hr^{-1}			cm hr^{-1}			cm hr^{-1}	
Mix Type										
CON†		108	147	127 a‡	57	150	103 a	33	125	79 a
BAFA		45	112	79 ab	53	145	99 a	31	196	114 a
BASFA		45	132	89 b	22	98	60 b	20	147	83 a
SFA		42	94	68 b	27	65	46 b	10	52	31 b
Average of all mixes		60 B§	121 A		40 B	114 A		24 B	130 A	
						ANOVA				
Source	df		184 DAS			367 DAS			543 DAS	
Mix	3		*			***			***	
Compaction	1		***			***			***	
Mix X Compaction	3		NS			NS			NS	

*,**,*** Significant at the 0.05, 0.01, and <0.001 probability levels, respectively.

† By volume: CON, 80% sand, 20% organic matter (OM); BAFA, 70% bottom ash (BA), 10% fly ash (FA), 20% OM; BASFA, 35% BA,35% sand, 10% FA, 20% OM; SFA, 70% sand, 10% FA, 20% OM.

‡ In columns, means followed by different lowercase letters signify differences by root mix, according to Tukey's Studentized Range Test ($\alpha = 0.05$).

§In rows, means followed by different uppercase letters signify differences by compaction, according to Tukey's Studentized Range Test ($\alpha = 0.05$).

the BAFA root mix. At 367 DAS, the CON and BAFA mixes possessed significantly greater K_s than the SFA and BASFA mixes.

As in the 184 and 367–d old columns, hydraulic conductivity rates of all compacted PG-mixes at 543 DAS were significantly reduced when compared to their uncompacted counterparts (Table 4). Regardless of compaction, the SFA mix did not maintain as rapid of conductivity as the CON, BAFA, or BASFA root mixes. Non–compacted root mixes showed rapid conductivity rates at the experiment conclusion. Saturated conductivity rates of all compaction–treated root mixes (sans the SFA mix) fell within the acceptable range for PG mixes (>15 cm hr^{-1}).[3]

Alteration of particle size distribution, particularly destruction of coarse particles, can adversely affect physical properties of root mixes. Particle size distribution (0–7 cm soil depth) was significantly affected by root mix and length of experimental period. Significant differences by root mix are expected because the original formulations were made volumetrically, while the components possessed varying particle densities (Table 2). Thus, root mix main effects on PSD will not be discussed. Furthermore, the sampling method used in PSD analysis was inadvertently confounded by application of compaction treatments. Applied compaction significantly reduced the conductivity rate of all root mixes (Table 3). Therefore, two separate linear models were performed; compacted putting greens and uncompacted putting greens. Analysis of variance indicated significant interaction between experimental period and root mix effects on particle size distribution in the upper 0–7 cm segment of the compacted PG root mixes (Table 5).

In the BAFA root mix, compaction resulted in a significant decrease of very coarse sand–sized particles (3.5 %). A similar effect due to cumulative compactions was observed in the BASFA root mix (Table 6). For the BASFA mixture, compaction treatments resulted in a 3.1% absolute percentage reduction (Table 6). These class shifts likely resulted from

TABLE 5. ANOVA table for percent gravimetric particle size distribution in the 0–7 cm depth segment, as affected by source

	Very coarse sand	Coarse sand	Medium sand	Fine sand	Very fine sand	Silt	Clay
			F-Values Compacted Columns				
Source							
Age	19.4***	1.8	0.3	1.3	10.7***	19.8***	1.0
Mix	74.3***	3.2*	199.9***	45.1***	54.0***	236.9***	29.4***
Age X Mix	7.8***	0.7	2.0	1.0	1.2	8.1***	0.7
			Non-compacted Columns				
Age	6.9**	1.2	0.2	0.9	10.0***	10.5***	1.1
Mix	77.5***	10.8***	195.4***	69.1***	36.6***	76.8***	15.9***
Age X Mix	2.2	1.2	1.0	1.9	2.2	1.7	1.3

*,**,*** Significant at the 0.05, 0.01, and <0.001 probability levels, respectively.

TABLE 6. Absolute percent mass differences in gravimetric particle size distribution recovered in the compacted 0–7 cm surfaces, as affected by age and cumulative compaction events

	Very coarse sand	Coarse sand	Medium sand	Very fine sand	Silt	Clay
Compacted Root mixes (only)			Δ Mass (%)			
Control (CON)						
6 and 12 month PSD versus 18 month PSD	−0.06	NS	NS	NS	+0.14	NS
BAFA						
6 and 12 month PSD versus 18 month PSD	−3.50**	NS	NS	NS	+7.09***	NS
BASFA						
6 and 12 month PSD versus 18 month PSD	−3.04**	NS	NS	NS	+0.54	NS
SFA						
6 and 12 month PSD versus 18 month PSD	+0.47	NS	NS	NS	+0.83	NS

*,**,*** Significant at the 0.05, 0.01, and <0.001 probability levels, respectively.

breakage of coarse BA particles. As quartz sand particles are known to resist alteration by compaction forces, a lesser reduction in very coarse, sand-sized particles would be expected in the mix with aggregates partially comprised of sand.[8]

As silt–sized particles were determined by subtraction, all gravimetric component percentages sum to 100. Therefore, a decrease in coarse particles must be countered by an increase in percentage of finer–sized particles. For the BAFA mix, compaction resulted in a significant increase in silt. This increasing trend was observed in the other CCP-mixes (Table 6), but was not significant.

At high FA application rates to mineral soil, the suspected mechanism responsible for reduced K_s is clay translocation and pore blockage.[21] In surface horizons of low ionic strength soils, intrinsic sodicity (high SAR) is more likely responsible for dispersion of clay–sized particles.Moderate additions of FA have been proven beneficial, particularly when salts associated with FA flocculate clay colloids. However, incremental FA additions from only 10 to 15% (w/w) have had significant offsetting repercussions on conductivity.[22]

Ishak et al. packed columns with alkaline fly ash–amended loamy sand A–horizons of the Appling series (clayey, kaolinitic, thermic family of typic Hapludults). Following 2 pore volumes of saturated flow, pressure head build–up at the column inlet doubled compared to the unamended control loamy sand. At 6 pore volumes of saturated flow, the pressure head tripled levels observed in the unamended control soil column.[23] Our results showed a similar effect on K_s when fly ash was the sole CCP amendment of a sand–based root mix (SFA), as saturated conductivity in the non-compacted SFA root mix diminished with continuing pore volumes.

Saturated hydraulic conductivity (K_S) is directly related to the quantity and connectivity of soil macropores in a soil (root mix). At the start of the experiment, the CON root mix treatment contained ~98.3% (w/w) particles of diameter >110 μm. In comparison, the BAFA, BASFA, and SFA experimental mixes contained 80.3, 85.1, and 90.5% (w/w) of these sized particles, respectively (data not shown). This is the likely reason why greater conductivity rates were observed in the CON mix at 184 DAS. Rawls et al. compiled K_s values from 1,323 soils collected over 32 states of the US, and concluded saturated hydraulic conductivity to be greatest in coarse–textured soils and least in fine–textured soils.[24] At 543 DAS (and after 18 compaction events), PSD of the CON mix remained unchanged (Table 6). Conversely, the compacted BAFA mix had a significant decrease in very coarse sand content. These data make it difficult to explain how the saturated conductivity of the BAFA mix remained statistically equal to the control mix following the 18–month experimental period. The structure of the BAFA root mix may have facilitated preservation of macropores, even having lost a significant number of very coarse sized particles.

Leachate Quality and Trace Element Transport

Concentrations of As, Cd, Cr, Cu, Mo, Ni, Pb, and Se in leachate were significantly affected by the interaction of leach events (time) and root mixes (data not shown). The oxyanions measured shared similar patterns of release from the putting green root mixes (Figure 1). Molybdenum concentration of the fly ash was 16 mg kg^{-1} (Table 1). Molybdate, a common form of Mo, is a plant-available anion present in trace concentrations in soil (0.1–7.2. mg kg^{-1}). It behaves similarly to P, being strongly adsorbed by Fe and Al oxides. Molybdenum concentration never exceeded 0.5 mg L^{-1} leachate in any given leaching event. The BASFA and SFA root mixes released significantly greater Mo than the control mix during the first leaching event. In the fourth pore volume, the SFA root mix column discharged 0.49 mg Mo L^{-1}, a significantly greater Mo concentration than observed of any other root-mix type throughout the experimental period (Figure 1). Contrast statements indicate average Mo levels in leachate of all CCP mixes significantly exceeded levels measured in discharge from the CON columns. The average

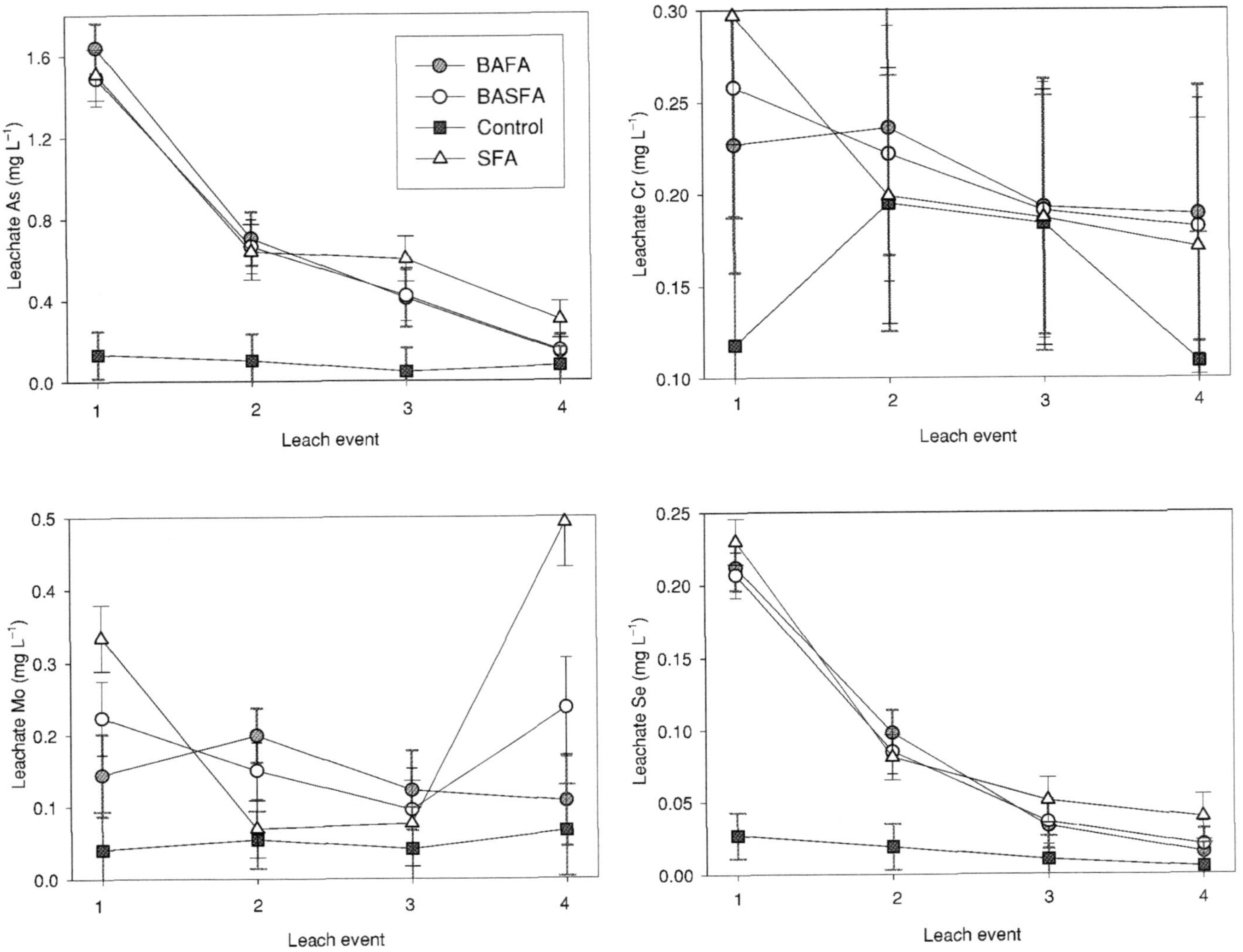

FIGURE 1. Leachate concentration of As, Cr, Se, and Mo (clockwise from top left); by putting green root mix and leach event (BA, bottom ash; FA, fly ash; S, sand).

Mo concentration of the SFA mix leachate exceeded that observed in all other mixes, while the BAFA and BASFA did not significantly differ from one another (data not shown).

Chromium concentrations in the CCP-amended root mixes did not exceed levels commonly encountered in mineral soils. The SFA PG mix discharged significantly more chromium than the control mix at the initial leaching, but concentrations did not significantly differ in subsequent drainage (Figure 1).

Arsenic and selenium occur as oxyanions in fly ash at concentrations greater than in most soils, and are associated with the surface layer of this by-product. It can be inferred from contrast statements (not shown) that fly ash is the major contributor to elevated oxyanion concentrations in leachate. For example, the average of all root mix columns containing CCP possessed significantly greater concentrations of As and Se in their leachate than the control (Figure 1). Moreover, when sand was used as the sand-sized aggregate, columns containing FA (SFA) leached significantly greater quantities of Se and As than the control root mix. It can be inferred that the coarse-textured aggregate utilized (BA or sand) had little bearing on As and Se concentrations in subsequent leachate (Figure 1).

This release of As (Fig. 1) very closely resembles data presented by Qafoku et al.[25] In that experiment, soil columns amended with 4% fly ash by mass (FA; 180 mg As kg^{-1}) were leached with a dilute solution of calcium sulfate. Arsenic concentrations in leachate reached a maximum of 80 μg As L^{-1} at 1 pore volume and diminished rapidly with subsequent pore volumes. In a similar column study, a soil was amended with the equivalent application of 41 Mg FA ha^{-1}. Leachate was collected and As concentrations reached a maximum level at approximately 1.7 mg As L^{-1} of 12 pore volumes.[26]

The chemistry of selenium is similar to As, with both being potentially toxic trace elements, both exhibiting varying

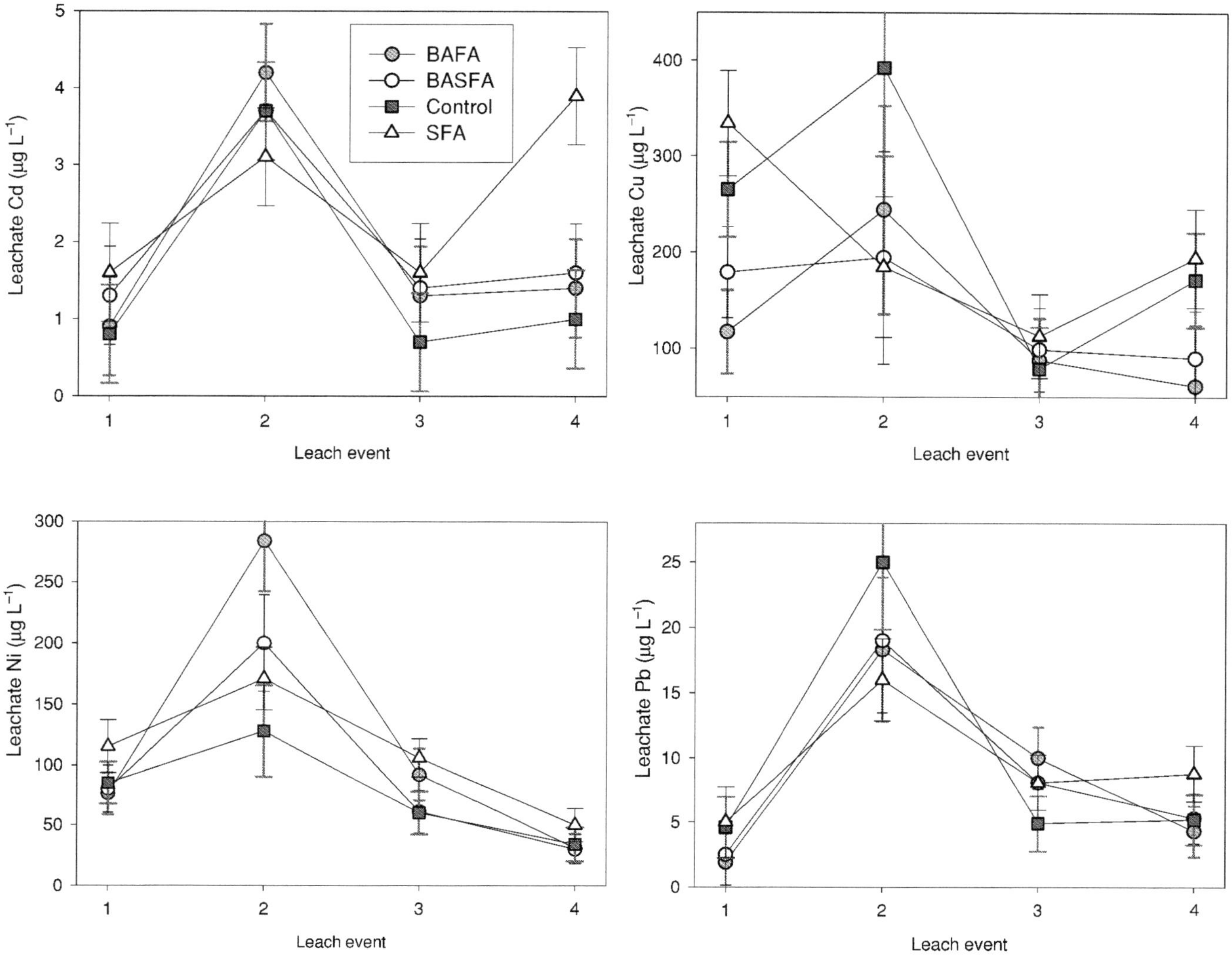

FIGURE 2. Leachate concentration of Cd, Cu, Pb, and Ni (Clockwise from top left); by putting green root mix and leach event (BA, bottom ash; FA, fly ash; S, sand).

oxidation states, and occurring as oxyanions. In the mentioned study, fly ash applications made at 41 Mg ha^{-1} resulted in approximately 0.3 mg Se L^{-1} of leachate after 9 pore volumes.[26] The exhibited release curve was similar to that of As (Fig. 1).

Metal concentrations in leachate (Figure 2) were low and decreased with cumulative leaching. This effect was likely due to increased plant uptake following sodding. At no time during the 100 day leaching period did Cd, Cu, Ni, or Pb concentrations in leachate exceed 500 parts per billion. The more-significant release of As and Se from the CCP-amended root mixes may limit potential usage. Putting greens generally comprise 1 ha of area on a typical golf course. If every newly-constructed green were to receive a 10-cm rainfall event (nearly one pore volume of putting green root mix per unit area), approximately 2.5-cm of leachate would drain from the greens. This 2.5 cm of leachate would comprise approximately 250,000 L and contain a cumulative total of approximately 0.5 kg of arsenic and 74 grams of selenium. Such discharge to a surface impoundment could prove detrimental to the environment under conventional drainage systems. Arsenic is currently the focus of scrutiny by the USEPA, leading to a recent reduction in the drinking water standard from 50 to 10 parts per billion (ppb).[27]

Conclusions

The simple connotation 'turfgrass' implies use. Uninterrupted enjoyment of high quality turfgrass cover, preventing the need for disruptive renovation procedures, requires a root mix that is resistant to compaction, and continually well–drained with ample air–filled porosity. This attempt to utilize CCP bottom ash and fly ash as substitute aggregates in golf course putting green construction was novel, but hardly original. Coal cinders have historically been utilized in the construction of golf course putting greens,[28] as well as drainage layers in bunker construction (J. Jennings, pers. comm., 2001). Bulk density of BA-amended root mixes was significantly less than observed in the control mix (Table 2).

At this time, utilization of fly ash in putting green root mix construction at the rates investigated in this study is not recommended. Bottom ash appears to have greater future application to root mix construction than fly ash. Bottom ash amended root mixes maintained rapid infiltration and hydraulic conductivity throughout the 18-month experimental period. However, bottom ash did not demonstrate physical stability under subjection to compactive forces. This tendency adversely affected percolation of water within the experimental period. Considering the expected functional period of golf course putting greens exceeds 20 years, our results prompt further investigation of suitable inorganic amendments for golf course putting green root mix inclusion.

Acknowledgements. This research was supported by the Electric Power Research Institute (EPRI), Palo Alto, California. The authors are grateful for the technical assistance provided by Stan Dudka, Gil Landry Jr., Robert Carrow, Larry Morris, Malcolm Sumner, Gene Weeks, Keith Harris, Lamar Larrimore, Chris Vanags, Melanie Parker, and Cameron J. Tribble.

References

1. McCoy, E.L., Sand and organic amendment influences on soil physical properties related to turfgrass establishment, *Agron. J.* 90, 411, 1998.
2. Spomer, L.A., Prediction and control of porosity and water-retention in sand-soil mixtures for drained turf sites, *Agron. J.* 72, 361, 1980.
3. United States Golf Association (USGA) Green Section Staff., Recommendations for a method of putting green construction, *USGA Green Section Record.* March/April, 1, 1993.
4. McCoy, E.L., Stehouwer, R.C., Water and nutrient retention properties of internally porous inorganic amendments in high sand content root zones, *J. Turf. Manag.* 2, 49, 1998.
5. Bigelow, C.A., Bowman, D.C., Cassel, D.K., Nitrogen leaching in sand-based rootzones amended with inorganic soil amendments and sphagnum peat, *J. Amer. Soc. Hort. Sci.* 126, 151, 2001.
6. Huang, Z.T., Petrovic, A.M., Clinoptilolite zeolite influence on nitrate leaching and nitrogen use efficiency in simulated sand putting greens, *J. Environ. Qual.* 23, 1190, 1994.
7. Richardson, M.D., Karcher, D.E., Addition of inorganic amendments to a mature, sand-based putting green, *Int. Turf. Soc. Res. J.* 9, 610, 2001.
8. Wasura, J.P., Petrovic, A.M., Physical stability of inorganic amendments used in turfgrass root zones, *Int. Turf. Soc. Res. J.* 9, 636, 2001.
9. Holiday, R., D.R. Hodgson, D.R., Townsend, W.N., Wood, J.W., Plant growth on fly ash, *Nature* 181, 1079, 1958.
10. Adriano, D.C., Weber, J.T., Influence of fly ash on soil physical properties and turfgrass establishment, *J. Environ. Qual.* 30, 596, 2001.
11. Pathan, S.M., Aylmore, L.A.G., Colmer, T.D., Fly ash amendment of sandy soil to improve water and nutrient use efficiency in turf culture, *Int. Turf. Soc. Res. J.* 9, 33, 2001.
12. Patterson, J.C., Jr., Henderlong, P.R., Turfgrass soil modification with sintered fly ash, in Proc. 1st Int. Turfgrass Res. Conf., Harrogate, England. 15–18 July 1969. Sports Turf Res. Inst., Bingley, England.
13. Schlossberg, M.J., Miller, W.P., Coal combustion by-product (CCB) utilization in turfgrass sod production, *HortScience.* 39, 408, 2004.
14. Gangloff, W.J., Ghodrati, M., Sims, J.T., Vasilas, B.S., Impact of fly ash amendment and incorporation method on hydraulic properties of a sandy soil, *Water Air Soil Poll.* 119, 231, 2000.
15. Campbell, D.J., Fox, W.E., Aitken, R.L., Bell, L.C., Physical characteristics of sands amended with fly ash, *Aust. J. Soil Res.* 21, 147, 1983.
16. Mattigod, S.V., D. Lai, D., Eary, L.E., Ainsworth, C.C., Geochemical factors controlling the mobilization of inorganic constituents from fossil fuel combustion residues. I. Review of the major elements, *J. Environ. Qual.* 19, 188, 1990.
17. Klute, A, Dirksen, C., Hydraulic conductivity and diffusivity: Laboratory methods, in *Methods of soil analysis, I. Physical and mineralogical methods—Monograph no. 9*, Klute, A., Ed., Amer. Soc. Agron.—Soil Sci. Soc. Amer., Madison, WI, 1986, 687.
18. Day, P.R., Particle fractionation and particle size analysis, in *Methods of soil analysis, Part I. Agronomy 9*, Black, C.A. et al., Eds. Amer. Soc. Agron.—Soil Sci. Soc. Amer., Madison, WI, 1965, 545.
19. Burt, R., Reinsch, T.G., Miller, W.P, A micro-pipette method for water dispersible clay, *Commun. Soil Plant Anal.* 24, 2531, 1993.
20. Gee, G.W., Bauder, J.W., Particle-size analysis, in *Methods of soil analysis, Part 1. Physical and mineralogical methods—Monograph no. 9*, Klute, A., Ed., Amer. Soc. Agron.—Soil Sci. Soc. Amer., Madison, WI, 1986, 383.
21. Sharma, S., Fulekar, M.H., Jayalakshmi, C.P., Fly ash dynamics in soil-water systems, *Crit. Rev. Environ. Control.* 19, 251, 1989.
22. Seaman, J.C., Aburime, S.A., Jackson, B.P., Punshon, T., Moisture retention and hydraulic conductivity of coarse-textured soils amended with coal combustion fly ash, in *Chemistry of trace elements in fly ash,* Sajwan K.S. et al. Eds., Kluwer Academic/Plenum, New York, 2003, 346.
23. Ishak, C.F., Seaman, J.C., Miller, W.P., Sumner, M., Contaminant mobility in soils amended with fly ash and flue-gas gypsum: Intact soil cores and repacked columns, *Water Air Soil Poll.* 134, 287, 2002.
24. Rawls, W.J., Brakensiek, D.L., Saxton, K.E., Estimation of soil water properties, *Trans. ASAE* 25. 1316, 1982.
25. Qafoku, N.P., Kukier, U., Sumner, M.E., Miller, W.P., Radcliffe, D.E., Arsenate displacement from fly ash amended soils, *Water, Air, and Soil Poll.* 114, 185, 1998.

26. Qafoku, N.P., Dudka, S., Sumner, M.E., Miller, W.P., Arsenic, boron, selenium and molybdenum displacement and transport in a fly ash amended soil leached with calcium phosphate solution, *Commun. Soil Sci. Plant Anal.* 32, 1499, 2001.
27. USEPA (U.S. Environ. Protection Agency), Federal Register, http://www.epa.gov/fedrgstr/EPA-WATER/2003/March/Day-25/w7048.htm, 2003.
28. Hurdzan, M.J., *Evolution of the modern green, part I.* American Society of Golf Course Architects (ASGCA), 221 North LaSalle St., Chicago, IL, 1985.
29. USEPA (U.S. Environ. Protection Agency), Acid digestion of sediments, sludges, and soils, USEPA SW-S846; Ch 3.2 method 3050A; U.S. Environ. Protection Agency, Gov, Print Office, Washington D.C., 1990.

14
Solute Leaching from Fly Ash Amended Soil Under Varying Degrees of Saturation

J. M. Hutchison[1], J. C. Seaman[1], B. P. Jackson[1], and S. A. Aburime[2]
[1]*Advanced Analytical Center for Environmental Sciences, Savannah River, Ecology Laboratory, The University of Georgia, Aiken, SC 29802, USA*
[2]*Department of Engineering, Clark Atlanta University, 233 James P. Brawley Drive, Atlanta, GA 30314, USA*

1. Abstract

Most of the fly ash produced in the US is stockpiled or disposed of in settling ponds where solute leaching can pose an environmental and health concern. Therefore, a series of saturated and unsaturated column experiments were conducted to evaluate the impact of saturation and pore solution residence time (i.e., equilibration time) on the leaching of solutes from fly ash when incorporated within surface horizon material from a loamy sand soil. Repacked soil columns were leached at various moisture contents using an Unsaturated Flow Apparatus (UFA), a modified centrifuge for conducting steady-state leaching experiments. Additional column experiments were conducted under saturated conditions to isolate the effects of residence time from that of water content. Addition of 10 percent fly ash (by weight) significantly increased water-holding capacity in the loamy sand. Leachate concentration of As and Se increased as residence time increased. Most Se leaching took place in the first few pore volumes, indicating that weathering the fly ash may help alleviate some of its phytotoxic effects. Since saturation was related to residence time in these experiments, the effect of water content could not be isolated, though there appeared to be a delay in the "leaching front" as the soil became desaturated. However, comparisons of unsaturated leaching studies to saturated leaching studies are difficult due to the variation in water content under unsaturated conditions.

2. Introduction

In 2001, over 61 million metric tons of fly ash were produced in the US as a result of the burning of coal.[1] Of that amount, about 32% was utilized in applications such as cement and concrete production, mining, road base and sub-base materials, structural fill and waste stabilization. Most fly ash is stockpiled, landfilled or stored in settling basins. Not only are sites for landfills becoming expensive and less available, but the weathering of fly ash can directly impact the environment through surface runoff and seepage to groundwater.

Fly ash has been shown to have many beneficial qualities as a soil amendment, including increased water-holding capacity for coarse textured soils and improved drainage for fine textured soils due to the dominance of silt-sized particles.[2–6] Fly ash can also act as an effective liming agent, reducing toxic effects of Al and Mn in acidic soils.[3,7] However, the application of fly ash in sufficient quantities to see these results may increase the potential for plant toxicity problems due to boron and environmental hazards associated with other trace elements.[8]

Fly ash contains many trace elements including As and Se that may have deleterious effects on soil and water quality and plant and animal life.[3,9–12] Selenium concentrations in plant tissues have been shown to consistently increase with fly ash treatment, being roughly proportional to either the rates of fly ash application or the Se content of the fly ash.[13–15] Though Se is not necessarily toxic to plants, these elevated concentrations could pose a risk to livestock as well as wild species such as woodchucks[16] and honey bees.[17] Aquatic species have also been affected by the elevated concentrations of As, Se and other trace elements in the leachate from fly ash basins.[9,18–21]

It is therefore important to understand the leaching dynamics of solutes in fly ash and how they behave under variable moisture conditions. Many column experiments are conducted under saturated conditions. However, in order to more adequately reflect natural conditions, unsaturated conditions were simulated in this study to approximate situations in which fly ash is incorporated into surface soils. The objective of this study was to examine the impact of saturation on the leaching of As and Se from repacked soil columns containing a mixture of soil and fly ash.

3. Materials and Methods

A coarse-textured surface soil (89% sand) from the Orangeburg series was collected on the Department of Energy's Savannah River Site near Aiken, SC (Table 1). Samples were stored in a field-moist state at 4 °C prior to characterization

TABLE 1. Select physical and chemical properties of the surface soil and coal combustion fly ash used in this study

	Soil	FA*	10% FA + soil
Sand (%)#	89.44 ± 1.92		87.17 ± 2.14
Silt (%)#	3.08 ± 0.38		7.48 ± 2.11
Clay (%)#	7.48 ± 2.30		5.35 ± 3.20
Organic carbon (g 100g $^{-1}$)^	1.35 ± 0.29		
pH-deionized water+	4.90 ± 0.01	12.15 ± 0.06	6.51 ± 0.02
pH-ARW+	4.92 ± 0.02	12.01 ± 0.05	6.60 ± 0.07
pH-1 *M* KCl+	3.84 ± 0.06	12.36 ± 0.06	5.88 ± 0.24
EC (μS cm^{-1})+	30.5 ± 5.47	2857 ± 194	203 ± 90
CBD-Fe (g kg^{-1})**	2.04 ± 0.04		
CEC (cmol$_c$ kg^{-1})##	15.94		
Clay mineralogy^^	k, HIV		
WDC, % (w/w)++	1.63 ± 0.04		1.45 ± 0.65
Extractable metals* **		(μg kg^{-1})	
As	1.73 ± 0.05	227.43 ± 9.12	2.37 ± 0.49
Se	0.49 ± 0.24	678.93 ± 50.17	39.98 ± 1.28

* FA = fly ash
\# Hydrometer method[22], three replicates
^ Dry combustion method[23]
\+ pH and EC taken in 1:2 soil:solution ratio for deionized water, ARW (artificial rainwater) or 1 *M* KCl after 30 minutes of intermittent stirring, three replicates
** CBD, citrate-dithionite-bicarbonate extraction[24]
\## Ammonium acetate method[25–26]
^^ Clay mineralogy determined by XRD: k = kaolinite, HIV = hydroxyl interlayered vermiculite
\++ WDC = Water dispersible clay, Micropipette method[27]
*** 1:2 soil:solution ratio, mixed intermittently for 30 minutes in artificial rainwater solution, particles allowed to settle, supernatant sampled and As and Se determined by ICP-MS

and column experiments. The material had a pH in deionized water of 4.90, a citrate-dithionite-bicarbonate extractable iron content of 2.04 g kg^{-1} and a cation exchange capacity of 15.94 cmol$_c$ kg^{-1}. The dominant clay minerals were kaolinite and hydroxyl-interlayered vermiculite.

The coal combustion fly ash (FA), obtained from a southeastern US power plant, was alkaline with potential liming capacity (pH = 12.15 in deionized water, Table 1). The treatment used in this study consisted of a 10% by weight addition of fly ash to the soil (thoroughly mixed) and will be termed "fly ash mixture" for simplicity throughout this paper. The fly ash was not added to the soil until immediately prior to the initiation of characterization, matric potential measurements or column experiments. Repacked columns were deemed appropriate to simulate fly ash incorporation into the soil after application as an amendment.

Column experiments were conducted using a surrogate artificial rainwater (ARW) solution, based on precipitation samples collected on the Savannah River Site.[28] The ARW contained the following (mg L^{-1}): 0.082 Ca^{2+}, 0.042 K^+, 0.033 Mg^{2+}, 0.199 Na^+, 0.139 NH_4^+, 0.351 Cl^-, 0.711 NO_3^- and 1.29 $SO_4{}^{2-}$.

After the pH and EC were measured, the supernatant solution in this 1:2 soil:solution ratio mixture was collected for chemical analysis by Inductively Coupled Plasma-Mass Spectrometry (ICP-MS). This provided an estimate of readily-leachable solutes to give an indication of what to expect in column effluents. Since the column experiments were conducted in ARW, only the ARW-extractable As and Se results are shown in Table 1. The extractable As content for the soil was 1.73 μg kg^{-1}. The extractable As for the fly ash was much higher (227.43 μg kg^{-1}), with a 10% addition of fly ash increasing the extractable soil As to 2.37 μg kg^{-1}. The fly ash was high in Se (678.93 μg kg^{-1}) and a 10% addition increased soil Se from 0.49 to 39.98 μg kg^{-1}. This data may indicate that the soil had a greater affinity for As than Se.

3.1. Matric Potential Comparison

A centrifuge method was used to perform the matric potential comparisons of the soil and fly ash mixture.[29] The centrifuge apparatus, initially described by Nimmo et al.[30–31], was modified by researchers at Pacific Northwest Laboratory[32] and is referred to as an Unsaturated Flow Apparatus (UFA, model L8-UFA, Beckman Coulter, Inc., Fullerton, CA, Figure 1). The UFA, described in detail elsewhere[33–35], consists of two infusion pumps (AVI 200A, 3M, St. Paul, MN) and a modified Beckman J6-MI centrifuge (UFA Ventures, Richland, WA), including a rotating seal, rotor and column assembly. The infusion pumps make it possible to deliver the leaching solution at controlled rates. The column assembly consists of the sample holder and the effluent collection cup. The packed soil columns, contained in the sample holder, are 5 cm in length with an internal diameter of 3.3 cm. The effluent collection cup

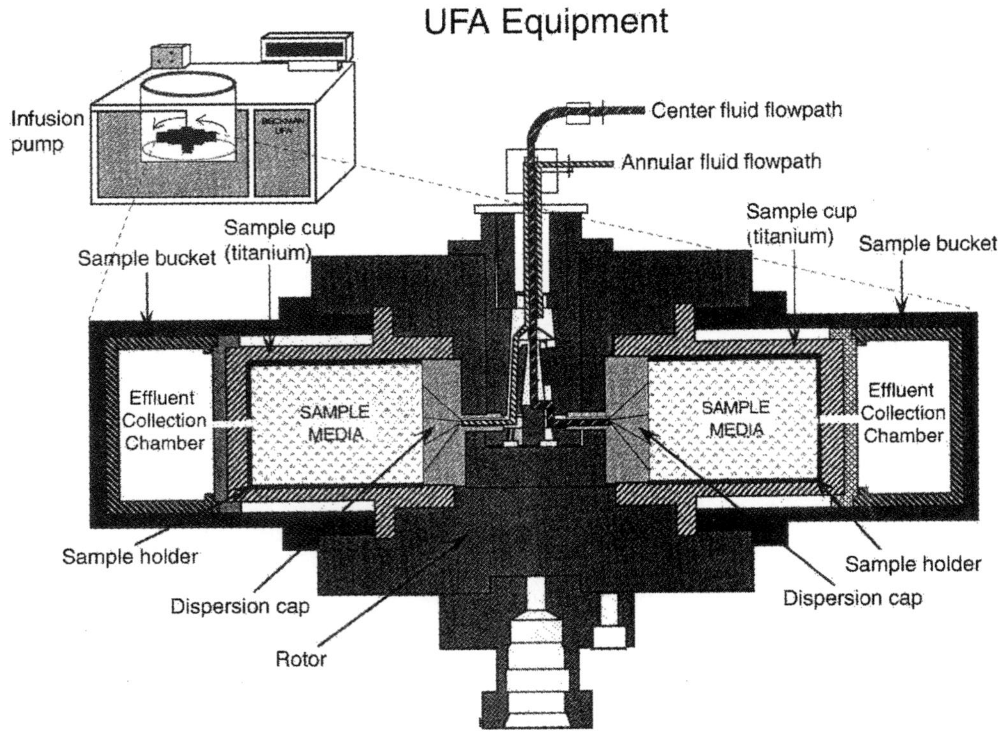

FIGURE 1. Diagram of the Unsaturated Flow Apparatus components[34].

is placed at the end of the sample holder and serves to collect the column effluent during centrifugation.

Field moist soil or fly ash mixture was packed into columns, saturated by slowly immersing the column in a beaker of ARW and weighed to determine initial water content. The columns were centrifuged at progressively higher rotation rates and durations to extract water under different equivalent tensions. At predetermined intervals for a given rotational speed, the centrifuge was stopped and the columns were removed and weighed to determine the amount of soil moisture that had been extracted. At the completion of the final centrifuge run, the columns were removed from the UFA and the final water content determined gravimetrically after heating the soil overnight at 60 °C. The calculated water tension, in bars, was then plotted as a function of water content to produce the characteristic matric potential curve.

3.2. Saturated Experiments

Saturated experiments were conducted prior to unsaturated experiments to isolate the impact of retention time on the leaching of solutes. Unlike electrostatic anion sorption processes[36] that are likely to be quite rapid and, therefore, somewhat insensitive to solution residence times, dissolution of ash-derived contaminants is likely to be highly sensitive to solution residence time, a parameter that can be controlled in the saturated columns by adjusting the inlet flow rate. The saturated system consisted of a column packed with fly ash mixture (length = 7.6 cm, internal diameter = 5 cm), a peristaltic pump delivering ARW and two inline electrodes measuring the pH and EC of the column effluent before collection by a fraction collector. Ottawa sand was packed into both ends of the saturated columns to adjust the diameter to length ratio to equal that of the UFA columns (0.66).

Two saturated experiments were completed. The first experiment (Sat1) was 98% saturated with a flow rate of 0.95 mL min^{-1} (Table 2). The average solute retention time for this column was 84 minutes, calculated as the volume of water-filled pore space divided by the volumetric flow rate. The bulk density of the fly ash mixture was 1.24 g cm^{-3}, calculated as the weight of solid added to the column (adjusted for moisture content) per column volume. The second column (Sat2) was 100% saturated, with a flow rate of 0.22 mL min^{-1}, giving a solute retention time of 281 minutes. The bulk density was 1.55 g cm^{-3}. Columns were leached for 24–30 pore volumes.

3.3. Unsaturated Experiments

Centrifuge techniques were also employed to simulate unsaturated flow in this study. This technique has the advantage of producing stable, low water contents in a relatively short time. The UFA, described above, contains a rotating seal at the center of the rotor that acts as a conduit through which fluid is delivered by the infusion pumps to the columns by two separate flow paths. This enables two columns to be run simultaneously with different inlet solutions. The solute residence time is controlled to some degree in the UFA by varying the

TABLE 2. Characteristics of saturated and unsaturated column experiments using surface soil and 10% fly ash*

Expt	bkgd	ρ_b (g·cm⁻)	F (cm^3·min^{-1})	rpm	S (%)	S range	RT (min)	As (μg 10 PV)	Se (μg 10 PV)	As (μg 20 PV)	Se (μg 20 PV)
Sat1	ARW	1.24	0.95	—	98	N/A	84	24.2	7.3	57.2	11.4
Sat2	ARW	1.55	0.22	—	100	N/A	281	24.9	14.2	59.8	25.4
Unsat1	ARW	1.50	0.25	600	60	68—58	45	3.1	2.1	—	—
Unsat2	$CaCl_2$	1.54	0.25	600	59	67—59	42	4.1	2.3	—	—
Unsat3	ARW	1.44	0.25	1100	21	28—22	18	2.4	1.3	4.9	2.4
Unsat4	ARW	1.42	0.25	1100	22	27—21	17	2.2	1.5	4.5	2.6
control	ARW	1.43	0.25	600	54	80—54	43	0.2	0.2	—	—

* Abbreviations are as follows: Expt, experiment; bkgd, background leaching solution; ρ_b, bulk density; F, flow rate; rpm, centrifuge revolutions per minute; S, percent saturation of column (percent of pore space filled with water); S range, range of saturation over course of experiment; RT, average solute residence time; PV, pore volume

inlet flow rate at a given centrifuge speed, assuming that the solution path remains similar at all saturation levels. The process is complicated somewhat further by the fact that fly ash will increase water-holding capacity in the coarse-textured soil and thus increase solute residence time to some degree.

Four columns were packed with fly ash mixture, resulting in an average bulk density of 1.47 g cm^{-3} (range: 1.42 to 1.54 g cm^{-3}; Table 2). Columns were then saturated by slowly immersing the column in a beaker of ARW. The length of saturation time was consistent among all columns and the residual ARW solution that was not retained in the column was collected and analyzed for solutes to maintain mass balance calculations. Columns were then centrifuged at predetermined speeds and flow rates (Table 2). The higher flow rate used in the saturated system (0.95 mL min^{-1}) resulted in significant backpressure and column plugging, so a slower flow rate was adopted (0.25 mL min^{-1}). The flow of ARW was not initiated until the centrifuge reached 75% of its maximum speed in an attempt to maintain consistent flow conditions.[37] Since the effluent collection cups can only hold about 20 mL, centrifugation was interrupted for sample collection at specified intervals. Each time a sample was taken, the water content of the columns was determined gravimetrically.

When conducting breakthrough experiments using the UFA, columns would normally be flushed with the leaching solution for 2 hours at the experimental flow rate and centrifuge speed to reach steady-state water contents.[36] However, when conducting these fly ash leaching studies, it is important to collect all column effluent since trace elements like As and Se may readily leach during the equilibration stage. Therefore, these columns were not equilibrated prior to the initiation of the experiment and therefore, the water content fluctuated over time. The reported saturation value for each unsaturated leaching experiment was taken at the point where the water content became stable, but the variation in saturation is given in Table 2 and changed as much as 6–26% during the course of the experiment.

Since two columns can be run on the centrifuge at one time, experiments Unsat1 and Unsat2 were run simultaneously. However, Unsat2 was run with 0.001 M $CaCl_2$ for comparison and to reduce possible clay dispersion and swelling that may occur under low ionic strength conditions. Unsat3 and Unsat4 are replicates to examine variation between experiments. One control experiment using only field moist soil without fly ash was also run on the centrifuge system. Arsenic and Se in all experiments was analyzed by ICP-MS.

4. Results and Discussion

The effect of adding fly ash to the soil can be seen in particle size, pH and electrical conductivity (EC) data in Table 1. Due to the dominance of silt-sized particles in fly ash, the 10% addition of this material increased the silt fraction of the soil from about 3 to 7%, raised the pH (in deionized water) from 4.90 to 6.51, increased the EC from about 31 to 203 μS cm^{-1} and appeared to decrease the water-dispersable clay fraction, though the difference was not significant.

4.1. Matric Potential Comparison

The addition of fly ash (10% by weight) increased the water-holding capacity of the soil (Figure 2). This was most likely a result of the addition of silt-sized particles to this sandy surface soil.[2–3,38] At any given matric potential, the soil amended with fly ash held a greater amount of water than the soil alone. This fly ash could be beneficial as an amendment to sandy or heavy clay soils to improve water holding capacity and drainage.

4.2. Saturated Experiments

The results of the saturated experiments are shown in Figure 3. These experiments were conducted to examine the effect of solute residence time on the leaching of solutes from fly ash amended soil. The pH of the column effluent ranged from 5.96 to 6.26 for Sat1 and 6.17 to 6.35 for Sat2 during the course of the experiment (data not shown). It is evident from the graphs that more As and Se leached from the columns as residence time increased (281 min. vs. 84 min.). For Se (Figure 3B), a large fraction of the leached solute eluted within the first few pores volumes, indicating that this fraction is readily soluble. This trend has been shown for other trace elements in saturated

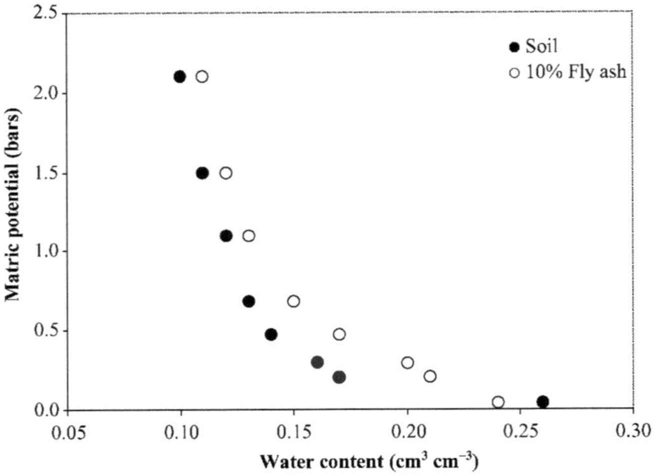

FIGURE 2. Matric potential as determined by the UFA method for coarse-textured surface soil and 10% by weight addition of fly ash to surface soil.

column leaching experiments.[5] However, with As (Figure 3A), leaching occurs over a longer period of column flushing, suggesting that the As is in a less soluble form than Se. For the longer residence time, the As concentration in the effluent actually increased with subsequent column leaching. This effect could not be explained by flucuations in effluent pH, but may be due to the soil's preference for As over Se (as seen in pH extracts, discussed previously, Table 1).

In order to compare columns, Table 2 includes the amount of As and Se leached in ten pore volumes of effluent. For

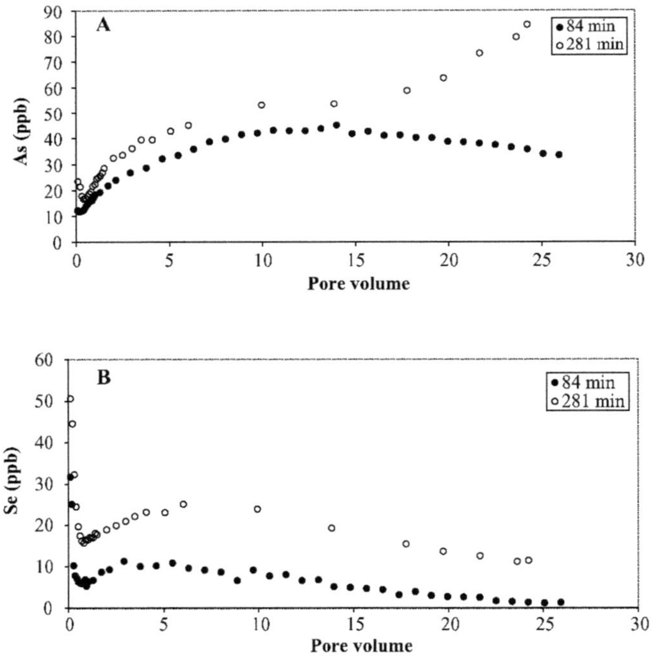

FIGURE 3. Saturated column experiments: comparison of residence time on leaching of As and Se.

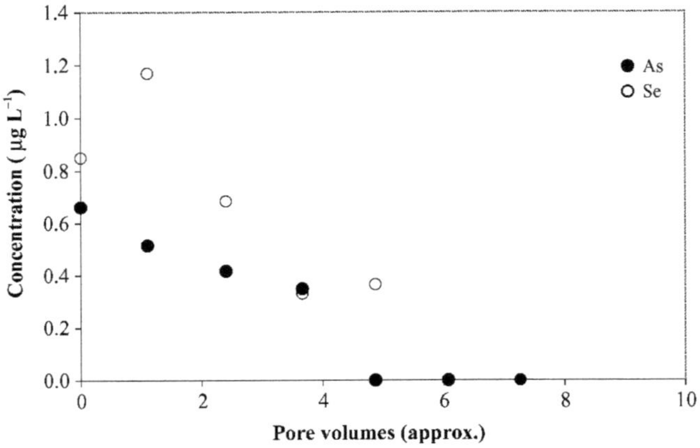

FIGURE 4. As and Se in effluent of control unsaturated column run using UFA.

the saturated experiments, As leaching was not affected by residence time. However, about twice as much Se leached from the longer residence time column (281 min., Sat2) than the shorter one (84 min., Sat1). These results are similar to those for data to 20 pore volumes, though not all columns were leached to 20 pore volumes (Table 2).

4.3. Unsaturated Experiments

The leaching patterns for As and Se in the control unsaturated experiment is shown in Figure 4. The pore volumes are labeled as approximate due to the previously mentioned problem with water content fluctuations during the course of the experiment. Most of the As leached out of the column in the first few pore volumes, with a peak concentration of 0.66 ppb. Selenium in the column effluent reached a peak concentration of 1.2 ppb in the first pore volume before dropping below the detection limit after about five pore volumes (approximate). Overall, the amount of As and Se in the control (only soil) unsaturated experiment was relatively low. The cumulative As leached in the control column was about 0.4 μg kg^{-1} (ppb), similar to the ARW extract value of 1.73 μg kg^{-1} given in Table 1. The cumulative Se leached from the control column was 0.2 μg kg^{-1}, similar to the ARW extract value of 0.49 μg kg^{-1} given in Table 1. The similarities in the data may be due to the fact that the control column solute residence time (43 min.) was similar to the time spent taking pH measurements and collecting supernatant (about 30 min.).

The results of the unsaturated experiments using fly ash mixture are shown in Figures 5, 6 and 7. Figure 5 shows the effect of ionic strength on the leaching of As and Se. For As, it appears that more solute leached out with the higher ionic strength solution, 0.001 *M* $CaCl_2$. Table 2 lists the amount leached (3.1 μg vs. 4.1 μg) in ten approximate pore volumes of leachate, though this difference was not significant. This may be because the experiments were not leached for sufficient pore volumes. The concentration of As in the column effluent increased with leaching, as seen in the saturated columns. The concentration

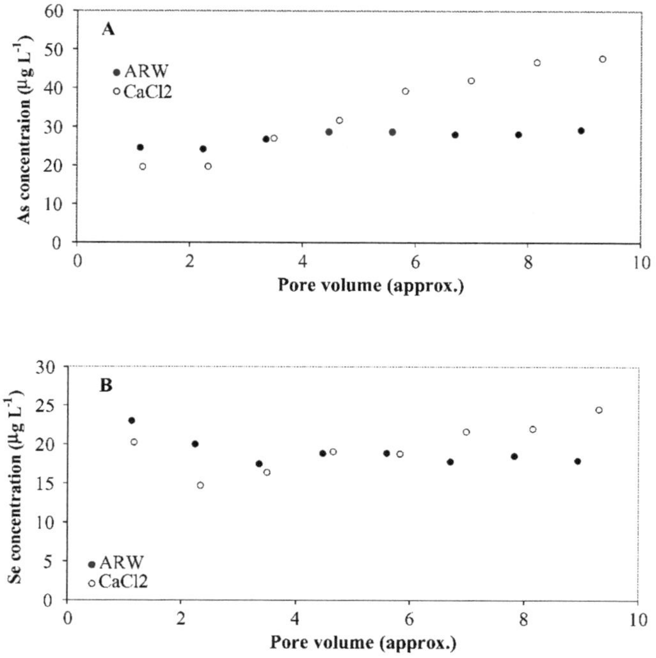

FIGURE 5. Comparison of the ionic strength of the leaching solution on As and Se leaching in unsaturated fly ash amended columns.

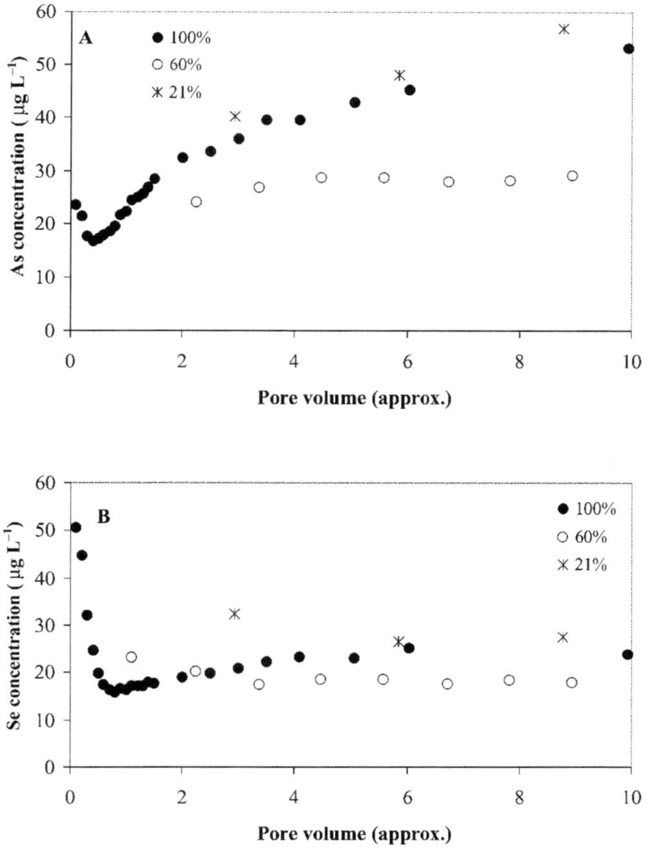

FIGURE 6. Comparison of As and Se leaching in columns of different saturation levels.

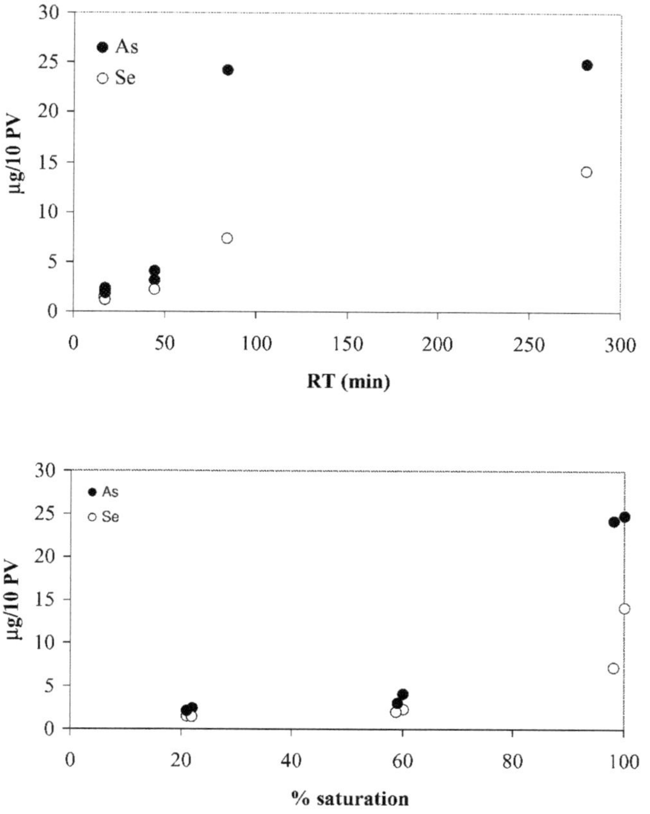

FIGURE 7. The amount of As and Se leached from columns after ten pore volumes (PV) of effluent with average solute residence time (RT) and saturation level.

of Se in the column effluent did not appear to decrease with leaching, unlike the saturated experiments where the Se concentration peaked in the first few pores volumes then decreased. This delay in the "leaching front" could be a consequence of desaturation as it is not seen in the saturated columns (Figure 3B). However, these observations could also be the result of insufficient leaching of the columns and the fact that the column leached with 0.001 *M* $CaCl_2$ was initially saturated with ARW.

Figure 6 shows the results of the effect of water content on the leaching of As and Se from fly ash mixture. For As, the saturated column (100%, Sat2) had a higher concentration over the range of measured pore volumes than the 60% (Unsat1) saturation column (Figure 6A). This was substantiated with the data in Table 2 which showed the total amount of As leached from Sat2 to be greater than Unsat1. However, with a further decrease in saturation to 21%, the initial concentration of the leachate was higher than the saturated case. Since the unsaturated columns held a much smaller amount of solute and soil, the values in Table 2 show As leaching to be greatest under saturated conditions. There may be a delay in the "leaching front" with desaturation which was more evident in the Se data (Figure 6B).

In the saturated column, a large fraction of the leached Se eluted in the first few pore volumes. The 60% saturated column usually had lower Se concentrations over the range of measured pore volumes than the 100% saturation columns. However, in the first few pore volumes, the 60% saturation column had higher Se concentrations than in the saturated columns. This delayed "leaching front" was even more pronounced for the lowest saturation column (21%). Therefore, even though more As and Se was leached out of the saturated columns (which have longer residence times and more soil to solution contact), the leaching curves sometimes showed higher concentrations for As and Se because not all the As and Se had been flushed in the first few pore volumes. However, since the unsaturated columns had smaller pore volumes than the saturated columns and effluent could not be sampled as often using the UFA, unsaturated data had less resolution, making it difficult to compare data from these two systems.

Figure 7A shows the amount of As and Se leached from the columns (μg per 10 pore volumes) with average solute residence time. Though results for As and Se differ, there is an increase in the amount of solute leached when the residence time is increased. Figure 7B shows the amount of As and Se leached from the columns with saturation. There was a noticeable increase in the amount of solute leached with increasing saturation. Since leaching varied similarly with both residence time and saturation, the effect of saturation on leaching could not be isolated.

5. Conclusions

Most of the leaching effects in this study can be attributed to residence time (equlibration time) differences. However, there did appear to be a delay in the "leaching front" of the solute with desaturation, though more column studies would have to be undertaken to examine this effect. Also, the results of this study are complicated by the variation in water content of the unsaturated columns during the course of the experiment. These problems would have to be overcome before the specific effect of saturation on leaching of solutes from fly ash amended soil could be isolated and examined.

References

1. Kalyoncu, R. S. [2003] Coal combustion products: U. S. Geological Survey Minerals Yearbook 2001, vol. 1, available online at http://minerals.usgs.gov/minerals/pubs/commodity/coal/coalmyb01.pdf
2. Chang, A. C., Lund, L. J., Page, A. L. and Warneke, J. E. Physical properties of fly ash-amended soils, *J. Environ. Qual.*, 6, 267, 1977.
3. Adriano, D. C., Page, A. L., Elseewi, A. A., Chang, A. C. and Straughan, I., Utilization and disposal of fly ash and other coal residues in terrestrial ecosystems: A review, *J. Environ. Qual.*, 9, 333, 1980.
4. Jacobs, L. W., Erickson, A. E., Berti, W. R. and MacKellar, B. M., Improving crop yield potentials of coarse textured soils with coarse fly ash amendments, *Proc. Ninth International Ash Use Symposium*, 3, 59, 1991.
5. Ghodrati, M., Sims, J. T. and Vasilas, B. L., Evaluation of fly ash as a soil amendment for the Atlantic Coastal Plain: I. Soil hydraulic properties and elemental leaching, *Water, Air, and Soil Pollution*, 81, 349, 1995.
6. Punshon, T., Adriano, D. C. and Weber, J., Restoration of eroded land using coal fly ash and biosolids, TR-113940, EPRI, 1999.
7. Keefer, R. F., Chapter 1: Coal ashes—Industrial wastes or beneficial byproducts?, in *Trace elements in coal and coal combustion residues*, R. F. Keefer and K. S. Sajwan, Eds., Lewis Publishers, Ann Arbor, 1993.
8. Carlson, C. C. and Adriano, D. C., Environmental impacts of coal combustion residues, *J. Environ. Qual.*, 22, 227, 1993.
9. Cherry, D. S. and Guthrie, R. K., Toxic metal in surface waters from coal ash, *Water Res. Bull.*, 13, 1227, 1977.
10. Adriano, D. C., Woodford, T. A. and Ciravolo, T. G., Growth and elemental composition of corn and bean seedlings as influenced by soil application of coal ash, *J. Environ. Qual.*, 7, 416, 1978.
11. Page, A. L., Elseewi, A. A. and Straughan, I., Physical and chemical properties of fly ash from coal-fired power plants with reference to environmental impacts, *Residue Rev.*, 71, 83, 1979.
12. Elseewi, A. A. and Page, A. L., Molybdenum enrichment of plants grown on fly ash–treated soils. *J. Environ. Qual.*, 13, 394, 1984.
13. Furr, A. K., Kelly, W. C., Bache, C. A., Gutenmann, W. H. and Lisk, D. J., Multielement uptake by vegetables and millet grown in pots on fly ash amended soil., *J. Agric. Food Chem.*, 24, 885, 1976.
14. Furr, A. K., Parkinson, T. F., Hinrichs, R. A., Van Campen, D. R., Bache, C. A., Gutenman, W. H., St. John, Jr., L. E., Pakkala, I. S. and Lisk, D. J., National survey of elements and radioactivity in fly ashes: Absorption of elements by cabbage grown in fly ash-soil mixtures, *Environ. Sci. Technol.*, 11, 1104, 1977.
15. Straughan, I., Elseewi, A. A. and Page, A. L., Mobilization of selected trace elements in residues from coal combustion with special reference to fly ash. P. 389–402 in D. D. Hemphill (ed.) *Trace substances in Environmental Health – XII, A Symp.* Univ. of Missouri, Columbia, 1978.
16. Fleming, W. J., Gutenmann, W. H. and Lisk, D. J., Selenium in tissues of woodchucks inhabiting fly ash landfills. *Bull. Environ. Contam. Toxicol.*, 21, 1, 1979.
17. De Jong, D., Morse, R. A., Gutenman, W. H. and Lisk, D. J., Selenium in pollen gathered by bees foraging on fly ash-grown plants, *Bull. Environ. Contam. Toxicol.*, 18, 442, 1977.
18. Gutenmann, W. H., Bache, C. A., Youngs, W. D. and Lisk, D. J., Selenium in fly ash, *Science* (Washington, D. C.), 174, 966, 1976.
19. Evans, D. and Giesy, J., Trace metal concentrations in a stream-swamp system receiving coal ash effluent, p. 782–790 in M. Wali (ed.) *Ecology and coal resource development*, Pergamon Press, New York, 1979.
20. Forbes, A. M., Magnuson, J. J. and Harrell, D. M., Effects of habitat modifications from coal ash effluent on stream macroinvertebrates: A synthesis. P. 241–249 in L. A. Krumholz (ed.) *Warmwater streams symposium*, Am. Fisheries Soc., Lawrence, KS, 1981.
21. Lemly, A. D., Toxicology of selenium in a freshwater reservoir: Implications for environmental hazard evaluation and safety, *Ecotoxicol. Environ. Safety*, 10, 314, 1985.

22. Gee, G. W. and Bauder, J. W., Chapter 15: Particle size analysis, in *Methods of soil analysis, Part 1: Physical and mineralogical methods, 2nd edition*, 9, A. Klute (ed.), American Society of Agronomy, Madison, WI, 383, 1986.
23. Nelson, D. W. and Sommers, L. E., Total carbon, organic carbon, organic matter, in *Methods of Soil Analysis*, 2, Page, A. L., Miller, R. H. and Keeney, D. R. (eds.), American Society of Agronomy, Madison, WI, 1982.
24. Jackson, M. L., Lin, C. H., and Zelazny, L. W., Oxides, hydroxides, and aluminosilicates. P. 101–150, in A. Klute (ed.) *Methods of soil analysis*, Part 1. SSSA Book Ser. 5 SSSA, Madison, WI, 1986.
25. Rhoades, J. D., Cation exchange capacity, p. 149–157, in A. L. Page (ed.) *Methods of soil analysis, Part 2*, SSSA Book Ser. 5, SSSA Madison, WI, 1982.
26. Thomas, G. W., Exchangeable cations, p. 159–164, in A. L. Page (ed.) *Methods of soil analysis, Part 2*, SSSA Book Ser. 5, SSSA, Madison, WI, 1982.
27. Miller, W. P. and Miller, D. M., A micro-pipette method for soil mechanical analysis, *Commun. Soil Sci. Plant Anal.*, 18, 1, 1987.
28. Strom, R. N. and Kaback, D. S., SRP baseline hydrogrologic investigation: Aquifer characterization groundwater geochemistry of the Savannah River Site and Vacinity (U), Westinghouse Savannah River Company, Environmental Sciences Section, Aiken, SC, 1992.
29. ASTM D 6527-00, Standard test method for determining unsaturated and saturated hydraulic conductivity in porous media by steady-state centrifugation, ASTM International, 2003.
30. Nimmo, J. R., Rubin, J. and Hammermeister, D. P., Unsaturated flow in a centrifugal field: Measurement of hydraulic conductivity and testing of Darcy's Law, *Water Resour. Res.*, 23, 124, 1987.
31. Nimmo, J. R., Akstin, K. C. and Mello, K. A., Improved apparatus for measuring hydraulic conductivity at low water content, *Soil Sci. Soc. Am. J.*, 56, 1758, 1992.
32. Conca, J. L. and Wright, J., Hydrostratigraphy and recharge distributions from direct measurement of hydraulic conductivity using the UFA method, PNL-9424, Pacific Northwest Laboratory, Richmond, WA, 1992.
33. Conca, J. L. and Wright, J., Diffusion coefficients in gravel under unsaturated conditions, *Water Resour. Res.*, 26, 1055, 1990.
34. Khaleel, R., Relyea, J. F. and Conca, J. L., Evaluation of van Genuchten-Mualem relationships to estimate unsaturated hydraulic conductivity at low water contents, *Water Resour. Res.*, 31, 2659, 1995.
35. Gamerdinger, A. P. and Kaplan, D. I., Application of a continuous-flow centrifugation method for solute transport in disturbed, unsaturated sediments and illustration of mobile-immobile water, *Water Resour. Res.*, 36, 1747, 2000.
36. Hutchison, J. M., Seaman, J. C., Aburime, S. A. and Radcliffe, D. E., Chromate transport and retention in variably saturated soil columns, *Vadose Zone J.*, 2, 702, 2003.
37. UFA Ventures, Inc., Chapter 4: Hydraulic conductivity, UFA technical procedures, Richland, WA, 1996.
38. Adriano, D. C. and Weber, J. T., Influence of fly ash on soil physical properties and turfgrass establishment, *J. Environ. Qual.*, 30, 596, 2001.

15
Solution Geochemistry Gradients in an Acid Mine Drainage Wetland Substrate

A.D. Karathanasis, J.E. Robinson, and E.M. D'Angelo
Department of Agronomy, University of Kentucky, N-122K Ag. Science-North, Lexington, KY 40546-0091, USA

1. Abstract

The geochemistry and treatment capacity of an abandoned multistage acid mine drainage (AMD) wetland receiving high metal and acidity loads was evaluated as a function of treatment stage, substrate depth, and seasonal variations. In spite of excessive overloading for its small size and the lack of maintenance for the past 6 years, the wetland treatment significantly decreased Fe, SO_4^{2-}, and acidity with successive treatment stage. Extensive Fe^{3+}-hydrolysis reactions, combined with the influx of oxygenated water from the preceding oxidation pond into the wetland cell substrate, maintained high Eh levels (>350 mV) throughout the wetland, thus inhibiting alkalinity contributions through dissimilatory SO_4^{2-}reduction. Only limited buffering occurred through the dissolution of the limestone gravel substrate, which was insufficient to overcome the secondary acidity produced from Fe^{3+} hydrolysis reactions. The main Fe removal mechanism was the precipitation of Fe-hydroxysulfate and Fe-oxyhydroxide minerals, which was promoted by microbial oxidation of Fe^{2+} in lower substrate depths and Fe^{3+} hydrolysis in the surface. Sulfate retention was accomplished through Fe-hydroxysulfate and gypsum precipitation throughout the wetland. Manganese precipitation was inhibited by high Fe^{2+} and Eh levels, while Al was retained in the form of alunite or jurbanite, primarily in lower substrate depths. Metal, SO_4^{2-}, and acidity loads increased in the spring months due to higher hydraulic loads and increased pyrite oxidation rates through microbial activity. Solution equilibrium speciations suggested greater saturation with respect to Fe- and Al-minerals during the fall months, when precipitation reactions were enhanced by lower hydraulic loads, longer residence times, and expanded oxidized rhizosphere zones by the hydrophytic vegetation.

2. Introduction

The dissolution of iron sulfide minerals, commonly as pyrite, associated with Appalachian coal seams in the United States has created a long-term pollution source for waterways receiving drainage from underground mines. Acid mine drainage (AMD) is defined by a low pH (<4), Fe and Mn concentrations above 6 and 2 mg L^{-1} respectively, increased sulfate concentrations, and the release of other metals including aluminum.[1,2,3] The dissolution of iron sulfides and other sulfide metal combinations initiates the production of insoluble ferric hydroxides and acidity according to the reaction[4]:

$$2FeS_{2(s)} + 7.5O_2 + 7H_2O \Leftrightarrow 2Fe(OH)_{3(s)} + 4SO_4^{-2} + 8H^+.$$

Since the 1980's, constructed wetlands have been used as remediation tools for acid mine drainage problems because of their low maintenance and cost requirements. However, site availability, hydrology, temperature, aging, loading rate, and microbial rate fluctuations could pose some limitations.[2,5,6,7,8]

The two major types of wetland systems used to treat acid mine drainage utilize either surface or subsurface water flow with oxidation/hydrolysis reactions being the main metal precipitation and removal processes. Subsurface systems have the added ability to generate alkalinity and enhance treatment functions through chemical, physical, and microbial substrate processes. Metal retention within the substrate may be associated with precipitation, adsorption to substrate sites, or complexation by organic residues.[8,9] The incorporation of aerobic basins and limestone substrates in subsurface flow systems maximizes metal precipitation and buffers the associated acidity increases. Oxidation basins promote hydrolysis reactions and enhance metal precipitation. Organic substrates provide carbon, energy, and surfaces that support microbially-induced metal precipitation, with strong influences from hydrology, temperature, retention times, pH, and Eh.[3] It has been reported that the presence of crystalline iron minerals, such as lepidocrocite, hematite, and goethite, in upper oxygenated sections of the wetland substrate, and amorphous ferrihydrite and Fe-oxyhydroxysulfates in lower depths.[9] The presence of a limestone substrate may also induce $FeCO_3$ mineral precipitation, thus removing available alkalinity buffering from the system.

Due to the low pH of these systems, microbial reactions are necessary for iron removal, while increased partial pressures of CO_2 due to respiration help to maximize the dissolution of limestone.[3,10] It has also been reported that microbial effects upon metals as being either through direct contact or through the release of organic acids and metabolites.[11] Aluminum and Mn solubility are mainly controlled by pH, with manganese removal at pH > 6 and Al removal at pH > 4[12].

The redox state of the wetland system is important in controlling the types of precipitates forming within the system as well as potential sulfide production. The oxidation of Fe^{2+} to Fe^{3+} is the rate-controlling removal step[5], but the dissolution of amorphous Fe-minerals under certain conditions may also be an important process.[11,13] Dissolution of precipitated reactive Fe-hydroxides and Fe-hydroxysulfates, particularly those of poor crystallinity, may transform the wetland from a sink into a net source of iron. Removal of Mn is difficult due to the complexity of redox interactions, including a tendency of dissolved Mn^{4+} to oxidize Fe^{2+} .[6,11,12,14,15]

The generation of sulfide through sulfate reduction, the corresponding alkalinity generation, and the precipitation of metal (except for Mn) sulfides are also important and rapid metal removal processes.[8,11,16] However, low sulfide levels may result from suppression of sulfate reducing bacteria due to overloading, redox > −100 mV, and short retention times.[7,16,17] It has also been reported low sulfide levels in a renovated wetland despite the removal of sulfate from the system, primarily through the formation of gypsum and Fe-oxyhydroxysulfate precipitation.[18]

Most studies of AMD treatment systems have focused upon influent and effluent chemical interactions considering the wetland substrates as "black box" systems. Monitoring and evaluating the geochemical gradients within the wetland substrates provides a better understanding of the multiple complex processes prevailing in these systems. The objective of this study was to determine geochemical gradients as a function of substrate depth within the wetland cells of an AMD wetland based upon water chemistry. The parameters studied included treatment stage, depth zone, season, and the combined effects of these parameters.

3. Materials and Methods

The study site, located in McCreary County, KY, consists of an abandoned wetland containing an initial settling lagoon, two parallel anoxic limestone drains, and a series of alternating aerobic/anaerobic cells (Fig. 1). Aerobic cells are essentially oxidation basins promoting hydrolysis processes and have no organic matter or vegetation. From there, water flows through subsurface pipes into anaerobic wetland cells where it percolates upward through a limestone gravel base overlaid by a hay/mushroom compost layer, which is planted with *Typha latifolia* (cattails). Due to topographic limitations of the site, the wetland is divided into two fields. The upper field consists of three alternating aerobic/anaerobic sequences that are connected via a flume with a second field containing two more treatment sequences. However, due to clogging and lack of maintenance the lower field no longer receives flow from the upper field of the treatment system. Therefore, sampling for this study focused on the three vegetated anaerobic wetland cells (Cells 1–3) of the upper wetland section covering a surface area of about 700 m^2. The first wetland cell contains only remnants of cattail vegetation due to the harsh acidity and high metal load conditions.

Water column and substrate pore-water samples were collected in winter 2002, spring 2003, and fall 2003. Pore-water equilibrators[19,20,21] of 35 cm length consisting of 8 mL slots at 1cm intervals in a polycarbonate sheet were used to sample the water-substrate column. The equilibrators were filled with deoxygenated, de-ionized water, covered with a 0.45 μm

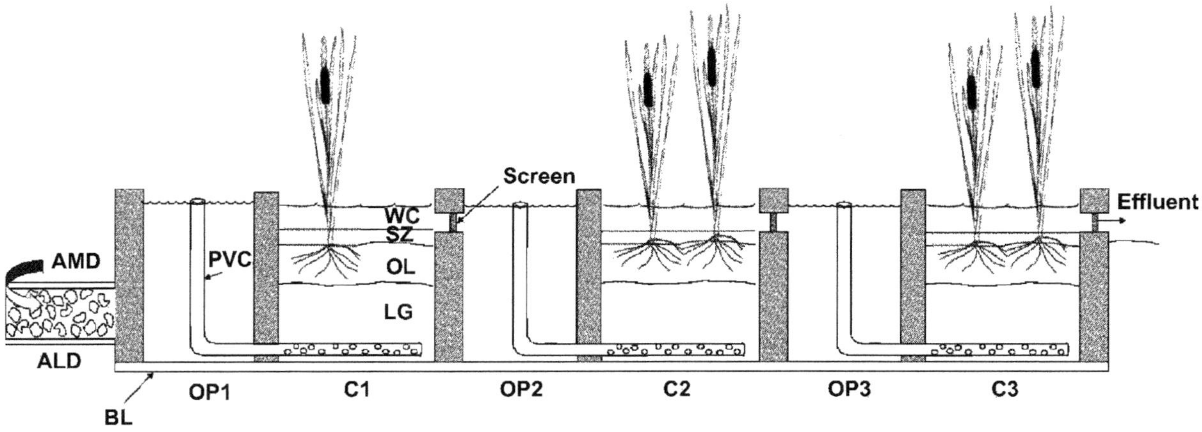

† AMD = acid mine drainage; ALD = anoxic limestone drain; BL = bentonite layer; OP = oxidation pond; PVC = polyvinyl chloride pipe; C = wetland cell; LG = limestone gravel; OL = organic layer; SZ = surface zone; WC = water column.

FIGURE 1. Cross sectional view of the wetland.

filter and protective 1mm Nitex mesh cover. Duplicate equilibrators were installed in each cell with approximately 5 cm in the water column and 30 cm below the substrate surface. The equilibrators remained in the substrate for 21 days to equilibrate with the interstitial substrate solution. Upon removal, each of the 35 samples per equilibrator was divided into three subsets. One mL of the water sample was stored in a N_2-purged glass vial for gas analysis, 1mL was stored in a syringe purged of air and kept in a rubber stopper for pH determination, and the remaining sample was acidified with concentrated HCl to a pH <2 for metal and sulfur analyses. All samples were stored at 4°C until analysis. The pH was measured immediately after returning from the field. Ferrous and total iron, gases, and sulfide were analyzed within 3 days. The remaining analyses were completed within 4 weeks.

All analytical determinations were conducted in accordance with the guidelines of Standard Methods for the Examination of Water and Wastewater.[22] The redox state of the cell substrate was assessed in situ with an Accumet AP62 multi-meter equipped with platinum tipped sample- and calomel reference-electrodes. The pH was measured using a Corning semi-micro pH probe attached to an Orion Model SA-520 pH meter. Alkalinity could not be determined by titration due to limited sample volume, so it was calculated as mg/L $CaCO_3$ equivalents from the CO_2 gas content of the 1 mL sealed vial sample. The gas samples were determined using a Shimadzu Gas Chromatograph GC-8A for CO_2, and Shimadzu Gas Chromatograph GC-14A for CH_4 and N_2O analysis[21].

A Solaar M5 Thermal Elemental flame atomic adsorption spectrometer was used to determine calcium, magnesium, aluminum, manganese, sodium, and potassium. Ferrous iron, ferric iron, and sulfide were measured colorimetrically with a Versa Max microplate reader. Ferrous iron and ferric iron were analyzed using the ortho-phenanthroline method[22], and sulfide-S was determined by the method[23] adapted as a plate reader technique. Sulfate-S was determined turbidimetrically using the BaCl method. Acidity calculations were made using the following equation:

$$\text{Acid (calc)} = 50[2Fe^{2+}/56 + 3Fe^{3+}/56 + 3Al/27 + 2Mn/55 + 1000(10^{-pH})],$$

where concentrations are in mg/L and the acidity term is the equivalent in mg/L of $CaCO_3^2$. Statistical analyses were performed using the Statgraphics Plus 5.0 software program using Duncan's multiple range test with an alpha level of 0.05. Mineral saturation index determinations were made using the Visual Minteq 2.23 software program.

4. Results and Discussion

4.1. Treatment Stage Effects

Table 1 shows concentration gradients and sequential treatment effects through Cells 1, 2, and 3 on the overall wetland geochemistry, combining all substrate depths. Iron (0.54–1.6 g/L), sulfate (1.9–4.2 g/L) and acidity (1.0–2.9 g/L) levels were among the highest ever reported in the literature for AMD composition, indicating the overwhelming metal and acidity load imposed on the wetland. More than 99% of the total iron was in the Fe^{2+} form. In spite of the overload, the wetland treatment resulted in drastic decreases in SO_4^{2-} (55%), Fe^{2+}(66%), and net acidity (76%) from Cell 1 to Cell 3, and gradual decreases in Mn, Al, and Ca concentrations. Considering an average hydraulic load of 20 L/min, this amounts to ~30.5 kg/day of Fe, and ~67.5 kg/d of SO_4^{2-} being retained, and ~54.7 kg/day of acidity being neutralized by the wetland. However, pH levels remained low, because the extremely high acidity generated through Fe^{2+} hydrolysis reactions was only partially buffered by the dissolution of the limestone substrate.

The pH drop in cell 2 is probably the result of increased iron hydrolysis reactions occurring in Cell 1 effluents entering the lower substrate of Cell 2. The main mechanism for the decrease of Fe^{2+} and SO_4^{2-} with increasing treatment stage is attributed to precipitation of Fe-oxyhydroxides such as ferrihydrite,

TABLE 1. Mean concentrations ±1 SD for AMD quality parameters and redox potential for the three wetland cells (combined substrate depths) representing advancing treatment stage

AMD parameter	Cell 1	Cell 2	Cell 3
pH	4.0 ± 0.9^b	3.7 ± 0.8^a	4.3 ± 1.2^c
SO_4^{2-}(mg/L)	4242.4 ± 1505.0^c	2641.6 ± 1093.8^b	1899.1 ± 1448.6^a
Fe^{2+}(mg/L)	1589.8 ± 683.6^c	861.5 ± 427.5^b	540.7 ± 490.2^a
Fe^{3+}(mg/L)	4.8 ± 4.2^a	9.3 ± 32.9^b	10.9 ± 37.5^b
Mn^{2+}(mg/L)	9.0 ± 3.6^c	4.8 ± 1.9^b	3.8 ± 2.3^a
Al (mg/L)	7.1 ± 7.2^b	4.2 ± 2.1^a	4.1 ± 2.4^a
Ca^{2+}(mg/L)	379.7 ± 93.8^c	323.6 ± 104.9^b	276.1 ± 130.0^a
Acidity* (mg/L)	2920.4 ± 1204.7^c	1594.3 ± 733.4^b	1011.9 ± 864.7^a
Alkalinity* (mg/L)	881.2 ± 268.5^b	535.7 ± 287.8^a	524.3 ± 374.8^a
Net Acidity* (mg/L)	2040.8 ± 1160.9^c	1062.7 ± 555.6^b	487.7 ± 729.8^a
Eh (mV)	590	480	360
*$CaCO_3$ equivalent			

§ Values followed by the same letter are not statistically significant at the a = 0.05 probability level

TABLE 2. Saturation indices (log Q/K) calculated by Visual Minteq for potential minerals precipitating within the wetland cells (combined substrate depths)

Minerals	Cell 1	Cell 2	Cell 3
Alunite	+1.5	−1.0	+2.5
Ferrihydrite	+3.4	+0.6	+0.2
Gibbsite	−1.5	−2.4	−0.5
Goethite	+6.1	+3.3	+2.9
Gypsum	−0.1	−0.2	−0.3
Jurbanite	−0.5	−1.0	−0.4
K-Jarosite	+12.2	+3.9	+0.5
Lepidocrocite	+5.3	+2.4	+2.0
Rhodochrosite	−4.3	−5.4	−4.2
Schwertmannite	+113.8	+69.9	+58.2
Siderite	−2.6	−3.6	−2.5

goethite, lepidocrocite, and Fe-oxyhydroxysulfates like jarosite and schwertmannite. Indeed, solution speciation analysis indicated supersaturation of the AMD with ferrihydrite, goethite, lepidocrocite, jarosite, and schwertmannite-type minerals (Table 2). Solution saturation with respect to the solubility of these minerals appeared to decrease with sequential treatment due to subsequent reductions in Fe and SO_4^{2-} loads. In addition to the precipitation of these minerals, further reductions in the SO_4^{2-} result from the precipitation of gypsum, particularly in lower carbonate-buffered substrate depths, where Ca and SO_4^{2-} levels were near equilibrium with gypsum. The decreasing Fe/SO_4 molar ratios with sequential treatment are consistent with the observed decreases in solution saturation indices for schwertmannite minerals, but they may also be partially attributed to precipitation of Fe-oxyhydroxysulfate minerals with higher SO_4/Fe stoichiometry. The increased alkalinity in cell 1 is attributed to the extra buffering capacity provided by the anoxic limestone drains to the AMD entering the first wetland cell. The alkalinity in cells 2 and 3 was relatively stable and consistent with the dissolution of the limestone substrate.

Gradual decreases in Mn in cells 2 and 3 are probably the result of sorption onto the substrate matrix sites rather than precipitation, as solution speciations indicated undersaturation with Mn-oxyhydroxides and rhodochrosite. The high Fe^{2+} concentrations of the AMD are also inhibiting Mn precipitation.[14,24] Aluminum reductions in cells 2 and 3 are attributed to sorption onto substrate matrix sites, but speciation analysis also suggested possible precipitation of alunite or jurbanite. While soluble Ca levels in the AMD are consistent with the dissolution of the limestone substrate, relatively high concentrations of Mg (106–150 mg/L), K (65–133 mg/L), and Na (118–148 mg/L) are attributed to the dissolution of silicate or carbonate minerals associated with the geology of the mine and impurities of the limestone substrate. The observed decreases in Ca (as well as in Mg, K, and Na) with increasing treatment stage may be attributed to increased plant-uptake since vegetation growth and density improved in Cells 2 and 3. No soluble sulfide species were detected, which is consistent with the relatively high Eh values (360–590 mV) measured throughout the wetland. Although these redox levels may appear somewhat atypical for wetland substrates, they are attributed to the unique design of this wetland, which allows the influx of oxygenated water into the wetland substrate from an oxidation pond preceding each wetland cell. The dominant Fe^{3+} hydrolysis reactions may also limit the system's ability to reach redox conditions suitable for sulfate reduction, thus making any alkalinity contributions from SO_4^{2-}reduction highly unlikely.

4.2. Substrate Depth Effects

Although the pore equilibrator sampling accounted for each cm of the wetland substrate depth, for the sake of clarity, the discussion of the results addresses four major depth zones, defined in terms of prevailing processes as (1) the water column, (2) the surface zone, (3) the organic layer, and (4) the limestone gravel layer. The water column (9 to 1 cm) represents an oxidized zone above the substrate surface with the lowest overall metal concentrations. Metal removal is predominantly controlled by iron hydrolysis reactions. Storm events directly influence this zone through dilution and mixing with adjacent flooded cells, while extensive dry periods cause higher soluble metal concentration gradients. The top 5 cm (0–5) of the substrate represents the surface zone within each wetland cell, where precipitation of Fe from water column and interstitial solution hydrolysis products occurs at a massive scale. In addition to Fe, some Mn immobilization may also occur in this zone through co-precipitation with Fe-oxyhydroxides. The third zone (6–20 cm) is dominated by organic constituents derived from the compost layer installed during the wetland construction and decomposed vegetation residues returned to the wetland at the end of the growth season each year. Interactions of the AMD with the organic substrate components are the major factors controlling pH, metal sorption and dissolution/precipitation processes through mediation by microbial activity and increased CO_2 pressure. Finally, the last depth zone (21–39 cm) is dominated by the limestone gravel substrate placed in the bottom of each cell during construction and constitutes the main source of alkalinity generated through dissolution processes. This alkalinity is expected to partially buffer the original acidity associated with the AMD, as well as secondary acidity produced during the Fe hydrolysis and precipitation reactions occurring within the wetland.

Table 3 shows average interstitial solution compositions of the four different substrate zones for all wetland cells combined. The results indicate significant concentration gradient stratification by depth zone, indicative of the different prevailing processes. Contrary to the trends observed in surface flow wetlands treating AMD, where the highest concentration gradients for Fe, SO_4^{2-}, and acidity are associated with the upper substrate, this wetland showed a significant increase in these parameters with increasing substrate depth. This is attributed to the high influx of Fe-hydrolysis products directly

TABLE 3. Mean concentrations ± 1 SD for AMD quality parameters and redox potential in the four substrate depth zones within the wetland

AMD parameter	Water column	Surface zone	Organic layer	Limestone gravel layer
PH	3.2 ± 0.8^{a}	3.6 ± 0.9^{b}	4.1 ± 1.0^{c}	4.2 ± 0.9^{c}
SO_4^{2-} (mg/L)	1767.3 ± 1182.2^{a}	2037.8 ± 1507.9^{a}	2922.2 ± 1690.3^{b}	3747.4 ± 1411.2^{c}
Fe^{2+} (mg/L)	440.3 ± 429.3^{a}	679.4 ± 616.2^{b}	1052.1 ± 708.0^{c}	1263.6 ± 638.1^{d}
Fe^{3+} (mg/L)	19.5 ± 34.5^{c}	9.6 ± 24.4^{b}	1.2 ± 7.5^{a}	9.1 ± 42.7^{b}
Mn^{2+} (mg/L)	7.1 ± 3.3^{c}	5.6 ± 2.7^{ab}	5.2 ± 2.3^{a}	6.5 ± 4.8^{bc}
Al (mg/L)	5.6 ± 6.0^{b}	4.8 ± 2.3^{ab}	5.7 ± 6.0^{b}	4.3 ± 2.5^{a}
Ca^{2+} (mg/L)	222.9 ± 117.2^{a}	246.5 ± 123.4^{a}	331.7 ± 105.5^{b}	392.6 ± 79.5^{c}
Acidity* (mg/L)	863.0 ± 747.0^{a}	1286.5 ± 1081.3^{b}	1934.8 ± 1259.5^{c}	2303.7 ± 1151.8^{d}
Alkalinity* (mg/L)	347.0 ± 303.0^{a}	490.5 ± 383.8^{b}	721.0 ± 332.6^{c}	725.0 ± 304.3^{c}
Net Acidity* (mg/L)	523.5 ± 491.6^{a}	796.0 ± 794.5^{b}	1212.4 ± 1074.5^{c}	1581.6 ± 1134.8^{d}
Eh (mV)	440	388	512	603
*$CaCO_3$ equivalent				

§Values followed by the same letter are not statistically significant at the a = 0.05 probability level

into the lower wetland substrate of each cell from the preceding oxidation pond. As the AMD infiltrated upward through the limestone gravel and the organic zone, precipitation of Fe-hydroxysulfates through Fe^{3+} hydrolysis and microbial oxidation reactions removed ample amounts of Fe and SO_4^{2-} from solution.[18,25] When the AMD reached the substrate surface Fe and SO_4^{2-} levels decreased further through the precipitation of Fe-oxyhydroxides as a result of the extra oxygen present in the free water column. This was demonstrated by the higher Fe^{3+} concentrations in the free water column zone. Decreases in total and net acidity of the AMD during this upward infiltration process are the result of gradual buffering occurring through the dissolution of limestone. This was demonstrated by pH increases with depth. However, increased alkalinity production in the lower substrate depths failed to compensate for the high acidity levels generated from Fe^{2+} hydrolysis reactions, thus resulting in increases of the overall acidity with depth.

Manganese and Al levels did not show much sensitivity to substrate depth, suggesting that they are mostly indifferent to the hydrolysis and microbial oxidation processes affecting Fe removal. High Fe^{2+} concentrations are known to inhibit Mn oxidation and even remobilize it through the contribution of electrons to precipitated Mn oxides.[14,24] Therefore, most of the Mn retention within the wetland occurs through co-precipitation with Fe-oxyhydroxides in the surface zone (0–5 cm) or sorption to the substrate matrix in lower depths.[6] Since pH levels within the wetland substrate are too low for Al-hydroxide precipitation, Al retention also occurs through sorption processes or precipitation of Al-sulfates. Indeed, solution speciation analysis supported the stability of jurbanite and alunite minerals. Surprisingly, soluble sulfide was not detectable at any substrate depth. Solution speciation analysis suggested considerable undersaturation with respect to FeS and pyrite in all water samples. This suggests that sulfate reducing bacteria were absent from the wetland substrate, and therefore, not contributing to extra alkalinity generation. Apparently, the extremely high SO_4^{2-} concentrations (>1,200 mg/L) and Eh levels (> −100 mV) present in the system inhibited dissimilatory sulfate reduction processes.[7,16,17] Calcium concentrations increased with depth reaching maximum values at the lowest substrate depth, which was dominated by the limestone gravel. Soluble Ca^{2+} and SO_4^{2-} levels were near equilibrium with gypsum throughout the wetland, particularly in the organic- and limestone-controlled substrate zones. These trends as a function of substrate depth were consistent within wetland cells, in spite of the decreasing influent and effluent loads with each subsequent treatment.

Solution speciation analysis by depth across all wetland cells suggested favorable environments for precipitation of goethite, lepidocrocite, and schwertmannite in all depth zones, with increasing supersaturation with increasing depth (Table 4). In contrast, the precipitation of ferrihydrite, jarosite, and alunite was favored mainly in the organic- and the limestone-controlled zones. Although still somewhat undersaturated, all depth zones were near equilibrium with gypsum, while Al and SO_4^{2-} levels in the two lowest depths approached equilibrium with jurbanite. Solution speciation trends by depth were somewhat different in the three wetland cells,

TABLE 4. Saturation indices (log Q/K) calculated by Visual Minteq for potential minerals precipitating within the four wetland substrate depth zones

Minerals	Water column	Surface zone	Organic layer	Limestone gravel layer
Alunite	−3.6	−1.3	+1.9	+1.9
Ferrihydrite	−1.8	−1.3	+2.4	+4.0
Gibbsite	−3.6	−2.5	−1.1	−1.1
Goethite	+0.9	+1.4	+5.1	+6.7
Gypsum	−0.5	−0.4	−0.2	−0.1
Jurbanite	−1.3	−1.0	−0.5	−0.5
K-Jarosite	−2.2	−1.7	+8.3	+13.1
Lepidocrocite	0.0	+0.5	+4.2	+5.8
Rhodochrosite	−6.1	−5.5	−4.2	−4.0
Schwertmannite	+35.8	+39.9	+95.3	+121.0
Siderite	−4.8	−3.8	−2.4	−2.4

TABLE 5. Mean concentrations ±1 SD for AMD quality parameters of the wetland for the winter, fall, and spring seasons

AMD parameter	Winter 2002	Spring 2003	Fall 2003
pH	3.6 ± 0.5^a	3.6 ± 0.7^a	4.8 ± 1.1^b
SO_4^{2-} (mg/L)	2442.4 ± 1642.5^a	4141.9 ± 1572.3^b	2196.3 ± 1008.8^a
Fe^{2+}(mg/L)	827.6 ± 649.8^a	1398.1 ± 792.6^b	764.0 ± 419.6^a
Fe^{3+} (mg/L)	7.0 ± 33.7^a	8.6 ± 24.6^a	5.4 ± 28.8^a
Mn^{2+} (mg/L)	6.5 ± 5.0^b	5.8 ± 2.0^{ab}	5.4 ± 2.7^a
Al	6.5 ± 6.6^b	2.3 ± 0.6^a	6.5 ± 3.4^b
Ca^{2+} (mg/L)	286.3 ± 101.9^a	300.3 ± 86.4^a	392.7 ± 132.5^b
Acidity* (mg/L)	1546.2 ± 1172.9^a	2550.2 ± 1396.3^b	1417.8 ± 743.1^a
Alkalinity* (mg/L)	439.8 ± 234.3^a	780.4 ± 326.4^b	720.3 ± 387.8^b
Net Acidity* (mg/L)	1106.4 ± 1015.7^b	1771.1 ± 1161.5^c	694.8 ± 664.2^a
*$CaCO_3$ equivalent			

§Values followed by the same letter are not statistically significant at the a = 0.05 probability level.

apparently due to the variable concentration gradients. The first cell, which showed the highest concentration gradients, exhibited saturation with ferrihydrite, goethite, lepidocrocite, schwertmannite, jarosite, and alunite in all substrate depths. In the second cell, the precipitation of goethite and schwertmannite was favored throughout the substrate, but ferrihydrite, lepidocrocite, jarosite, and alunite were stable only in the lowest two depths. Finally, in the third cell only the organic- and the limestone–controlled zones were favorable for the precipitation of these minerals.

4.3. Seasonal Effects

Table 5 shows seasonal effects on the overall wetland geochemistry regardless of cell and depth. The drastic increases in Fe, SO_4^{2-}, and acidity load during the spring compared to the fall and winter composition of the AMD were striking. Spring rainfall was expected to increase dilution of these soluble components as a result of increasing hydraulic load. Apparently, greater hydraulic loads caused greater dissolution and oxidation of FeS and pyrite in the mine that overrode dilution effects. Microbially-mediated mobilization of Fe and SO_4^{2-} may have also occurred within the wetland substrate as a result of decomposition of organic constituents acting as sorption or precipitation sites in the substrate matrix. Further increases in Fe and SO_4^{2-} concentration gradients may result from the increased solubility of the Fe-minerals caused by higher temperatures, while net acidity increases are attributed to higher hydrolysis rates. Another factor contributing to higher Fe, SO_4^{2-}, and acidity loads within the wetland during the spring season is probably the shorter residence time of these parameters within the wetland due to higher hydraulic loads, which do not allow enough interaction time for their effective retention in the wetland. This is supported by the lowest overall levels of Fe, SO_4^{2-}, and acidity observed in the wetland during the fall season, when the hydraulic loads were the lowest and the residence time the highest. The reduced net acidity is also responsible for the observed significant pH increase in the wetland during the fall season. Sizeable increases in alkalinity during the spring and fall seasons compared to the winter months are probably the result of enhanced limestone dissolution in the presence of elevated CO_2 pressure caused by increased microbial respiration. This is consistent with respective increases in soluble Ca concentrations. Manganese and Al behavior was not affected significantly by seasonal variations. These trends were consistent between wetland cells, even though the concentration gradients decreased with each successive treatment.

Solution speciations suggested that the wetland was a favorable environment for the precipitation of ferrihydrite, goethite, jarosite, lepidocrocite, and schwertmannite throughout the year, with increasing supersaturation trends from winter, to spring, to fall seasons (Table 6). Enhanced precipitation of Fe-minerals during the fall may be influenced by the fully-grown vegetation through oxygen transport into the plants' rhizosphere and expanded microbial attachment and oxidation opportunities. The precipitation of alunite, gibbsite, and jurbanite was favored only during the fall when the hydraulic residence time was the longest. Solution compositions were near equilibrium with gypsum throughout the year and particularly during the warmer months. Solution

TABLE 6. Saturation indices (log Q/K) calculated by Visual Minteq for potential minerals precipitating within the wetland at different seasons

Minerals	Winter 2002	Spring 2003	Fall 2003
Alunite	−1.4	−2.8	+6.7
Ferrihydrite	+0.2	+0.3	+3.8
Gibbsite	−2.5	−3.2	+1.2
Goethite	+2.9	+3.0	+6.5
Gypsum	−0.3	−0.2	−0.2
Jurbanite	−0.9	−1.4	+0.3
K-Jarosite	+2.6	+3.7	+10.5
Lepidocrocite	+2.0	+2.2	+5.6
Rhodochrosite	−5.4	−5.3	−2.8
Schwertmannite	+65.0	+67.8	+111.1
Siderite	−3.8	−3.3	−1.1

undersaturation with respect to rhodochrosite and siderite decreased during the fall, but it was still significant enough to prevent precipitation, except for wetland cell 3, which reached near equilibrium levels with respect to siderite in the fall. Seasonal saturation indices decreased from wetland cell 1 to cell 2, but were high enough to support the stability of all Fe-oxyhydroxide and Fe-hydroxysulfate minerals mentioned above throughout the year. In contrast, ferrihydrite and jarosite were unstable in wetland cell 3 during the spring and winter months. As mentioned earlier, the precipitation of Al-sulfates and Al-hydroxides was favored only during the fall, particularly in wetland cells 1 and 3 with some preference for alunite over jurbanite or gibbsite. Although seasonal saturation indices generally increased with depth, there were no particular trends for Fe-minerals, but the precipitation of jurbanite, gibbsite, and siderite was favored mainly in the lowest substrate depths.

5. Conclusions

In spite of the abandoned state of the wetland, AMD passing through a sequence of multiple oxidation ponds and vegetated cells showed drastic reductions in Fe, SO_4^{2-}, and acidity loads with advancing treatment stage. Metal and SO_4^{2-} retention was accomplished mainly through microbially-mediated precipitation of Fe-oxyhydroxysulfates and Fe-oxyhydroxides, but the secondary acidity produced from the ensuing Fe^{3+} hydrolysis reactions could not be buffered by the dissolution of the limestone gravel substrate. Gypsum precipitation contributed to SO_4^{2-} retention. Lower substrate zones were more favorable environments for precipitation reactions, particularly during the fall season, due to enhanced microbial activity, increased buffering capacity, lower hydraulic loads, longer residence times, and greater oxygen diffusion in the plants' rhizosphere.

References

1. Brodie, G.A., Hammer, D.A., and Tomljanovich, D.A. Treatment of acid drainage from coal facilities with man-made wetlands. *In* D.A. Hammer (ed.) *Constructed wetland for wastewater treatment: Municipal, industrial and agriculture.* Proc. from the 1st Int. Conf. On Constructed Wetlands for Wastewater Treatment, Chattanooga, TN. June 13-17, 1998. Lewis Publ., Chelsea, MI. 1989, 903.
2. Hedin, R.S., Nairn, R.W., and Kleinmann, R.L.P. Passive treatment of coal mine drainage.Bureau of Mines Information Circular 9389. US Bureau of Mines, Pittsburgh, PA. 1994, 35.
3. Mitsch, W.J., and Wise, K.M. Water quality, fate of metals, and predictive model validation of a constructed wetland treating acid mine drainage. *Water Resources.* 32, 1888, 1998.
4. Bigham, J.M., Schwertmann, U., and Carlson, L. Mineralogy of precipitates formed by the biogeochemical oxidation of Fe(II) in mine drainage. *In* H.C.W. Skinner and R.W. Fitzpatric (Ed.), *Biomineralization.* Cremlingen-Destedt, Germany: Catena Verlag. 1992.
5. Singer, P.C., and Stumm, W. Acidic mine drainage: the rate-determining step. *Science.* 167, 1121, 1970.
6. Hem, J.D. Redox coprecipitation mechansism of manganese oxides. *In* M.C. Kavanaugh and J.O. Leckie (ed). *Particles in water: Characterization, fate, effects, and removal.* Advances in Chemistry Series no. 189. American Chemistry Society. Washington, DC. 1980, 45.
7. Postgate, J.R. *The Sulfate Reducing Bacteria.* 2nd ed. Cambridge University Press, Cambridge. 1984.
8. August, E.E., McKnight, D.M., Hrncir, D.C., and Garhart, K.S. Seasonal variability of metals transport through a wetlands impacted by mine drainage in the Rocky Mountains. *Environmental Technology.* 36, 3779, 2002.
9. Barton C.D., and Karathanasis, A.D. Aerobic and anaerobic metal attenuation processes in a constructed wetland treating acid mine drainage. *Environmental Geosciences.* 5, 43, 1998.
10. Stumm, W., and Morgan, J.J. *Aquatic Chemistry.* 3rd ed. J. Wiley and Sons, New York. 1995.
11. Tarutis, W.J., Unz, R.F., and Brooks, R.P. Behaviors of sedimentary Fe and Mn in a natural wetland receiving acid mine drainage, Pennsylvania, USA. *Applied Geochemistry.* 7, 77, 1992.
12. USDA-NRCS-USEPA. *A Handbook of Constructed Wetlands: Volume 4 Coal Mine Drainage.* Chester, PA. 1996.
13. Munch, J.C., and Ottow, J.C.G. Preferential reduction of amorphous to crystalline iron oxides by bacterial activity. *Soil Science Society of America Journal.* 129, 15, 1980.
14. Hem, J.D. Rates of manganese oxidation in aquatic systems.*Geochimica Cosmochimica Acta.* 45, 1369, 1981.
15. Nairn, R.W., Hedin, R.S., and Watzlaf, G.R. A preliminary review of the use of anoxic limestone drains in the passive treatment of acid mine drainage. *In* Proceedings of Twelfth Annual West Virginia Surface Mine Drainage Task Force Symposium, Morgantown, WV, April 3–4, 1991.
16. McIntire, P.E., and Edenborn, H.M. The use of bacterial sulfate reduction in the treatment of drainage from coal mines. *In*: Proceedings of the West Virginia Mining and Reclamation Conference. West Virginia University, Morgantown, WV.1990, 409.
17. Hilton, B.L., and Oleszkiewicz, J.A. Sulfide induced inhibition of anaerobic digestion. *Journal of Environmental Engineering.* American Society of Civil Engineers. 114, 1377, 1998.
18. Barton C.D., and Karathanasis, A.D. 1999. Renovation of a failed constructed wetland treating acid mine drainage. *Environmental Geology.* 39, 39, 1999.
19. Carignan, R., Rapin, F., and Tessier, A. Sediment porewater sampling for metal analysis: A comparison of techniques. *Geochimica et Cosmochimica Acta.* 49, 2493, 1985.
20. Brandl, H., and Hanselmann, K.W. Evaluation and application of dialysis porewater samplers for microbiological studies at sediment-water interfaces. *Aquatic Sciences.* 53, 55, 1991.
21. D'Angelo, E.M., and Reddy, K.R. Diagenesis of organic matter in a wetland receiving hypereutrophic lake water: I. Distribution of dissolved nutrients in the soil and water column.*Journal of Environmental Quality.* 23, 928, 1994.

22. American Public Health Association. *Standard Methods for the Examination of Water and Wastewater*. 19th ed. APHA, Washington, DC. 1996.
23. Cline, J.L. 1969. Spectrophotometeric determination of hydrogen sulfide in natural waters. *Limnology and Oceanography*. 14, 454, 1969.
24. Burdige, D.J., Dhakar, S.P., and Nealson, K.H. Effects of Mb oxide mineralogy on microbial and chemical Mn reduction. *Geomicrobiology Journal*. 10, 27, 1992.
25. Wieder, R.K. 1993. Ion input/output budgets for wetlands constructed for acid coal mine drainage treatment. *Water, Air, Soil Pollution*. 71, 231, 1993

16
Removal of Trace Elements from Aqueous System: Comparison of Two Fly Ash Materials

K.S. Sajwan[1], S. Paramasivam[1], M.C. Potts[1], G.S. Ghuman[1], J. Afolabi[1], and A.K. Alva[2]
[1]*Department of Natural Sciences and Mathematics, Savannah State University, P.O. Box 20600, Savannah, GA 31404, U.S.A.*
[2]*USDA-ARS, Pacific West Area, Vegetable and Forage Crops Research Unit, 24106 N. Bunn Road, Prosser, WA 99350, U.S.A.*

Abstract

A comparative study was initiated to assess the feasibility of using fly ash FA for the removal of selected trace elements from wastewater and industrial effluents. The specific objectives of this study were to (i) determine the feasibility of using FA to remove Zn, Cu, Cd, Cr, Pb, and Mn from aqueous medium by using mixed trace element solution and (ii) to identify the optimum conditions for the maximum removal of Zn, Cu, Cd, Cr, Pb, and Mn by FA from mixed trace element solution. Two FA materials with contrasting characteristics from a coal-fired power plant in Port Wentworth PWW-FA, and McIntosh McInt-FA in Savannah, GA were used in this study. Results of this study indicated that both FA materials were very efficient ($\sim$100%) in removing most of the elements studied when 5 mg L^{-1} mixed trace element solution was equilibrated with 200 g L^{-1} FA material for 8 hours. However, when the concentrations of mixed elements were increased up to 25 mg L^{-1}, PWW-FA was very efficient in removing about 100% of Cr, and Pb and about 45% of Cu at 8 h contact time from aqueous medium. Similar increasing trend was observed only for Pb with 60% removal but not for Cr and Cu with adsorbent McInt-FA. Results also indicated that FA material with greater carbon content depicted greater removal of trace elements from aqueous solution containing mixture of trace elements. This study suggests that FA could be successfully used as an adsorbent to remove wide variety of trace elements, which are commonly found in contaminated aquatic habitats.

Introduction

The chemical composition of processed wastewater produced by the wastewater treatment plants in cities around the globe varies with location and source of influent wastewater. Influent wastewater collected from various locations for purification typically contains high levels of N, P, K, and certain trace elements and organic compounds. Among the various trace elements (Zn, Cu, Fe, Mn, Ni, Cd, Cr, Pb, and Se) commonly present in influent wastewater, Cd is widely recognized as the most hazardous elements[1,2]. However, other elements are also equally important and hazardous to natural flora and fauna, animals and humans when used in the water purification process. Furthermore, the degree of toxicity depends on the concentration of these elements in purified effluent water as well as other factors such as microorganisms associated with the water.

Though conventional adsorbents are available and used for treatment by wastewater treatment plants, activated carbon is commonly used and considered as the most promising technique for removing the majority of organic contaminants and selected elements from aqueous systems[3]. However, due to its high cost and limited availability, it becomes necessary to identify and investigate alternative adsorbent sources. FA is, an inexpensive and abundantly available byproduct, from coal-fired power plants, and consists of finely dispersed particulate matter[4,5]. The quantity of coal burnt in USA during 2000 is about 860 million metric tones[6] and resulted in the production of very large amount of coal combustion byproducts [CCB]. Currently, FA is the most abundant of the CCB and coal-fired power plants produce a huge and potentially unmanageable quantity of FA. The composition of FA is similar to clay in that it contains large amount of silica, aluminum and some unburnt carbon[7]. A potential use of FA, often overlooked, is as an adsorbent, capable of removing many organic contaminants because of its large surface area per unit volume and its high residual carbon content. The concentration of residual carbon contained within the FA determines its effectiveness as an adsorbent. In early 90s, interest arose in the investigation of materials that could be used in non-conventional methods for scavenging trace element ions from industrial wastewaters[8]. Singh and Rawat[9] studied the adsorption of Cu from an aqueous system on bituminous coal and found that adsorption was dependent on concentration, pH and temperature. They also studied the kinetics and the mechanism of Cu sorption in this system. Several studies have shown that the efficacy of trace elements such as Cd, Pb, Mn, Zn, As, and Se and organic compound adsorption on FA could be enhanced by a CO_2 infusion process[10,11,12,13]. It was also demonstrated[14,15] that the sorptive behavior of trace elements on FA in aqueous systems is partly determined by the pH of the system, the trace elements As, Cd, Cr, Cu, Pb, Ni, and Zn. The highest efficiency is attained at their lowest solubility in solutions when the pH is between 8 and 9. They attributed this observation to increased adsorption and precipitation processes. In addition, significant increase in solubility of these trace elements were observed when the pH dropped from 12 to 9 and also below 6 in the presence of alkaline FA in aqueous system[14]. In our previous study, we demonstrated that FA could be successfully used to remove Cd from wastewater containing a mixture of various inorganic pollutants[16].

Therefore, there is considerable interest in investigating the ability of FA to remove trace elements including Cd from wastewater and contaminated aquatic systems. The objectives of this study were to assess the: (i) feasibility of using FA to remove Zn, Cu, Cd, Cr, Pb, and Mn from aqueous systems by using mixed element solution; (ii) effect of adsorbate concentration on removal of elements using varying rates of FA; (iii) effect of adsorbent concentration on the removal of elements using varying rates of mixed element solution; and (iv) effect of contact period on elemental removal from aqueous medium.

Materials and Methods

Adsorbent

The FA used for this experiment was collected from the Port Wentworth (PWW-FA) Power Plant located in Port Wentworth, Savannah, Georgia and the McIntosh (McInt-FA) Power Plant located in Rincon, Georgia. Basic characteristics of the FA are presented in Table 1.

Rates of Adsorbents

Varying concentrations (25 g L^{-1}, 50 g L^{-1}, 100 g L^{-1}, and 200 g L^{-1}) of adsorbents were used in this study. In this study, we used 50 mL of the adsorbate (mixed element solution) and appropriate amounts of two different adsorbents (1.25 g, 2.50 g, 5.00 g, and 10.00 g) to obtain the listed rates of adsorbents.

TABLE 1. Selected properties of FA materials used in this study.

Properties	Units	Source of fly ash: PWW-FA	Source of fly ash: McInt-FA
Al	%	0.41	1.20
Fe	%	1.35	0.50
Ca	%	0.85	0.19
Mg	%	0.47	0.09
K	%	0.06	0.22
Na	%	0.04	0.05
S	%	0.30	0.12
Cd	μg g^{-1}	2.80	1.33
Mn	μg g^{-1}	138	22
Si	μg g^{-1}	72	80
PH		6.9	6.9
C	%	18.5	3.6
Specific gravity	g mL^{-1}	1.2	1.3
Size	mm	0.1–0.2	0.1–0.3

Adsorbate

The combined salts on common inorganic contaminants of wastewater and aqueous systems (e.g. Cd, Cr, Mn, Pb, Zn, Cu) were dissolved to prepare mixed element solutions. Removal of trace elements is difficult when effluents contain a mixture of trace elements due to differences in their solubilities.[17]

Rates of Adsorbates in Mixed Metal Solution

Mixed element solution containing Cd, Cr, Mn, Pb, Zn, and Cu were prepared to have the following concentrations: 5.0 mg L^{-1}, 25.0 mg L^{-1}, and 50.0 mg L^{-1} to run the batch adsorption experiments.

Contact Time in Shaker

Adsorbent and adsorbate were placed in 50-mL centrifuge tube and kept in contact by shaking the tubes in reciprocal shaker (New Brunswick Scientific Co. Inc., Edison, New Jersey, USA) at 200 rpm for the following contact periods: 0.5 h, 1.0 h, 2.0 h, 4.0 h, and 8.0 h.

Experiments to Study Kinetics of Adsorption of Selected Trace Elements

The studies with element-fly ash systems were performed to assess the effect of elemental concentration in solution, quantity of FA mixed in the solution, and time of contact between the adsorbate and adsorbent.

The adsorption studies were performed at room temperature (25° ± 2 °C) as batch experiments with three replicates. Fifty mL of the (adsorbate) mixed element solution (5, 25, and 50 mg L^{-1}) were combined in a 100 mL centrifuge tube with the different rates of adsorbent FA (25, 50, 100, and 200 g L^{-1}). The tubes were agitated continuously in a reciprocal shaker (New Brunswick Scientific Co. Inc., Edison, New Jersey, USA) at 200 rpm for the following contact periods: 0.5 h, 1.0 h, 2.0 h, 4.0 h, and 8.0 h. Along with these, three control blanks (with deionized water as adsorbate) for each contact period and each adsorbent rates included. Control and treated samples at the end of specified contact periods were removed from the shaker, centrifuged at 3400 rpm and filtered through "Whatman" filter paper No. 42 and analyzed for pH and elemental concentration remained in the supernatant. The pH and concentrations of elements remained in the supernatant were measured using pH meter (Accumet Model 15; Fisher Scientific, Pittsburgh, PA) and Inductively Coupled Plasma Optical Emission Spectrophotometer (ICP-OES; Plasma RL 3100, Perkin Elmer Inc., Norwalk, CT) respectively.

Results and Discussion

Disappearance of trace elements from aqueous medium in the presence of FA can occur due to various types of reactions, which take place in the aqueous medium. Adsorption, microbial immobilization, and precipitation are some of them for us to consider. Each of these processes is generally governed by certain sets of conditions prevailing at a particular point in time in aqueous medium. Among the various process, the adsorption of solute onto the adsorbent presumably occurred in three consecutive steps: (a) solute from solution moved to the exterior surface of the adsorbent, (b) the solute moved through the pores of the adsorbent by intraparticle diffusion, and (c) solute was adsorbed at the specific site on the surface of the adsorbent particle. The adsorption of trace elements and their kinetics are generally dependent on concentration of adsorbent and adsorbate, pH and temperature.

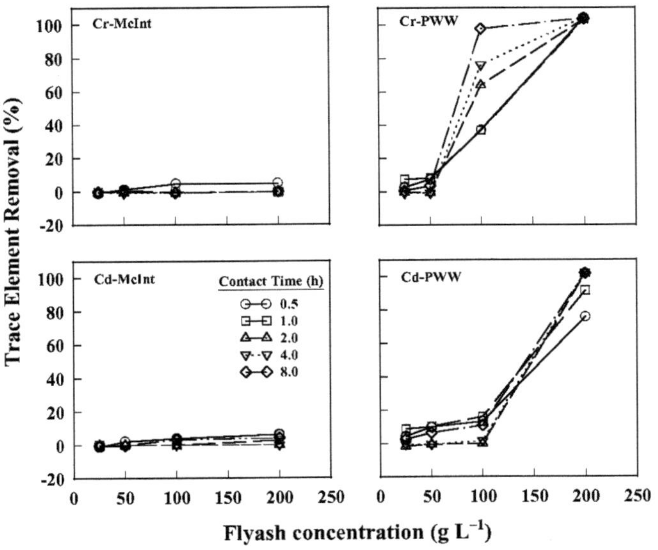

FIGURE 1. Effect of adsorbent FA concentration, adsorbent source and contact time on the percentage removal of Cd and Cr from 5 mg L^{-1} mixed trace elements solution.

Effect of Adsorbent Concentration

Effect of adsorbent concentration and contact time on the removal of various elements from a mixed aqueous solution system is presented in Figures 1–3. In this study, two types of adsorbents (PWW-FA and McInt-FA) used were collected from two different coal-fired power plants in Savannah, GA. These two adsorbents have contrasting chemical properties and also have substantially different amount of carbon content (Table. 1). Presence of high content of activated carbon could play a significant role in removing various elements from aqueous medium. The PWW-FA removed greatest percentage of trace elements, which had greater percentage of carbon in it compared to that of McInt-FA from McIntosh power plant in Savannah, GA. Substantially low percentage (0 to 3%) of Cd and Cr removal was observed when 5-mg L^{-1} mixed-trace element solution was equilibrated with McInt-FA irrespective of rates of adsorbent and contact times. In contrast, almost all the Cd (~100%) was removed when 5-mg L^{-1} mixed-trace element solution was equilibrated with PWW-FA for 2 h. (Fig. 1). Results also indicated that increase in contact time beyond 2.0 h did not have any observable effect on the removal of Cd by PWW-FA. However, the percentage removal of Cd from aqueous medium by PWW-FA decreased with decreasing contact time below 2.0 h. It was also interesting to note that the contact time did not have any impact on the removal of Cr with PWW-FA at 200 g L^{-1} rate. Irrespective of differences in contact time, almost 100 % of Cr in mixed element solution was removed by PWW-FA.

The removal of Mn from solution phase by adsorbent McInt-FA was insignificant. In contrast, almost all the Mn (98%) was removed from solution by PWW-FA at contact period of 8.0 h at 200 g L^{-1} adsorbent rate (Fig. 2). Percentage removal of Mn by PWW-FA decreased with decreasing contact time and it was about 18% when the contact time was 0.5 h.

It was interesting to note that Pb behaved somewhat differently compared to all other trace elements studied (Fig. 2). Almost 100% Pb from the mixed element solution was removed by PWW-FA at 200 g L^{-1} rate at all the contact times studied. In contrast, the other adsorbent McInt-FA removed only 78–94% Pb from the mixed element solution at the similar 200 g L^{-1} adsorbent rate. Percentage of Pb removal from aqueous medium increased with both increasing contact time as well as with increasing rate of adsorbent irrespective of source of adsorbent.

Effect of rate and source of adsorbent and contact time on the removal of Cu and Zn is somewhat similar to what was

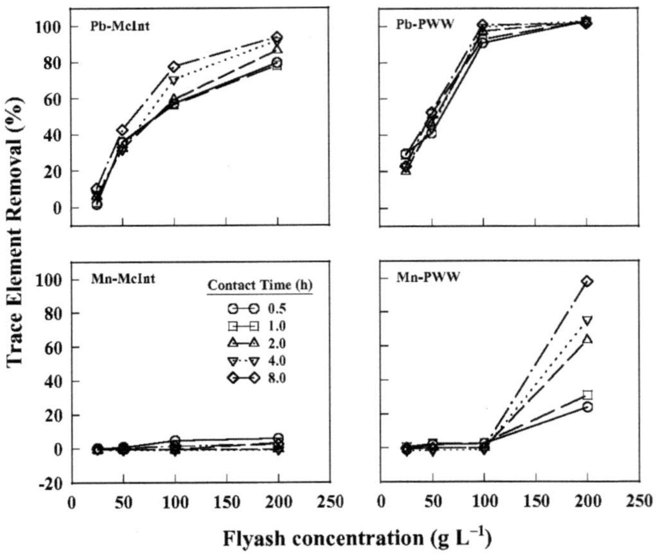

FIGURE 2. Effect of adsorbent FA concentration, adsorbent source and contact time on the percentage removal of Mn and Pb from 5 mg L^{-1} mixed trace elements solution.

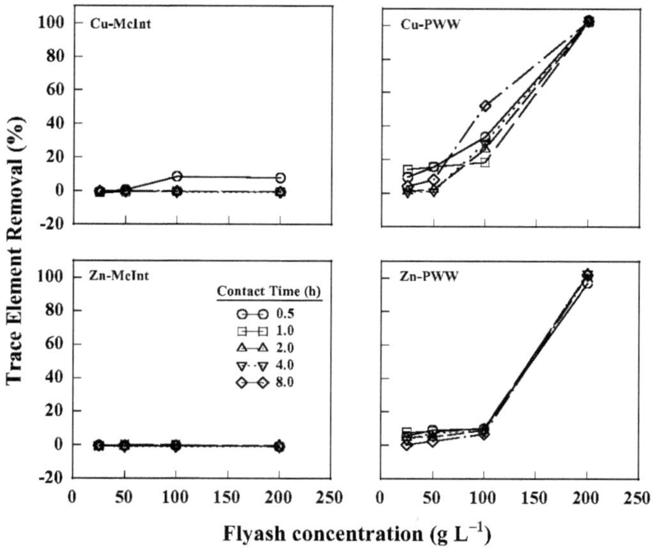

FIGURE 3. Effect of adsorbent FA concentration, adsorbent source and contact time on the percentage removal of Zn and Cu from 5 mg L^{-1} mixed trace elements solution.

observed Cd and Cr. The removal of Zn and Cu from mixed metal solution by adsorbent McInt-FA was insignificant. In contrast, almost 100% of Zn and Cu from mixed element solution were removed by PWW-FA at the adsorbent rate of 200 g L^{-1} at all contact times (Fig. 3).

Effect of Adsorbate [Trace Element] Concentration

Effect of concentration of adsorbate and contact time on the removal of various elements from a mixed aqueous solution system is presented in Figures 4–6. At 5.0 mg L^{-1} concentration level of mixed element solution, almost all the elements disappeared (were removed) with PWW-FA. This observation was true for all contact times used in this study. With the exception of Pb, there was no noticeable significant removal of other elements at the 50.0 mg L^{-1} rate even at prolonged contact time of 8 h. However, at the medium adsorbate concentration (25 mg L^{-1}), substantial removal of Cr, Pb and Cu was observed from solution phase when the mixed element solution was equilibrated with adsorbent with greater carbon content PWW-FA (Fig. 4–6).

With the exception of Pb, percentage removal of Cd, Cr, Mn, Cu, and Zn was observed with adsorbent McInt-FA, which has substantially less carbon content. This observation was true at all the concentration range of mixed element solution and contact times studied.

Removal of Pb from mixed element solution (adsorbate) concentration at 5.0 mg L^{-1} by McInt-FA was about 88% at 0.5–1.0 hr contact times. However, when the contact time was extended to 2.0–8.0 h, percentage removal Pb increased to 92–100%. In contrast, almost 100% Pb removal was observed when 5.0 mg L^{-1} of adsorbate equilibrated with PWW-FA for all the contact times studied (Fig. 5).

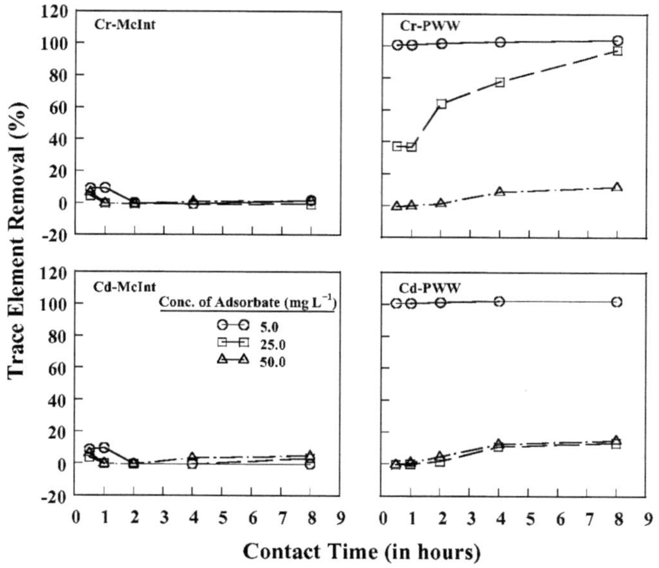

FIGURE 4. Effect of adsorbate concentration, and contact time on the percentage removal of Cd and Cr when mixed trace elements solution was equilibrated with 200 g L^{-1} of adsorbent from two different sources.

Effect of Contact time and Adsorbent Concentration

Effect of contact Time and adsorbent concentrations on the removal of various trace elements were studied with 5 mg L^{-1} mixed-element solution. With the exception of Pb, practically there were no measurable amounts of Cd, Cr, Zn, Cu and Mn removal observed when varying rates of adsorbent McInt-FA

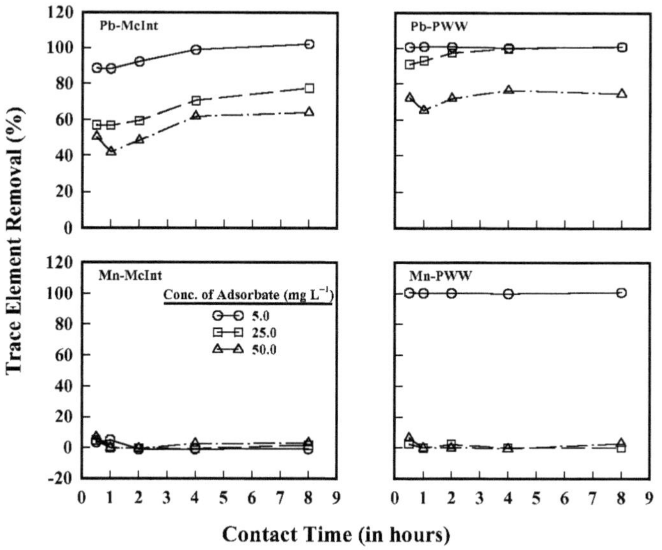

FIGURE 5. Effect of adsorbate concentration, and contact time on the percentage removal of Mn and Pb when mixed trace elements solution was equilibrated with 200 g L^{-1} of adsorbent from two different sources.

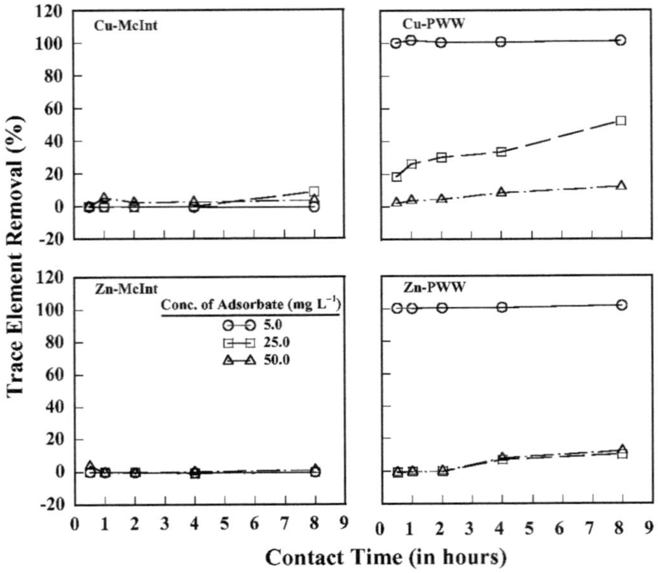

FIGURE 6. Effect of adsorbate concentration, and contact time on the percentage removal of Zn and Cu when mixed trace elements solution was equilibrated with 200 g L^{-1} of adsorbent from two different sources.

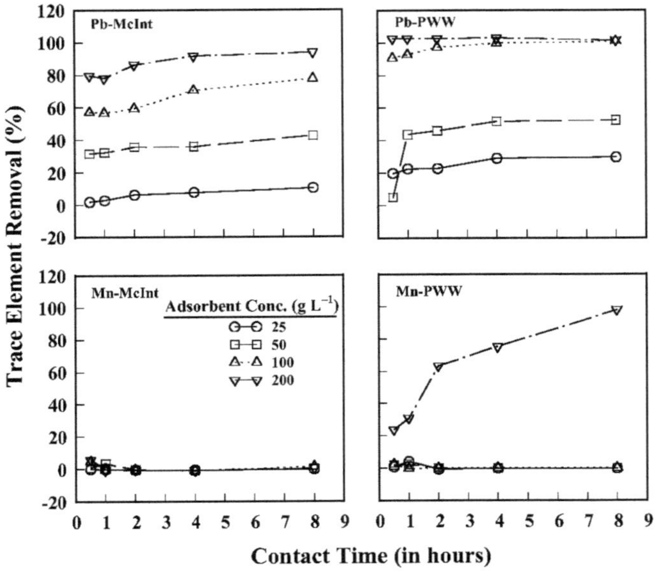

FIGURE 8. Effect of adsorbent concentration, and contact time on the percentage removal of Mn and Pb when 5 mg L^{-1} mixed trace elements solution was equilibrated with adsorbent from two different sources.

was used. Removal of Cd by PWW-FA at 200 g L^{-1} rate was 76% at 0.5 h contact time and then gradually increased to reach the 100% and remained constant with the increase of contact time (Fig. 7). Removal of Cr by PWW-FA was about 100% and it was not affected by the contact time at the rate of 200 g L^{-1}. However, Cr removal by PWW-FA increased substantially with increasing contact time when the PWW-FA was at 100 g L^{-1} rate. Removal of Cr by PWW-FA reached almost 100% when the contact was 8 h.

Contact time and adsorbent concentration did not have any significant impact on Mn removal when adsorbent McInt-FA was used (Fig. 8). However, about 93% of Pb was removed by McInt-FA at 8.0 hr contact time with 200 g L^{-1} adsorbent rate. In contrast, Mn removal increased with increasing contact time and reached about 98% with adsorbent PWW-FA at 8.0 hr equilibrium period with 200 g L^{-1} adsorbent rate. But percent removal of Pb by adsorbent PWW-FA at 200 g

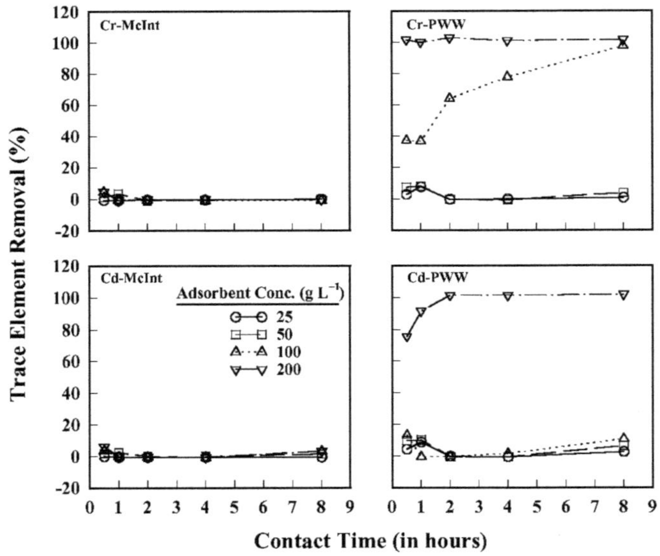

FIGURE 7. Effect of adsorbent concentration, and contact time on the percentage removal of Cd and Cr when 5 mg L^{-1} mixed trace elements solution was equilibrated with adsorbent from two different sources.

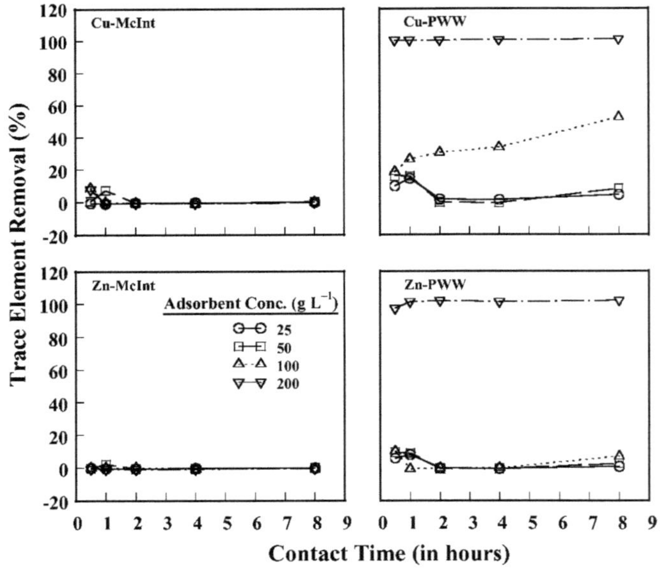

FIGURE 9. Effect of adsorbent concentration, and contact time on the percentage removal of Zn and Cu when 5 mg L^{-1} mixed trace elements solution was equilibrated with adsorbent from two different sources.

L^{-1} was about 100% for all the contact times. The percent removal of Pb was almost closer even at 100 g L^{-1} rate and varied from 91% to 100%. When the rate of adsorbent decreased to 50 and 25 g L^{-1}, the percent removal of Pb was substantially lower and accounted 40% and 20% respectively. It was interesting to note that; removal pattern of Pb was somewhat similar when the adsorbent was McInt-FA. However, the percent removal of Pb was about 20% less at each rates of adsorbent compared to what was observed with adsorbent PWW-FA.

Contact time and adsorbent concentration did not have significant effect on the removal of Zn and Cu with adsorbent McInt-FA. However, about 97% and 100% of Zn and Cu were removed from the mixed element solution when adsorbent PWW-FA was used at 200 g L^{-1} for all the contact times. Other rates of adsorbent did not have significant effect on the removal of Zn from mixed element solution.

Conclusions

Results of this study indicated that removal of elements increased with increasing concentration of adsorbent up to 200 g L^{-1}. Results of this study also indicated that adsorbent PWW-FA at the rate of 200 g L^{-1} was the most effective in removing trace elements from aqueous medium compared to other adsorbent McInt-FA. Maximum percentage removal of various trace elements varied with type of trace elements. We observed almost 100% adsorption for selected trace elements only at concentration level 5 mg L^{-1}. Furthermore, various trace elements studied behaved differently with respect to adsorbate concentration. Interestingly, adsorption or removal of elements such as Cr, Pb increased with increasing concentration up to 50 mg L^{-1}. Similarly, removal of various elements increased with increasing contact time. However, the contact time needed for maximum removal from aqueous medium varied with various trace elements. From this study, it was apparent that about 8hr of equilibrium would be sufficient to obtain maximum adsorption or removal of trace elements from aqueous solution.

Results of this study further indicated that both FA materials were very efficient ($\sim$100%) in removing most of the elements studied when 5 mg L^{-1} of mixed elements equilibrated with 200 g L^{-1} for 8 hr. However, when the concentrations of mixed elements were increased up to 25 mg L^{-1}, PWW-FA was very efficient in removing about 100% of Cr, and Pb and about 45% of Cu at 8 h contact time from aqueous medium. Similar increasing trend was observed only for Pb with 60% removal but not for Cr and Cu with adsorbent McInt-FA. Results also indicated that fly ash material with greater carbon content depicted greater removal of trace elements from aqueous solution containing mixture of trace elements. This study revealed that fly ash could be successfully used as an adsorbent to remove wide variety of trace elements, which are commonly found in contaminated aquatic systems.

Acknowledgements. This work was supported by the U.S. Department of Energy and the U.S. Environmental Protection Agency Grant Award No. DE-FG09-02SR22248.

References

1. Alloway, B.J., Heavy metals in soils. Blackie Publishers, London, 1995.
2. Jackson, A.P., and Alloway, B.J., The transfer of cadmium from agricultural soils to the human food chain. *in Biogeochemistry of Trace Metals,* D.C. Adriano (ed.) Lewis Publishers, Chelsea, Michigan, 109, 1992.
3. Weber, W. J., Jr. and Morries, J. C., Kinetics of adsorption on carbon from solutions, *J. Sanitary Engineering, Div. Am. Civil Engr.* 89, (2), 31, 1963.
4. Sajwan, K.S., Alva, A.K., and Keefer, R.F., Biogeochemistry of trace elements in coal and coal combustion by-products, Kluwer academic/Plenum Publishers. New York, 359, 1999.
5. Khandekar, M.P., Bhide, A.D., and Sajwan, K.S., Trace Elements in Indian coal and coal fly ash. *in Biogeochemistry of trace elements in coal and coal combustion by products,* K.S. Sajwan et. al. (eds.) Kluwer Academic/Plenum Publishers, New York, 99, 1999.
6. Kalyoncu, R. Coal Combustion Products, 2000. http://minerals.usgs.gov./minerals/pubs/commodity/coal/874400.pdf
7. Adriano, D.C., Page, A.L., Elseewi, A.A., Chang, A.C., and Straughan, I., Utilization and disposal of fly ash and other coal residues in terrestrial ecosystems: A Review, *J. Environ. Qual.* 9, 333, 1980.
8. Rao, A. J., Verma, N., and Kaur, A., Bottom Ash for Adsorption of Nickel Metal Ion from Industrial Wastewater, *Indian J. Environ. Health*, 32, (3), 280, 1990.
9. Singh, D., and Rawat, N. S., Removal of Copper from Aqueous System by Sorption on Bituminous Coal, *Asian Environ.*, 56, 1992.
10. Frye, G.C., and Thomas, M.M., Adsorption of organic compounds on carbonate minerals. 2: Extraction of carboxylic acids from recent and ancient carbonates, *Chem. Geol.*, 109, 215, 1993.
11. van, Proosdij, E.M.H., and Reddy, K.J., Immobilization of contaminants with in-situ calcite precipitation, A preliminary evaluation, *in. Contaminated Soils. 3rd International Conference on Biogeochemistry of Trace Elements*, R. Prost (ed.), Paris, France, 1997.
12. Zavarin, M., and Doner, H.E., Selenium, Nickel, and Manganese interactions with calcite, *in Sorption and Desorption of Trace Elements, Proceedings of 4th International Conference on the Biogeochemistry of Trace Elements,* I.K. Iskandar et al. (eds.), U.S. Army Cold Regions Research and Engineering Laboratory, Hanover, New Hapshire, 1997.
13. Reddy, K.J., Coal fly ash chemistry and carbon dioxide infusion process to enhance its utilization, *in Biogeochemistry of trace elements in coal and coal combustion by-products*, K.S. Sajwan et al. (eds.), Kluwer Academic/Plenum Publishers, New York, 133, 1999.
14. Theis, T.L., and Wirth, J.L., Sorptive behavior of trace elements on fly ash in aqueous systems, *Environ. Sci. Technol*, 11, 1096, 1997.

15. Rai, D., Ainsworth, C.C., Eary, L.E., Mattigod, S.V., and Jackson, D.R., Inorganic and organic constituents in fossil fuel combustion residues, Electric Power Research Institute, EA-5176, 1987.
16. Hajarnavis, M., Sajwan, K.S., Paramasivam, S., Chetty, C. S., and Reddy, G.R., Potential use of fly ash to remove cadmium from aqueous systems, *in Biogeochemistry of trace elements in coal and coal combustion by-products,* K.S. Sajwan et al. (eds.), Kluwer Academic/Plenum Publishers, New York, 177, 2003.
17. Weber, W. J., Physico-Chemical Processes for Water Quality Control. Wiley-Interscience, New York, NY, 1972.

V
Use of Coal Ash as Agriculture Soil Amendment

17
Transport and Plant Uptake of Zn in an Oxyaquic Haplustalf Amended with Coal Ash and Sewage Sludge: A Field Study

U.M. Khodke[1], S. Tripathy[2], R.K. Panda[3], H. Veeresh[4], and K.S. Sajwan[5]

[1]*College of Agricultural Engineering, Marathwada Agricultural University, Parbhani, Maharashtra, 431 402, India*
[2]*Department of Geology and Geophysics, Indian Institute of Technology, Kharagpur, West Bengal, 721 302, India*
[3]*Department of Agricultural and Food Engineering, Indian Institute of Technology, Kharagpur, West Bengal, 721 302, India*
[4]*Department of Soil Science and Agricultural Chemistry, UAS, GKVK, Bangalore, Karnataka, 560 065, India*
[5]*Department of Natural Sciences and Mathematics, Savannah State University, Savannah, GA 31404, USA*

Abstract

Field experiments were conducted to study the movement and uptake of Zn in an Oxyaquic Haplustalf (pH = 5.2) amended with coal ash (pH = 7.5) and sewage sludge (pH = 4.3) at low application rates. The plots received cumulative loadings of sludge (39.5 Mg ha^{-1}), ash (78 Mg ha^{-1}) and mixture of ash and sludge (1:2 i.e. 19.5:39 Mg ha^{-1}) over two years. Irrigated peanut (*Arachis hypogaea L.*) was the test crop. Irrigations were scheduled at two levels viz. 15% (I_1) and 50% (I_2) maximum allowable depletion (MAD) of available soil water (ASW) in the root zone. Soil samples were collected from 0–15, 15–30, 30–45 and 45–60 cm depths during the experiment and analyzed for pH and water extractable concentration (WEC) of Zn. Pre and post experiment profile soil samples were also analyzed for total Zn concentration.

The water extractable concentrations (WEC) of Zn were generally higher in sludge treated plots than the ones receiving mixture and ash. This is in accordance to the order of Zn loading by the amendments. Cyclic variation in WEC of Zn was perceptible within the top 15 cm layer of the soil compared to that at higher depths in all sampling periods across the treatments; but was more distinct in the case of plots treated with only sludge. The increase in WEC of Zn below 15 cm layer in sludge treated plots indicated the possibility of leaching. Noticeable increases in total Zn concentrations in the surface layers of sludge and mixture treated plots were observed. The total Zn concentration in the profile soil from sludge applied plots indicated significant movement of Zn within 15–30 cm and thereafter a marginal increase up to 45cm signifying the possibility of Zn movement at lower depths. Plant accumulation of Zn was the highest in sludge treated plots (150 mg kg^{-1}) followed by the ones treated with mixture (99 mg kg^{-1}) and coal ash (56 mg kg^{-1}) as compared to the control (25 mg kg^{-1}). On the other hand, the ratio of concentrations of Zn in the plant to its total in surface soil followed the order -: control < ash < mixture < sludge indicating that the plant accumulation is governed by its loading rates.

1. Introduction

Coal fly ash contains most of the beneficial plant nutrients[1] and has been used to alleviate micronutrient deficiencies in soil[2–3] such as B[4–5], Mg[6], Mo[7], S[6,8] and Zn[9] for favorable crop growth and productivity. It has also been used as soil amendment for field crops[4,10–12] and on acid strip mine soils to improve growth and development of crops.[13] Significant increases in growth parameters and crop yield have been observed with application of fly ash at both lower[14] and at higher rates.[15] Fly ash is known to neutralize soil acidity to a level suitable for agricultural crops[16] and makes easy the availability of plant nutrients except N.[17–19] Its alkaline pH has beneficial effects both in the cases of acidity and metal availability in the soil thereby lessening phytotoxocity.[20–21] Moreover, the physical properties of soil is also improved due to the presence of predominantly silt size particles in ash.[22]

However, phytotoxic constituents, trace elements, soluble salts and high pH of un-weathered fly ash can have direct and indirect detrimental effects on plant establishment, microbial activity and nutrient balances in soils.[23–27] In addition, environmental risk is also associated due to the potential buildup of metal concentrations in soil, plant and ecosystem beyond toxic limits following land application.[28–29] The possible mechanisms by which fly ash is likely to damage crops are B toxicity,[30–33] induced P deficiency[31], salt injury[22,30], and heavy metal toxicity.[2,16,30,34–35]

Sewage sludge is a rich source of nitrogen, phosphorus, organic matter and other essential nutrients required for plant growth[36–38] and improves crop yield.[39] Presence of high organic matter and substantial N and P in sludge makes it preferentially a regenerator of soil[40]. However indiscriminate use of sludge has the potential to cause contamination and phytotoxicity.[41–44] Long-term field observations have shown that sludge-applied metals can remain sufficiently available in non-acid soils to harm sensitive crops.[45] The reduction in yield and retardation in plant growth are generally caused by high concentrations of metals in sludge,[46] induced salinity,[47] high application rates[48] and by high cumulative loading rates.[49–50] In addition, problems may arise if toxic metals are mobilized in the soil solution and are available either to be taken up by plant or transported through drainage water.

Combining coal ash and sewage sludge can produce a mixture with more balanced properties and less potential for environmental contamination than either of them applied individually[28]. Therefore, it can offer a viable and distributed recycling alternative to landfill with potential for agronomic use as substitutes for lime, fertilizer or simply as soil amendment[3].

Studies report little movement of metals below the zone of incorporation in sludge treated soils.[51–58] However, long term and high rates of sludge application could result in significant movement of metals[59–62] in highly structured[63] or fine structured soils[64] via macropore or preferential transport.[65–66] Even long-term wastewater sludge application on soil surface showed movement of heavy metals to a depth below 30 cm.[39,67–68] Some studies have also shown little potential for trace metal movement below the soil profile and hence groundwater contamination[69–71] while others suggest that metals could move downward in the soil profile under specific circumstances[72]. However, disparities on the movement of metals may be expected owing to their nature of occurrences and differences in the experimental conditions.[73]

Similarly, while relatively little movement of metals below the depth of cultivation or sludge application have been observed;[47,74] more pronounced movement of metals within or below the profiles of amended soils[75,76] even with lower application rates[77] have also been reported. Therefore, feasibility of using ash, sludge and their mixtures at low application rates needs to be assessed from environmental point of view.

Soil solution concentrations has been regarded as indicators of mobile pool of metals in a soil system[78] since potentially toxic heavy metals become more hazardous on entering into the liquid phase in soil. Assessment of water extractable or water-soluble forms of metals in the vadose zone helps in understanding the potential risk associated with treating soil with either ash or sludge or both.

Plant metal uptake under different soil moisture regimes, especially under the influence of ash and sludge application has not been addressed yet. Under submerged conditions, the production of organocomplexes and reductions of Fe and Mn in the soil generally tend to enhance the availability of Zn and Cu. On the other hand, submergence of acid non-calcareous soils generally tends to increase the availability of Cu and Mo and depresses Zn due to increased pH and formation of sulfides.[79] However, the behaviour of Zn and Cu under submerged conditions seems to be complex since both decrease and increase in their readily available forms have been reported.[80] On the contrary, water stress, in general, was found to reduce nutrient uptake and transfer from the root to the shoot within the plant because of restricted transpiration rates.[81] In contrast with these extreme conditions, little is known as yet about metal uptake by plant under different soil moisture regimes in case of waste amended soil.

This study attempts to quantify the movement of Zn in an Oxyaquic Haplustalf treated with low rates of application of ash and sewage sludge in variably saturated soil water regimes under field conditions. The response of a leguminous crop to the amendment with respect to Zn uptake was also looked in.

2. Materials and Methods

2.1. Field Experiment

Field experiments were conducted during February through June 2000 and February through May 2001 in a factorial randomized block design with leveled basin plots of size 5 m × 2 m, bound by dikes on all sides. The soil in the field was pulverized and flat beds were prepared. Coal ash and sewage sludge either alone or in combination was mixed thoroughly into the top 15 cm soil layer as per the treatments and sowing was carried out one week after the treatment application.

Treatments consisted of application of coal fly ash, sludge individually and their mixture in 2:1 proportions (Table 1) and control (untreated). Experiments included plots under both un-cropped and cropped conditions with similar treatments. Peanut (*Arachis hypogaea L.*), cultivar AK-12-24 was grown in both the years. The inter-row spacing was 30 cm and plant-to-plant spacing was 15 cm.

In the first year 26 Mg ha^{-1} of coal ash, 13 Mg ha^{-1} of sewage sludge and 13: 6.5 Mg ha^{-1} of ash-sludge mixture were incorporated into the top 15 cm soil (Table 1). However, in the second experiment the application rates were doubled for the corresponding plots without any change in the field layout. Check basin method of irrigation was adopted for all the plots with a buffer strip separating the basins from each other. Irrigations were scheduled at two levels viz., 15% (I_1) and 50% (I_2) maximum allowable depletion (MAD) of available soil water (ASW) in the root zone. Measured irrigation water was applied directly to the plots as soon as the desired MAD limit was attained.

TABLE 1. Experimental details.

Treatments	Quantity applied (Mg ha^{-1})
Control	0
FA	26
FA + SS (2:1)	13 FA + 6.5 SS
SS	13
Irrigation	I_1 (15%MAD)
schedule	I_2 (50%MAD)
Ground Cover	Cropped
	Uncropped

FA-Fly ash; SS-Sewage sludge
MAD-Maximum allowable depletion of available water

TABLE 2. Layer-wise physical properties of soil at the experimental site.

Soil depth (cm)	Particle size distribution ($g\ kg^{-1}$)			Bulk density ($g\ cm^{-3}$)	Saturated hydraulic conductivity ($cm\ day^{-1}$)
	Clay	Silt	Sand		
0–15	138	262	600	1.59	11.5
15–30	204	225	571	1.55	6.72
30–45	234	218	548	1.50	0.89
45–60	278	205	517	1.52	0.74
60–90	294	224	482	1.58	0.34

2.2. Sampling Procedure and Analysis

The base soil samples were collected at 15 cm increments to a depth of 75 cm using an auger of 2.8 cm diameter from the experimental field and were air-dried, mixed and screened through 2 mm sieve prior to analysis. Soil is sandy loam in texture and taxonomically grouped under the order Alfisol (Oxyaquic haplustalf). The layer-wise physical properties of soil are given in Table 2. Composite samples of approximately 50 g (dry basis) taken from respective layers were used for all the chemical analyses. The chemical properties of the representative surface soil (0–15 cm), ash and sludge used in the study were determined (Table 3) by standard procedures and techniques. Profile soils were collected from different soil layers of 0–15, 15–30, 30–45 and 45–60 cm depths throughout the experiment and were analyzed for water extractable concentration (WEC) of Zn (Table 4). Pre and post experiment profile soil were also analyzed for total Zn concentration (Table 4). Maximum care was taken to minimize the contamination of samples by the soil material of overlying layers. Soil samples were kept in Ziploc plastic bags immediately after sampling for further use. Each time after collecting the soil samples, the auger was cleaned with deionized water and dried properly before re-insertion.

The pH of the soil solution (at 1:2.5 (w/v) soil/solution ratio[82]) prepared with deionized water and the electrical conductivity (EC) of the same at a soil/solution ratio of 1:5 (w/v)[82] were measured. Cation exchange capacity (CEC) was determined by ammonium acetate method.[82] The organic carbon (OC), total nitrogen and available nitrogen were estimated through Walkley-Black procedure[83], micro-kjeldahl technique[84] and alkaline permanganate method[85] respectively.

Total Zn concentration was determined on aqua regia digests of soil samples, ash and sludge[86–87] ground to $<150\ \mu m$ and plant available element contents were determined on Mehlich 3 extraction[88] followed by measurements employing inductively coupled plasma atomic emission spectrometer (ICP OES), Perkin-Elmer PE-3300DV. Selective major elements in soil were analyzed by using Philips PW 2404 X-ray Fluorescence spectrometer.[89]

The water-soluble concentration of Zn in the soil was extracted using deionised water (Millipore 18.2 M cm) at a 1:5 (w/v) soil/solution ratio. The soil suspensions were shaken for 30 minutes at $25 \pm 1°C$ in an incubator shaker and then

TABLE 3. Some pertinent chemical properties of soil, coal ash and sewage sludge used in the experiment.

Parameter	Soil (0–15 cm)	Fly ash (FA)	Sewage sludge (SS)	USEPA[1]	EEC[2]	MOE[3]
pH (1:2.5)	5.3	7.5	4.3			
EC ($\mu S\ cm^{-1}$)	42	125	2995			
CEC ($cmol\ kg^{-1}$)	5.4	4.5	19.7			
OC ($g\ kg^{-1}$)	4.5	8.6	144			
Carbonates ($g\ kg^{-1}$)	6.7	88.8	3.5			
	Total concentrations ($g\ kg^{-1}$)					
N	0.5	0.2	10.3			
P	0.57	1.49	7.81			
K	0.42	0.69	2.2			
Ca	1.45	4.4	24.8			
Mg	3.9	11.5	33			
Fe	15.6	21.4	89.9			
Mn	0.06	0.28	0.49			
	Available concentrations ($mg\ kg^{-1}$)					
N	76.9	0	2625			
P	33.0	360	96.0			
K	92.0	120	150			
	Total concentrations ($mg\ kg^{-1}$)					
Zn	25.1	98.0	3183	7500	2500	840

1. US Environmental protection agency: Ceiling limits for use of sludge in agriculture
2. European commision DG, Environment waste management unit working document, 1999: Limit values for regulated constituents in biosolids
3. Ministry of Environment, Ontario: Maximum permissible metal concentrations in dried biosolids

TABLE 4. The total and water extrcatable concentrations of Zn in the soil profile prior to experiment.

	Soil Depth (cm)			
Zn concentration	0–15	15–30	30–45	45–60
Total, mg kg^{-1}	25.11 ± 0.92	23.02 ± 0.65	25.30 ± 0.81	25.70 ± 0.65
Water Extractable, mg kg^{-1}	0.56 ± 0.12	0.42 ± 0.15	0.47 ± 0.09	0.45 ± 0.02

Number of observations = 12; * mean ± standard deviation

centrifuged at 3000 rpm for 15 min, followed by filtration of the supernatant through 2.5 μm (Whatman Grade 42, ash 0.007%) filter paper to collect particulates in the acid clean LDPE bottle. The remaining soil suspension in the centrifuge tube was again mixed with another 25 ml of distilled water and the process was repeated to collect a total of 50ml solution.[90] The solution was analyzed for Zn using ICP-OES.

Plant samples of the peanut crop were collected during different stages of growth. At each harvest, four plants from each plot were randomly removed, thoroughly washed with deionized water and roots were separated from the shoot before drying in a hot air oven at 70 °C to a constant weight. The plant material was then weighed for dry matter determination. Oven dried shoot and shell samples were ground in a Pulverisette (P-14, Fritsch) with titanium blade and sieved through 80-mesh while kernel samples were ground using an agate mortar and pestle and subsequently passed through 2 mesh sieve. Ground samples were stored in airtight plastic vials. The powdered plant samples were digested using tri-acid mixture[82] and concentration of Zn was determined using Atomic Absorption Spectrophotometer (Perkin Elmer; AAnalyst 300 with HGA-850).

3. Results and Discussion

3.1. Temporal Variation of WEC of Zn

In general, the soil is acidic in nature, low in organic carbon, cation exchange capacity, nitrogen and phosphorus, and medium in potassium content (Table 3). The ash used in the present study was slightly alkaline (pH = 7.5) with low salinity (EC = 125 μScm^{-1}) whereas the sludge was highly acidic (pH = 4.3) with high salinity (EC = 2995 μScm^{-1}), OC (144 g kg^{-1}) and N (2625 mg kg^{-1} available N and 10.3 g kg^{-1} total N) contents.

During the first experiment, the temporal variation in WEC of Zn under both irrigation schedules was marginal and did not indicate any clear trend and hence not discussed further. This may be attributed to long sampling interval and low application rates used in the first year. It is understood that the water extractable forms of metals recurrently can undergo desorption, readsorption and redistribution in a given time period depending on their concentrations in each form[91] and prevailing conditions. The temporal variations in WEC of Zn were more pronounced in the second year when the additional amounts of amendment (double the amount used in the first year) were incorporated and the soil samples were collected more frequently (Figure 1 & 2) i.e. 3, 7, and 18 DAA and subsequently at an interval of 15 days throughout the experiment that mostly was a day after the irrigation. However, to evaluate the effect of amendment on the temporal variation of WEC of Zn in the soil profile, only the plots under I_1 (15% MAD) were considered. Due to wide variation in the magnitude of Zn concentrations, different scales on Y-axis have been used in the figures for better visual comparison. Initial WEC of Zn on 0 DAA shown in these figures is the actual observed concentration before the start of second year experiment.

Variations in WEC of Zn were more pronounced in the surface (0–15 cm) layers than the deeper ones (Figure 1 & 2). A noticeable increase in WEC of Zn was observed initially between 3 to 18 days after amendment compared to the remaining period of the experiment. The range was within 4 to 8 mg kg^{-1} till the 34th day and then gradually declined. The increase in WEC of Zn during the initial period can be attributed to higher rate of Zn solubility owing to decomposition of both or any one of the amendment components after application into the soil.[92–93] The gradual changes in the WEC of Zn thereafter could be attributed to its re-adsorption in soil and retransformation into sparingly soluble chemical forms over a period of time.[91] The marginal cyclic variations in WEC of Zn observed during rest of the period may be due to its simultaneous adsorption and release in the soil. In consonance with the surface layer, lower layers especially 15–30 cm and 30–45 cm exhibited marginal variations in WEC of Zn and thus suggest insignificant leaching. However, the potential for leaching of Zn from such amendment is high immediately after the waste application and corroborates with earlier findings.[92–93] On the contrary, metal leaching through soil profile via macropore or preferential transport can also not be ruled out and is likely to occur without substantial increase in lower layer concentrations.[66,94]

The crop cover did not show any significant effect on the WEC of Zn and exhibited similar variations as in the case of uncropped ones. In mixture treated plots, the effect of ash on increasing the soil pH and associated decrease in metal solubility in the presence of sludge was not clearly noticed. Lower WEC of Zn in mixture treated plots compared to those treated with sludge might be due to relatively lesser Zn loading. Marginal variation in WEC of Zn was also observed temporarily in the ash treated plots (Figure 2) since the ash contained significant amounts of Zn (Table 3).

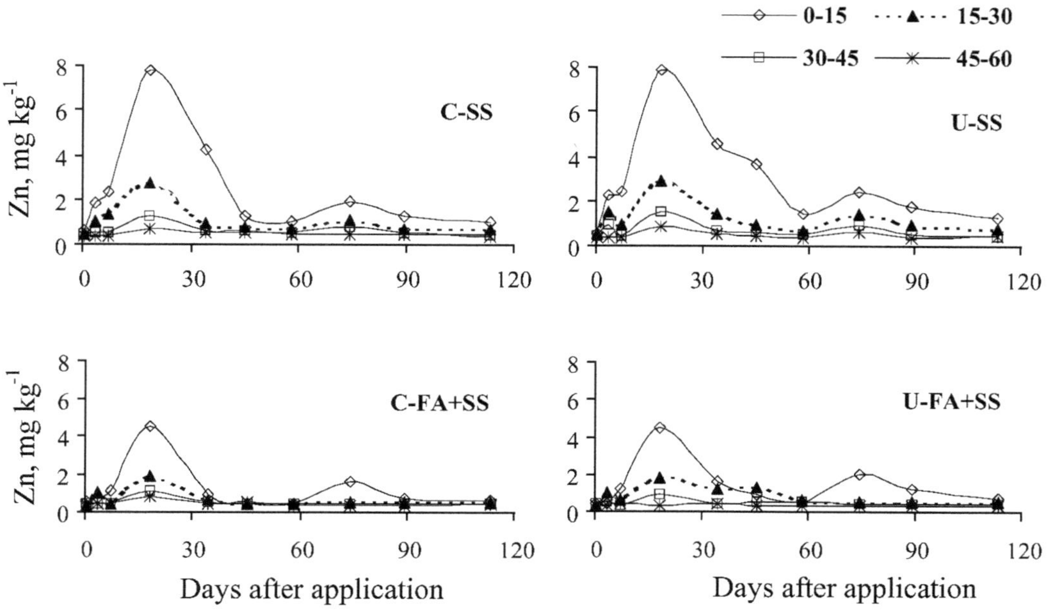

FIGURE 1. Temporal variation of water extractable Zn in different soil layers of sludge and mixture treated plots.

3.2. Distribution of Total Zn in Soil Profile

Distribution of total concentration of Zn in the soil profile under different treatments at the end of second experiment is shown in Figure 3.

Total concentration of Zn in the soil profile increased on application of ash and sludge, either alone or in combination in proportion to the metal loadings (Table 5). A marginal increase in the value of total soil Zn to a depth of 30–45 cm, especially in sludge treated plots (Figure 3) indicated possible movement at greater depths. This is ascribed to higher amounts of total Zn in sludge (3183 mg kg^{-1}) and its potential mobility even in the short term at low loading rates[77,95] in oxyaquic haplustalf. In the short term, relatively less downward movement in the

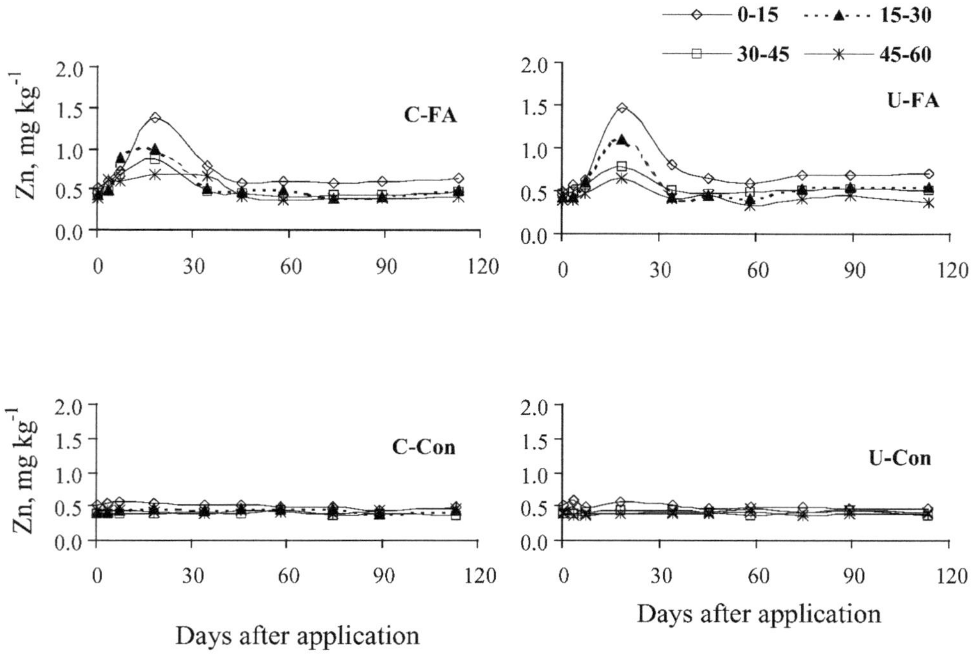

FIGURE 2. Temporal variation of water extractable Zn in different soil layers of fly ash treated and control plots (C-Cropped; U-Uncropped; SS-Sewage Sludge; FA-Fly ash; Con-Control).

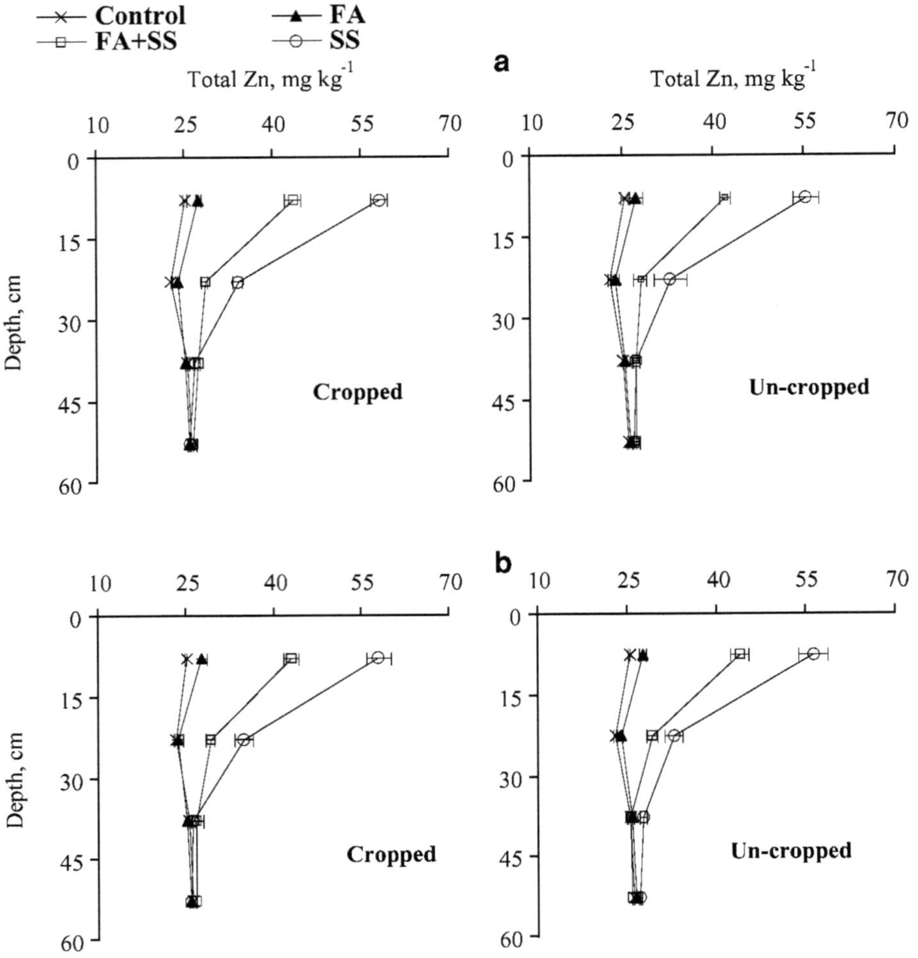

FIGURE 3. Total concentration of Zn at various soil depths in the plots under different treatments (a.15% and b. 50% maximum allowable depletion of available water).

distributions of metals in profiles of waste amended soils below the depth of application has been noticed.[45,74] However, more pronounced movement of sludge-borne Zn, in particular, below 40 cm in different soils have also been observed.[76]

Presence of crop cover did not alter the vertical distribution of Zn (Table 5) since uptake by crop is very low in terms of the fraction of total metal removed from the soil.[39,51,96,97] The effect of irrigation schedule on the vertical movement also could not be noticed (Table 5). Leaching or downward movement of Zn could be expected during initial period due to higher rate of Zn solubility due to decomposition of the amendment However, the irrigation schedules were maintained only after initial establishment of crop (32 DAA in first year and 39 DAA in the second year), during which the plots under both schedules received similar irrigation water. Further between two experiments (June 2000 to Feb 2001) all the plots, irrespective of the ground cover, received the equal amount of water through rainfall, which was comparatively higher than the amount of irrigation. In addition to water, metal movement in waste treated soil might depend on many other factors such as prevailing soil conditions,[98] metal adsorption affinity of soil, chemical forms of metals and microbial biomass activities which were not considered in this study.

3.3. Plant Uptake of Zn

The average concentrations of Zn in the dry matter of shoot, shell and kernel of peanut determined after each crop harvest are summarized in Table $6_{a\&b}$. The concentration range of Zn was comparatively higher in all the plant parts and is attributed to the acidic nature[71,77] of the soil itself (pH = 5.3).

In general, the effect of irrigation schedule on plant uptake of Zn was marginal for kernel but significant for shoot and shell, showing higher concentrations under I_1 than I_2 in both the years. This indicated that uptake was reduced to a certain extent due to water stress under I_2 schedule. Both increase and decrease in plant concentrations of essential nutrients under water stress conditions has been reported.[79] Unlike irrigation schedules, the application of amendment had more distinct

TABLE 5. Total Zn concentrations (mg kg^{-1}) in the soil profiles of different treatments.

			Soil depth, cm			
Ground Cover	Irrigation Schedule	Treatment	0–15	15–30	30–45	45–60
Cropped	I_1	Control	24.93	22.65	25.45	26.09
		FA	27.32	23.84	25.14	25.74
		FA + SS	43.32	28.36	27.34	26.25
		SS	58.02	33.98	26.42	24.54
	I_2	Control	25.01	23.11	25.32	26.12
		FA	27.58	23.57	25.18	25.82
		FA + SS	42.92	29.04	26.78	26.65
		SS	57.83	34.73	25.88	25.74
Un-cropped	I_1	Control	25.48	22.97	25.12	26.12
		FA	27.25	23.73	25.41	26.18
		FA + SS	42.08	28.02	27.32	27.25
		SS	54.11	32.95	27.15	26.84
	I_2	Control	25.33	22.85	25.42	25.95
		FA	27.45	23.72	25.71	26.46
		FA + SS	43.85	28.95	25.32	25.76
		SS	56.26	32.88	27.48	26.95

Effect of treatment:

	Soil depth, cm			
Treatment	0–15	15–30	30–45	45–60
Control	25.19a	22.90a	25.33a	26.07
FA	27.40a	23.72a	25.36a	26.05
FA + SS	43.04b	28.59b	26.69b	26.48
SS	56.56c	33.64c	26.73b	26.02
SEm (±)	0.80	0.49	0.35	0.27
LSD (0.05)	2.30	1.41	1.00	NS

SEm and LSD shown are for the effect of treatment at 0.05
Values in columns followed by same letter are non-significant (effect of treatment) at 0.05
Effect of ground cover, irrigation schedule and interactions are non-significant.
I_1-15%MAD; I_2-50%MAD; FA-Fly Ash; SS- Sewage Sludge

effect on Zn uptake and accumulation in different plant parts. The increase in Zn concentration of plants grown on treated soils is supported by the increased total concentrations of soil at higher rates of amendment application in the second year (Figure 3). Application of sludge significantly increased Zn in peanut plant parts over the control in both the experiments; the degree of increase being more in the second year owing to higher quantity of its addition. In ash and mixture treated plots, although there were significant increases in plant metal concentrations in both the experiments, the magnitude was more pronounced during the second year, obviously due to higher metal loadings. The results also indicated that the presence of sludge had more control than the ash on Zn uptake in mixture treated plots due to decrease in the soil pH slightly, which resulted in increased metal mobility and consequently higher plant accumulation.[99]

3.4. Distribution of Zn in Plant Parts

The distribution of Zn within peanut plant parts was estimated by calculating its ratio with respect to "total" (sum of the concentrations in shoot, shell and kernel) plant concentration and is expressed as percentage and its distribution in various plant parts is illustrated in Figure 4. It is seen that Zn is principally localized in the shoot than shell and kernel where it is distributed in similar manners. Further, the concentrations of Zn in plant parts were systematically higher in the sludge treated plots followed by mixture and ash treated ones.

The relationships between Zn concentration in the plant parts and its "total" concentrations in soil are shown in Figure 5. There were noticeable differences in the magnitudes of Zn concentration in shoot (vegetative tissues) and the kernel (storage organs). It is seen that within the ranges of observed Zn concentration in soils, its relationship with plant concentration was linear and significant (Table 7). The slope of the linear regression in Table 7 symbolized the transfer efficiency (TE) of Zn from soil to the plant.[100] Greater slopes in the case of vegetative tissues than the storage tissues (kernel) indicated the presence of physiological barrier inhibiting the transfer of metals.

The order of the slopes for the plant parts was shoot > shell > kernel and is similar to the metal distribution patterns in the plant parts (Figure 4). The soil pH seems to be the major factor regulating the mobility and transfer of Zn within

TABLE 6a. Effect of irrigation schedule and solid waste application on concentration of Zn in peanut plant parts (Experiment-1)

	Zn concentration, mg kg^{-1}								
	Shoot			Shell			Kernel		
Treatment	I_1	I_2	Average	I_1	I_2	Average	I_1	I_2	Average
Control	31.8	29.9	30.9a	12.8	11.6	12.2a	10.6	10.7	10.7a
FA	40.9	34.0	37.5a	16.5	14.5	15.5b	16.5	14.9	15.7b
FA + SS	70.5	62.3	66.4b	19.7	18.1	18.9b	24.3	18.9	21.6c
SS	114	95.7	105c	25.7	19.7	22.7c	21.6	23.7	22.7c
Average	64.3‡	55.5†	SEm(I) = 2.62 LSD(I) = 7.95 SEm(D) = 3.71 LSD(D) = 11.5	18.7‡	15.9†	SEm(I) = 0.83 LSD(I) = 2.53 SEm(D) = 1.18 LSD(D) = 3.58	18.3†	17.1†	SEm(I) = 0.74 LSD(I) = 2.28 SEm(D) = 1.06 LSD(D) = 3.22

TABLE 6b. Effect of irrigation schedule and solid waste application on concentration of Zn in peanut plant parts (Experiment-2)

	Zn concentration, mg kg^{-1}								
	Shoot			Shell			Kernel		
Treatment	I_1	I_2	Average	I_1	I_2	Average	I_1	I_2	Average
Control	25.9	23.5	24.7a	14.9	11.5	13.2a	12.7	11.8	12.3a
FA	56.2	55.6	55.9b	25.5	24.1	24.8b	28.1	25.5	26.8b
FA + SS	113	84.3	98.7c	36.6	36.5	36.6c	34.6	33.2	33.9c
SS	182	150	166d	55.3	46.6	51.0d	43.2	40.6	41.9d
Average	94.3‡	78.4†	SE(I) = 3.82 LSD(I) = 11.6 SEm(D) = 5.40 LSD(D) = 16.4	33.1‡	29.7†	SE(I) = 1.07 LSD(I) = 3.26 SEm(D) = 1.52 LSD(D) = 4.61	29.7†	27.8†	SE(I) = 1.11 LSD(I) = 3.38 SEm(D) = 1.57 LSD(D) = 4.78

I = Irrigation schedule and D = Treatment; Effect of interactions are non-significant
Values in columns followed by same letter are non-significant (effect of treatment) at 0.05
Values in rows followed by same symbol are non-significant (effect of irrigation schedule) at 0.05
FA—Fly Ash; SS—Sewage Sludge

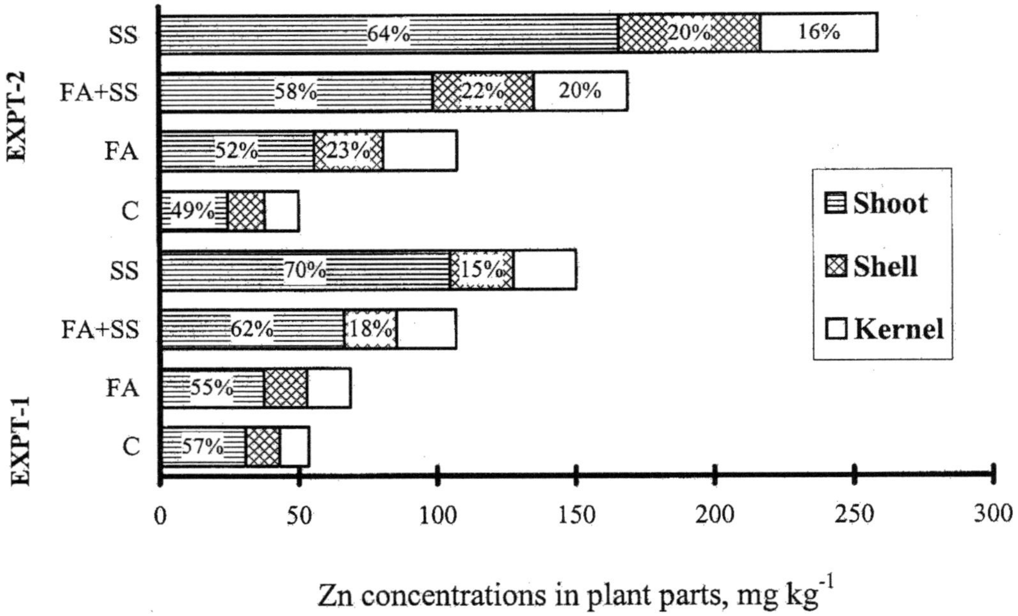

FIGURE 4. Partitioning of Zn in different plant parts of peanut in both the experiments.

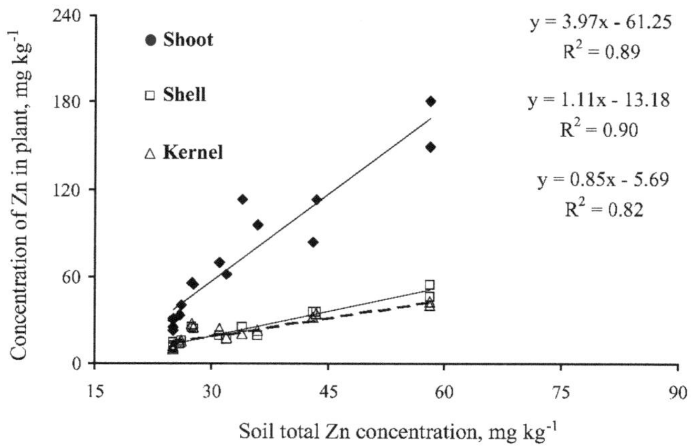

FIGURE 5. Relationship between total concentrations of metals in soil.

the plant in addition to its loading as clearly demonstrated by comparatively higher accumulation of Zn in plants grown in the treated plots receiving sewage sludge than those either with ash or with mixtures.

4. Conclusions

It can be concluded that even under low application rates of ash and sludge Zn is able to move within the profiles of the amended soil. The concentration of water extractable Zn increased immediately after the amendment application and remained transient for the rest of the period. Despite the presence of ash in the mixture, its effect in increasing the soil pH and immobilizing Zn below the surface layer in the mixture amended soil is not witnessed. While accumulation of Zn in peanut plant parts showed that the uptake is governed by the loading rates, albeit reduced under water stress conditions, the transfer ratios revealed that accumulation is more in the vegetative parts than the storage organs.

Acknowledgements. This work has been supported by projects, Land Restoration Through Waste Management and Fly Ash Management in India. The authors thank India Canada Environment Facility, New Delhi, India, Canadian International Development Agency and International Development Research Center, Canada. The authors also thank M K Panigrahi for critically going through the manuscript.

TABLE 7. Range of Zn concentrations (mg kg^{-1}) in peanut plant parts and parameters of linear regression between plant and soil concentration.

	Range	Intercept	Slope	R^2
Plant Part		Zn		
Shoot	23.5–182.0	−61.25	3.97	0.89 **
Shell	11.5–55.3	−13.15	1.11	0.90 **
Kernel	10.6–43.2	−5.69	0.85	0.82 **

** Significant at 1%

References

1. Negi, B.S. and Meenakshi, V. 1991. Characterization of coal ash from thermal power plants in India. p. 143–152 In: K.C. Sahu (ed.) Environmental Impact of coal utilization. Internat Conf IUIT Bombay, India.
2. El-Mogazi, D., Lisk, D.J. and Weinstein, L.H. 1988. A review of physical, chemical and biological properties of fly ash and effects on agricultural ecosystems. Sci Total Environ 74: 1–37.
3. Schumann, A.W. and Sumner, M.E. 1999. Plant nutrient availability from mixtures of fly ashes and biosolids. J Environ Qual 28: 1651–1657.
4. Plank, C.O. and Martens, D.C. 1974. Boron availability as influenced by application of fly ash to soil. Soil Sci Soc Am Proc 38: 974–977.
5. Ransome, L.S. and Dowdy, R.H. 1987. Soybean growth and boron distribution in a sandy soil amended with scrubber sludge. J Environ Qual 16: 171–175.
6. Hill, M.J. and C.A. Lamp. 1980. Use of pulverized fuel ash from Victorian brown coal as a source of nutrients for a pasture species. Aust Exp Agric Animal Husb 20: 377–384.
7. Doran, J.W. and Martens, D.C. 1972. Molybdenum availability as influenced by application of fly ash to soil. J Environ Qual 1: 186–189.
8. Elseewi, A.A., Bingham, F.T. and Page, A.L. 1978. Availability of sulphur in fly ash to plants. J Environ Qual 7: 69–73.
9. Schnappinger, M.G., Martens, D.C. and Plank, C.O. 1975. Zinc availability as influenced by application of fly ash to soil. Environ Sci Technol 9: 258–261.
10. Plank, C.O., Martens, D.C. and Hallock, D.L. 1975. Effect of soil application of fly ash on chemical composition and yield of corn and on chemical composition of displayed soil solution. Plant Soil 42: 465–478.
11. Martens, D.C. and Beahm, B.R. 1976. Growth of plants in fly ash amended soils. p. 657–664 In: Proc. Fourth Int Ash Utilizations Symp, US Bureau of Mines St. Louis, US.
12. Adriano, D.C., Page, A.L., Elseewi, A.A., Chang, A.C. and Straughan, I. 1980. Utilization and disposal of fly ash and other coal residues in terrestrial ecosystems: A review. J Environ Qual 9: 333–343.
13. Wochok, Z.S., Fail, J.I. and Hosmer, M. 1976. Analysis of plant growth in fly ash amended soil. Proc. Fourth Internat Fly Ash Symp St. Louis US.
14. Kuchanwar, O.D. and Matte, D.B. 1997. Study of graded doses of fly ash and fertilizers on growth and yield of groundnut Arachis hypogaea L. J Soils Crops 7(1): 36–38.
15. Jacobs, L.W., Erickson, A.E., Berti, W.R. and McKeller, B.M. 1991. Improving crop yield potentials of coarse textured soils with fly ash amendment. In: Proc. 9th Int. Ash Use Symp. Vol. 3. EPRI GS-7162. Am Coal Ash Assoc Alexandria, VA.
16. Moliner, A.M. and Street, J.J. 1982. Effect of fly ash and lime on growth and composition of corn (Zea mays L.) on acid sandy soil. Soil Crop Sci Soc Proc 41: 217–220.
17. Elseewi, A.A., Grimm, S.R., Page, A.L. and Straughan, I.R. 1981. Boron enrichments of plants and soils treated with coal ash. J Plant Nutr 3: 409–427.
18. Druzina, V.D., Miroshrachenko, E.D. and Chettov, O.D. 1983. Effect of industrial pollution on nitrogen and ash content in meadow phytoenotic plants. Bot Zh 68: 1583–1591.

19. Wong, M.H. and Wong, J.W.C. 1989. Germination and seedling growth of vegetable crops in fly ash amended soil. Agric Ecosyst Environ 26: 23–35.
20. Fang, M., Wong, J. W. C., Li, G. X. 1998. Changes in biological parameters during co-composting of sewage sludge and fly ash residues. Biores Technol. 64: 55.
21. Wong, J.W.C. and Su, D.C. 1997. Reutilization of coal fly ash and sewage sludge as an artificial soil-mix: Effects of pre-incubation on soil physico-chemical properties. Biores Technol 59: 97–102.
22. Carlson, C.L. and Adriano, D.C. 1993. Environmental impact of coal combustion residues. J Environ Qual 22: 227–247.
23. Elliot, I.F., Tittenmore, D., Papendick, R.D., Cochren, V.I. and Bezdicek, D.F. 1982. The effect of Mount Helen's ash on soil microbial respiration and numbers. J Environ Qual 11: 164–166.
24. Arthur, M.A., Zwick, T.C., Tolle, D.A. and Van Voris, P. 1984. Effects of fly ash on microbial CO2 evolution from agricultural soil. Water Air Soil Pollut 22: 443–453.
25. Wong, M.H. and Wong, J.W.C. 1986. Effects of coal ash on soil microbial activity. Environ Pollut 40: 127–144.
26. Cervelli, S.G., Petruzzelli, G. and Pema, A. 1987. Fly ashes as amendments in cultivated soils. I. Effect on mineralization and nitrification. Water Air Soil Pollut 33: 331–338.
27. Pitchel, J.R. and Hynes, J.M. 1990. Influence of fly ash on soil microbial activity and population. J Environ Qual 19: 593–597.
28. Alva, A.K., Paramasivam, S., Prakash, O., Sajwan, K.S., Ornes, W.H., and Van Clief, D. 1999. Effects of fly ash and sewage sludge amendments on transport of metals in different soils. p. 207–222 In: K.S. Sajwan, A.K. Alva, and R.F. Keefer (eds.) Biogeochemistry of Trace Elements in Coal and Coal combustion Byproducts. Kluwer Academic/ Plenum Publ., New York.
29. Clark, R.B., Zeto, S.K., Baligar, V.C. and Ritchey, K.D. 1999. Nickel, lead, cadmium, and chromium concentrations in shoots of maize grown in acidic soil amended with coal combustion byproducts. p. 259–267 In: K.S. Sajwan, A.K. Alva, and R.F. Keefer (eds.), Biogeochemistry of trace elements in coal and coal combustion byproducts. Kluwer Academic Publ, New York, US.
30. Collier, G.F. and Greenwood, D.J. 1977. Potential phytotoxic components of pulverized fuel ash. J Sci Food Agric 28: 137–144.
31. Adriano, D.C., Woodford, T.A. and Ciravalo, T.G. 1978. Growth and elemental composition of corn and bean seedlings as influenced by soil application of coal ash. J Environ Qual 7: 416–421.
32. Kukier, U., Sumner, M.E. and Miller, W.P. 1994. Boron release from fly ash and its uptake by corn. J Environ Qual 23: 596–603.
33. Adriano, D.C. and Weber, J.T. 2001. Influence of fly ash on soil physical properties and turfgrass establishment. J Environ Qual 30: 596–601.
34. Tolle, D.A., Arthur, M.F. and Van Vorris, P. 1983. Microcosm/ field comparision of trace element uptake in crops grown in fly ash amended soil. Sci Total Environ 31: 243–261.
35. Shukla, K.N. and Mishra, L.C. 1986. Effect of fly ash extract on growth and development of corn and soybean seedlings. Water Air Soil Pollut 27: 155–167.
36. Hue, N.V., Silva, J.A. and Arifin, R. 1988. Sewage sludge-soil interactions as measured by plant and soil chemical composition. J Environ Qual 17(3): 384–390.
37. Walter, I., Miralles, R., Funes, E., Gorospe, M.J. and Bigeriego, M. 1990. Effect of sewage sludge used as fertilizer in the central region of Spain. p. 304–309 In: L'Hermite (ed.). Use of Sewage Sludge and Liquid Agricultural Waste. Elsevier Applied Sci.
38. Wong, J.W.C., Lai, K.M., Fang, M. and Ma, K.K. 1998. Effect of sewage sludge amendment on soil microbial activity and nutrient mineralization. Environ International 24: 935–943.
39. Berti, W.R. and Jacobs, L.W. 1998. Distribution of trace elements in soil from repeated sewage sludge application. J Environ Qual 27: 1280–1286.
40. Scancar, J., Milacic, R., Strazar, M., Burica, O., Total metal concentrations and partitioning of Cd, Cr, Cu, Fe, Ni and Zn in sewage sludge, Sci. Total Environ., 250, 9, 2000.
41. Sauerbeck, D.R., Plant element and soil properties governing uptake and availability of heavy metals derived from sewage sludge, Water Air and Soil Pollut., 57–58, 227, 1999.
42. Heckman, J.R., Angle, J.S., Chaney, R.L., Residual effects of sewage sludge on soybean: I. Accumulation of heavy metals, J. Environ. Qual., 16, 113, 1987.
43. Smith, S.R. 1994a. Effect of soil pH on availbility of metals in sewage treated soils. I Nickel, copper and zinc uptake and toxicity to ryegrass. Environ Pollut 85: 321–327.
44. Smith, S.R.1994b. Effect of soil pH on availability to crops of metals in sewage sludge-treated soils. II Cd uptake by crops and implications for human dietery intake. Environ Pollut 86: 5–13.
45. Alloway, B. and Jackson, A. 1991. The behaviour of heavy metals in sewage sludge amended soils. Sci Total Environ 100: 151–176.
46. Moreno, J.L., Garcia, C., Hernandez, T. and Ayuso, M. 1997. Application of composted sewage sludges contaminated with heavy metals to an agricultural soil. Soil Sci Plant Nutr 43: 565–573.
47. Viller, M.C., Beloso, M.C., Acea, M.J., Cabaneiro, A., Gonzalez-Prieto, S.J., Carballas, M., Diaz-Ravina, M. and Carballas, T. 1993. Physical and chemical characterization of four composted urban refuses. Biores Technol 45: 105–113.
48. Brallier, S., Harrison, R., Henry, C. and Dongesn, X. 1996. Liming effects on availability of Cd, Cu, Ni and Zn in a soil amended with sewage sludge 16 years previously. Water Air Soil Pollut 86: 195–206.
49. Ibekwe, A.M., Angle, J.S., Chaney, R.L. and Van Berkum, P. 1995. Sewage sludge and heavy metal effects on nodulation and nitrogen fixation of legumes. J Environ Qual 24: 1199–1204.
50. Berti, W.R. and Jacobs, L.W. 1996. Chemistry and phytotoxicity of soil trace elements from repeated sewage sludge applications. J Environ Qual 25: 1025-1032.
51. Kelling, K.A., Keeney, D.R., Walsh, L.M. and Ryan, J.A. 1977. A field study of the agricultural use of sewage sludge. III Effect on uptake and extractability of sludge borne metals. J Environ Qual 6: 352–358.
52. Chang, A.C., Page, A.L., Foster, K.W. and Jones, T.E. 1982. A comparison of cadmium and zinc accumulation by four cultivars of barley grown in sludge-amended soils. J Environ Qual 11: 409–412.

53. Emmerich, W.E., Lund, L.J., Page, A.L. and Chang, A.C. 1982. Movement of heavy metals in sewage sludge-treated soils. J Environ Qual 11: 174–178.

54. Chang, A.C., Warneke, J.E., Page, A.L. and Lund L.J. 1984. Accumulation of heavy metals in sewage sludge-treated soils. J Environ Qual 13: 87–91.

55. Williams, D.E., Vlanis, J., Pukite, A.H. and Corey, J.E. 1980. Trace element accumulation, movement and distribution in the soil profile from massive application of sewage sludge. Soil Sci 129: 119–132.

56. Williams, D.E., Vlanis, J., Pukite, A.H. and Corey, J.E. 1984a. Metal movement in sludge-treated soils after six years of sludge addition. I. Cadmium, copper, lead, and zinc. Soil Sci. 137: 351–359.

57. McGrath, S.P. 1987. Long term studies in metal transfer following application of sewage sludge. In: Coughtrey et al. (eds.) Pollutant transport and fate in ecosystem. Special Publication No. 6, British Ecol. Soc. Blackwell Scientific Publ. London.

58. McGrath, S.P. and Lane, P.W. 1989. An explanation for the apparent losses of metals in a long-term field experiment with sewage sludge. Environ Pollut 60: 235–256.

59. Logan, T.J. and Chaney, R.L. 1983. Utilization of municipal wastewaters and sludge on land-metals. p. 235–326. In. A.L. Page et al. (eds.) Utilization of municipal wastewaters and sludge on land. Univ of Calif Riverside, CA.

60. Dowdy, R.H., Latterell, J.J., Hinesley, T.D., Grossman, R.B. and Sullivan, D.L. 1984. Trace metal movement in a silt loam soil. In: D.W. Nelson et al. (eds.) Chemical mobility and reactivity in soil systems. SSSA Spec Publ No. 11. ASA and SSSA, Madison, WI.

61. Schirado, T.L., Vergara, T.I., Schalascha, E.B. and Pratt, P.F. 1986. Evidence for movement of heavy metals in a soil irrigated with untreated wastewater. J Environ Qual 15: 9–12.

62. Lund, L.J., Page, A.L. and Nelson, C.O. 1976. Movement of metals below sewage disposal ponds. J Environ Qual 5: 330–334.

63. Dowdy, R.H., Latterell, J.J., Hinesly, T.D., Grossman, R.B. and Sullivan, D.L. 1991. Trace metal movement in an Aeric Ochraqualf following 14 years of annual sludge application. J Environ Qual 20: 119–123.

64. Yingming, L. and Corey, R.R. 1993. Redistribution of sludge-borne cadmium, copper, and zinc in a cultivated plot. J Environ Qual 22: 1–8.

65. McBride, M.B., Richards, B., Steenhuis, T, Russo, J.J. and Sauvé, S. 1997. Mobility and solubility of toxic metals and nutrients in soil fifteen years after sludge application. Soil Sci 162: 487–500.

66. Richards, B.K., Steenhuis, T.S., Peverly, J.H. and McBride, M.B. 1998. Metal mobility at an old, heavily loaded sludge application site. Environ Pollut 99: 365–377.

67. Streck, T. and Richter, J. 1997a. Heavy metal displacement in a sandy soil at the field scale: I Measurements and parameterization of sorption. J Environ Qual 26: 49–56.

68. Baveye, P., McBride, M.B., Bouldin, D., Hinesly, T.D., Dahdoh, M.S.A. and Abdel-sabour, M.F. 1999. Mass balance and distribution of sludge-borne trace elements in a silt loam soil following long-term applications of sewage sludge. Sci Total Environ 227: 13–28.

69. McGrath, S.P., Chang, A.C., Page, A.L. and Witter, E. 1994. Land application of sewage sludge: scientific perspectives of heavy metal loading limits in Europe and the United Sates. Environ Rev 2: 108–118.

70. Kabata-Pendias, A. and Adriano, D.C. 1995. Trace metals. p. 139–167 In: J.E. Rechcigl (ed.) Soil amendments and environmental quality. CRC Press/Lewis Publ. Boca Raton, FL.

71. Smith, S.R. 1996. Agricultural recycling of sewage sludge and the environment. CAB International, UK.

72. McBride, M.B. 1995. Toxic metal accumulation from agricultural use of sludge: Are USEPA regulations protective? J Environ Qual 24: 5–18.

73. Nemeth, T., Bujtas, K., Csillag, J., Partay, G., Likacs, A. and van Genuchten, M.Th. 1999. Distribution of ecologically significant fractions of selected heavy metals in the soil profile. p. 251–270 In: H.M. Selim, and I.K. Iskander (eds.) Fate and transport of heavy metals in vadose zone. Lewis Publ Boca Raton, FL.

74. Seaker, E. 1991. Zinc, copper, cadmium and lead in mine-spoil, water and plants from reclaimed mine land amended with sewage sludge. Water Air Soil Pollut 57–58: 849–859.

75. Sidle, R.C. and Kardos, L.T. 1977a. Transport of heavy metals in a sludge treated forested area. J Environ Qual 6: 431–437.

76. Darwish, O. and Ahmed, F. 1997. Avialability, movement and distribution of sludge-borne elements in soil. Alex Sci Exch 18: 505–521.

77. Planquart, P., Bonin, G., Prone, A. and Massiani, C. 1999. Distribution, movement and plant availability of trace metals in soils amended with sewage sludge composts: application of low metal loadings. Sci Total Environ 241: 161–179.

78. Kabata-Pendias, A. and Adriano, D.C. 1995. Trace metals. p. 139–167 In: J.E. Rechcigl (ed.) Soil amendments and environmental quality. CRC Press/Lewis Publ. Boca Raton, FL.

79. Alam, S.M. 1994. Nutrient uptake by plants under stress conditions. p. 227–244 In: M. Pessarakli (ed.) Plant and Crop Stress. Marcel Dekker Inc., New York.

80. Iu, K.L., Pulford, I.D. and Duncan, H.J. 1982. Influence of water logging on subsequent plant growth and trace metal content. Plant Soil 66: 423–431.

81. Erlandsson, G. 1975. Rapid effects on ion and water uptake induced by changes in water potential in young wheat plants. Physiol Plant 35: 256–260.

82. Jackson, M.L. 1967. Soil Chemical Analysis. Prentice-Hall of India, Bombay.

83. Nelson, D.W., Sommers, L.E. 1982. Total carbon, organic carbon and organic matter, In Methods of Soil Analysis, Part II, 2nd Edition, Page, A, L., Miller, R. H. and Keeney, D. R., Eds., American Society of Agronomy, Madison, Wisconsin, 539.

84. Bremner, J.M., Mulvaney, C.S. 1982. Nitrogen -Total. In Methods of Soil Analyses, Part 2, 2nd Edition, Page, A.L., Miller, R.H. and Keeney, D.R., Eds., American Society of Agronomy, Madison, Wisconsin, 595.

85. Subbaiah, B.V., Asija, G.L. 1956. A rapid procedure for the estimation of available nitrogen in soils. Current Sci 25: 259.

86. Lynch, J. 1996. Provisional Elemental Values for Four New Geochemical Sil and Till Reference Materials, Till-1, Till-2, Till-3, and Till-4. Geological Survey of Canada contribution paper 1996, 701, 277.

87. Maier, E.A., Griepink, B., Muntau, H., Vercoutere, K. 1993. Certification of the total contents of the Aqua Regia soluble contents of Cd, Co, Cu, Pb, Mn, Ni, and Zn in a sewage sludge, Commission of the European Communities, Community bureau of reference, 60.

88. Tran, T.S., Simard, R. R. 1993. Soil sampling and Methods of Soil Analysis. Lewis Publ., Boca Raton, FL, 43.
89. Jenkins, R, Gould, R.W. and Dalegedcke 1998. Quantitative X-ray spectroscopy series. Vol. 20, Marcel Dekker Inc. New York, USA.
90. Ma, Y.B. and Uren, N.C. 1998. Transformations of heavy metals added to soil: Application of a new sequential extraction procedure. Geoderma 84: 157–168.
91. Leita, L. and De Nobili, M. 1991. Water-extractable fractions of heavy metals during composting of municipal solid waste. J Environ Qual 20: 73–78.
92. Jackson, B.P., Miller, W.P., Schumann, A.W. and Sumner, M.E. 1999. Trace element solubility from application of fly ash/organic waste mixtures. J Environ Qual 28: 639–647.
93. Jackson, B.P. and Miller, W.P. 2000. Soil solution chemistry of a fly ash–poultry litter -and sewage sludge-amended soil. J Environ Qual 29: 430–436.
94. Sidle, R.C., Kardos, L.T. and van Genuchten, M.Th. 1977b. Heavy metals transport model in a sludge treated soil. J Environ Qual 6: 438–443.
95. Olaniya, M., Bhoyar, R. and Bhide, A. 1991. Effects of solid waste disposal on land. Indian J Environ Health 34: 143–150.
96. Dowdy, R.H. and Larson, W.E. 1975. Metals uptake by Barley seedlings grown on soils amended with sewage sludge-treated soils. J Environ Qual 4: 229–233.
97. Bidwell, A.M. and Dowdy, R.H. 1987. Cadmium and zinc availability to corn following termination of sewage sludge application. J Environ Qual 16: 438–442.
98. Welch, J.E. and Lund, L.J. 1987. Soil properties, irrigation quality and soil moisture level influences on the movement of Ni in sewage sludge treated soils. J. Environ Qual 16: 403–410.
99. Zufiaurre, R., Olivar, A., Chamorro, P., Nerin, C. and Callizo, A. 1998. Speciation of metals in sewage sludge for agricultural uses. Analyst 123: 255–259.
100. McGrath, S.P., Zhao, F.J., Dunham, S.J., Crosland, A.R. and Coleman, K. 2000. Long-term changes in the extractability and bioavailability of zinc and cadmium after sludge application. J Environ Qual 29: 875–883.

18
Influence of Fly Ash and Other Amendments on Leaching of Soil Nutrients and Metals from a Calcareous Soil

Qingren Wang, Yuncong Li, and Waldemar Klassen

Tropical Plant Research and Education Center, Institute of Food and Agricultural Sciences, University of Florida, Homestead, FL, 33031, U.S.A.

Abstract

The influences on the leaching of soil nutrients and heavy metals by separate application to a calcareous soil of fly ash, biosolids, N-viro soil, co-compost, and yard waste compost in comparison to a control without an amendment were determined in a pot experiment. The results showed that the concentration of NO_3-N in leachate from biosolids was significantly higher than in leachate from other treatments. The levels of heavy metals found in the leachates from all amended soils were so low as to suggest these amendments may be used without risk of leaching significant amounts of these elements. Nevertheless the level of heavy metals in leachate from fly ash amended soil was substantially greater than in leachates from the other treatments. The results suggest that at heavy loading rates of biosolids, leaching of NO_3^- can be an important concern. Leaching of inorganic P can be increased significantly by application of both co-compost and biosolids, but significantly decreased by applying fly ash and N-viro soil. Biosolids, N-viro soil and fly ash applications significantly increased Ca and Mg concentrations in leachates. Copper concentration in leachate was increased by application of biosolids, while Fe concentration in leachates was increased by biosolids, fly ash and co-compost applications. The concentrations of Zn, Mo and Co in leachates were increased by application of fly ash. The concentrations of heavy metals in leachates were very low and unlikely to be harmful, although they were increased significantly by fly ash application.

Introduction

Fossil fuels are used widely in modern power plants to produce electrical energy. The inorganic residue from combustion of pulverized coal is called "Coal Combustion Byproducts (CCBs)". The amount of CCBs rapidly accumulated and caused enormous problems of disposal. In 2002 the United States produced 76.5 million tons of fly ash and Britain produced about 10 million tons. Australian coal-fired power stations produce over 8 million tons of fly ash ever year[1]. Some CCBs can be utilized effectively in construction applications such as pavement installation and concrete production. These uses of coal fly ash conserve energy by reducing the demand for typical pavement materials such as lime, cement and crushed stone. Each ton of fly ash used to replace a ton of cement can save the equivalent of nearly one barrel of imported oil, and also reduce greenhouse gas emissions, which could otherwise contribute to global warming. However, only a small proportion of fly ash can be utilized in the industry to replace cement[1]. For example, in Britain where 25 to 30% of concrete's cement content is replaced by pulverized fuel ash (PFA), less than one million tons of the 10 million tons produced each year are used in this manner. Likewise in Australia, only approximately 10% of fly ash is incorporated into cement and concrete, and the remainder is stored as unsightly ash dams[1]. Florida has tremendous requirements for electricity, and three of the 70 major U.S. coal-fired electrical utilities are located in this state[2]. According to American Coal Ash Association, over 117 million tons of coal combustion products (CCPs) including about 87 million tons of fly and bottom ash were produced in 2001, but less than 32% of coal ash produced was utilized[3].

Utilization of fly ash as a soil amendment in agriculture represents a potentially large market nationally and internationally. There are a number of benefits that result from the application of fly ash to agricultural soils including improved soil texture, modified soil pH, increased soil nutrient holding capacity, and supply of nutrients necessary for plant growth and development. Adriano et al. studied the effects of fly ash on soil chemical properties, and growth and mineral nutrition of corn and bush bean[4]. They reported that the application of fly ash increased concentrations of extractable K, Ca, Mg, Cu, Fe, Mn, and Zn in the soil and increased concentrations of K and Fe in corn shoots and of K, Ca, Mg, S, and Cu in bush bean shoots[4]. Similar results have been reported for a loamy sand soil[5]and for soil used for rice production[6]. Application of fly ash also increased the boron content of alfalfa[7] and of corn[8].

TABLE 1. Characteristics of the Krome gravelly loam soil used for the experiment.

PH (Soil : water =1 : 2.5)	$CaCO_3$ equivalent	Organic C ($g\ kg^{-1}$)	Total N ($g\ kg^{-1}$)	Extractable P ($mg\ kg^{-1}$)	Extractable K ($mg\ kg^{-1}$)
7.9	45%	2.8	1.14	22.7	129

Clearly fly ash application in agricultural production would also lead to reduced rates of inorganic fertilizer application and effectively decrease leaching and surface runoff of nitrogen (N), phosphorus (P) and other soil nutrients. Phosphorus contamination of ground and surface waters has become a serious world-wide concern. This concern is especially acute in Florida where the eutrophication in some lakes, canals and wetlands through non-point source pollution contributes to ecosystem degradation of the Everglades and marine bays[9]. An important goal in the Everglades Restoration Program is to reduce P in surface runoff from the agricultural area by 25%.

In spite of the above mentioned potential benefits, the agricultural use of coal fly ash is only 0.02 million tons per year, which is equivalent to 0.02% of fly ash production from coal[2]. The main deterrent to greater use of fly ash in agriculture is the environmental concern that persists even though the U.S. Environmental Protection Agency (EPA) has determined that power plant fly ash applied to agricultural soils is largely free of health and environmental risks and EPA does not regulate this agricultural use[10].

The agricultural use of other soil amendments, such as biosolids and compost, are hampered by similar concerns. These amendments can supply certain plant nutrients, improve soil physical and chemical properties, may totally or partially replace inorganic fertilizers, and, effectively improve crop growth and production. Nevertheless a major concern is that the heavy metal contents in those amendments may lead to unsafe accumulations in the soil as a consequence of heavy amendment loading rates or repeated applications over the long term. Therefore, the objective of the study was to elucidate the influence of the application of different soil amendments (coal fly ash, biosolids, and composts) on leaching of soil nutrients and metals from an agricultural soil.

Materials and Methods

Soil Background

The potting soil was Krome very gravelly loam (loamy-skeletal, carbonatic, hyperthermic Lithic Udorthents), containing 58.8% gravel (>2 mm), 19.9% sand, 12.5% silt and 8.8% clay with a high concentration of $CaCO_3$ and a high pH value (Table 1).

Soil Amendments

Soil amendments were coal fly ash, biosolids, yard waste compost (YW-compost), N-viro soil (1: 1 mixture of fly ash: biosolids), co-compost (3: 7 of biosolids: yard waste compost) with no soil amendment as the control. The characteristics of those soil amendments are displayed in table 2[11].

Experimental Design and Management

A randomized complete block design with 3 replicates was carried out in a greenhouse. Plastic pots (23 cm diameters, 20 cm high, each with a capacity of 8 kg soil) were used for the experiment. Each soil amendment at a rate equivalent to 50 t ha^{-1} was thoroughly mixed with the soil and then placed into the pots. Twenty okra (*Abelmoschus esculentus* L.) seeds

TABLE 2. Characteristics of soil amendments and element contents ($mg\ kg^{-1}$) used in the experiment[11].

	Fly ash	N-viro soil	Biosolids	Co-compost	YW-compost
pH	11.8	9.2	6.1	6.7	7.7
EC ($mS\ cm^{-1}$)*	5.4	6.4	1.6	1.0	1.7
Extractable N†	6.4	13.8	1461.2	145.9	29.0
Ortho-P	8.9	209.1	236.6	130.3	85.9
Total P	156	5,430	29,080	13,433	729
Total Ca	161,294	115,122	43,179	44,133	153,614
Total Mg	1,781	1,814	3.774	2,452	2,027
Total Fe	6,985	9,094	8,387	13,210	5,559
Total Zn	69	168	1,254	347	598
Total Mo	18.7	12.0	11.4	6.0	3.4
Total Co	41.5	16.6	2.8	2.6	4.5
Total Cd	1.6	1.7	3.6	1.8	1.8

* The ratio of soil: water was 1: 2.

†N was extracted by 2.0 *M* KCl; Ortho-P was extracted with ammonium bicarbonate-DTPA, and determined by means of AA3. Total content of elements was determined by means of ICP after samples digested with method EPA 3050A

were sown in each pot. Drip irrigation was set up to deliver water 2 l hr^{-1} to each pot and an automatic control timer was installed to control the irrigation frequency and period, which depended on plant growth stage and plant size. During okra flowering 3.2 g pot^{-1} of chemical fertilizer (6 N-6 P_2O_5-12 K_2O) was applied. The fertilizer consisted of 2% nitrate N, 1% ammonia N, 2% water soluble organic N and 1% water insoluble N; 6% of P_2O_5; 12% K_2O; 2% Mg; 0.06% B; 0.06% Cu; 0.14% Zn; 0.36% Fe and 0.004% Mo. Okra plants were harvested 2 months after planting (when floral buds were visible, i.e., pre-flowering stage) and dried at 70°C for 72 hrs to obtain the biomass dry weight.

Sampling and Analysis

Leachates were collected by placing the pot (which had a drain hole) on a 5 cm high PVC ring in a plastic saucer. Leachate collection was initiated at the beginning of the 3rd week after okra germination. Once per week, the amount of leachate was recorded and the samples were collected, filtered and NO_3-N, NH_4-N and inorganic P were analyzed by means of an autoanalyzer (AA3) and metals by ICP. The data were subjected to the analysis of variance (ANOVA) and Duncan's multiple range tests using SAS (version 8.1).

Results and Discussion

Leachate Volumes

All pots received the same amount of irrigation water throughout the okra growing period. However the volumes of leachate collected below the pots with soil amendment treatments and the control were significantly different. Thus the volumes of leachate were significantly reduced with the application of all soil amendments compared to the control. Most notably the volume of leachate from the biosolids treatment was reduced by 63.7% compared to that from the control (Table 3).

The leachate volumes from the biosolids and co-compost treatments were significantly lower than from the control. Water use by the plants might be a major factor for the relatively

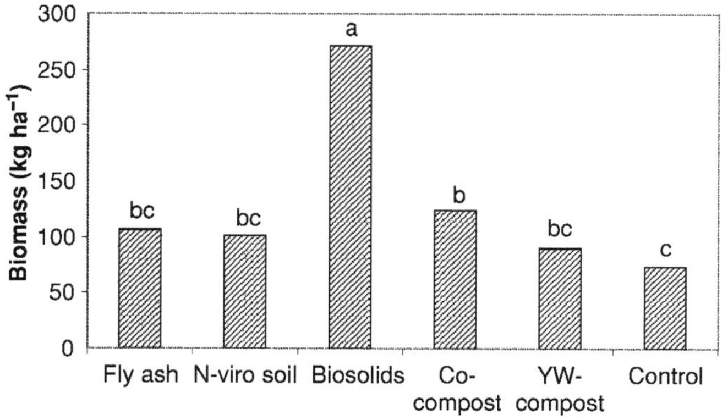

FIGURE 1. Okra biomass at visible floral bud (pre-flowering) stage produced on Krome very gravelly loam soil with various amendments. The same letter (s) on the bar represents no significant difference at $p \leq 0.05$.

low amounts of water leaching from these two treatments because the okra plants grew best with biosolids application, and significantly better in the co-compost treatment than in the control (Fig. 1). Large plants lead to a higher evapotranspiration rate, water usage and less amount of water leaching. The high organic matter content in biosolids may also have contributed to the higher water retention. However there were no significant differences observed in leachate volumes in the fly ash and N-viro soil treatments compared to the control since amounts of biomass produced with these treatments were relatively low. This result is not consistent with the findings of Ghodrati et al. that the moisture holding capacity in a soil amended with 30% fly ash was increased from 12% to 25%[5], but this inconsistency may be related to differences in application rates, types of soil and plant growth status.

Leaching of NO_3-N and NH-N

The concentration of NO_3-N in the leachate was significantly higher (15- to 68-fold) with application of biosolids than with the other soil amendments and the control, and the differences observed among all the other treatments and the control were not significant statistically (Table 3). The result indicates that high application rates of biosolids (e.g., equivalent to 50 t ha^{-1}) may cause an environmental risk of ground water pollution. On

TABLE 3. Volume of leachates, concentrations of NO_3-N, NH_4-N and Ortho-P in Krome very gravely loam soil with various amendments.

	Volume of leachate (ml pot^{-1})			NO_3-N (mg l^{-1})			NH_4-N (mg l^{-1})			Ortho-P (mg l^{-1})		
	Max	Min	Mean	Max	Min	Mean	Max	Min	Mean	Max	Min	Mean
Fly ash	473	415	428^{ab}*	32.1	0.3	19.2^{b}	3.2	0.01	0.56^{a}	0.52	0.02	0.06^{c}
N-viro soil	495	415	422^{ab}	45.5	0.2	10.5^{b}	2.5	0.02	0.43^{a}	0.48	0.01	0.08^{c}
Biosolids	250	50	192^{c}	604.4	35.5	288.5^{a}	3.8	0.1	0.68^{a}	4.21	0.27	0.54^{a}
Co-compost	430	110	343^{b}	15.2	1.2	4.2^{b}	2.6	0.1	0.26^{a}	3.28	0.18	0.63^{a}
YW-compost	480	230	451^{ab}	26.3	2.3	5.3^{b}	3.1	0.06	0.25^{a}	2.36	0.02	0.18^{bc}
Control	800	450	529^{a}	35.5	5.2	12.6^{b}	1.5	0.04	0.31^{a}	1.25	0.03	0.26^{b}

*Means following by same letters within a column represent no significant at $p \leq 0.05$.

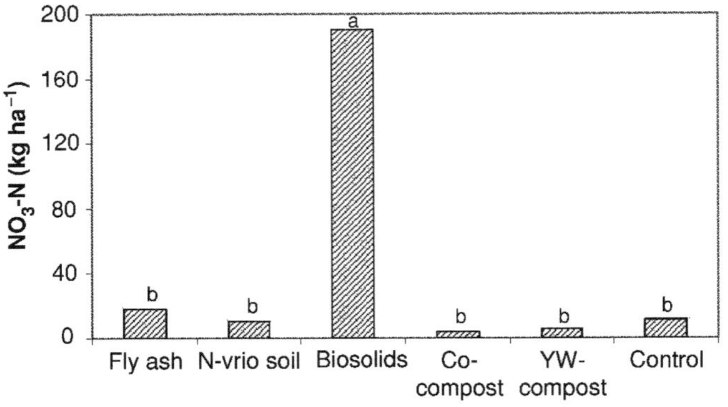

FIGURE 2. Amounts of NO_3-N leached from soil profiles treated with various amendments. The same letter (s) on the bar represents no significant difference at $p \leq 0.05$.

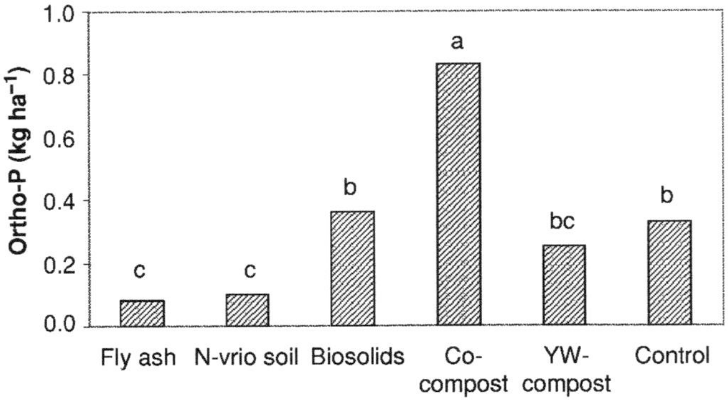

FIGURE 3. Amounts of Ortho-P leached from soil profiles treated with various amendments. The same letter (s) on the bar represents no significant difference at $p \leq 0.05$.

the other hand the concentrations of NH_4-N in the leachates from all treatments were low and statistically the same as in the control (Table 3). The levels of NH_4-N loss from the Krome calcareous soil were negligible with or without application of amendments, and this may be related to a high nitrification rate in this type of soil.

With respect to nutrient loss, on the average, an equivalent of 190 kg ha^{-1} of NO_3-N was leached from the biosolids treatment, while only 20 kg ha^{-1} or less was leached from each of the other treatments (Fig. 2). The large amount of NO_3-N leached from the biosolids treatment is the direct result of the high concentration of N in the biosolids. The KCl extractable N from biosolids was about 1,600 mg kg^{-1} and about 260 times higher than from fly ash (Table 2). This implies that the application of biosolids at a high rate (e.g., 50 t ha^{-1}) might cause NO_3-N leaching into ground water even though it can increase the plant growth dramatically. When biosolids or other high N soil amendments are applied, N leaching may be reduced by applying little or no inorganic fertilizer during crop growing season.

Leaching of P

Generally, the concentration of ortho- or inorganic P in the leachates with or without soil amendment was low (Table 3). This might be attributed to the calcareous soil used in this experiment, which has a high pH (7.9) and abundance of $CaCO_3$ (45%) (Table 1). Still the mean concentrations of ortho-P were significantly lower in treatments with fly ash containing soil amendments (e.g., fly ash and N-viro soil in this experiment). This result implies that even in a calcareous soil with a high pH, fly ash application can still effectively further reduce the P concentration in leachates, which suggests a promising environmentally important benefit for reducing the non-point source P pollution.

Total amounts of P leached from pots throughout 2-month okra growing period were calculated and presented as kg ha^{-1} for treatments with each soil amendment and a control (Fig. 3). Both co-compost and biosolids have high contents of extractable P and total P. Nevertheless the amount of P leached from soil treated with biosolids was significantly lower than that treated with co-compost. The relatively low amount of P leached from the soil treated with biosolids may be caused by the high P uptake by plants and the lower leachate volume from these pots because the treatment with biosolids produced a high amount of okra biomass (Fig. 1).

The application of fly ash or the fly ash-containing amendment, N-viro (50% fly ash), significantly reduced the amount of ortho-P leached compared to the amounts leached from the untreated control, biosolids- and co-compost treatments even though N-viro soil contains over 200 mg kg^{-1} of inorganic P compared to about 20 mg kg^{-1} in the control. The high retention of soil inorganic P by fly ash might result from strong adsorption and precipitation. It is relevant to note that 30–70% of fly ash particles on a mass basis are finer than 10 μm, and on a numerical basis, 40-60% of the particles range from 0–1 μm[12]. Cheung and Venkitachalam (2000) reported that the maximum P sorptions were 13,766 mg kg^{-1} and 3,082 mg kg^{-1} for unweathered and weathered fly ash, respectively[13]. The high pH value together with the abundant Ca in the fly ash (Table 1) might contribute to a high precipitation of soil inorganic P as well. The high retention of soil inorganic P by fly ash implies that fly ash can be used effectively to prevent soil P from leaching.

Leaching of Ca and Trace Metals

A significantly larger amount of Ca was leached from the fly ash treatment than from any of the other treatments and the control (Fig. 4). This is caused by the high Ca concentration in fly ash. Some soil microelements necessary for plant growth and development, e.g., Zn, Fe, Mo and Co, were significantly higher in the fly ash treatment leachate than in that of the control and several other treatments. Amounts of Cu, and Mo in the biosolids treatment were significantly higher than in any other treatment, and the amount of Fe in the biosolids treatment was significantly higher than in the control (Fig. 5). The

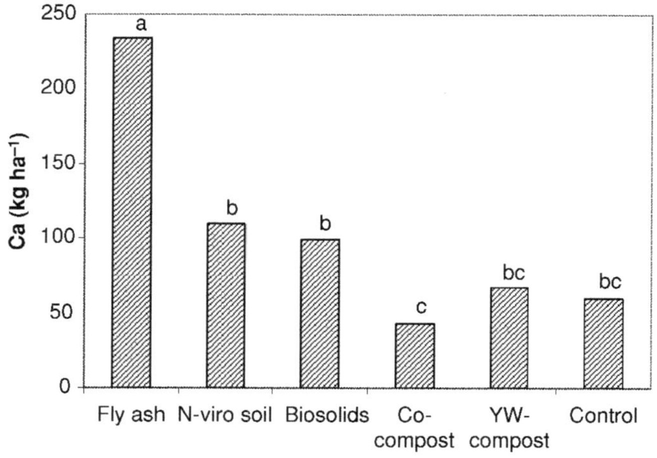

FIGURE 4. Amounts of Ca leached from soil profiles treated with various amendments. The same letter (s) on the bar represents no significant difference at $p \leq 0.05$.

elevated levels of these microelements in the leachates from the various soil amendment treatments may benefit plants, for example, certain amounts of soil-available Fe, Mo and Co facilitate development of rhizobia and N-fixation in leguminous crops.

The leachate from the fly ash treatment contained significantly higher concentrations of the heavy metals, As, Cd, Cr, Ni and Pb, than that of any other treatment (Fig. 6), which is related to the high concentration of those metals the fly ash contained (Table 2). Nevertheless the application to soil of fly ash or of the other soil amendments probably would not cause environmental problems because the concentrations of these heavy metals in the leachates proved to be far below those considered hazardous. Moreover the uptake of these heavy metal elements is pH sensitive, and in highly calcareous soils these heavy metal elements are quite stable and unavailable to plants.

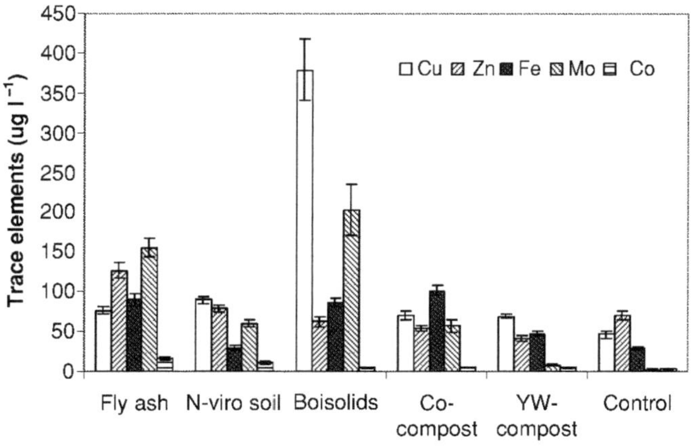

FIGURE 5. Mean concentrations of microelements in leachates from soil profiles treated with various amendments. Vertical bars represent standard error of the means ($n = 18$).

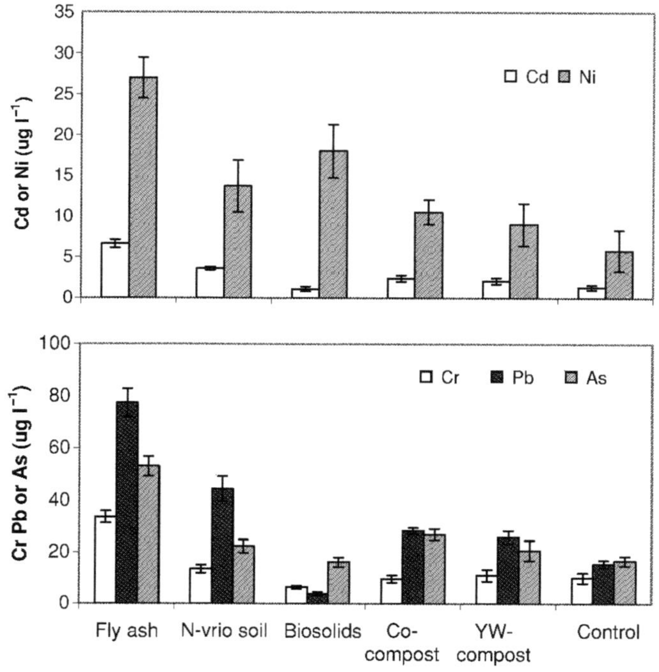

FIGURE 6. Mean concentrations of heavy metals in leachates from soil profiles treated with various amendments. Vertical bars represent standard error of the means ($n = 18$).

Conclusion

The application of some soil amendments, especially biosolids, to a calcareous soil can effectively increase soil water retention, because biosolids greatly enhance plant growth. Heavy application of biosolids can potentially cause high NO_3-N concentrations, because biosolids have a high N content. Fly ash application can effectively reduce leaching of soil inorganic P, and this application appears to have the potential to reduce the surface water eutrophication through non-point source P pollution. Also the application of fly ash can contribute the plant-essential microelements, Fe, Co, and Mo, to soil and to plants. The heavy metal concentrations in the fly ash are below those considered to be hazardous even though the levels of in fly ash of Cd, Cr, Pb, Ni, and As are considerably higher than in untreated Krome very gravelly loam calcareous soil.

Acknowledgments. This research was supported by the Florida Agricultural Experiment Station, and approved for publication as Journal Series No R-10671.

References

1. The Fly Ash Resource Center. 2004. Coal Combustion Byproducts (CCBs). http://www.geocities.com/Capecanaveral/Lauchpad/2095/flyash.html.

2. USDOE (U.S. Dept. of Energy). 2001. Annual Electric Generator Report, Electric Power Monthly, DOE/EIA-0226 (01/03), Energy Information Administration, Washington, DC.
3. ACAA (American Coal Ash Association). 2001. CCP Production and Use Survey—2000, American Coal Ash Association, Syracuse, NY.
4. Adriano, D. C., Woodford, T.A. and Ciravolo, T.G. 1978. Growth and elemental composition of corn and bean seedlings as influenced by soil application of coal ash. J. Environ. Qual. 7: 416–429.
5. Ghodrati, M., Sims, J.T., and Ciravolo, T.G. 1995. Evaluation of fly ash as a soil amendment for the Atlantic coastal plain: II. Soil hydraulic properties and elemental leaching. Water, Air and Soil Pollu. 81: 349–361.
6. Sikka, R. and Kansal, B.D. 1995. Effect of fly-ash application on yield and nutrient status of soils. Bioresource Tech. 51:199–203.
7. Mulford, F.R. and Martens, D.C. 1971. Response of alfalfa to boron in fly ash. Soil Sci. Soc. Am. Proc. 35:296.
8. Kukier, U. and Sumner, M.E. 1996. Boron availability to plants from coal combustion by-products. Water Air Soil Pollu. 87–93–110.
9. SFWMD (South Florida Water Management District). 2000. Everglades Consolidated Report. SFWMD, West Palm Beach, Fl.
10. USEPA (U.S. Environmental Protection Agency). 2000. Report to Congress: waste from the combustion of fossil fuels. EPA Docket #F-2000-FF2F-FFFFF. Office of Solid Waste, USEPA, Washington, D.C.
11. Wang, Q.R., Li, Y.C. and Klassen, W. 2003. Effects of soil amendments at a heavy loading rate associated with cover crops as green manures on the leaching of nutrients and heavy metals from a calcareous soil. J. Envir. Sci. Health, Part B. 38:865–881.
12. The fly ash resource center. 2004. Fly ash quality control. http://www.geocities.com/CapeCanaveral/Launchpad/2095/qualitycontrol.htm.
13. Cheung, K.C. and Venkitachalam, T.H. 2000. Improving Phosphate Removal of Sand Infiltration System Using Alkaline Fly Ash. Chemosphere, 41: 243–249.

19
Amendment of Fly Ash to Container Substrates for Ornamental Plant Production

Jianjun Chen[1] and Yuncong Li[2]
[1]*Mid-Florida Research and Education Center, IFAS, University of Florida, Apopka, FL 32703, USA*
[2]*Tropical Plant Research and Education Center, IFAS, University of Florida, Homestead, FL 33031, USA*

Abstract

Ornamental plants are largely produced in containerized soilless substrates that generally consist of sphagnum peat, bark, and other organic matter. Such substrates initially have a low pH (around 4.0) and limited amount of mineral nutrients. In order to raise pH and increase fertility, commercial ground limestone and chemical fertilizers have to be incorporated into the substrates. Fly ash, a coal combustion byproduct possessing alkalinity and containing essential mineral elements in silt-sized particles, could be an alternative to limestone amendments and nutrient sources for container substrates. This study examined chemical properties of four fly ash sampled from power plants in (1) North Carolina, (2) Michigan(a), (3) Florida, and (4) Michigan(b) and incorporated an appropriate amount of each fly ash or a commercial dolomite, respectively, into a soilless basal substrate, resulting in fly ash- or dolomite-amended substrates. After testing chemical properties of the substrates, seedlings of Pink Splash (*Hypoestes phyllostachya* Bak.), a herbaceous ornamental foliage plant, were planted in the basal, fly ash-, and dolomite-amended substrates. Substrate pH and plant heights were monitored during the course of plant growth. All fly ashes were able to raise the pH of the basal substrate from 3.8 to 7.4 except the one collected from Michigan(b) that was unable to change the pH regardless of the amount of fly ash used. Plants grown in the substrates amended with fly ash from North Carolina, Michigan(a), and Florida had comparable heights and similar fresh and dry weights as well as overall quality to those grown in the dolomite-amended substrate. Whereas, plants grown in the basal substrate and the substrate amended with Michigan(b) fly ash were significantly smaller and had less fresh and dry weights than those grown on the rest of substrates, thus, plants were not marketable. These results show that fly ash, after selection based on their chemical properties, can be used as alternative to commercial dolomite in soilless substrates for ornamental plant production. Utilization of fly ash as a container substrate amendment may represent a new market for the beneficial use of coal combustion byproducts.

Introduction

Fly ash is the fine residue captured from flue exhausts when coal is burnt in a power station. The physical and chemical properties of fly ash vary with the composition of the parental coal and conditions during coal combustion.[1,2] In general, fly ash is comprised primarily of fine sand- and silt-sized particles, thus, the principle compounds are SiO_2 and Al_2O_3.[3] Fly ash also contains varying amounts of C, Ca, Mg, K, Fe, P, and a small quantity of heavy metals[4]. The presence of carbonates and nutrient elements has led to the use of fly ash as a soil amendment in crop production. Fly ash amendment has been reported to improve soil texture, modify soil pH, and provide essential plant nutrients for increasing crop production.[5,6] These beneficial effects of fly ash have been demonstrated in corn (*Zea mays* L.) and bush bean (*Phaseolus vulgaris* L.)[7], barley (*Hordeum vulgare* L. var. 'Leduce')[8], rice (*Oryza sativa* cv. 'PR 106'), wheat (*Triticum aestivum* L.)[9], tomato (*Lycopersicon esculentum* Mill. cv. 'Pusa Ruby')[10], and alfalfa (*Medicago sativa* L.).[11]

According to the U.S. Department of Energy (USDOE)[12] power plants in the U.S. used over 976 million tons of coal in 2002. Over 129 million tons of coal combustion products including about 77 million tons of fly ash were produced in 2002[13], but less than 38% of coal ash was utilized as an ingredient in cement and other construction-related products. Agricultural use of coal ash was only 0.01 million tons per year.[13] Most fly ashes are disposed in landfills or slurry ponds.[14] Landfilling, however, serves as a major source of environmental pollution through fly ash erosion and leachate generation.[15,16]

One practical and potentially beneficial use of fly ash is to incorporate it into container substrates for ornamental plant production. Since the components of container substrates encompass sphagnum peat, pine bark, coir dust, perlite, or vermiculite, pH is generally low (around 4.0), and mineral nutrients are often limited. Liming materials, such as dolomite and calcitic limestone, and chemical fertilizers have to be used for adjusting pH and improving substrate fertility. As a result, thousands of tons of liming materials and chemical fertilizers are used each year as substrate amendments, which represents a great portion of total production costs. Ornamental plants now represent the sixth largest agricultural commodity group in the U.S., with farmgate cash receipts totaling $10.94 billion in 1997[17]. The ornamental plant industry is the fastest growing major segment of U.S. agriculture, averaging 7.1% annual growth during 1992–1997. Since fly ash is readily available as a disposal material, its amendment into container substrates could be another avenue of beneficial use in agriculture.

However, limited information is available on the use of fly ash as an amendment to container substrates for ornamental plant production. The objectives of this study were to examine chemical properties of selected fly ashes and ash-amended substrates and determine if fly ash-amended substrates could be used for marketable ornamental plant production.

Materials and Methods

Fly ash, dolomite, and components of container substrates. Three ashes: (1) fluidized bed fly ash, generated from a power plant in Michigan [Michigan(a)], (2) coal fly ash from North Carolina, and (3) coal fly ash from another plant in Michigan [Michigan(b)] were obtained from N-Viro International Corp., Toledo, Ohio. The fourth fly ash was obtained from a power plant in Florida. Heavy metal concentrations of As, Cd, Cr, Cu, Pb, Hg, Mo, Ni, Se, Zn, and Mn were analyzed using the EPA method 6010A by N-Viro International Corp.[18] A commercial ground dolomite, containing $CaCO_3$ 91.7%, $MgCO_3$ 3.1%, SiO_2 3.4%, $Al_2O_3$0.45%, and Fe_2O_3 0.55% with pH 8.5, was obtained from Ash Grove, Portland, Oregon. Container substrate components of sphagnum peat, pine bark, and perlite were provided by Fafard, Inc., Apopka, Florida.

Container substrate preparation and chemical analysis. A basal substrate, designated as SPP, comprised of 70% sphagnum peat, 15% pine bark, and 15% perlite based on volume was prepared. The pH of the basal substrate was 3.8. Based on the common container foliage plant production practices, 4.2 kg of the ground dolomite were incorporated into each cubic meter of the basal substrate, which resulted in a pH of 6.8 in the substrate (SPP-DO).[19] To determine the amount of fly ash needed to raise the pH of the basal substrate equivalent to that of the dolomite-amended substrate, each fly ash was incorporated into the basal substrate at 6, 9, 12, 15, 18, 21, and 36 kg per cubic meter. After measuring pH of the fly ash-amended substrates at each rate, 18 kg of fly ash per cubic meter of the basal substrate was selected. This resulted in substrates of SPP-NC, SPP-MI(a), SPP-FL, and SPP-MI(b) [basal substrate, SPP, amended with fly ash from North Carolina, Michigan(a), Florida, and Michigan(b), respectively]. Heavy metals of As, Cr, Cu, Pb, Mo, Ni, Se, Zn, and Mn as well as P, K, Ca, Mg, and Fe of the prepared substrates were analyzed. Total N and C were analyzed by a CNS Auto-Analyzer (VarioMAX, Elementar Americas, Inc. Mt Laurel, New Jersey). Substrate pH and electrical conductivity were tested by extracting solution from saturated substrates using the pour-through method.[20]

Containerized ornamental plant production. The prepared basal, dolomite, and four fly ash-amended substrates were respectively filled in 15-cm diameter plastic containers; seedlings of Pink Splash (*Hypoestes phyllostachya* Bak.) were planted with seven plants per container. One week after planting, five grams of an 18.0N-2.6P-10.0K controlled-released fertilizer with microelements (Osmocote 18-6-12, 8-9 month duration, The Scotts Co., Marysville, Ohio) was applied on the surface of the container substrates. Plants were grown on raised benches in a shaded and evaporated pad cooled greenhouse under a maximum photosynthetically active radiation of 285 $m^2 s^{-1}$. Temperatures ranged from 20 to 32 °C and relative humidity from 50 to 100%. Plants were overhead irrigated through sprinklers one to two times a week at 6 mm per application.

Plant growth was closely monitored, including weekly inspection for plant growth disorders. Initial plant height and widths were recorded. Root-zone solutions of each substrate were extracted monthly using the pour-through method[20], pH and electrical conductivity (EC) of the solutions were measured. Five months after planting, plants attained marketable sizes; plant height and widths were again measured. Plants were graded visually for overall quality, as described by Chen et al (2003a)[21], where: 1 = poor; 2 = substandard, unsalable; 3 = good, salable; 4 = very good; and 5 = excellent. Plant shoots were harvested by cutting at the substrate surface, shoot fresh weights were recorded, and dry weights were determined after drying in a forced-air oven for 48 hr at 80 °C. Shoot concentrations of mineral elements were analyzed by the soil test laboratory at the University of Florida, Gainesville, Florida.

Experimental design and data analysis. The experiment was a completely randomized design with 10 replications. Plant heights, fresh and dry weights, and monthly pH of each substrate were analyzed using the Statistical Analysis System (SAS Institute, Inc., 1996, Cary, NC.).[22] When significant differences ($P < 0.05$) of each measured parameter occurred among substrates, means were separated using Fisher's Protected Least Significant Differences (LSD) at the 5% level.

Results and Discussion

Chemical properties of fly ash. The pH of fly ashes from North Carolina, Michigan(a), Florida, and Michigan(b) were 11.9, 12.0, 11.8, and 8.9, respectively (Table 1). Electrical conductivity readings were 10.1, 8.4, and 9.6 dS m^{-1} for fly ashes from North Carolina, Michigan(a), and Florida, respectively, but only 2.0 dS m^{-1} for that from Michigan(b). The high pH and EC readings were associated with high concentrations of either Ca or Mg, or both, in fly ashes collected from Michigan(a), North Carolina, and Florida, although K concentration in fly ashes collected from North Carolina and Florida were more than doubled than that collected from North

TABLE 1. Electrical conductivity, pH, percentage of total carbon and nitrogen, and concentrations of phosphorus, potassium, calcium, magnesium, and iron in four fly ashes.

Fly ash[z]	EC (dS m^{-1})	pH	C (%)	N (%)	P	K	Ca	Mg	Fe
					(mg kg^{-1})				
NC	10.1	11.9	8.6	0.2	7713	13111	320040	5267	23533
MI(a)	8.4	12.0	5.4	0.7	4905	5227	349770	1769	44057
FL	9.6	11.8	6.5	0.1	2382	15961	266494	4260	46396
MI(b)	2.0	8.9	11.0	0.3	1738	4930	9761	1755	53285

[z]Fly ashes from power plants in North Carolina (NC), one site of Michigan [MI(a)], Florida, and the other site of Michigan [MI(b)], respectively.

Carolina and Michigan(b) (Table 1). The fly ashes also contained high concentration of Fe and P. Total C was 11% in fly ash of Michigan(b) but ranged from 5.4 to 8.6% in the rest of ashes. Total N was low, 0.1-0.7%, in all four fly ashes.

The fly ashes also contained various heavy metals (Table 2). Concentrations of Cd and Hg were negligible in all ashes. The highest concentrations of As, Cr, Cu, Pb, Mo, Ni, Se, and Zn, regardless of the source of fly ashes, were 48.3, 41.7, 83.2, 115.0, 15.0, 80.0, 28.9, and 194.0 mg kg^{-1}, respectively. These concentrations were generally well below the maximum pollutant concentration of individual metals in clean sewage and metal-loading limits for land application suggested by the U.S. Environmental Protection Agency (1993)[23], where concentrations of Cr, Cu, Mo, Ni, Pb, Se, and Zn were recommended as 1200, 1500, 18, 420, 300, 36, and 2800 mg kg^{-1}, respectively. However, As in fly ash collected from North Carolina was 48.3 mg kg^{-1}, which was slightly higher than the USEPA's 41.0 mg kg^{-1}. Concentration of Mn in all of the fly ashes ranged from 58.6 to 265.0 mg kg^{-1}.

Chemical properties of fly ash-amended container substrates. The pH of container substrates increased from initial 3.8 to 7.4 as the amount of fly ashes from North Carolina, Michigan(a), and Florida increased (Fig. 1). However, the pH of substrate amended with fly ash collected from Michigan(b) failed to respond to the increased addition of ash. This failure in raising pH may be explained by its lower pH and EC, correspondingly low Ca and Mg originally. Since container plant growth generally requires a pH range of 6.0–6.5, a rate of 18 kg of fly ash per cubic meter of the basal substrate was thus selected for neutralizing pH.

The pH readings were 6.8, 6.9, and 6.7 for SPP-MI(a), SPP-NC, and SPP-FL substrates, respectively (Table 3). These pH values were comparable to the pH of substrate amended with commercial dolomite (SPP-DO). However, pH values of the basal substrate (SPP) and substrate amended with fly ash collected from Michigan(b) [SPP-MI(b)] remained 3.8 and 3.9, respectively, which were not adequate levels for many containerized ornamental production. The EC readings of SPP-NC, SPP-MI(a), and SPP-FL before the application of the controlled-release fertilizer were 1.8, 1.5, and 1.4, respectively. These readings were also comparable to that of substrate amended with dolomite (SSP-DO). However, EC readings of substrates of SSP and SPP-MI(b) were only 0.3 and 0.4 dS m^{-1}, respectively. The C/N ratio of all substrates ranged from 19.7 to 26.7, indicating that they were within maturity range for containerized plant production. [24] Comparing elemental concentrations of each fly ash with those of corresponding fly ash-amended container substrates, P, K, Ca, Mg, and Fe were significantly reduced in fly ash-amended substrates (Table 1 vs Table 3); heavy metals in the substrates were much lower than fly ashes only (Table 2 vs Table 4), thus were not a concern of leaching in ornamental plant production.

Plant growth in basal, dolomite, and fly ash-amended container substrates. Plant growth was closely monitored during the five-month growth period. Plants produced in substrates of SPP and SPP-MI(b) were significantly smaller and shorter than those produced in SPP-NC, SPP-MI(a), and SPP-FL (Fig. 2). Old leaf margins of Pink Splash grown in SPP-MI(b) substrates were burnt. As a result, the overall quality of plants grown in SPP and SPP-MI(b) substrates were 2.5 to 2.0 (data not shown), respectively, suggesting that the plants were not marketable. However, overall quality of plants grown in the other substrates was graded above 4.0 (data not shown), and they all were marketable. The reduced growth in SPP and SPP-MI(b)

TABLE 2. Concentration of heavy metals in four fly ashes.

Fly ash[z]	As	Cd	Cr	Cu	Pb	Hg	Mo	Ni	Se	Zn	Mn
						(mg kg^{-1})					
NC	48.3	1.2	41.7	83.2	26.0	0.6	6.6	44.6	28.9	48.4	161.0
MI(a)	22.1	0.9	36.7	28.5	17.5	0.7	8.0	26.9	10.4	71.1	265.0
FL	ND[y]	3.0	ND	65.0	115.0	ND	15.0	80.0	ND	194.0	258.6
MI(b)	19.6	1.0	18.5	49.3	14.1	0.4	6.8	20.9	4.3	19.8	116.7

[z]Fly ashes from power plants in North Carolina (NC), one site of Michigan [MI(a)], Florida, and the other site of Michigan [MI(b)], respectively.
[y]ND = not determined.

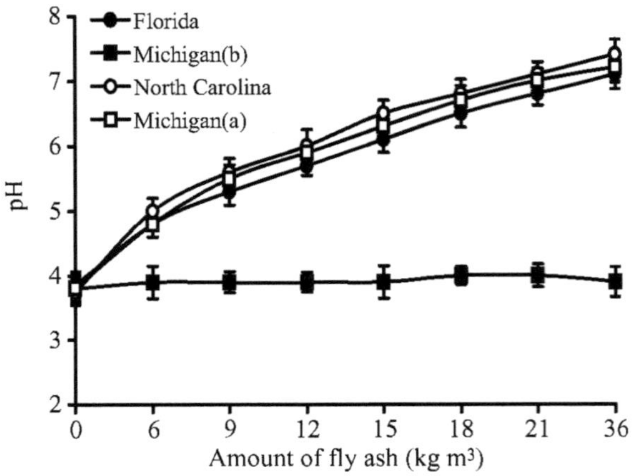

FIGURE 1. Changes in pH of the basal substrate (SPP) consisting of 70% sphagnum peat, 15% pine bark, and 15% perlite based on volume after amending with different rates of fly ash collected from North Carolina, Michigan(a), Florida, and Michigan(b).

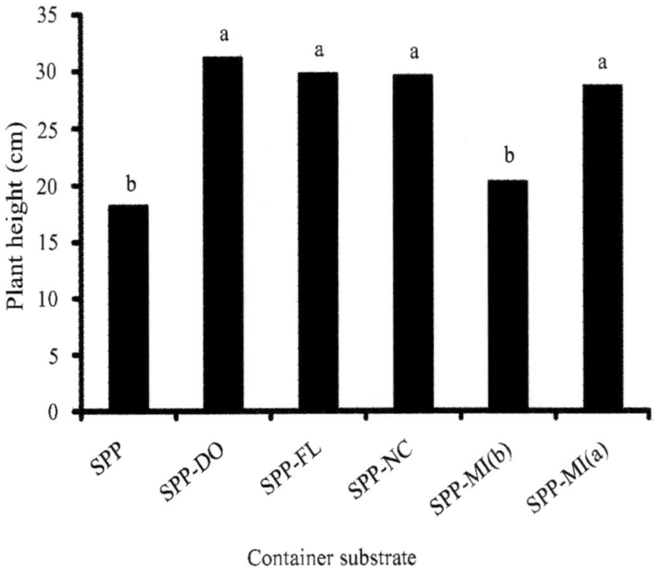

FIGURE 2. Pink Splash (*Hypoestes phyllostachya* Bak.) heights (cm) after growing in substrates of SPP (basal substrate consisted of 70% sphagnum peat, 15% pine bark, and 15% perlite based on volume), SPP-NC, SPP-MI(a), SPP-FL, and SPP-MI(b) [basal substrate amended fly ashes collected from North Carolina, Michigan(a), Florida, and Michigan(b)] as well as SPP-DO (basal substrate amended with a commercial dolomite) for five months. Means with different letters above the bars indicate significant plant height differences among substrates based on the LSD test ($P < 0.05$).

was probably attributed to the low substrate pH (Fig. 3). Pink Splash is a plant requiring substrate pH of 5.5 to 6.5 for healthy growth. The pH in both SPP and SPP-MI(b), however, were around 4.0 during the entire course of plant growth. Low pH, such as at 4.0 can cause decreased availability of substrate Ca and Mg but increased solubility of B, Fe, and Mn. In addition, low pH has been shown to directly affect the permeability of root cell membranes and impose leackage of various ions from the roots. [25] Plants produced in SPP and SPP-MI(b) had lower Ca than those produced in the other substrates (Table 5). In addition, margin burnt plants grown in SPP-MI(b) were likely caused by phytotoxicity of B, Fe, and Mn as these elements in shoots were much higher than plants grown in the other substrates (Table 5).

Plant sizes and heights produced from substrates of SPP-NC, SPP-MI(a), and SPP-FL were comparable to those produced from (SPP-DO). No growth disorder symptoms were observed from those plants either. Plant fresh and dry weights were also coincided with plant sizes and heights (Fig. 4), suggesting that Pink Splash growth in substrates amended with fly ashes from Michigan(a), North Carolina, and Florida produced the same amount of dry weights as that grown in dolomite-amended substrate. Menzies and Aitken (1996)[26] evaluated growth parameters of Boston fern [*Nephrolepis exaltata* (L) Schott], petunia (*Petunia* x *hybrida* Hort.), and tomato (*Lycopersicon esculentum* Mill.) in fly ash-amended substrates and also concluded that fly ash could be successfully used as a component of container substrates for ornamental plant production.

Potential use of fly ash as an amendment for containerized plant production. This study shows that fly ash can be used

TABLE 3. Electrical conductivity, pH, total carbon and nitrogen, carbon to nitrogen ratio, and concentrations of other nutrient elements in basal, dolomite-, and fly ash-amended container substrates.

Subsrate^z	EC ($dS\ m^{-1}$)	pH	C (%)	N (%)	C/N	P ($mg\ kg^{-1}$)	K	Ca	Mg	Fe
SPP	0.3	3.8	27.7	1.4	19.8	15	65	148	54	37
SPP-NC	1.8	6.9	25.4	1.1	23.1	769	1265	21799	676	1894
SPP-MI(a)	1.5	6.8	24.8	1.0	24.5	493	548	22721	225	4194
SPP-FL	1.4	6.7	24.0	0.9	26.2	279	1381	17869	548	2295
SPP-MI(b)	0.4	3.9	25.2	1.0	25.2	213	515	1050	209	4351
SPP-DO	1.2	6.8	26.5	1.2	22.1	392	365	16143	489	4437

^zSPP was the basal substrate containing 70% sphagnum peat, 15% pine bark, and 15% perlite based on volume; SPP-NC, SPP-MI(a), SPP-FL, and SPP-MI(b) were the basal substrate amended with 18 kg m^3 of fly ashes from North Carolina, Michigan(a), Florida, and Michigan(b); and SPP-DO was the basal substrate amended with 4.2 kg m^3 a commercial dolomite.

TABLE 4. Concentration of heavy metals in basal, dolomite-, and fly ash-amended container substrates.

Fly ash[z]	As	Cr	Cu	Pb	Mo	Ni	Se	Zn	Mn
					(mg kg^{-1})				
SPP	0.1	0.2	1.0	0.5	0.0	0.5	0.0	1.2	2.5
SPP-NC	6.8	7.7	19.6	4.2	1.2	8.4	4.6	25.1	14.8
SPP-MI(a)	3.5	5.9	8.5	2.6	1.8	6.5	2.2	38.0	15.9
SPP-FL	4.6	4.6	13.9	4.1	2.6	14.3	2.5	42.0	11.9
SPP-MI(b)	3.0	4.0	10.8	3.7	0.9	4.6	0.1	22.4	17.8
SPP-DO	2.0	2.6	4.7	2.3	1.0	7.9	1.8	28.5	4.7

[z]SPP was the basal substrate containing 70% sphagnum peat, 15% pine bark, and 15% perlite based on volume; SPP-NC, SPP-MI(a), SPP-FL, and SPP-MI(b) were the basal substrate amended with 18 kg m^3 of fly ashes from North Carolina, Michigan(a), Florida, and Michigan(b); and SPP-DO was the basal substrate amended with 4.2 kg m^3 commercial dolomite.

TABLE 5. Concentration of mineral elements in shoots of Pink Splash (*Hypoester phyllostachya* Bak.) after four month of growth in basal, dolomite-, and fly ash-amended container substrates.

Substrate[z]	P	K	Ca	Mg	B	Cu	Fe	Mn	Na	Ni	Pb	Zn
		(%)						(mg kg^{-1})				
SPP	0.76	3.3	0.8	0.4	91.1	3.5	107.1	196.5	849.0	0.3	0.4	50.0
SPP-NC	0.25	3.9	4.1	0.4	134.9	3.4	82.1	120.7	809.5	0.3	0.3	29.6
SPP-MI(a)	0.30	4.0	3.0	0.9	133.2	2.2	120.6	120.6	785.2	0.4	0.5	36.7
SPP-FL	0.24	3.8	3.8	0.5	119.0	3.6	70.5	113.3	820.5	0.6	0.3	40.2
SPP-MI(b)	0.57	4.0	1.1	0.4	217.4	9.2	192.2	279.6	894.1	2.8	0.8	59.9
SPP-DO	0.63	3.3	2.1	1.3	76.7	2.6	96.9	179.3	772.4	0.3	0.4	38.8

[z]SPP was the basal substrate containing 70% sphagnum peat, 15% pine bark, and 15% perlite based on volume; SPP-NC, SPP-MI(a), SPP-FL, and SPP-MI(b) were the basal substrate amended with 18 kg m^3 of fly ashes from North Carolina, Michigan(a), Florida, and Michigan(b); and SPP-DO was the basal substrate amended with 4.2 kg m^3 commercial dolomite.

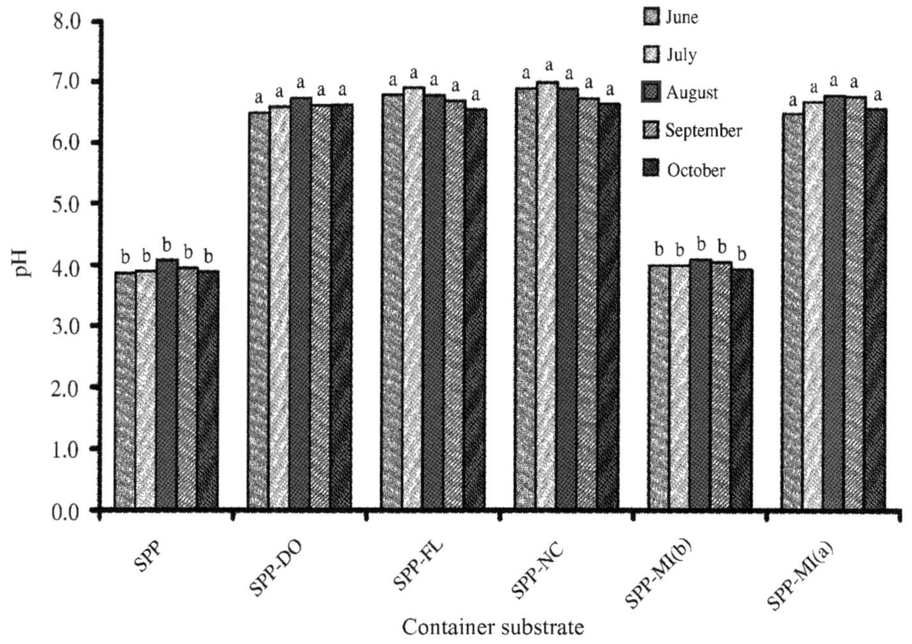

FIGURE 3. The pH of SPP (basal substrate consisting of 70% sphagnum peat, 15% pine bark, and 15% perlite based on volume), SPP-NC, SPP-MI(a), SPP-FL, and SPP-MI(b) [basal substrate amended fly ashes collected from North Carolina, Michigan(a), Florida, and Michigan(b)] as well as SPP-DO (basal substrate amended with a commercial dolomite) during each month of Pink Splash (*Hypoestes phyllostachya* Bak.) growth. Means with different letters above the bars indicate significant pH differences among substrates at each month based on the LSD test ($P < 0.05$).

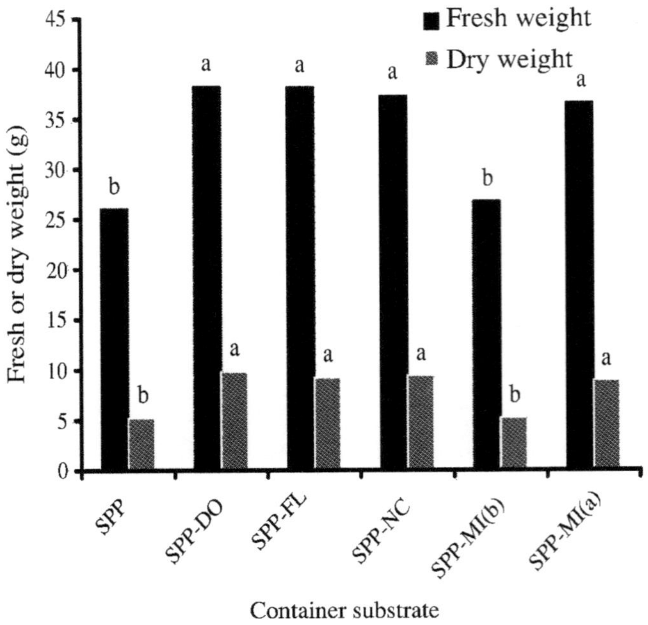

FIGURE 4. Fresh and dry weights of Pink Splash (*Hypoestes phyllostachya* Bak.) after growing in SPP (basal substrate consisting of 70% sphagnum peat, 15% pine bark, and 15% perlite based on volume), SPP-NC, SPP-MI(a), SPP-FL, and SPP-MI(b) [basal substrate amended fly ash collected from North Carolina, Michigan(a), Florida, and Michigan(b)] as well as SPP-DO (basal substrate amended with a commercial dolomite) for five months. Means with different letters above the bars indicate significant fresh or dry weight differences among substrates based on the LSD test ($P < 0.05$).

as an amendment for neutralizing container substrate pH for ornamental plant production. However, not all fly ashes serve this purpose, such as the fly ash collected from Michigan(b). Some criteria need to be met for this beneficial use, which include the ability of neutralizing substrate pH. Such ability can be practically determined by pH changes in response to increased amounts of fly ash amendment to a basal substrate. Heavy metals in fly ashes should not be excess the maximum pollutant concentrations in clean sewage and metal-loading limits for land application suggested by the U.S. Environmental Protection Agency (1993)[23]. Additionally, the suitability of fly ash-amended substrates for ornamental plant production can be assessed by the chemical properties, such as pH, EC, and mineral element concentrations of the substrates against the established optimal ranges of the properties.

The substrates amended with fly ashes from Michigan(a), North Carolina, and Florida had chemical properties well within the established ranges of commercial container substrates.[27] Substrate pH during the course of plant growth were stable (Fig. 3), and no growth disorders appeared on plants. Although this study only tested one ornamental plant species, the chemical properties of SPP-MI(a), SPP-NC, and SPP-FL, indicated that these substrates could also be used for producing other species. The use of fly ash in containerized plant production will open another market for a large-volume utilization of fly ash in agriculture. Since the ornamental plant industry is the fastest growing major sector of the U.S. agriculture, millions of tons of ground limestone and chemical fertilizers have been used to increase pH and fertility while preparing container substrates for ornamental plant production. Once they attain marketable sizes, ornamental plants with their container substrates are sold to consumers. Consequently, there is a continuous demand for replacement of container substrates in ornamental plant production. Correspondingly, there will be a continuous need for fly ash to replace limestone. This will represent a significant portion of fly ash being beneficially used each year and a great reduction of fly ash in landfills that often pose environmental concern due to heavy metal contamination of soils and groundwater.

Acknowledgments. We thank N-Viro International Corporation, Toledo, OH for providing fly ashes, Russell D. Caldwell for technical assistance of this study, and Kelly Everitt for critical reading of this manuscript. This research is supported in part by the Center for Natural Resources, IFAS, University of Florida and approved for publication as journal series R-10255 by the Florida Agricultural Experiment Station.

References

1. Adriano, D.C., Page, A.L., Elseewi, A.A., Chang, A.C., and Straughan, I. Utilization and disposal of fly ash and other coal residues in terrestrial ecosystems: A review, *J. Environ. Qual.* 9, 333, 1980.
2. Pathan, S. M., Aylmore, L.A.G., and Colmer, T. D. Properties of several fly ash materials in relation to use as soil amendments, *J. Environ. Qual.*, 32, 687, 2003.
3. Iyer, R.S., and Scott, J.A. Power station fly ash: A review of value-added utilization outside of the construction industry. *Resour. Conser. Recycl.* 31:217, 2001.
4. Carlson, A.C., and Adriano, D.C. Environmental impacts of coal combustion residues, *J. Environ. Qual.* 22, 227, 1993.
5. Bilski, J.J., Alva, A.K., and Sajwan, K.S., Fly ash, in *Soil Amendments and Environmental Quality*, Reckcigl, J.E., Ed., CRC Press, Inc., Boca Raton, Fla., 1995, 327.
6. Korcak, R.F. Utilization of coal combustion by-products in agriculture and horticulture, in *Agricultural Utilization of Urban and industrial By-Products*, Karlen, D.L., Wright, R.J., Kemper, W.D., Eds., American Society of America (Special Publication No. 58), Madison, WI, 1995, 107.
7. Adriano, D.C., Woodford, T.A., and Ciravolo, T.G., Growth and elemental composition of corn and bean seedlings as influenced by soil application of coal ash, *J. Environ. Qual.* 7, 416, 1978.
8. Sale, L.Y., Naeth, M.A., and Chanasyk, D.S. Growth response of barley on unweathered fly ash-amended soil, *J. Environ. Qual.*, 25, 684, 1996.
9. Sikka, R., and Kansal, B.D. Effect of fly-ash application on yield and nutrient composition of rice, wheat and on pH and available nutrient status of soils. *Bioresour. Technol.* 51, 199, 1995.
10. Khan, M.R., and Khan, M.W. The effect of fly ash on plant growth and yield of tomato. *Environ. Pollution.* 92,105, 1996.

11. Mulford, F.R., and Martens, D.C. Response of alfalfa to boron in fly ash. *Soil Sci. Soc. Am. Proc.* 35, 296, 1991.
12. USDOE (United States Department of Energy). *Annual Coal Report, 2002.* DOE/EIA-0584, Energy Information Administration, Washington, DC. 2002.
13. ACAA (American Coal Ash Association), *2002 Coal Combustion Product (CCP) Production and Use Survey*, American Coal Ash Association, Syracuse, NY, 2003.
14. USDOT (United States of Transportation). Fly ash facts for highway engineers. FHWA-SA-94-081. 1995.
15. Page, A.L., Elseewi, A.A., and Straughan, I.R. Physical and chemical properties of fly ash from coal-fired power plants with reference to environmental impacts. *Residue Rev.* 71, 83, 1979.
16. Gupta, D.K., Rai, U.N., Tripathi, R.D., and Inouhe, M. Impacts of fly-ash on soil and plant responses. *J. Plant Res.* 115, 401, 2002.
17. USDA (United States Department of Agriculture), National Agricultural Statistics Service (NASS). *Census of Agriculture-1997.* US Dept. Agriculture, Washington, D.C. 1999.
18. USEPA. Test methods for evaluating solid waste, physical/ chemical method. EPA SW-846, 3rd ed. USEPA, Washington, DC. 1997.
19. Poole, R.T., and C.A. Conover. Changing medium pH with hydrated lime. University of Florida, IFAS, *CFREC-Apopka Res. Rep. RH-92-1.* 1992.
20. Yeager, T.H., Wright, R.D., and Donohue, S.J. Comparison of pour-trough and saturated pine bark extract N, P, K, and pH levels. *J. Amer. Soc. Hort. Sci.* 108, 112, 1983.
21. Chen, J., Beeson, R.C. Jr., Yeager, T.H., Stamps, R.H., and Felter. L.A. Evaluation of captured rainwater and irrigation runoff for greenhouse foliage and bedding plant production. *HortScience* 38, 228, 2003a.
22. SAS Institute. *SAS User's Guide*, SAS Institute, Cary, NC., 1996.
23. USEPA (United States Environmental Protection Agency). Standards for the use or disposal of sewage sludge: final rules 40 CFR parts 257, 403, and 503. Fed Reg. 58 (32)9284–9415. USEPA, Washington, DC. 1993.
24. Chen, J., McConnell, D.B., Robinson, C.A., Caldwell, R.D., and Huang, Y. Rooting foliage plant cuttings in compost-formulated substrates. *HortTechnology* 13, 110, 2003b.
25. Yan, F., Schubert, S., and Mengel, K. Effect of low root medium pH on net proton release, root respiration, and root-growth of corn (*Zea Mays* L.) and broad bean (*Vicia faba* L). *Plant Physiol.* 99, 415, 1992.
26. Menzies, N.W., and Aitken, R.L. Evaluation of fly ash as component of potting substrates. *Sci. Hort.* 67, 87, 1996.
27. Nelson, P.V. *Greenhouse operation and management.* 6th ed. Prentice Hall, Upper Saddle River, NJ. 2003.

20
Influence of Coal Combustion Flue Gas Desulfurization Waste on Element Uptake by Maize (*Zea Mays L.*)

Anna S. Knox[1], John D. Knox[2], Domy C. Adriano[3], and Kenneth S. Sajwan[4]

[1] *Savannah River National Laboratory, Westinghouse Savannah River Company, 773-42A, Savannah River Site, Aiken, SC 29808, USA*
[2] *Columbia County Board of Education, Appling, GA 30802, USA*
[3] *The University of Georgia, Savannah River Ecology Laboratory, Aiken, SC 29802, USA*
[4] *Department of Natural Sciences and Mathematics, Savannah State University, Savannah, GA 31404, USA*

Abstract

A greenhouse study was conducted to determine the effect of coal combustion flue gas desulfurization (FGD) waste from a coal combustion electric power facility, located in Cope, SC on elemental uptake by maize (*Zea mays L.*). Unweathered FGD was applied to an Orangeburg series soil (Typic Paleudult) with an initial soil pH_{salt} of 4.90. The FGD was added at 0, 1, 2, 4, 6, 8, and 10% by weight. The test plant, maize, was harvested after 6 weeks of growth. Within 56 days of the FGD application, all rates of FGD significantly increased pH in the soil and the soil leachate above 6.0. The elemental concentration of the maize tissues indicated a characteristic elevation of B, Se, Mo, and As. However, no visual symptoms of toxicity of B or other elements in plants were observed. Increasing level of FGD caused a steady decline in biomass dry weight, with the highest treatment producing plants, which had approximately half the biomass of the control plants. Due to elevated concentrations of B and other elements (Se, Mo, and As) and due to adverse yield effects measured on plants, we do not recommend using FGD materials as a soil amendment for the purpose of growing agronomic and horticultural crops until sufficient time is allowed to substantially leach soluble salts including B and Se. This will necessitate monitoring the electrical conductivity and B and Se contents of the FGD waste to acceptable levels.

1. Introduction

According to the American Coal Ash Association, 105 million metric tons of coal combustion by-products (CCBs) were produced in the United States by the power generating utilities in 1997. Of that total, 1.68 million tons were applied on lands affected by mining operations.[1] The CCBs can be used in mine reclamation for acid mine drainage (AMD) prevention and treatment, subsidence control and surface restoration.[2] Class C Fly ash and Class F fly ash mixed with lime exhibit self-cementing properties and can be used to cap surfaces, line pavements and isolate acidic materials in the backfill to prevent AMD formation. In addition, highly alkaline CCBs, such as FGD and FBC residues, are used to directly neutralize acidic materials.

In the United States combustion of more than 800 million metric tons of coal annually results in approximately 75 million tons of solid residues. Even though these solid waste residues represent only about 2% of all the solid by-products now being produced in the US, these combustion residues are the sixth most abundant material for resource recovery. In addition, coal combustion residues account for about 90% of all fuel combustion wastes produced in the USA. Presently, only about 20% of these wastes are utilized, with the remainder deposited in landfills or surface impoundments.[3] Landfills designed for disposing of these wastes will be costly. Therefore major uses for these residues need to be discovered to abate the potential environmental consequences of storage and its associated cost.

Combustion of coal produces a variety of residues, including fly ash, bottom ash, flue gas desulfurization waste (scrubber sludge), fluidized bed boiler waste, and coal gasification ash.[4] Fly ash is the residue from coal combustion that enters the flue gas stream. Bottom ash is the residue from coal combustion that remains in the boiler. Flue gas desulfurization (FGD) waste, often called as scrubber sludge, results from the addition of limestone and/or dolomite to the coal either before (in the case of fluidized bed combustion) or after (in the case of flue gas desulfurization) combustion. These wastes are typically a combination of ash and various Ca, Mg, and S compounds.[5,6]

In order to meet air pollution control laws, the industry has to either retrofit existing facilities or install desulfurization devices to capture SO_x compounds from the flue gas, which otherwise would enter the atmosphere and contribute to the acid precipitation.

Because of the large amounts of coal ash generated each year, a great deal of research has been conducted to identify and determine the feasibility of utilizing these wastes in agriculture, road and building construction, and industry. The majority of this research has concentrated on the potential for using ash as an amendment to agricultural soils due to its potential to improve the physical and chemical properties of, infertile, degraded soils.[7] Amendment of agricultural soils with fly ash can improve soil tilt and texture, increase pH (i.e., for acidic soils) and elevate the concentrations of some macro- and micronutrients. [7,8] However, this amendment may also result in excessive soluble salt concentrations, excessive B, and increased concentrations of other potentially toxic trace elements, reduction in the availability of soil N and P, and elemental imbalances due to excessively high pH. Because of inconsistent chemical properties of fly ash and other residues, farmers are shunning these products as an amendment. But the most limiting factor in inhibiting greater usage of this product on land is the economics of transportation and application.

To meet air pollution regulations, a facility of the South Carolina Electric and Gas Co. installed a flue gas desulfurization (FGD) facility at their Cope Plant, near Orangeburg, SC. The environmental impacts associated with the disposal of FGD waste are similar in many ways to those associated with ordinary fly ash. Particle sizes of FGD waste are generally in the range of 5 to 50 μm.[9] Trace elements are present in these wastes as well, primarily in the ash component of the waste.[9,10] Like fly ash, FGD sludge contains high concentrations of soluble salts (Miller, 1987) and is characterized by high pH. As with fly ash, the principal concerns with scrubber sludge disposal include groundwater contamination from leachate, and elevated concentrations of trace elements in plants and animals in the vicinity of the disposal area.[9] An additional problem, unique to FGD waste, is the presence of high concentrations of sulfite, a phytotoxic compound. Under anoxic conditions, this can result in the production of H_2S gas.[11]Under aerobic conditions, high levels of sulfite can substantially increase the oxygen demand, due to the oxidation of sulfite to sulfate, in surface and groundwater systems affected by these wastes. [9] Several investigators have suggested that FGD waste could be used as a source of essential nutrients, including B and Se, for soils deficient in these elements.[12] Because of the lack of data, few conclusions can be drawn concerning the potential impact of FGD waste on the environment; however, it is evident that more research is needed in this area.

This study investigated the influence of flue gas desulfurization (FGD) waste from Cope Plant, (South Carolina Electric and Gas Co.) as a soil amendment and with the following objectives: i). To evaluate the feasibility of unweathered (fresh) FGD residue on plant growth and elemental uptake; ii). To evaluate the effects of FGD waste on the soil solution chemistry, and iii) to delineate the more potentially toxic and deleterious elements that can cause direct phytotoxicity or in soil the quality of food chain.

2. Materials and Methods

The Ap horizon of an Orangeburg Series-Typic Paledult soil was collected from Jim Traywick's farm, Cope, South Carolina. The soil was air dried, passed through a 2-mm sieve to remove gravel and coarse debris. The flue gas desulfurization (FGD) material was collected from Cope Coal Fired Power, Cope, South Carolina (Figure 1). Polyethylene pots were used for this study. Drainage holes were drilled in the bottom of each polyethylene pot, and the pots were lined with permeable nylon fabric. The pots were the filled with 7 kg soil blended with FGD material treated soil and placed in a 2.5-cm high plastic plate to collect drainage water (leachate). The experiment had seven treatments, which included the following rates of FGD: 0% (control), 1%, 2%, 4%, 6%, 8% and 10% by weight added to soil separately. The pots were arranged in a completely randomized design with each rate of application replicated four times. Five maize (*Zea Mays*, var. Pioneer 3165) seeds were sown in each pot and thinned to one after a week of germination. Maize plants were grown for six weeks. The plants were grown in a greenhouse under fluorescent lighting programmed to provide 16 hours of daylight. The

FIGURE 1. Flue gas desulfurization (FGD) waste was received from Cope Plant, (South Carolina Electric and Gas Co.) Orangeburg, SC.

day and night temperatures of the greenhouse were maintained at 30 ± 3°C, respectively. All the plants were watered daily and moisture content was maintained approximately to field capacity through the growing period. Throughout the equilibration periods of eight weeks the leachate from each treatment was collected in collection plates six times and was analyzed for pH, EC (electrical conductivity), and elemental concentrations. Plants were harvested at the end of six week. The harvest tissues from each pot were carefully rinsed in four successive deionized distilled water baths, oven dried to 60°C to constant weight, weighed to the nearest 0.01g for biomass production, and ground in Wiley Mill to pass a 22-mesh (841 μm) sieve. The dried plant samples were wet ashed in a nitric-perchloric acid mixture, and analyzed for elemental concentrations. Soil cores were also collected from each pot at the time of plant harvest. The soil samples were air dried, sieved to pass a 2-mm screen, and digested with HF + Aqua Regia and extractable with 0.1 *M* $HNO_3$for elemental analysis. For quality control, National Bureau of Standards tomato leaves (NBS, Washington, DC, NBS No. 15730) and spiked soil samples were used through all analysis. Both pants and soil-digested samples were analyzed through inductively coupled plasma optical emission spectrophotometer (ICP-OES) for elemental concentrations.

TABLE 1. Effect of FGD on pH of soil leachate over a period of 8 weeks.

Treatments	Week 1	Week 2	Week 3	Week 5	Week 7	Week 8
0% FGD	5.65	5.42	5.5	5.49	5.52	5.64
1% FGD	5.89	5.85	5.89	5.85	5.94	6
2% FGD	6.28	6.37	6.39	6.37	6.27	6.31
4 % FGD	7.05	6.9	7.33	7.15	7.03	7.21
6% FGD	7.37	7.43	7.55	7.62	7.53	7.57
8% FGD	7.5	7.57	7.68	7.7	7.64	7.65
10% FGD	7.74	7.74	7.77	7.86	7.75	7.83

TABLE 2. Effect of FGD on Electric Conductivity (mS/cm) of soil leachate over a period of 8 weeks.

Treatments	Week 2	Week 3	Week 5	Week 7	Week 8
0% FGD	0.28	0.235	0.217	0.195	0.19
1% FGD	2.41	1.045	1.132	1.25	0.935
2% FGD	3.27	1.45	1.47	1.66	0.97
4% FGD	4.38	1.57	1.95	1.92	1.7
6% FGD	4.72	1.91	1.99	2.15	1.98
8% FGD	5.47	2.05	2	2.1	2.01
10% FGD	5.33	1.95	2.11	2.16	2.02

3. Results and Discusion

3.1. Effect of FGD on Soil Leachate Property

Almost all doses of added FGD to the soil resulted in increased pH and EC values of the leachate. Pots were leached 6 times in total. Addition of 1% of FGD was the only treatment where the pH values of leachate did not change significantly over a period of 8 weeks (Table 1). For all FGD treatments EC values were elevated above the 0% FGD addition, indicating a proportional relationship between FGD application and increase in leachate salinity (Table 2). Progressive leaching demonstrated in this case that recovery of the background EC values accrued faster in comparison to pH. Also the characteristic increase in soluble salts followed by a gradual decline is indicative of unweathered ash.[7]

Concentration of tested elements in leachates from FGD amended soil increased progressively with FGD addition, and the highest concentrations were obtained in the 10% FGD treatment. For example, As increased from 0.0055 to 3.93 μg/g, respectively, for control and 10%FGD. Selenium concentrations increased by a factor of 22, in 10%FGD soils compared to control soil (Table 3). Elemental composition of the leachate also indicates a high soluble Ca concentration; Ca concentration increased a hundred fold following the addition of 10% FGD material.

Addition of FGD to the soil not only increased total concentration of elements but also the bioavailable pool of these elements. For example, the extractable As and Se levels were significantly higher in the treatment with 10% of FGD than in the untreated soil (Figure 2 and 3, respectively).

3.2. Effect of FGD on Maize Growth and Element Uptake

The influence of FGD waste on plants is indicated by maize yield and element concentrations in plant tissues. Yield of six-week-old maize started to decrease with the second dose of

TABLE 3. Concentrations of elements in leachate from soil treated with FGD (mg/kg or μg/kg*) over an eight-week equilibrium period.

Treatment	B	Na	Mg	K	Ca	As*	Se*	Mo*	Sb*	Hg*	Cd*
0%FGD	4.06	2.04	0.74	2.82	1.8	0.055	0.505	0.055	0.01	0.108	0.092
1%FGD	4.41	2.43	5.75	4.44	59.7	0.313	1.44	0.497	0.142	0.122	0.44
2% FGD	4.52	2.36	6.62	4.04	86.5	0.78	2.61	1.515	0.503	0.152	0.464
4% FGD	4.95	2.67	7.02	4.52	95.7	0.756	3.85	4.179	0.769	0.134	0.546
6% FGD	5.78	3.26	8.05	5.36	97.2	1.5	7.51	6.182	1.47	0.15	0.489
8% FGD	5.81	3.22	9.21	6.17	99.5	2.54	7.84	16.33	1.962	0.206	0.523
10% FGD	7.52	4.76	10.9	7.22	111	3.93	11.28	31.06	3.99	0.48	0.58

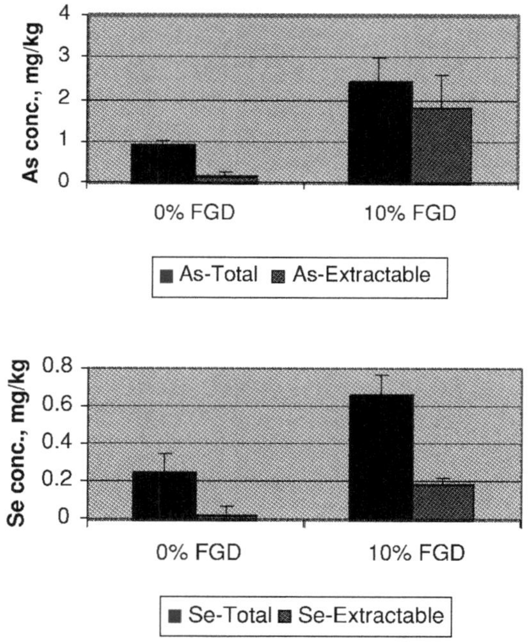

FIGURE 2. Total and extractable concentration of As and Se in soil with 0 and 10% of FGD. Soil was analyzed immediately after maize harvest (total –HF and aqua regia and extractable with 0.1 *M* HNO_3).

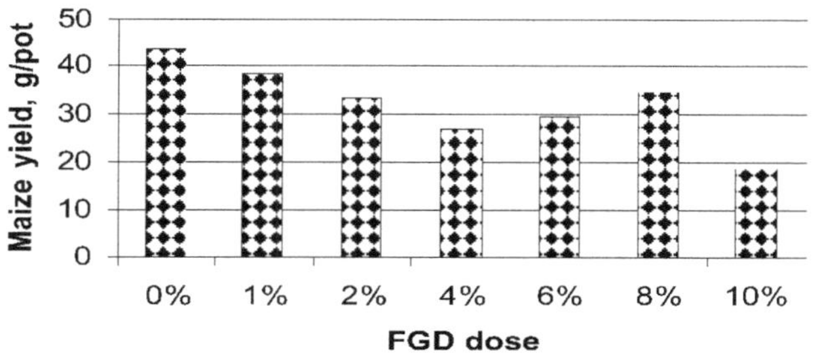

FIGURE 3. Effect of FGD on maize yield (g/pot).

TABLE 4. Effect of FGD application on the concentrations of Cd, Cr, Ni, Pb and Sb in maize leaves.)

Element (mg/kg)	0%	1%	2%	4%	6%	8%	10%
Cd	0.168a*	0.214a	0.208a	0.217a	0.196a	0.198a	0.220a
Cr	0.382a	0.457a	0.465a	0.360a	0.491a	0.409a	0.465a
Ni	1.14a	1.20a	0.916a	0.698b	1.027a	0.899b	1.027a
Pb	0.526a	0.630a	0.480a	0.455a	0.638a	0.681a	0.694a
Sb	0.016a	0.006a	0.013a	0.009a	0.016a	0.012a	0.025a

* Means followed by letters a and b are significantly different between doses of FGD at $P < 0.05$.

FGD and the lowest yield was obtained in treatment with the highest dose of FGD (i.e., 10%) (Figure 3). Increasing level of FGD caused a steady decline in biomass dry weight, with the highest treatment producing plants that had approximately half the biomass of the control plants. However, no visual symptoms of metal toxicity in plants were observed (Figure 4). Other pot studies using cola combustion products (CCPs) amendments have shown a similar reduction in overall biomass.[7,13]

From the twenty studied elements in maize tissues (Al, As, B, Ba, Ca, Cd, Cr, Cu, Fe, K, Mg, Mn, Mo, Ni, Na, Pb, Sb, Se, Tl, and Zn) only five elements (Cd, Cr, Cu, Pb, and Sb) were not significantly influenced by FGD material (Table 4). Arsenic, B, Mo and Se are the elements of the greatest concern in this study because the excess of these elements can cause severe soil contamination resulting in plant toxicity and potential detrimental effects to live stock. Trace amounts of arsenic is essential for animal nutrition, but not for plant growth.[14] Molybdenum is essential for both plant and animal nutrition, while Se is essential only for animal growth. The concentrations of these elements were significantly elevated by each addition of FGD (Figures 5, 6, 7 and 8). Arsenic in maize tissues increased 5 times in treatment with 10% FGD, compared to the control treatment (Figure 5). Kabata-Pendias[15] reported that sufficient and normal As concentrations in

FIGURE 4. Effect of FGD on maize growth.

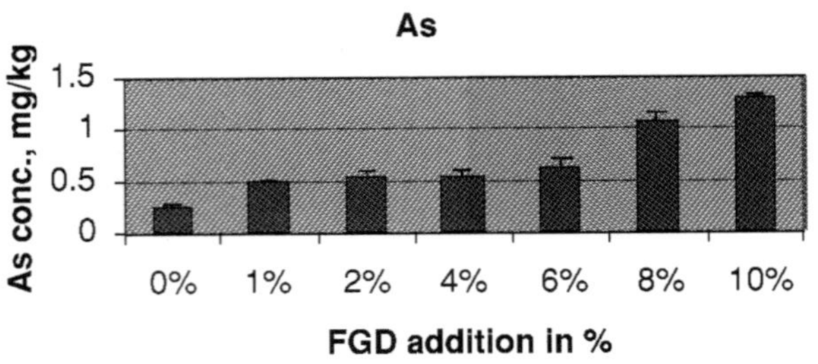

FIGURE 5. Effect of FGD on As concentration in maize leaves.

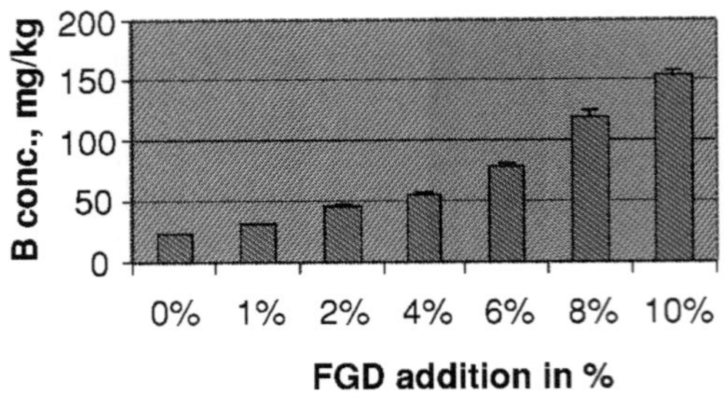

FIGURE 6. Effect of FGD on B concentration in maize leaves.

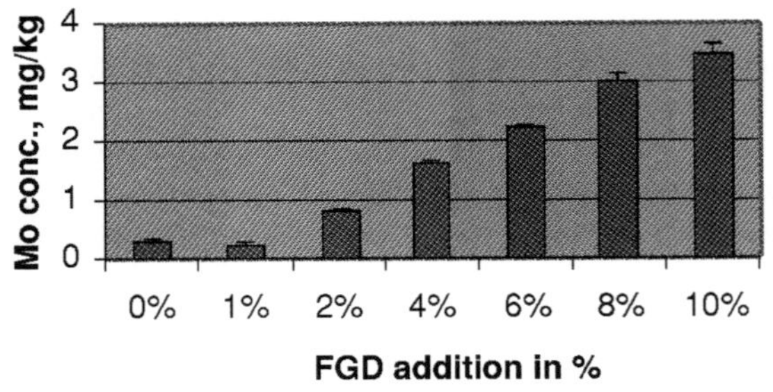

FIGURE 7. Effect of FGD on Mo concentration in maize leaves.

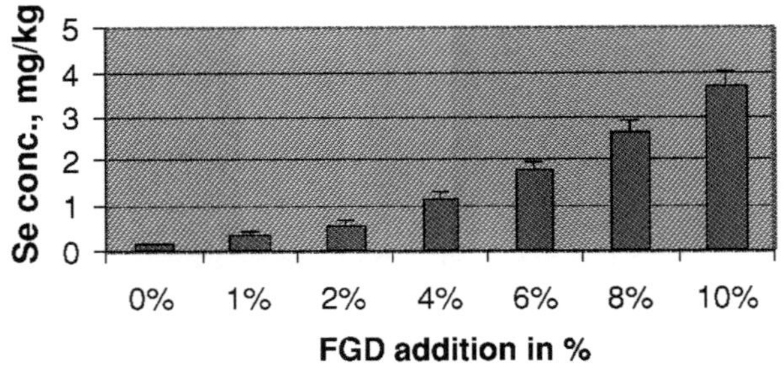

FIGURE 8. Effect of FGD on Se concentration in maize leaves.

mature leaf tissue for various species (in mg/kg DW) are between 1 and 1.7 mg/kg, however, a tolerable level of As in agronomic crops is 0.2 mg/kg. The tolerable level of As in agronomic crops was exceeded with the first FGD rate (1%). Boron concentrations in maize tissues increased significantly with each dose of FGD, and in the treatment with 10% FGD plants had concentrations (155 mg/kg) considered to be toxic for maize (Figure 6). According to Kabata-Pendias[15], the sufficient or normal B concentration in mature leaf tissues range from 10 to 100 mg/kg, and this level was exceeded with 8%FGD addition. Kukier and Sumner[16] also found excessive rate of fly ash increased B in corn (*Zea mays* L.) tissue to phytotoxic levels. Molybdenum concentration in plants in the control treatment was 0.279 mg/kg but in the treatment with the highest dose of FGD the concentration of this element increased to 3.48 mg/kg (Figure 7). Addition of each dose of FGD to the soil significantly increased the uptake of some macronutrients by the plant (Figure 8), for example, Ca concentration in maize tissues increased from 4975 mg/kg (control soil) to 8340 mg/kg (treatment 10% of FGD). It is a common belied that fly ash is rich Ca and Mg and therefore, can be used as a soil amendment for liming purpose and enhance the bioavailability of Ca and Mg. While it has been generally observed that fly ash application, including eastern U.S. ashes, improves the extractable Ca level in soil and plants[7,8], this study indicates that that may not be true for Mg nutrition. Applications of FGD can create an imbalance in Ca/Mg nutrition, inducing Mg deficiency and perhaps K deficiency (Figure 9). Some micronutrients such as Mn and Zn concentrations in maize tissues decreased with increasing dose of FGD (Figure 10).

Application of FGD to agricultural lands could have both advantages and disadvantages. The pH enhancement may cause plant nutrient imbalance, particulary P and Mg deficiency and antagonistic reactions among elements because of excessive Ca, K, and S. Some other researchers observed similar antagonistic reaction among elements.[7,16]

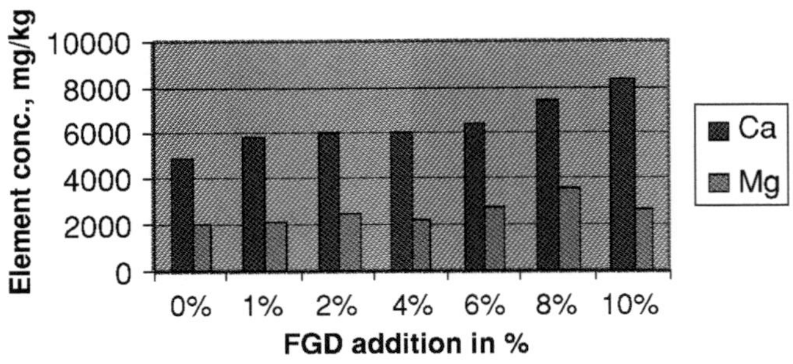

FIGURE 9. Effect of FGD on Ca and Mg concentrations in maize leaves.

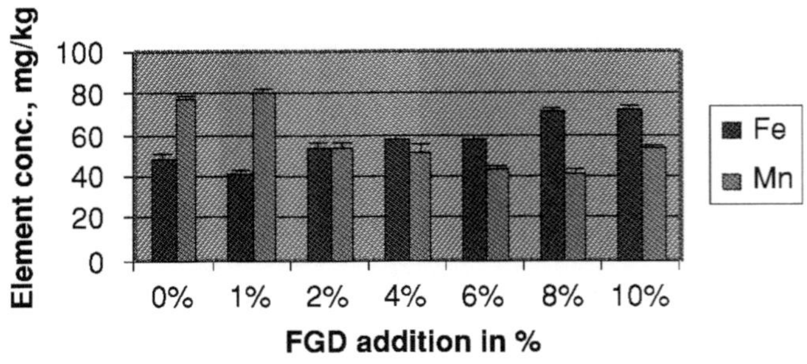

FIGURE 10. Effect of FGD on Fe and Mn concentrations in maize leaves.

Summary and Conclusions

From the twenty studied elements only concentrations of 5 elements (Cd, Cr, Cu, Pb, and Sb) in maize tissues were not influenced by the FGD addition. All rates of FGD increased uptake of some macronutrients (Ca, Mg, and K) by 6 week-old maize plants. Concentration of some micronutrients (Mn and Zn) in maize leaves decreased with increasing dose of FGD, very likely due to antagonistic reactions with other elements such as Ca or Fe. FGD can serve as a supplementary source of certain essential elements for plants, such as B, Se, and Mo. However, these micronutrients could be phytotoxic when at high levels in soils. Although, for example B in soil is fairly soluble and leachable, this phytotoxic effect is only temporary in unweathered ash. Boron and soluble salts in unweathered ash by-products might restrict seed germination and establishment of plants especially at high rates of application, unweathered materials should be allowed to "weather" to enable leaching these harmful constituents. This implies that the timing of planting is important to avoid potential salt-related problems. Soil tests for electrical conductivity and metal level (e.g., B) can be conducted to determine when to plant. This study suggest that because of elevated concentrations of B and other elements and due to adverse yield effects measured on plants, unweathered FGD would not be a suitable amendment for 6-week old maize on this soil.

Acknowledgement. Financial Assistance Award Number DE-FC09-96SR185 from the DOE to the University of Georgia Research Foundation supported the work.

References

1. American Coal Ash Association. *Innovative Applications Of Coal Combustion Products (CCPs).* Alexandria, Virginia: American Coal Ash Association, Inc. 1998.
2. Butalia, T.S., and William E. W. Market Opportunities For Utilization Of Ohio Flue Gas Desulfurization (FGD) and Other Coal Combustion Products (CCPs), Volume 1 Executive Summary. 2000. http://ccohio.eng.ohiostate.edu/ccpohio/Marketing/Volume1.PDF
3. U.S. Environmental Protection Agency. Waste from the combustion of coal by electric utility power plants. USEPA Rep. 530-SW-88-002. USEPA, Washington, DC. 1988.
4. Keefer, R.F., and Sajwan, K. S. *Trace Elements in Coal and Coal Combustion Residues.* Lewis Publishers, Boca Raton, FL. 1993.
5. Sajwan, K.S., Keefer, R.F., and Alva, A.K. *Biogeochemistry of Trace Elements in Coal and Coal Combustion Residues*, Kluwer Academic/Plennum Publishers, New York, NY. 1999.
6. Sajwan, K.S., Alva, A.K., and Keefer, R.F. *Chemistry of Trace Elements in Fly Ash.* Kluwer Academic/Plennum Publishers, New York, NY. 1993.
7. Carlson C. L., and Adriano, D.C. Environmental impacts of coal combustion residues. J. *Environ. Qual.* 22, 227, 1993.
8. Adriano, D.C., Page, A.L., Elseewi, A.A., Chang, A.C., and Straughan, I. 1980. Utilization and disposal of fly ash and other coal residues in terrestrial ecosystems: A review. *J. Environ. Qual.*, 9, 333, 1980.
9. Santhanam, C.J., Lunt, R.R., Johnson, S.L., Cooper, C.B., Thayer, P.S. and Jones, J.W. 1979. Health and environment impacts of increased generation of coal ash and FGD sludges. *Environ. Health Perpect.* 33, 131, 1979.
10. Miller, J.P., Jr. Environmental impact of fly ash disposal. In S.K. Majumdar et al. (ed.) Environmental consequences of energy production: Problems and prospects. *The Pennsylvania Acad. of Sci.*, Easton, PA. 1987, 233.
11. Adams, D.F., and Farwell, S.O. Sulfur gas emissions from stored flue gas desulfurization sludges. *J. Air Pollut. Control Assoc.* 31,557, 1981.
12. Ransome, L.S., and Dowdy, R.H. Soybean growth and boron distribution in sandy soil amended with scrubber sludge. *J. Environ. Qual.* 16:171, 1987.
13. Waker, W. J., and Dowdy, R. H.. Elemental composition of barley and ryegrass grown on acid soils amended with scrubber sludge. *J. Environ. Qual.* 9(1), 27, 1980.
14. Adriano, D.C. 2001. Trace elements in terrestrial environments—Biogeochemistry, bioavailability and risk of metals. 2nd ed., Springer, New York. 2001.
15. Kabata-Pendias, A. *Trace Elements in Soils and Plants.* CRC Press, Boca Raton, FL. 2001.
16. Kukier, U., Sumner, M. E., and Miller, W.P. Boron release from fly ash and its uptake by corn. *J. Environ. Qual.* 23, 596, 1994.

21
Amelioration of Soil Acidity with Class-C Fly Ash: A Field Study

Maxim J. Schlossberg[1], F. Clint Waltz Jr.[2], and William P. Miller[1]
[1]*Dept. of Crop and Soil Sciences, University of Georgia, Athens, GA 30602-7272, U.S.A.*
[2]*Dept. of Crop and Soil Sciences, Georgia Agricultural Experiment Stn., 119 Redding Bldg., Experiment, GA 30212, U.S.A.*

Abstract

Coal combustion products (CCP) include fly ash and bottom ash and are generated nationally at rates of 10^8 Mg per year. Fly ashes (FA) comprise the majority of CCP production, and can possess widely-variable physicochemical properties. Current consumption/utilization of FA in the US does not approach production levels, and results in FA stockpiling. Class-C fly ash is generated from combustion of sub-bituminous and lignite coal products. This class of FA often possesses an alkaline pH, resulting from calcium, magnesium, and potassium oxide inclusions. Some Class-C FA have been reported to contain as much as 60% calcium carbonate equivalency (CCE), prompting investigations of FA as a lime substitute in agricultural/horticultural applications. Furthermore, FA often possesses small concentrations of exchangeable micronutrients. Thus, considering availability of Class-C FA, its potential beneficial use as a liming agent, and the expansive area of low-to-medium maintenance turfgrass systems currently afflicted by suboptimal soil pH levels; our objective was to evaluate the acid-neutralizing efficacy of a Class-C FA when substituted for pulverized limestone (PL) in field application. This study was conducted on a severely-acidic bermudagrass (*Cynodon* spp. L.) rough of a Georgia golf course during 2001 and 2002. Following material characterization, $CaCO_3$ equivalent was applied at a prescribed rate of 3.8 Mg $CaCO_3$ ha^{-1}, using either PL or Georgia-produced FA (CCE = 45.5%). Soil samples collected one year following showed FA to have neutralized significantly greater acidity in the upper 8 cm of soil than the PL. Furthermore, exchangeable P, Mg, and Zn levels in the FA-treated 0–8 cm of soil exceeded levels observed in the PL-treated or control plots. Due to the disparity in CCE, requisite FA application greatly exceeded that of PL, reducing economic advantage of FA-substitution. However, in locations where Class-C FA is plentiful, high-grade PL is costly, and soil pH suboptimal for important crops; results of this study indicate soil liming with Class-C FA to be a beneficial use of this CCP. Comprehensive characterization of CCP and proper application rate (not exceeding agronomic requirements) are essential components of beneficial use.

Introduction

Fly ashes (FA) generated from the combustion of sub-bituminous and lignite coal commonly possess <50% alumino-, ferric-, and silica-oxides by mass; and are generally categorized as Class-C FA.[1] Neutralization of soil acidity by land application of class-C FA has been proven effective. Matsi and Keramidas amended two acidic surface soils (Ultic Haploxeralf and Typic Haploxeralf) with two alkaline FA at rates equal to 5, 20, and 50 g kg^{-1} soil. These amendment rates (equivalent to 12.4, 49.4, and 123.6 Mg ha^{-1}) resulted in significant plant-growth increases over control soils.[2] Additionally, cumulative plant uptake of both P and B significantly increased with addition of FA. Phosphorous uptake significantly increased with FA additions even though the FA contained less extractable-P than did the mineral soils. This was attributed to healthier root systems of the ryegrass (*Lolium* spp. L.) in more neutral pH soil, as these amendments resulted in significant increases in soil pH. At the highest FA application rate, soil pH of the less buffered soil increased 3 units and the pH of the more-clayey soil 2 units.[2]

A greenhouse study examined the effect of 5 and 10 % (by mass) FA additions versus traditional liming of three acidic fine sands. Pots, filled with soil or amended soil, were sowed with corn (*Zea mays* L.) seedlings. The higher rate of FA significantly raised soil pH above the control in all the fine sand as did the lower rate in 2 of 3 fine sands. An agricultural-grade limestone application made at 0.3% (by mass) also elevated soil pH levels above the control.[3] Positive correlations ($r = 0.92$) between the calcium carbonate equivalency (CCE) of fly ash and the final pH of soils receiving an 80 Mg ha^{-1} application have been reported.[4] Efficacy of acid neutralization relies on the addition of a 'high CCE' FA to an inherently acidic soil.

Turfgrasses vary in their responses to soil acidity and several papers report the optimum ranges for growth.[5,6,7] Despite the well-documented adverse effects of sub-optimal pH on nutrient availability, root growth, and Al toxicity; further result of acidic conditions specific to turfgrass management is excessive thatch accumulation. Maintenance of neutral soil pH through lime applications has been shown to slow thatch development in long-term studies.[8,9]

Several types of agricultural lime are available to turfgrass managers. These liming materials vary in their CCE, nutrient concentration, and fineness. Value of a particular turfgrass area generally dictates the intensity of management applied to its culture. Golf course putting greens, lawn bowling greens, football fields, baseball infields, and tennis courts generally possess the greatest value of all turfgrass stands and receive the highest degree of management/culture. Conversely, highway right-of-ways, industrial parks, parking lots and other utility turfgrass applications receive comparably minimal inputs. These utility turf areas receive little usage and are commonly vast in area. These traits of utility turves make them ideal candidates for CCP disposal at safe rates, particularly Class-C FA. Ubiquitous Class-C FA containing measurable CCE could be used to treat soil acidity of utility turves where application of traditional lime is normally cost prohibitive.

Class-C FA materials have been used to amend and treat agricultural production soils with good results. In the southern Piedmont, many bermudagrass fairways are constructed on relatively inhospitable soils/subsoils. Golf courses possess an average of 25 acres of fairways, approximately 20% of their total area. In the Southeastern United States (SE US), acidic soils prevail in large portions of Georgia, The Carolinas, Alabama and Mississippi. This region contains over 2,500 golf course facilities, and represents a *volume market*. On the basis of lime substitution, potential annual demand for Class-C FA in this region exceeds 65,000 Mg annually.

Traditionally, the best opportunity for effective lime treatment of an acid soil is at tillage.[10] Considering nearly all turfgrass systems regenerate perennially; tillage is an uncommon procedure in turfgrass culture and conducted in installations or renovations almost exclusively. Core aerification is the process that removes soil 'cores' from the 0-8 cm upper-soil profile. This process serves multiple purposes; one being the removal of thatch (a dense organic layer of undecomposed plant biomass), and a second being increased soil/air interface. Broadcasting lime, fertilizers, or amendments immediately following aerification improves efficacy by stabilizing these materials in contact with the soil, minimizing risk of surface transport.

Bermudagrass (*Cynodon* spp. L.) is a prolific and deep-rooted perennial turfgrass, but is adversely affected by low soil pH and high aluminum levels. These acidity factors are so problematic in some regions, research toward developing Al-tolerance bermudagrass germplasm as commenced on several continents.[11,12] The importance of optimizing soil pH and minimizing Al solubility when managing grasses is well documented.[8,9,13] As power plants in the SE US produce large quantities of FA possessing fairly high pH (10-12), CCE of FA can reach 60% and also possess high levels of Fe, Mn, Cu, B, Zn and Mo.[4,14] Thus, our experimental objective was to evaluate the acid-neutralizing efficacy of a Class-C FA when substituted for PL in field application.

Materials and Methods

Class-C FA was collected from a coal-fired generation plant operated by the Geogria Power Company (Atlanta, GA). Agricultural grade pulverized limestone (PL) was purchased at a local farm supply store (Athens, GA). Duplicate samples of FA and PL (Table 1) were digested using HF/*aqua regia*, analyzed by inductively-coupled plasma mass spectrometry (ICP-MS), and corrected for initial moisture content.[15]

In Feb. 2001, a preliminary site visit was made to evaluate fairway soil pH levels at The Standard Club in Duluth, GA. Due to the diligent agronomic practices conducted by the golf course superintendent, no fairways at The Standard Club required lime treatment. However, an acidic, low-use bermudagrass rough area was sampled. Conditions of this Cecil series (fine, kaolinitic, thermic, Typic Kanhapludults) were representative of numerous SE US fairways under typical management, yet lacking recent lime application (Table 2).

In March 2001, a core aerification procedure was performed on the experimental sites. A single pass was made over two 9-m^2 areas, creating a uniform matrix of 5 cm-deep and 1.6 cm-diam. holes on 5-cm centers. This procedure vacated 8% of the surface area or ~3% of the hectare furrow slice

TABLE 1. Chemical characteristics of liming materials.

	Fly Ash (FA)	Pulverized Limestone (PL)
pH (1:1)	12.2	8.6
	%	
CCE†	45.5	98.7
P	0.5	<0.1
Ca	10.1	32.3
Mg	3.0	3.5
K	0.3	<0.1
Al	5.7	<0.1
Fe	3.4	0.3
Na	0.7	<0.1
	mg kg^{-1}	
As	24	<1
B	780	7
Cd	1	<1
Cr	48	1
Cu	180	1
Mo	12	1
Pb	19	1
Se	6	<1
Zn	97	13

† CCE, calcium carbonate equivalency.[20]

TABLE 2. Baseline soil pH, extractable nutrients, and reserve acidity, March 2001.

	P	K	Ca	Mg	Zn	Mn	pH	pHb†
	kg ha^{-1}						1:2	
				0-15 cm				
Site 1	‡15 ± 2	132 ± 4	621 ± 72	94 ± 5	1 ± 0	33 ± 0	5.3 ± 0.1	7.66 ± 0.0
Site 2	7 ± 2	104 ± 12	320 ± 5	69 ± 3	1 ± 0	17 ± 2	5.0 ± 0.1	7.60 ± 0.0

† pHb, Buffer pH for lime requirement.[16]
‡ $\mu \pm$ standard error (SE).

(15-cm deep), on an empirically-based volumetric basis. Theoretically, this provided 39 m^3 ha^{-1} of total available void space available for replacement by soil amendments. Based on initial soil test results, site 2 required a 3.8 Mg ha^{-1} $CaCO_3$ application (Table 2) to raise the soil pH to an optimal level for bermudagrass culture (6.5). Site 1 required a 3.4 Mg ha^{-1} $CaCO_3$ application to raise the soil pH to 6.5. These application rates were derived from Adams-Evans buffer pH measurement and recommendations, excluding the 1.5 conversion factor.[16] Equivalent applications of PL or FA providing 3.8 Mg $CaCO_3$ ha^{-1} were made to both sites, March 2001.

The PL used as a traditional treatment possessed a CCE of 98.7% and a bulk density of 2.5 g cm^3 (Table 1). Thus, this treatment required only 1.54 m^3 ha^{-1}. The mass of FA required to meet the equivalent neutralizing strength of 3.8 Mg pure lime was 8.35 Mg FA (CCE of FA = 45.5%). This mass of FA resulted in an application volume of 6.28 m^3 ha^{-1}. To minimize any error or soil surface/lime interactions associated with differences in the volume of material added, we backfilled the aerification holes of all plots with 39 m^3 ha^{-1} of a quartz sand/liming material mixture. The control plots were backfilled with quartz sand at an equivalent rate of 39 m^3 ha^{-1}. On a mass/hectare furrow slice basis, PL was added to treated plots at 0.2% mass of the ha-furrow-slice. Fly ash, providing equivalent $CaCO_3$ neutralizing power, was applied at a rate providing 0.4% mass of the ha-furrow-slice. Once treatments were applied, manual irrigation was gently applied to wash all soil amendments into the aerification holes, ensuring stability of the materials and initiating the dissolution step of the liming reaction.

The experimental design was comprised of 2 independent Latin squares with 3 treatments (columns) and three replications (rows) in each. The PROC GLM subroutine facilitated analysis of all data (SAS Institute, Cary, NC). Pre-determined multiple comparisons were conducted on significant main treatment effects ($p \leq 0.05$) using Tukey's Studentized Range procedure ($\alpha = 0.05$).

In the one year following experiment initiation, sites were irrigated and mowed following standard golf course maintenance practices. In April 2002, two core subsamples (3.8-cm diam.) were collected from each square meter plot, segmented into 0–8 cm and 8–15 cm portions, and returned to the lab for analysis. Soils were dried, cleaned of biomass and debris, and ground to pass a 2-mm sieve. Soil pH was measured using a 1:2 soil to water dilution. Nutrients were extracted, then quantified by ICP-MS analysis.[17]

Results

Preliminary analysis conducted in 2001 clearly characterizes the experimental site as inhospitable and sub-optimal for culture of bermudagrass under intensive management (Table 2). Site 2 clearly possessed less optimal soil chemical conditions than site 1 (Table 2). No differences were observed between the preliminary analysis of the 0–8 cm deep soils made in 2001 and the control plots 0–8 cm deep soils made in 2002 (Tables 2 & 3). From this it can be inferred that there was no transport/contamination of adjacent plots by the liming materials following application. Following one year, nutrient and chemical properties in the top 8-cm of the plots treated with both the experimental and traditional liming agents were markedly improved (Table 3). One year after treatment, soil pH levels of the fly ash treated plots (0–8 cm depth) were significantly greater than the control plots at both sites (Table 3). Comparatively, traditional PL application significantly increased soil pH (0–8 cm depth) in only one of the two sites. At equal application rates, lesser efficacy observed on site 2 may be attributed to its greater reserve of exchangeable acidity (Table 2). These results were not dissimilar from the outcome of a recent, similar study. Aldous and Balfour applied 1, 2, 3, and 4 Mg of an alkaline waste lime-sludge ha^{-1} to a turfgrass practice fairway near Melbourne, Victoria (Australia).[18] The waste-lime sludge was a by-product of the paper-processing industry and contained 36% Ca and 1.1% Mg (pH~11). After 115 d, soil pH significantly increased in both lime and by-product treated plots.

Levels of exchangeable Ca, K (site 1 only), and Mn (site 2 only) increased similarly among the fly ash and lime treatments. However, levels of exchangeable P, Mg, and Zn in the FA-treated plots were greater than PL-treated and control plots. This represents a significant agronomic advantage in soils commonly possessing such deficiencies. Considering both liming materials possessed similar total Mg (~3%), Mg availability/solubility in FA appears greater than PL. Enhancement of exchangeable P and Zn in the FA treated plots was likely a direct result of loading rates. Concentrations of P and Zn in FA exceeded those measured in the PL (Table 1). Equivocally

TABLE 3. Soil pH and extractable nutrients by depth and treatment, April 2002.

	P	K	Ca	Mg	Zn	Mn	pH
	kg ha^{-1}						1:2
Site 1				0–8 cm			
Fly Ash (FA)	46 a†	167	4648 a	623 a	6 a	35	6.6 a
Control	17 b	153	755 b	119 b	2 b	30	5.3 c
Pulverized Lime (PL)	17 b	161	4211 a	216 b	2 b	34	6.3 b
LSD ($\alpha = 0.05$)	3	NS	437	122	1	NS	0.1
				8–15 cm			
Fly Ash (FA)	9	97	450 b	76	2	11	5.2
Control	15	102	432 b	64	1	15	5.4
Pulverized Lime (PL)	10	105	679 a	90	1	15	5.5
LSD ($\alpha = 0.05$)	NS	NS	132	NS	NS	NS	NS
Site 2				0-8 cm			
Fly Ash (FA)	58 a	122	2981 a	433 a	3 a	32 a	6.0 a
Control	7 b	99	533 b	80 b	1 b	24 b	4.9 b
Pulverized Lime (PL)	9 b	124	4262 a	180 b	1 b	30 a	5.6 ab
LSD ($\alpha = 0.05$)	10	NS	1281	110	1	2	1
				8–15 cm			
Fly Ash (FA)	3	76	248 ab	60	0 b	9	4.9
Control	3	85	221 b	52	1 a	10	5.0
Pulverized Lime (PL)	3	91	302 a	59	0 b	11	4.9
LSD ($\alpha = 0.05$)	NS	NS	65	NS	<1	NS	NS

† Different letters in a column signify statistical differences by Tukey's Studentized Range.

similar to the soil P and Zn results, availability of soil B, Cu, and/or Mo in the FA-treated plots may also stand elevated due to the loading effect.

Significant alteration of soil pH and nutrient levels at soil depths greater than 8 cm were generally not observed 1 year following application. However, Ca levels in deep soil depths (8–15 cm) of the PL-treated plots significantly increased over control levels (sites 1 & 2) and the FA-treated plots (site 1 only). This was likely the result of a significantly greater mass of Ca applied in the PL treatments. The 3.8 Mg ha^{-1} application of PL provided 1.23 Mg of Ca. Conversely, the 8.35 Mg ha^{-1} application of FA provided only 0.84 Mg Ca.

Conclusions

These results show applicability of substituting Class-C FA for traditional liming materials in treatment of soil acidity in mid to low-intensity turfgrass culture. The potential market for penetration of Class-C FA as a combined soil acidity ameliorant and/or micronutrient fertilizer is great. This list of turfgrass sites includes (but is not limited to): golf courses, parks, athletic fields, industrial parks, turf parking lots, cemeteries, and roadside right-of-ways. The additional benefit of increased soil P, Mg and Zn is significant, as these nutrients are often limiting in soils of the SE US.

There were no adverse environmental implications associated with FA replacement of traditional agricultural lime. The 8.35 Mg FA ha^{-1} application resulted in an annual load rate of 0.2 kg As, 0.4 kg Cr, and 0.05 kg Se ha^{-1}. These annual load rates are well below the USEPA regulations for land application of biosolids.[19] Classical economic cost/benefit analysis of this application may expose less-favorable aspects. The FA used in this study possessed a bulk density of 1.33 g cm^{-3} and a CCE of 45.5%. These physicochemical characteristics require a FA application twice that of traditional limestone to achieve the equivalent acid-neutralizing power. Clearly, any beneficial use of Class-C FA in liming soil will depend on product availability, associated shipping costs, P and micronutrient sufficiency of the target soil, and location of the potential user/site.

References

1. Mattigod, S.V., Rai, D., Eary, L.E., Ainsworth, C.C., Geochemical factors controlling the mobilization of inorganic constituents from fossil fuel combustion residues: I. Review of major elements, *J. Environ. Qual.* 19, 188, 1990.
2. Matsi, T., Keramidas, V.Z., Fly ash application on two acid soils and its effect on soil salinity, pH, B, P, and on ryegrass growth and composition, *Environ. Poll.* 104, 107, 1999.
3. Molliner, A.M., Street, J.J., Effect fly ash and lime on growth and composition of corn (*Zea mays* L.) on acid sandy soils, *Proc. Soil Sci. Soc. Fla.* 41, 217, 1982.
4. Schumann, A.W., Sumner, M.E., Plant nutrient availability from mixtures of fly ashes and biosolids, *J. Environ. Qual.* 28, 1651, 1999.
5. Beard, J.B., Turfgrass: Science and Culture, Prentice-Hall, Englewood Cliffs, NJ, 1973.
6. Musser, H.B., Turf management, Revised ed. McGraw-Hill Book Co., New York, 1962.

7. Spurway, C.H., Soil reaction (pH) preferences of plants, *Michigan Agric. Exp. Stn. Spec. Bull.* 306, 1941.
8. Engel, R.E., Alderfer, R.B., The effect of cultivations, lime, nitrogen, and wetting agent on thatch development in 1/4-inch bentgrass turf over a ten-year period, *NJ Agric. Exp. Stn. Bull.* 818, 32, 1967.
9. Murray, J.J., Juska, F.V., Effect of management practices on thatch accumulation, turf quality, and leaf spot damage in common Kentucky bluegrass, *Agron. J.* 69, 365, 1977.
10. Woodhouse, W.W., Effect of placement and rate of phosphorous, potash, and limestone on the growth of alfalfa and lespedeza, *Soil Sci. Soc. Amer. Proc.* 20, 15, 1956.
11. Ma, J.M., Furukawa, J., Recent progress in the research of external Al detoxification in higher plants, *J. Inorg. Biochem.* 9, 46, 2003.
12. Matsumoto, H., Cell biology of aluminum toxicity and tolerance in higher plants, *Int. Rev. Cytol.* 200, 1, 2000.
13. Hoveland, C.S., Achievements in management and utilization of southern grasslands, *J. Range Manage.* 53, 17, 2000.
14. Pathan, S.M., Aylmore, L.A.G., Colmer, T.D., Properties of several fly ash materials in relation to use as soil amendments, *J. Environ. Qual.* 32, 687, 2003.
15. Jones, R.L., Dreher, G.B., Silicon, Methods of soil analysis, SSSA Book Ser. 5. SSSA, Madison, WI, 1996.
16. Adams, F., Evans, C.E., A rapid method for measuring lime requirement of red-yellow podzolic soils, *Soil Sci. Soc. Amer. Proc.* 26, 65, 1962.
17. Mehlich, A., New extractant for soil test evaluation of phosphorus, potassium, magnesium, calcium, sodium, manganese, and zinc, *Commun. Soil Sci. Plant Anal.* 9, 477, 1978.
18. Aldous, D.E., Balfour, D., Evaluation of an industry lime by-product on a golf course fairway in Southeastern Victoria. *Int. Turf. Soc. Res. J.* 9, 817, 2001.
19. United States Environmental Protection Agency (USEPA), *40 CFR Parts 257, 403, and 503. Fed. Reg.* 58, 9248, 1993.
20. Barber, S.A., Liming materials and practices, Soil acidity and liming, ASA Monograph 12, Madison, WI, 171, 1984.

22
Phytoavailability of Trace Elements from a Landfill Containing Coal Combustion Waste

Sally Maharaj[1], Christopher Barton[1*], Bon-Jun Koo[1], and Lee Newman[2]

[1] *University of Kentucky, Department of Forestry, Thomas Poe Cooper Bldg., Lexington, KY 40546-0073, USA*

[2] *University of South Carolina, Arnold School of Public Health, 800 Sumpter St., Columbia, SC 29208, USA*

1. Abstract

The 488-D Ash Basin (488-DAB) is an unlined, earthen landfill on the U.S. Department of Energy's Savannah River Site, SC that contains approximately one million tons of coal combustion wastes (CCWs). Pyrite that is co-mingled with the CCWs has undergone oxidation and formed sulfuric acid, which has dissolved metals and trace elements in the CCWs and facilitated their mobility. The acid leachate contributes to groundwater deterioration in the area and threatens biota on, and adjacent to, the landfill. A study was undertaken to examine CCWs and vegetation on the 488-DAB to assess the potential for phytoavailability of these elements and to determine if a secondary contamination source exist. Results indicated that trace element concentrations of the CCWs were higher than those of native soils in the area. Mean pH (1.79 ± 0.75) and As concentrations (64.7 ± 43.0 mg kg^{-1}) for the CCWs were in the range of critical plant toxicity. Sequential extractions of the CCWs indicated that Mn, Zn, Ni, and Cr were organically bound and in exchangeable fractions in the material and likely phytoavailable, while Pb and Se were tightly bound in the crystalline and acid extractable (residual) fractions. The fractionation patterns for Al, As, Cd and Fe were very similar with most bound to the poorly crystalline fraction, which suggest that their solubilities are likely controlled by a common precipitate or mineral in the CCWs. Tissue analysis indicated that Mn and Zn were accumulated by species growing on the 488-DAB, as predicted. However, Ni and Cr were not accumulated at levels above that of the CCWs. Uptake of Se and Cd in tissue was observed at levels over that of the CCWs in several species, but not all. Given these conditions, results suggest that periodic monitoring of plant species growing on the 488-DAB should continue to ensure that toxicity and secondary contaminant problems do not arise.

2. Introduction

The degradation of water resources from coal mining and coal combustion activities is a problem of global significance. In the U.S., 860 million metric tons (Mg) of coal were burned in 2000[1], producing 98 million Mg of coal combustion wastes (CCWs). In general, CCWs are comprised of fly ash (FA), bottom ash (BA), boiler slag (BS) and flue gas desulfurization residue (FGD). Fly ash is the most abundant CCW worldwide with a production of over 500 million Mg per year[2]. In 2000, only 29% of the total CCWs produced in the U.S. was reused for beneficial purposes, leaving 69 million Mg for disposal[1,3]. Lagoons, disposal mounds, and landfills are commonly used as sites for CCW disposal[3,4]. Unfortunately, the leaching of excessive soluble salts and potentially toxic trace metals from these sites poses a significant threat to the environment[5]. Moreover, uptake of these elements by biological means and transport through the food web may extend the zone of contamination well outside the disposal area.

Plant roots can uptake metals and trace elements from soils or contaminated sediments and transport them to aboveground tissue where storage may occur. Accumulation of trace elements in plant tissue is influenced by the type of contaminant, concentration range of the contaminant, and physicochemical characteristics of the soil matrix. In addition, elemental accumulation differs among differing plant types[6,7]. Plant analysis may be utilized as a toxicity index for identifying accumulation of heavy metals and providing insight on their mobilization sequences and phytoavailability[8]. The presence of relatively distinct chemical fractions of trace metals in soils indicates that soils may form associations with metals that differ in strength (or reversibility) and, therefore, in their susceptibility to plant uptake, leaching, and extractability[9,10].

The 488-D Ash Basin (488-DAB), located at the U.S. Department of Energy's Savannah River Site (SRS) in South Carolina, is an unlined landfill containing approximately one million Mg of CCWs. Pyrite (FeS_2) that is co-mingled with the CCWs has resulted in the formation of acidic drainage (AD) as described by the following reaction:

$$FeS_{2(s)} + 3.75\ O_2 + 3.5\ H_2O \rightarrow Fe(OH)_{3\ (s)} + 2H_2SO_4 \quad (1)$$

where iron sulfide and other mixed-metal sulfides decompose upon exposure to the atmosphere and produce sulfuric acid.

Subsequently, the dissolution of materials in the landfill (coal, ash, soil etc.) by the sulfuric acid has resulted in the formation of a metal enriched solution phase that has contributed to groundwater deterioration in the area and poses a threat to biota on, and adjacent to, the basin[11]. As such, a study was undertaken to obtain information pertaining to metal and trace element distribution on the 488-DAB and to assess the potential for uptake by the various plant species growing on the landfill.

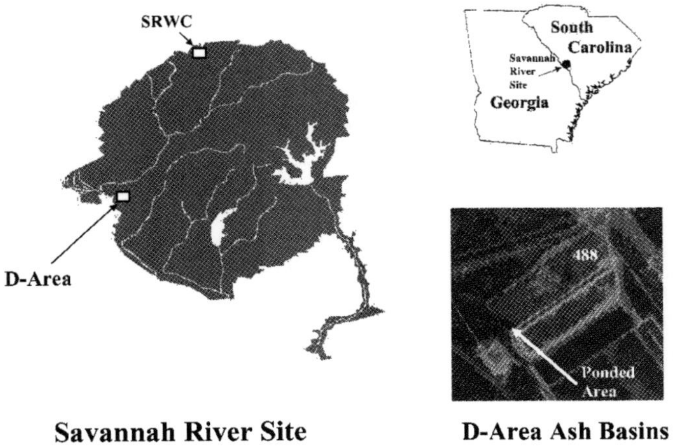

FIGURE 1. Study area at the Savannah River Site.

3. Materials and Methods

3.1. Site Description and Background

The 488-DAB received sluiced FA, dry FA and CR from the early 1950's to the mid-1990's. Non-hazardous wastes deposited in the basin contain metals typical of fly ash and coal (As, Co, Cr, Cu, Fe, Mn, Ni and Zn). The 488-DAB is $\sim 550 \times 180 \times 5.5$ m in size and contains $\sim 5.4 \times 10^5$ m^3 of waste material. The basin was constructed on the existing land surface at ~35 m (msl) and is ~9 m above the Savannah River. The present surface of the basin slopes gently to the west and is filled with waste, except in the far western reach where surface waters collect during rainfall events (Fig. 1).

3.2. Sample Collection, Processing and Analysis

A $2 \times 1 \times 1$ m deep trench was excavated in the center of 488-DAB to examine the distribution of materials in the rooting zone of the landfill. Six horizons were noted in a cross-section of the trench. Of these, a coal rubble zone (0–5 cm) and FA residue zone (20–50 cm) were clearly unique and relatively homogenous, while intermediate horizons appeared to be mixtures of the two outlying horizons. The shallow horizon consisted of poorly-sorted, pebble-sized pieces of coal and pyrite, while the deeper horizon contained a relatively uniform fine-grained (ash-like) material. Incremental coring at 5-cm depths was performed to collect samples for analysis. Cores from each depth were collected in triplicate, mixed in a bucket, and transferred to a plastic bag. The bags were transported to the laboratory, dried and sieved. In addition, soil samples were collected from an uncontaminated site at SRS for comparison purposes.

Three sub-samples from each of the 488-DAB horizons were sequentially extracted to determine the water-soluble (WS), exchangeable (EX), organically bound (OB), poorly crystalline (PC), crystalline (C) and residual (R) forms of metals bound to the substrate. A metal fractionation method was developed using techniques described by Miller and McFee[12], Henrot and Wieder[13], and McDaniel and Buol[14] and is outlined in Table 1. Analysis of the various fractions was performed using inductively coupled plasma-mass spectrometry or optical emission spectrometry (ICP-MS, ICP-OES).

An extraction for total recoverable elements was performed on separate 488-DAB samples and the SRWC native soil samples using the HNO_3-HCl microwave digestion method[15] followed ICP-MS and ICP-OES analyses. Quality assurance-quality control protocols were followed for all analytical procedures as outlined in USEPA Method 6020[16].

Vegetation samples were collected on the 488-DAB from areas where natural seeding had occurred. Efforts were taken to obtain samples from all dominant non-woody species that

TABLE 1. Sequential extraction procedure for CCWs from the 488-DAB.

Step	Fraction	Solution	Soil (g)	Solution (mL)	Condition*
1	Water Soluble	DDI water	1.0	10	Shake 30min
2	Exchangeable	1M KNO_3	1.0	25	Shake 16hrs (overnight)
3	Organically Bound	0.1M $Na_4P_2O_7$	1.0	15	Shake 16hrs (overnight)
4	Poorly Crystalline	0.2M $(NH_4)_2C_2O_4H_2O$	1.0	15	Shake 4hrs in the dark
		0.2M$H_2C_2O_4$ (pH 3.0)			
5	Crystalline	CBD: 0.3M $Na_3C_6H_5O_7.2H_2O$	1.0	15	Heat at 80°C in a water bath for 1hr.
		0.1M $NaHCO_3$			
		1g $Na_2S_2O_3$			Stir occasionally
6	Residual	(1:4) HCl	1.0	10	Reflux on hot plate for 30min.
		(1:1) HNO_3		4	

*After extraction with a reagent, the suspension was centrifuged, decanted and analyzed. Between extraction phases, the CCWs were washed twice with 10mL DI-water, centrifuged, water decanted and discarded.

existed "naturally" on the basin. Species that had been planted for experimental purposes were not included in this survey[11].

The plant specimens were washed with distilled water to remove dust and particulate material, then dried at 60°C. The dried plant tissue was ground using a Tekmar Analytical Mill and digested following USEPA methodology[15]. Metal analysis was performed by ICP-OES. The accuracy and precision of the plant tissue digestion and analysis was determined by the analysis of a spiked NIST standard reference material (SRM 1515- apple leaves). Average recovery for all metals was 105% with the exception of cadmium, which was consistently high. A method blank was included for every 10 samples.

Analysis of variance (ANOVA), general linear models (PROC GLMs), and PROC Univariate were performed to determine significant differences in elemental composition of the plant tissues. All statistical models were performed using SAS[17]. Plant tissue concentrations are considered statistically different at $p < 0.05$.

4. Results and Discussion

4.1. Characterization of Native Soil and CCWs

Chemical characteristics of the basin materials and native soil are presented in Table 2. The pH of the native soil was slightly acidic, while the 488-DAB material was highly acidic and potentially phytotoxic. The 488-DAB CCWs was greatly limited with respect to nitrate and P concentrations, and exhibited very low K content (<3 mg kg^{-1}). The 488-DAB material was also characterized by elevated Fe and Al concentrations, which were over those of background levels for Coastal Plain soils in South Carolina[18], but not in the critical range for plant growth and survival. However, the As concentration for the 488-DAB CCWs was within the critical range for plant toxicity (20–50 mg kg^{-1})[19]. The CCWs also exhibited a relatively high C content, which may be attributed to coal particles within the waste material. Cadmium, Cr, Ni, and Se showed no measurable concentrations (BDL, below detection limit) in the native soil. Calcium levels were high in the basin material and may have derived from a sulfur emission control system at the power plant. Manganese levels, on the other hand, were relatively low in the CCWs but elevated in the native soil. High Mn concentration in the native soil was likely derived from the parent material.

TABLE 2. Average chemical characteristics of substrate materials*.

Parameter	Native soil	CCW (488-DAB)
pH (1:1)	5.07 (0.62)	1.72 (0.75)
EC (1:5) (mS cm^{-1})	0.07 (0.08)	5.92 (1.86)
NO_3-N	10.3 (3.0)	BDL
P†	6.8 (2.9)	BDL
K†	17.2 (10.0)	2.3 (1.5)
Mg†	48.7 (34.3)	174.3 (82.2)
Ca†	208.2 (118.3)	526.0 (47.0)
Al‡	4193.8 (146.3)	1767.9 (495)
Fe‡	3510.5 (195.5)	20476.4 (13827.1)
Mn‡	222.3 (8.89)	7.18 (12.3)
Zn‡	14.3 (11.1)	1.96 (6.00)
Cd‡	BDL	0.1 (0.58)
Pb‡	8.15 (1.1)	28.4 (12.4)
As‡	9.88 (0.72)	64.7 (43.0)
Ni‡	BDL	26.8 (10.72)
Cr	BDL	43.36 (5.01)
Se‡	BDL	8.88 (5.80)
C (%)	1.29 (1.22)	11.6 (2.25)
N (%)	0.08 (0.05)	0.21 (0.04)

* mg kg^{-1} except where noted otherwise; standard error in parenthesis.
† Mehlich-1 method (HNO_3-H_2SO_4).
‡ U.S. EPA method 200.2 (HNO_3-HCl).
BDL = below detection limit.

4.2. Soil Fractionation

Sequential extraction of CCW samples at the three different depths (0–5, 10–15, and 20–25 cm) revealed that the WS and EX fractions contained minimal concentrations of Fe, As, Cd, and Cr (Fig. 2, 3 and 4).

Sequential extraction of 0–5 cm materials showed that approximately 70% of the Fe was extracted in the PC fraction, with a slight contribution ($<10\%$) from the OB fraction (Fig. 2). Substrates with little OB iron can attribute the PC fraction to ferrihydrite ($5Fe_2O_3 \cdot 9H_2O$), which is soluble in ammonium-oxalate[20]. Iron in the OB and R fractions increased at the 10–15 cm zone (Fig. 3), then showed a large reduction in the 20–25 cm zone (Fig. 4). The deeper zone also showed an increase in the C fraction over that of the shallower horizons (Fig. 4). The loss of the OB phase in the deeper zone may be attributed to a lack of organic matter at this depth. Sparse vegetation and shallow roots were present at the surface, but did not extend into the deeper profile where sampled. The increase in citrate-bicarbonate-dithionite extracted (crystalline) Fe in the deeper zone is likely linked to the elution of minerals such as goethite (α-FeOOH), hematite (α-Fe_2O_3), and lepidocrocite (γ-FeOOH), which are oxalate insoluble[21].

Manganese fractionation exhibited substantially different patterns than those observed for Fe (Fig. 2, 3 and 4). Exchangeable forms of Mn were prevalent in the 0–5 cm material, while the mid (10–15 cm) and deep (20–25 cm) zones were dominated by the PC fraction. High EX Mn fractions have been noted elsewhere in systems that are influenced by CCWs and AD[22,23]. The replacement of EX by PC forms in the deeper zone is probably the result of an increase in pH. Although a pH > 8.0 is generally needed for Mn precipitation, Mn in areas with high Fe concentrations may undergo co-precipitation with Fe at much lower pH levels[24]. It has also been shown that Mn^{2+} can be removed by adsorption to reactive surfaces of crystalline Fe minerals at low pH values. Thus, the PC and C Mn phases may be controlled by the type and amount of Fe present in the sample[24].

In all three zones, Al and As were primarily observed in the PC phase with a smaller amount found in the OB phase

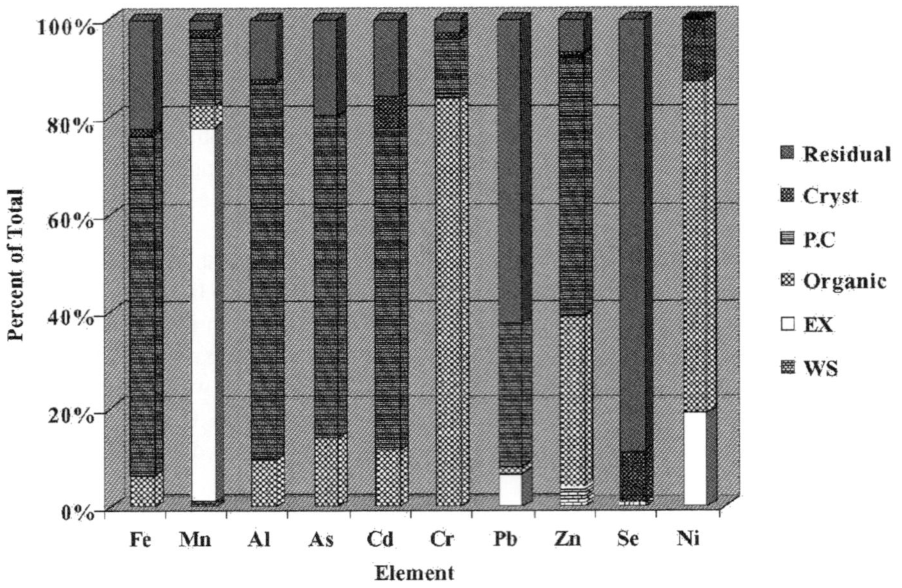

FIGURE 2. Fractionation of elements with respect to percent total at 0–5 cm depth soil profile.

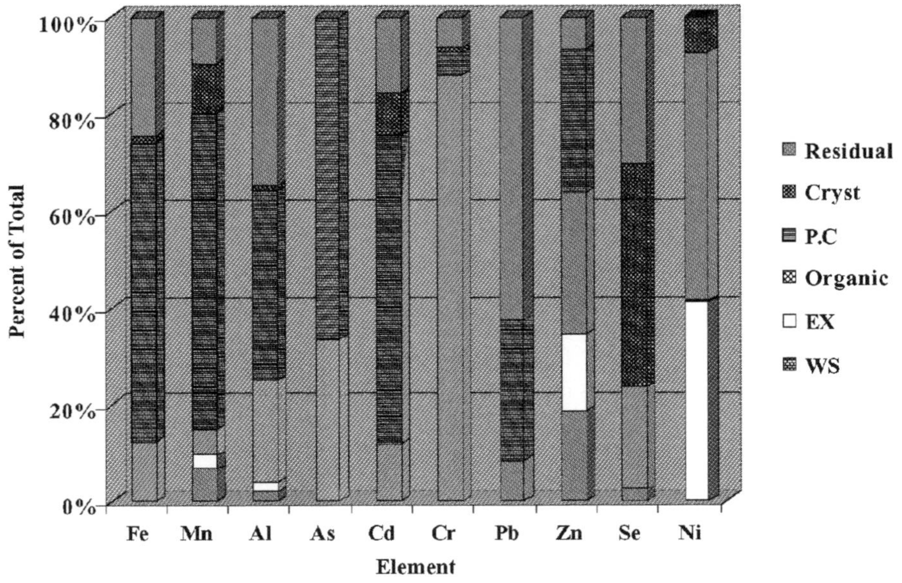

FIGURE 3. Fractionation of elements with respect to percent total at 10–15 cm depth soil profile.

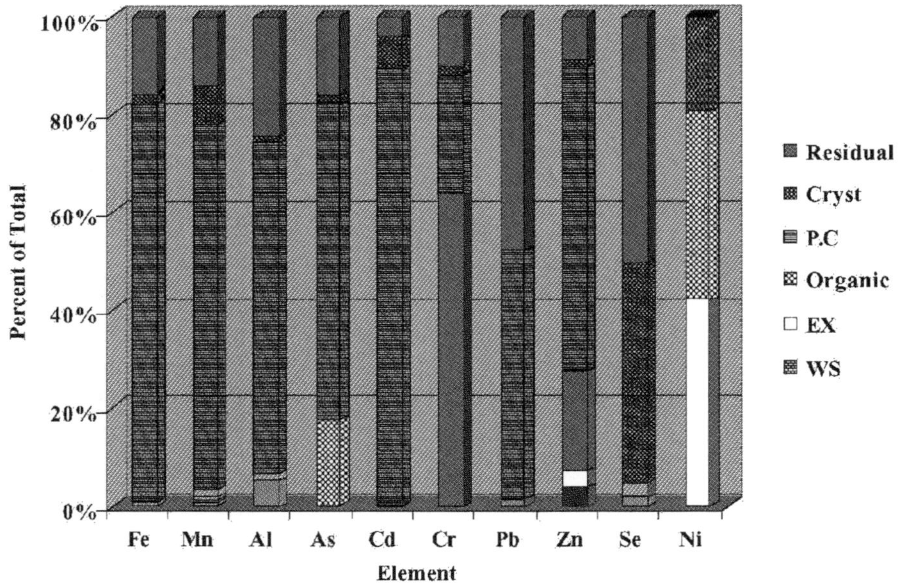

FIGURE 4. Fractionation of elements with respect to percent total at 20–25 cm depth soil profile.

(Fig. 2, 3 and 4). The fractionation patterns of Al and As are very similar to that of Fe, suggesting that their solubility may be controlled by an Fe dominated mineral or precipitate. Zinc was found in both the OB and PC fractions in appreciable quantities. It is interesting to note that the WS and EX fractions increased dramatically in the 10–15cm zone (Fig. 3). It has been shown that Zn is retained on oxyhydroxides by surface complexation and thus can be solubilized through mineral dissolution[25]. The primary control on Zn solubility may be explained by the formation and dissolution of secondary amorphous Fe precipitates within the basin. The OB Zn phase showed a decrease with depth from 35 to 18% (Fig. 2 and 4). Zinc has been shown to have a weak association with organic matter and is readily released during this extraction phase[26].

Selenium was limited to the C and R phases in all three zones (Fig. 2, 3 and 4). This partitioning behavior is in stark contrast with the other elements described earlier and implies that Se is strongly associated with either a primary mineral or a silicate. Therefore, the mobility and phytoavailability of Se is low as compared to those of the other elements.

Lead showed similar characteristics to Se in that it was limited to the PC and R phases (Fig. 2, 3 and 4). However, in the 0–5cm and 10–15cm zones, a small concentration ($<10\%$) was found in the EX and OB phase. This suggests that Pb has a limited potential for phytoavailability, which has been previously observed[8,27,28].

The recovery of Cr was highest in the OB fraction at all three depths (Fig. 2, 3 and 4) and the order of distribution followed OB$\gg$ PC$\gg$C>R. Cadmium was observed primarily in the PC phase with a small percentage being detected in the OB phase at the 0–5cm and 10–15cm depths. The 20–25cm depth showed $>80\%$ of Cd being present in the PC phase. Minimal concentrations were also detected in the C and R phases. Nickel was found in the EX fraction which increased with depth, the highest percentage detected in the OB fraction (Fig. 2, 3 and 4). This decreased in the deeper zones with subsequent increases in the EX fraction. This suggests that Ni may be readily phytoavailable.

4.3. Plant Tissue

Analysis of the plant tissue indicated significantly different tissue concentrations for Al, Mn, Fe, Zn, Cr and Ni among species (Table 3). Big blue stem (*Andropogon gerardii*) exhibited the highest concentration of Al (86.1 mg kg^{-1}) and the lowest concentration of Zn (20.9 mg kg^{-1}) from the species examined.

TABLE 3. Elemental concentrations for plant tissue collected on the 488-DAB*.

	Elements (mg kg^{-1})									
Plant species	Al	Mn	Pb	Fe	As	Zn	Se	Cd	Cr	Ni
Panicum capillare	23.39	138.082	4.576	1.8	20.447	77.5	14.19	3.9189	6.272	8.7735
(Witchgrass)	(CDEF)	(BCD)	(A)	(CD)	(A)	(B)	(A)	(A)	(AB)	(ABCDE)
Andropogon gerardii	86.1	22.933	3.394	210	9.349	20.87	4.678	2.2533	7.542	5.7439
(Big Blue Stem)	(A)	(FG)	(A)	(B)	(A)	(C)	(A)	(A)	(A)	(ABCDEFG)
Agrostis perennans	20.28	41.134	2.637	77	2.611	39.94	2.661	0.023	0.934	7.4486
(Bentgrass)	(DEF)	(FG)	(A)	(CDE)	(A)	(C)	(A)	(B)	(C)	(ABCDEF)
Lespedza bicolor	35.64	45.671	16.891	119	6.593	30.19	10.344	0.0431	0.549	9.7234
(shrubby lespedza)	(BCDEF)	(EFG)	(A)	(C)	(A)	(C)	(A)	(B)	(C)	(ABCD)
Specularia perfoliata	68.32	91.452	2.422	418	4.972	77.99	5.14	0.2455	3.163	9.7263
(Venus Looking Glass)	(A)	(DE)	(A)	(A)	(A)	(B)	(A)	(A)	(BC)	(ABCD)
Eupatorium compositifolium	36.78	38.875	3.819	121	4.622	74.01	10.762	1.6922	2.39	6.0324
(Dogfennel)	(BCDEF)	(FG)	(A)	(C)	(A)	(B)	(A)	(A)	(C)	(ABCDEFG)
Lolium perenne	35.56	98.207	1.99	103	5.722	46.26	9.404	0.2874	1.878	1.5361
(Ryegrass)	(BCDEF)	(DE)	(A)	(CD)	(A)	(C)	(A)	(A)	(C)	(DEFG)
Solidago stricta	26.78	75.156	3.076	79	4.261	46.26	6.111	0.143	0.627	3.6304
(Goldenrod)	(BCDEF)	(EF)	(A)	(CDE)	(A)	(C)	(A)	(A)	(C)	(BCDEFG)
Baccharis halimifolia	45.26	182.491	3.521	79	1.746	108.16	7.844	0.7734	0.263	6.8962
(Eastern baccharis)	(BCD)	(B)	(A)	(CDE)	(A)	(A)	(A)	(A)	(C)	(ABCDEFG)
Paspalum notatum	17.76	114.526	1.229	74	11.927	38.06	7.824	0.1647	2.819	1.7362
(Bahia grass)	(DEF)	(CDE)	(A)	(CDE)	(A)	(C)	(A)	(A)	(BC)	(EFG)
Panicum (sp)	20.2	166.885	1.28	100	7.794	97.93	6.627	0.2871	0.913	5.5234
(Not Known)	(DEF)	(BC)	(A)	(CDE)	(A)	(AB)	(A)	(A)	(C)	(ABCDEFG)
Panicum milaceum	43.25	325.976	3.122	122	3.964	74.75	2.796	0.4224	1.71	4.6808
(Broomcorn millet)	(BCDE)	(A)	(A)	(C)	(A)	(B)	(A)	(A)	(C)	(ABCDEFG)
Smilax bona-nox	24.02	171.277	3.275	64	2.149	40.4	4.138	0.1502	1.363	2.046
(Greenbrier)	(CDEF)	(B)	(A)	(CDE)	(A)	(C)	(A)	(A)	(C)	(CDEFG)
Rubus argustus	29.47	89.366	3.714	107	2.542	36.66	8.301	0.3185	0.784	6.7408
(Blackberry)	(BCDEF)	(DE)	(A)	(CD)	(A)	(C)	(A)	(A)	(C)	(ABCDEFG)

* Scheffe's Test using the GLM Procedure in the SAS System done to determine significant differences between species ($p = 0.05$). Means with the same letter are not significantly different.

Manganese uptake exhibited wide variability amongst species with broomcorn millet (*Panicum milaceum*) having the highest concentration (326 mg kg^{-1}). Plant tissue Mn and Zn concentrations were higher than CCWs levels for all species sampled on the 488-DAB (Tables 2 and 3). Both elements are essential micronutrients for plants and uptake is likely influenced by this factor. In addition, sequential extractions had indicated a likelihood for Mn phytoavailability with >70% bound to the EX fraction (Fig. 2.).

Even though the Fe concentration was high in the CCWs, plant Fe concentrations were negligible (Tables 2 and 3). Iron uptake by the non-woody species may have been inhibited by the PC and C Fe forms which were dominant in the CCWs. As such, Fe was not readily available for uptake. Antagonism between Zn and Fe may contribute to the low Fe tissue concentration levels. High Zn levels have been shown to inhibit Fe and Cd uptake by both root and foliar systems[29]. However, Cd concentrations in witchgrass (*Panicum capillarye)*, big blue stem, dogfennel (*Eupatorium compositifolium*) and eastern baccharis (*Baccharis halimifolia*) were higher than the mean CCWs concentration + standard error Cd levels, which suggests that antagonism may not be a major factor for uptake in this system.

Although As was not significantly different among species, witchgrass accumulated concentrations that could potentially be phytotoxic (10–50 mg kg^{-1})[19]. Most of the species exhibited phytotoxic levels of Se (5–30 mg kg^{-1})[29] (Table 3). Shrubby lespedeza (*Lepedeza bicolor*), witchgrass and dogfennel exhibited Se levels in excess of 10 mg kg^{-1}, which was higher than the mean concentration of the CCWs. The concentration of Ni, Cr and Pb in the plant tissues were higher than one would expect to find from plants growing on a non-contaminated site, but the levels were below that of the CCWs, which leads us to believe that phytoaccumulation of these elements is unlikely.

In all, metal and trace element concentrations of the species examined were found to be within the regulatory limits for solid waste disposal established by the U. S. Environmental Protection Agency[30] and therefore do not pose contamination risk at this time. However, species such as witchgrass and dogfennel were shown to accumulate some elements at levels above that of the CCWs, which suggest that periodic monitoring of plant species growing on the 488-DAB should continue to ensure that toxicity and secondary contaminant problems do not arise.

5. Conclusions

In an effort to evaluate the potential for phytoavailability of trace elements from a landfill containing CCWs, sequential extractions of the CCWs and tissue analysis from plants growing on the basin were performed. Results indicated differences in bonding fractionation between the elements examined and in elemental concentration amongst the differing plant species. Plant uptake of Mn and Zn was predicted by sequential extraction, and accumulation was observed in the tissues of all species examined. Although the fractionation procedure suggested otherwise, Ni and Cr were not accumulated in the plants at levels above that of the CCWs. The sequential extraction patterns for Al, As, Cd and Fe were very similar suggesting that their solubilities may be controlled by a common precipitate or mineral in the CCWs. Given that these four elements were bound primarily in the poorly crystalline fraction, it was not surprising that uptake was limited to levels below that of the CCWs. However, concentrations of these elements were by no means low and phyto-toxic levels of As were observed. Similar findings were also observed with respect to Se. Although phytoaccumulation of trace elements does not appear to be a significant problem at this time, periodic screening of vegetation on the 488-DAB is recommended to ensure that a secondary contaminant problem does not arise as new species colonize the basin and/or physicochemical changes to the CCWs occur.

Acknowledgements. This work was funded in part by the U.S. Department of Energy—Savannah River Operations Office through the U.S. Forest Service-Savannah River (Interagency Agreement DE-IA09-00SR22188) and the Savannah River Ecology Laboratory through the University of Georgia Research Foundation, Inc (Financial Assistance Award DE-FC09-96SR18546). Support was also provided by the U.S. Forest Service—Southern Research Station, Center for Forested Wetlands Research; and the Environmental Research and Training Laboratory (ERTL) at the University of Kentucky.

References

1. Kalyoncu, R., Coal Combustion Products, 2000. http://minerals.usgs.gov/minerals/pubs/commodity/coal/874400.pdf.
2. Kikuchi, R., Application of coal ash to environmental improvement. Transformation into zeolite, potassium fertilizer, and FGD absorbent, *Resour. Conserv. and Recycl.*, 27, 333, 1999.
3. Punshon, T., Seaman, J. C., and Sajwan, K. S., The Production and Use of Coal Combustion Products, in *Chemistry of Trace Elements in Fly Ash*, Sajwan, K. S., Alva, A. K., and Keefer, R. F., Eds., Kluwer Academic/ Plenum Publishers, New York, 2003, 1–11.
4. Twardowska, I., Tripati, P. S. M., and Das, R. P., Trace elements and their mobility in coal ash-fly ash from Indian power plants in view of its disposal and bulk use in agriculture, in Proceedings of 6th International Conference on the Biogeochemistry of Trace Elements. Guelph, Canada, 2001.
5. Jackson, B., Shaw-Allen, P., Mills, G., Hopkins, W., and Jagoe, C., Trace-element protein interactions in fish from a fly-ash settling basin. A study using size exclusion chromatography coupled to ICP-MS, in Proceedings of 6th International Conference on the Biogeochemistry of Trace Elements, Guelph, Canada, 2001.
6. Weis, J. S., Windham, L., and Weis, P., Patterns of metal accumulation in leaves of the tidal marsh plants *Spartina alterniflora* Loisel and *Phragmites australis* Cav. Trin ex Steud. over the growing season, *Wetlands*, 23, 459, 2003.

7. Koo, B-J., Adriano, D. C., Bolan, N. S., and Barton, C.D., Root exudates and microorganisms, in *Encyclopedia of Soil in the Environment*, 2004. in press.
8. Murphy, A. P., Coudert, M., and Barker, J., Plants as biomarkers for monitoring heavy metal contaminants on landfill sites using sequential extraction and inductively coupled plasma atomic emission spectrophotometry (ICP-AES), *J. Environ. Monit.*, 2, 621, 2000.
9. Viets, F. G. Jr., Chemistry and availability of micronutrients in soils, *J. Agric. Food Chem.*, 10, 174, 1962.
10. Miller, W. P., McFee, W. W., and Kelly, J. M., Mobility and retention of heavy metals in sandy soils, *J. Environ. Qual.*, 12, 579, 1983.
11. Barton, C. D., Marx, D., Adriano, D.C., Koo, B.-J., Newman, L. A., Czapka, S., and Blake, J., Phytostabilization of a landfill containing coal combustion waste, *Environ. Geosci.*, 11(2), 2004. in press.
12. Miller, W. P., and McFee W. W., The distribution of Cd, Zn, Cu, and Pb in soils of industrial northwestern Indiana, *J. Environ. Qual.*, 12, 29, 1983.
13. Henrot, J., and Weider, R. K., Processes of iron and manganese retention in laboratory peat microcosms subjected to acid mine drainage, *J. Environ. Qual.*, 19, 312, 1990.
14. McDaniel, P. A., and Buol, S. W., Manganese distributions in acid soils of the North Carolina Piedmont, *Soil. Sci. Soc. Am. J.*, 55 152, 1991.
15. USEPA, Methods for the determination of metals in environmental samples, Method 200.2 EPA/600/R-94/111, Washington, DC, United States Environmental Protection Agency, 1994a.
16. USEPA, Inductively coupled plasma-mass spectrometry, Method 6020, Rev 0, *in* SW-846, Test Methods for Evaluating Solid Waste, Physical/Chemical Methods, Washington, DC, United States Environmental Protection Agency Office of Solid Waste, 1994b.
17. SAS, *SAS/STAT User's Guide*, Version 8, SAS Institute Inc., Cary, NC, 1999.
18. Canova, J. L., Elements in South Carolina inferred background soil and sediment samples, *South Carolina Geol.*, 41, 11, 1999.
19. Alloway, B. J., *Heavy Metals in Soils*, John Wiley & Sons, Inc., London, 1990.
20. Schwertmann, U., and R. M. Taylor, R. M., Iron oxides, in *Minerals in the Soil Environment*, 2nd Ed., Soil Science Society of America, Dixon, J. B., and Weed, S. B., Eds., Madison, WI, 1989, 379–438.
21. Barton, C. D., and Karathanasis, A. D., Aerobic and anaerobic metal attenuation process in a constructed wetland acid mine drainage, *Environ. Geosci.*, 5, 43, 1998.
22. Weider, R. K., Linton, M. N., and Heston, K. P., Laboratory mesocosm studies of Fe, Al, Mn, Ca and Mg dynamics in wetlands exposed to synthetic acid coal mine drainage, *Water Air Soil Pollut.*, 51, 181, 1990.
23. Anderson, M. A., Bertsch, P. M., Feldman, S. B., and Zelazny, L. W., Interactions of acidic metal-rich coal pile runoff with a subsoil, *Environ. Sci. Technol.*, 25, 2038, 1991.
24. Stumm, W., and Morgan J. J., *Aquatic Chemistry*, 2nd Ed., John Wiley & Sons, Inc., New York, 1981.
25. Calvet, R., and Bourgeois, S., Some experiments on extraction of heavy metals present in soils, *Int. J. Environ. Anal. Chem.*, 39, 31, 1990.
26. Narwal, R. P., Singh, B. R., and Salbu, B., 1999, Association of cadmium, zinc, copper, and nickel with components in naturally heavy metal rich soils studied by parallel and sequential extraction, *Commun. Soil Sci. Plant Anal.*, 30, 1209, 1999.
27. Lasat, M. M., Phytoextraction of toxic metals: A review of biological mechanisms, *J. Environ. Qual.*, 31, 109, 2002.
28. Gleyzes, C., Tellier, S., and Astruc, M., Sequential extraction procedures for the characterization of fractionation of elements in industrially-contaminated soils, in *Methodologies for Soil and Sediment Fractionation Studies*, Quevauviller, P., Ed., The Royal Society of Chemistry, Cambridge, London, 2002, 66–99.
29. Kabata-Pendias, A., and Pendias, H., *Trace Elements in Soils and Plants*, CRC, Press, Florida, 2001.
30. U.S. EPA, Appendix A: Generic SSLs for the residential and commercial/industrial scenarios, 2001. http://www.epa.gov/superfund/resources/soil/ssgmarch01.pdf

23
Potential Uses of Fluidised Bed Boiler Ash (FBA) as a Liming Material, Soil Conditioner and Sulfur Fertilizer

Hailong Wang[1], Nanthi Bolan[2], Mike Hedley[2], and Dave Horne[2]

[1] *Forest Research, Private Bag 3020, Rotorua, New Zealand*

[2] *Institute of Natural Resources, Massey University, Palmerston North, New Zealand*

Abstract

To meet the Clean Air standard, many North American and European power plants have adopted fluidized bed combustion techniques. These systems require lower capital investment, to reduce the sulfur-dioxide (SO_2) emissions in flue gases, than the wet flue gas desulfurization process. In the former system, limestone chip mixed with the fine coal is burnt in a bed suspended by compressed air. Sulfur dioxide released as the coal is burnt, reacts with CaO generated in the furnace therefore minimizing SO_2 emissions from the stack. The resulting ash from the boiler bed and trapped fly ash, known as fluidized bed boiler ash (FBA), contains $CaSO_4$ and unreacted CaO. Mixing this ash with water, to overcome dust problems, subsequently converts CaO to $Ca(OH)_2$. The chemical composition of FBA is highly dependent on the efficiency of the boilers and the nature of the fuel and limestone sources.

As many countries including Australia and USA are rich in coal resources, coal plays a significant role in supplying energy. But, some of the coal resources are rich in sulfur (S) and with increasing concern for environmental pollution, low cost techniques, such as fluidized bed combustion, are required to reduce SO_2 emission from the burning of such high S coal. Consequently, large quantities of FBA or similar by-products would be expected. Although the resulting large volumes of FBA materials are generally disposed of in landfills, limited landfill space and increased costs, however, have stimulated investigations in many countries to develop and demonstrate agricultural and environmentally safe uses for FBAs, thereby reducing the cost of SO_2 scrubbing.

Agricultural utilization of FBA overcomes a waste problem and at the same time provides a liming material and an inexpensive form of S source. Furthermore, FBA also has potential as a micro-nutrient fertilizer, providing elements such as boron and selenium. In this chapter the chemical characteristics of FBA and its potential uses as a liming material, soil conditioner and S fertilizer are reviewed.

1. Introduction

Fluidized bed combustion is a technology utilized in the burning of sulfur-enriched coal to limit the amount of S released into the atmosphere.[1] Coal is burnt in a bed of inert ash and crushed limestone. The bed is fluidized (held in suspension) by injecting air through a perforated floor. Sulfur dioxide (SO_2) released during combustion is converted to calcium sulfate through its reaction with the calcined limestone. Such boilers thus have potential in coal-fired power plants for meeting air quality standards without the high costs of installing scrubbers to remove SO_2 from stack gases. Fluidized bed furnaces burn coal at lower temperatures (between 815 to 870°C)[2] than do conventional coal fired furnaces (1400 to 1600°C)[3] because the lower temperatures result in optimum S capture, minimizing the limestone usage and reducing nitrogen oxide formation.[1,2] The chemical reactions involving the limestone charge in a fluidized bed boiler can be described by the following equations:

$$CaCO_3 \rightarrow CaO + CO_2 \quad (1)$$

$$CaO + SO_2 \rightarrow CaSO_3 \quad (2)$$

$$2CaSO_3 + O_2 \rightarrow 2CaSO_4 \quad (3)$$

Because of its relatively low cost, the fluidized bed combustion technique has become widely used for reducing SO_2 emissions from electricity-generating plants, especially in the United States. The fluidized bed combustion techniques are generally smaller in scale and require less capital investment than the wet desulfurization of flue gases.[1,4] Existing coal fired power plants can be modified to fluidized bed combustion, and therefore present a simple option for bringing older plants into compliance with clean air legislation.[4] One disadvantage, however, is that approximately 200 kg of fluidized bed boiler ash (FBA) is generated per 1000 kg of coal burned.[5] Thus relatively large amounts of FBA waste are produced in fluidized bed combustion power plants. It is estimated that a 1000 megawatt fluidized bed power plant without regeneration of the bed material would produce about 1.8×10^6 kg of FBA per day.[6] Therefore,

TABLE 1. Chronological summary of chemical characterization of fluidized bed boiler ashes (FBA).

Sorbent	$CaCO_3$ %	pH	Elemental analysis[1]	Reference
Calcite	53–58	12.5	52% $CaSO_4$, 23.6% Ca, 0.5% Mg, 36–112 $\mu g\ g^{-1}$ B, H.M. trace	3, 12, 13, 14, 15, 16, 17, 18, 19, 20, 21, 22
Calcite	51		27% Ca, 7.4% S, 1.1% Mg, 88 $\mu g\ g^{-1}$ B, H.M.$\leq$20 $\mu g\ g^{-1}$	23
Dolomite	68		41% $CaSO_4$, 12.9% Ca, 9% Mg	24
Calcite	65		28% $CaSO_4$, 24.3% Ca, 1.1% Mg, 106 $\mu g\ g^{-1}$ B, H.M. $\leq$715$\mu g\ g^{-1}$	24, 25
Calcite	50		28% Ca, 3.6% S, 0.5% Mg, H.M.$\leq$11 $\mu g\ g^{-1}$	26, 27
Calcite			32% Ca, 10% S, 5.5% Mg, H.M.$\leq$25 $\mu g\ g^{-1}$	28
Calcite & Dolomite	105		4% $CaCO_3$, 31% $CaSO_4$, 29% Ca, 11% Mg, H.M.$\leq$41 $\mu g g^{-1}$	29, 30
Dolomite			27% Ca, 5.9% S, 6.1% Mg, 74 $\mu g\ g^{-1}$ B, H.M.$\leq$11 $\mu g\ g^{-1}$	31
Dolomite	60	10.5	18% Ca, 5.2% S, 11% Mg, 171 $\mu g\ g^{-1}$ B, 5.6 $\mu g\ g^{-1}$Se, H.M. $\leq$52 $\mu g\ g^{-1}$	4, 32
Calcite	97	11.9	39% Ca, 3.8% S, 0.5% Mg	33, 34
Calcite	46	12.5	39% Ca, 16% S, 0.1% Mg	35, 36, 37
Calcite	52	12.4	25% Ca, 6.2% S, 0.6% Mg, 3.6% Fe, 1.8% Al, 0.7% Na, 0.3% K, 1360 $\mu g\ g^{-1}$ B, 1250 $\mu g\ g^{-1}$ Sr, 875 $\mu g\ g^{-1}$ P, 88 $\mu g\ g^{-1}$ Mn, H.M. $\leq$6.5 $\mu g\ g^{-1}$	11
Calcite	31		16% Ca, 0.3% S, 0.9% Mg, H.M. $\leq$ 6.8 μg g-1	38
Calcite	55.2	12.4	41% Ca, 16% S, 0.4% Mg, 0.2% Al, 24 $\mu g\ g^{-1}$ K, 171 $\mu g g^{-1}$ B, 117 $\mu g\ g^{-1}$ P, 192 $\mu g\ g^{-1}$ Mn	39
Dolomite	50	12.4	13% Ca, 2.3% S, 7.5% Mg, 2.1% Fe, 0.5% Al, 0.04% Na, 0.3% K, 369 $\mu g\ g^{-1}$ B, 312 $\mu g\ g^{-1}$ P, 71 $\mu g\ g^{-1}$ Mn, H.M. $\leq$52 $\mu g\ g^{-1}$	40
Dolomite	60	12.2	45% Si, 24% Ca, 8.6% S, 16% Mg, 2.1% Fe, 2.8% Al, 1% Na, 0.1% K, 206 $\mu g\ g^{-1}$ B, 100 $\mu g\ g^{-1}$ P, H.M. $\leq$913 $\mu g\ g^{-1}$	7

[1]H.M. = heavy metals.

in addition to installation and operating expenses, significant costs for disposal of FBA material are incurred. Development of beneficial uses for FBA material that are environmentally benign or even beneficial would significantly reduce the costs of controlling SO_2 emission.[4]

Fluidized bed boiler ash is a mixture of conventional coal combustion ash (either bed or fly ash), the SO_2 reaction product (primarily anhydrite, $CaSO_4$) and unspent sorbent (generally lime, limestone or dolomite).[2,4] Therefore, the fuel and the limestone source dictate the composition of the FBA. In many countries including the United States, FBA is generally disposed of in landfills. Limited landfill space and increased tipping fees result in the development and demonstration of beneficial and environmentally safe uses for FBA.[7] Fluidized bed boiler ash is usually highly alkaline because of the presence of unspent sorbent, and also contains anhydrite, thus has potential to be used as liming material and S fertilizer.

2. Physical and Chemical Properties of FBA

The physical and chemical characteristics of FBA depend on a variety of factors, including the composition of the parent coal, SO_2 sorbent and specific boiler configuration used.[8] Fluidized bed boiler ash is composed predominantly of small particles, with particle sizes ranging from 2 μm to 4 mm.[3,4] The ash also exhibits pozzolanic properties, i.e. it can react with water to form cement due to its similar composition to cement.[4,9] It has been observed that the slaked FBA becomes a rock-like material after piling up for a period of several months. Researchers use FBA as a low strength concrete for cattle walkways and dairy cattle feedlots. Fluidized bed boiler ash can also be utilized to make synthetic aggregates, to control acid mine drainage and to remove CO_2 from multi-component gas (natural gas) streams, and as a substitute for cement in road construction and mine reclamation. In addition, FBA can also potentially be used as a soil liming amendment with gypsum for improving soil physical and chemical properties, stabilization of soil, cover on landfills, a liner for creation of solid waste disposal sites, alkaline stabilization of biosolids and temporally stabilizing hazardous spills.[10]

2.1. Agronomically Useful Elements

The major elements in FBA include Si, Al, Fe, Ca and S, together with lesser amounts of Na, Mg, B, Sr and K.[11] Table 1 shows the chemical analysis of some FBA samples. The components (elemental analysis) important to agriculture are divided into four groups: liming material, essential plant nutrients, heavy metals and phytotoxic elements.[5]

Fluidized bed boil ash has a Ca content ranging from 13% to 39% (Table 1). Magnesium content of FBA varies from 0.1% [35,36] to 16% [7] depending on whether the sorbent is calcite or dolomite (Table 1). Due to the presence of unspent sorbent, FBA is usually highly alkaline (pH_{water} 10.5–12.5) with significant neutralization potential (Table 1). It has been reported that the best agricultural use of FBA is as a liming source to overcome the problems of associated with soil acidity.[3,5,13,14,23]

The S content of FBA is also relatively high, being mainly in the form of $CaSO_4$ (anhydrite). Therefore, FBA is a source of

TABLE 2. Summary of effect of fluidized bed boiler ashes (FBA) on plant elemental composition.

Plant species	Trial type	Plant tissue analysis relative to the control[1]	Reference
Apple	Sand culture	Leaf: Ca, S, Sr, ↑; N, K, n.c.; P, Mg, B, Mn, Zn, Cu, Fe, ↓; Peel & flesh: n.c.	15
Apple	Pot	Leaf: Ca, ↑; Mn, Zn, ↓; K, N, P, S, B, n.c.	14
Apple	Pot	Root: Ca, ↑; P, Mn, Cu, Al, n.c.	17
Apple	Field (Control aglime)	Leaf: Ca, n.c.; Mg, ↓; Peel & flesh: n.c.	16
Apple	Field	Leaf: Ca, ↑; Mg, ↓; Peel & flesh: n.c.	18
Peach	Pot	Leaf: Mg, ↑; Mn, Zn, Al, ↓	24
Pecan	Pot	Leaf: Ca, P, K, ↑; Mn, Zn, Al, ↓	25
Vegetables, corn, wheat, peanuts	Field	Total content: Ca, P, Mg, N, K, Zn, B, Cu, Cd, Mn, Fe, n.c.	48
Red clover, Tall fescue, Oat, Buckwheat	Pot (Control: $Ca(OH)_2$)	Mg, S, ↑; Zn, ↓	21
Tall fescue, Lotus	Field (Control: dolomite)	Ca, Fe, S, Zn↑; Mg, P, ↓	26
Tall fescue	Field	Ca, Fe, S, ↑; Mn, Zn, ↓	27
Maize, wheat	Field (Control: dolomite)	Ca, Fe, ↑; Mg, ↓ K, Mn, Cu, Zn, n.c.	28
Tobacco, Corn	Pot (Control aglime)	Leaf: Cl, P, S, ↑; N, K, Cd, Ni, Zn, ↓; Shoot tissue: Cl, Cu, ↑; Pb, ↓;	29
Orchardgrass, white clover	Field	Ca, Mg, S, ↑; N, P, K, Mn, Zn, Al, n.c.	31
Tall fescue	Pot	Ca, Mg, S, K, ↑; N, P, B, Mn, Zn, Al, Cd, Cr, Cu, Ni, Se, ↓	32
Alfalfa	Field	S, ↑; N/S, ↓; Heavy metals, n.c.	38
Maize	Field	B, ↑	49
Maize	Pot	Ca, Mg, S, Fe,↑; Mn, Zn, ↓; P, n.c.	50

[1] ↑, increase; ↓, decrease; n.c., no change, relative to the untreated control unless the control is stated.

both Ca and S for plant nutrition in addition to having liming value.[3,12,21,24,25] Anhydrite in FBA does not neutralize acidity as does lime but it may ameliorate subsoil acidity in leaching environments.[29,30,41,42,473,44,45,46]

Mineralogical analysis indicated that anhydrous FBA is mainly composed of calcium oxide (CaO), anhydrite and quartz (SiO_2).[3,24] In comparison, slaked FBA contains gypsum ($CaSO_4.2H_2O$), ettringite [$Ca_6Al_2(SO_4)_3(OH)_{12}\cdot 26H_2O$], portlandite [$Ca(OH)_2$] and quartz.[11] Ettringite has very low solubility in water.[9,11] Aqueous extraction indicated that approximately 50% of S in a slaked FBA was water soluble gypsum, and the rest was ettringite, which has potential to reduce leaching loss of S that can be a major problem in some soils with low S retention capacity.[11]

Compared with Ca and S, the amounts of K and P in FBA are negligible. Magnesium and K appear as oxides, hydroxides, or sulfates in FBA. Phosphorus probably occurs as a form of calcium phosphate.[5] Since these elements are usually applied to soil in the form of lime or fertilizer, their presence in FBA is desirable.

The micronutrient content in FBA is low except for boron (B) that ranges from 36 to 1360 $\mu g\ g^{-1}$ in FBAs (Table 1). Boron is an essential mineral nutrient for all vascular plants. The functions of B are primarily extracellular, which relates to lignification and xylem differentiation.[47] FBA also contains small amounts of selenium (Se, 0.16–5.6 $\mu g\ g^{-1}$).[4,5] Selenium is an essential nutrient for animals but not for plants.[47] Due to the oxidizing conditions to which these elements are exposed during combustion, Fe, Mn, Cu and Zn probably occur as oxides in FBA, and B and Mo as borates and molybdates.[5] With more intensive cropping systems and the increasing use of purer, high analysis fertilizers, the need for micronutrient fertilization is becoming more apparent, and the presence of these nutrients in FBA may be beneficial.

2.2. Potentially Phytotoxic Elements

The amounts of micronutrients in FBA, except for B, are within the range of these elements usually found in soils.[5] Korcak & Kemper[20] observed no phytotoxic effects of micronutrients when FBA was used as a lime source even at very high rates (Table 2). However, the availability of micronutrients to plants depends not only on the amounts applied in FBA, but on the soil properties and the specific crop being grown. Therefore, their entrance into the human food chain via FBA application should be carefully monitored.[5]

Similar to other coal combustion by-products, B concentrations in FBAs (Table 2) are higher than that found in most soils, which average at 10 $\mu g\ g^{-1}$ (ranging from 2–100 $\mu g\ g^{-1}$).[51] Boron is not toxic in most agricultural soils, unless supplied in excessive amounts by fertilizers.[5] Typical B toxicity symptoms on mature leaves are marginal and/or tip chlorosis and necrosis. They reflect the distribution of B in shoots, following the transpiration stream.[47] Wang[52] reported that high rate application of FBA (up to 19185 kg ha^{-1}) in a field trial had not induced visual symptom of B toxicity in a white clover and ryegrass dairy pasture in a peat soil. However, Clark et al.[53]

reported that a soil mixed with more than 0.5% FBA resulted in significant dry matter reduction of maize in a pot experiment due to B toxicity. Crews & Dick[40] observed B toxicity symptoms in plant tissue when red oak is grown in soil applied with FBA at two times or higher the lime requirement rate. Therefore, care should be taken when applying FBA with high levels of B to sensitive crops such as cherry, peach, lupin, and kidney bean, especially when these crops are growing on sandy soils.[5]

Stout et al.[22] reported that the heavy metal content in FBA is comparable to that of more traditional agricultural soil amendment products, such as lime and gypsum. In addition, the oxide form of heavy metals in FBA makes them much less available to plants than the organic forms in sewage sludge.[22] A number of studies have confirmed that the accumulation of heavy metals in food crops or herbage grown on FBA treated soils has not caused any adverse effects on the health of animals consuming these diets.[21,26,27,28,31,48] However, any increase in the quantity or availability of these metals in soils should be carefully monitored using appropriate soil tests.[5]

3. Crop Responses to FBA Application

Substantial and highly significant plant growth or yield responses have been reported in a number of experiments involving either broadcast application or incorporation of FBA into the soil. The crops studied include corn,[3,7,29] peanuts,[38,46] fescue,[10] tobacco,[10,29] soybean,[30] apples,[12,17,20] peaches,[24] pecan,[25] red clover, tall fescue, oat, buckwheat,[10,21] white clover and ryegrass pasture.[52] Typically, yield increases were shown mainly to be associated with reductions in soil acidity and Al phytotoxicity[3,4,10,14,22,25] and improvements in Ca and S nutrition.[3,12,15,16 52] Improved rooting with depth has also been reported.[30,46]

Dick et al.[10] suggested that FBA application has the potential to improve growth and yield of pH sensitive crops such as alfalfa on acid soils. With a crop such as corn that is less pH sensitive, the potential yield benefit from FBA application may be less.[10]

4. FBA as a Liming Material

4.1. Effect on Soil Acidity

Research has focused on the potential for using FBA as liming material to reduce the acidity of low pH soils because it is an alkaline material with $CaCO_3$ equivalence ranging from 31% to 105% (Table 1). A number of studies have shown that addition of FBA can increase the pH of acid soils.[3,7,14,18,20,24,29,30,33,40] Application of FBA at the recommended liming rate effectively neutralized acidity in the zone of incorporation[10,14,15] and in some cases increased the pH of underlying soil.[30]

Marsh and Grove[29] evaluated the efficacy of an FBA derived from burning petroleum coke (dolomite as sorbent) in reducing soil acidity on a silt loam soil (pH_{water} 4.75) compared with limestone and $Ca(OH)_2$. Soil pH rose with increasing rates of application for all liming materials in the experiment. The increase in soil pH was less in the limestone treatment, primarily because the carbonates of the limestone reacted more slowly than the oxides-hydroxides associated with the equivalent liming weights of the other materials. Generally soil pH measurements are not accurate measures of the amount of soil acidity neutralized because soil pHs measured in water are sensitive to soil salt concentrations. Where higher rates of $Ca(OH)_2$ and FBA were used pH values may be lower than expected because of the high ionic strength in soil solution. Soil solution electrical conductivity (EC) remained constant across rates of limestone, but increased with rate for $Ca(OH)_2$, and especially for the FBA. This necessitated overcoming the salt effect by using 1 M KCl as the measurement medium to obtain pH values permitting comparisons among the amendments.[29]

Korcak and Kemper[20] investigated the long term effects of FBA applied at disposal levels on soil chemical properties. In 1980 FBA was applied as a soil 'cap' about 50 mm thick within the rows of an established, young apple orchard at rates up to 1.12×10^5 kg ha^{-1}. Eleven years after application, soil profile pH was significantly elevated by the FBA treatment to a depth of 660 mm. They suggested that horticultural utilization of FBA at relatively high rates of application may be a practical consideration for disposing of FBA. In orchard settings, FBA can be surface applied and remain in place for long periods. Results from their original apple study showed overall good growth and productivity of the trees.[18]

In an incubation study, Wang[52] found that FBA was as efficient as lime as a soil acidity ameliorant for a peat soil. Regression analysis indicated that the increase in soil pH in the incubated peat samples was linearly related to the application rates of FBA and lime ($R^2 > 0.9$, Figure 1). In a field

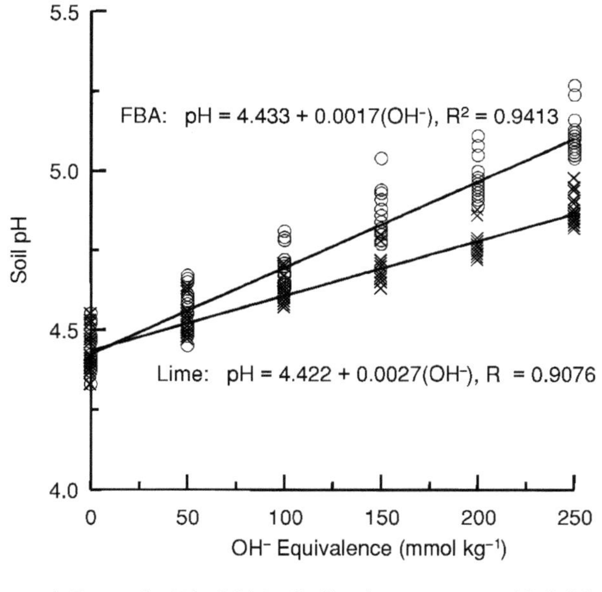

FIGURE 1. Peat soil pH in 0.01 M $CaCl_2$ after treatment with fluidized bed boiler ash (FBA) (O) and lime (×) in an incubation experiment.[52]

trial on peat soil, the surface applied FBA and lime treatments (up to equivalence of 10000 kg ha^{-1} $CaCO_3$) significantly increased the soil pH_{water} of only the surface 0–25 mm soil layer ($P< 0.01$) and pH in 0.01 M $CaCl_2$ at 25–50 mm layer, but had no effect on soil pH below 50 mm.[52] This indicated that there was a negligible movement of the bases generated from the surface application of FBA and lime. These results further indicated that although surface applied FBA was not able to increase the subsurface soil pH in acid peat, it had a similar liming value to lime. The amounts of FBA products needed to increase soil pH to a certain level, however, will vary from one FBA material to another, as FBA materials vary considerably in their neutralization potential (Table 1).[5]

Not only the $CaCO_3$ equivalence but also the particle size will affect the time required for FBA to be an effective liming material. Terman et al.[3] compared the neutralizing ability of FBA of different sizes with that of fine calcite. In an incubation experiment, fine $CaCO_3$, and coarse (< 3.4 mm) and fine (< 0.1 mm) FBA (33.1% CaO) were mixed with 1 kg of soil (Typic Paleudult) at rates of 0, 1, 2, 4 and 8 g kg^{-1} soil and incubated for five weeks. After one week, the soil pH response to increasing rates of fine $CaCO_3$ was linear reaching a maximum at pH 7.7–7.8. More than 95% of the maximum pH change was achieved with the 4 g kg^{-1} of fine $CaCO_3$. With a rate of 8 g kg^{-1}, coarse FBA increased soil pH linearly from 4.4 to 4.9 and fine FBA to 7.4. They suggested that the immediate liming value of coarse and fine FBA was about 8 and 47%, respectively, that of fine $CaCO_3$.

Edwards et al.[24] also confirmed that particle size of FBA affected its effectiveness in raising soil pH. Their incubation study indicated that FBA from a calcitic sorbent source (ground to pass a 0.15 mm sieve) was more effective in altering soil pH than FBA from a dolomitic source (90% passing through a 2 mm sieve and 25% passing through a 0.15 mm sieve). Both FBA sources, however, were adjusted to equal Ca contents. They suggested that the smaller particle size of the calcitic FBA was more reactive than the bigger sized dolomitic FBA. Thus when comparing FBA liming effectiveness, FBA particle size must be considered as well as the nature of the fuel and sorbent source.

4.2. Effect on Other Soil Chemical Properties

Besides increasing pH, FBA has been shown to modify other soil chemical properties. Changes in soil chemical composition result both from soil enrichment with soluble salts and major and trace elements present in the FBA plus from changes in soil pH and soil element solubility. In general FBA application causes water soluble concentrations of Al, Mn and Fe to decrease and the base status of acid soils to increase. The base status of soil below the zone of incorporation is commonly increased as well owing to transport of Ca^{2+} and Mg^{2+}. Mobility of Ca and Mg (when dolomite is used as the SO_2 sorbent) in FBA is the result of the greater solubility of $CaSO_4$ and $MgSO_4$ compared with that of $CaCO_3$/$CaMg(CO_3)_2$.

Korcak[16] reported that FBA treatments (at one and two times of the soil lime requirement on weight basis) greatly elevated extractable soil Ca compared to the untreated control. Soil extractable metals (Mn, Zn, Cu, Pb and Cd) were not significantly different, while the KCl extractable soil Al was significantly reduced by FBA treatment and the effect lasted throughout the four-year experimental period.[16] Similarly, of the studies to date there has been no evidence that land application at the recommended liming rate of the FBA would lead to elevated levels of potentially toxic trace elements in soil or water. Metal toxicity is likely to occur only if the lime sorbent or the fuel contains excessive levels of toxic elements.

Utilization of high application rates of FBA materials to agricultural land at disposal level is limited by the high alkalinity produced when the material is mixed within the soil.[3] However, Korcak[18] applied rates of FBA of up to 1.12×10^5 kg ha^{-1} as a within-row cap in an established apple orchard. Over six years, cumulative yields were increased in three of four cultivar/rootstock combinations. Both EC and exchangeable Ca in amended soil were significantly elevated, whereas exchangeable Mg decreased due to the low level Mg in FBA1 (0.8% MgO, Table 1). Similarly, extractable Fe, Mn, Zn and Cu plus exchangeable K were significantly decreased with increasing FBA. On a percent-saturation basis, the low and high FBA rates resulted in about 90% Ca saturation of the soil exchange capacity, compared to about 10% for the control soils. Concurrently, exchangeable Mg was significantly reduced by FBA application. Soil Ca/Mg was thus increased from 4.3 for the control soils to 24.3 at the high FBA rate. Foliar Mg levels decreased with time following high FBA application, indicative of the greatly increased soil Ca status and the potential to leach Mg from the surface horizons.

These apple plots were re-examined eleven years after the initial application and five years after the plots were ploughed.[20] X-ray diffraction patterns of remnant cemented pieces of the applied FBA showed that most of the original CaO has converted to calcium carbonate and some quartz, gypsum and ettringite were present. The formation of calcium carbonate with time is expected and it leads to the maintenance of a relatively high soil pH. Surface pH values from these plots after eleven years is about 7.6. Soil pH remained significantly higher in the top 660 mm of the treated sites compared to unamended sites. This has been accompanied by significant increases in extractable and total Ca, B and S with a concomitant reduction in extractable Mg. The higher B levels were not reflected in increased apple tissue concentrations of B throughout the first six years after application.[18] Therefore, application of FBA at relatively high rates, 1.12×10^5 kg ha^{-1}, can have long lasting effects on the soil environment. A potential area of concern is the imbalance created in the Ca to Mg ratio when evaluated on an extractable nutrient basis.[20] This situation can be corrected by application of $MgSO_4$ as a soil amendment or, for orchards, as a foliar spray. Alternatively, the use of dolomitic limestone as part of the absorbent during coal combustion would greatly increase the Mg content of the residue and lessen the potential for Ca:Mg imbalances.[20]

Therefore, for utilization in horticultural systems, FBA appears to be one of the better coal combustion products due to lower metallic concentration.[19] When surface applied and not incorporated, the pH and alkalinity effects are reduced, allowing the use of higher rates. When the situation is not suitable for surface applications that can remain in place for several years, application rates need to be at or near the lime requirement of the soil to prevent detrimental effects on crop growth.[19]

In addition to the application of FBA to soils, FBA has potential to be added to mixtures containing organic materials such as sewage sludges or manure composts. The high pH of the FBA materials would assist in reducing the bioavailability of heavy metals such as Cd. A greenhouse experiment on apple seedlings was conducted using FBA at application rates of one or two times the soil lime requirement mixed with composted sewage sludge at 0, 2.5 and 5.0×10^4 kg ha^{-1}.[13] The mixtures reduced Mn uptake from an indigenously high-Mn soil and increased apple tissue Ca concentration after 15 weeks.[19]

4.3. Effect on Plant Elemental Composition

Application of FBA not only changes the chemical properties of the soil but affects plant root uptake of nutrients from soil. The results from a number of studies showed that when FBA is applied to soil at the rate of its lime requirement, FBA treatment increases concentrations of Ca and S in plant tissue, but has no detrimental effect on plant growth.

When relatively large quantities of FBA were disposed of on an established apple (*Malus domestica* Borkh) orchard at two rates (9.2 and 36 kg m^{-2}) no nutrition related problems were noted for any of four tree types used.[18] Although leaf Ca concentrations were significantly increased by FBA treatment, little if any influence of applied FBA was noted on peel and flesh Ca concentrations throughout the study. One of the concerns of applying a high Ca product to orchard soils is the effect on the Mg status of the trees. The applied FBA, due to the calcitic limestone used in combustion, had a relatively low Mg content (0.8% MgO) and Korcak[18] found that leaf Mg was significantly lowered by the high FBA rate for the last two growing seasons. Both peel and fruit flesh Mg were little affected by treatments throughout the study.[18] Leaf Mg was generally within the acceptable range of 0.15–0.25% for apple trees.[54] Foliar concentrations of N, P, K, S, Mn, Fe, Ca, B, Al and Zn were generally little affected by treatment or apple tree type throughout the study and found to be within normal concentration ranges.[54] Similar results were noted for elemental concentrations in both fruit peel and flesh.[16,18]

Using FBA as a mine spoil amendment, Stehouwer et al.[32] found high rates application of FBA of up to 120 g kg^{-1} increased tall fescue growth, and chemical analysis of plant tissue indicated that concentrations of the main nutrients were generally within sufficiency ranges. Although tissue Ca, Mg and S were increased by FBA treatment, tissue concentrations of most trace elements were decreased.

Stout et al.[21] compared FBA with $Ca(OH)_2$ as a lime source in a greenhouse study with red clover (*Trifolium pratense* L.), tall fescue (*Festuca arundinacea* Schreb.), oat (*Fagopyrum sagittatum* Gilib.) and buckwheat (*Avena sativa* L.). The FBA treatment was applied at a rate to give sufficient alkali to adjust the pH of Westmoreland silt loam (Ultic Hapludalf, mixed, mesic) up to 6.5. Fluidized bed boiler ash treatments increased Ca levels in red clover by more than the $Ca(OH)_2$ treatment. This did not occur in the non-legumes. All species had higher tissue Mg and S concentrations when grown with FBA rather than $Ca(OH)_2$. In spite of the increased Zn loading, Zn levels in herbage decreased relative to the control as soil pH increased as a result of FBA application. Cochran et al.[26] also reported that application of FBA (6760 kg ha^{-1}) to a *Festuca arundinacea/Lotus corniculatus* sward on reclaimed coal mine spoil increased herbage Ca, Fe, Zn, S and Cu and lowered herbage Mg and P compared with dolomitic limestone (3380 kg ha^{-1}). In a similar experiment, Vona et al.[31] found there were no differences in micronutrient concentrations in herbage due to treatment. The above experiments and a number of other studies on the effect of FBA treatment on plant tissue elemental composition are summarized in Table 2. As expected, therefore, when Ca and Mg rich alkaline materials are added to soil-plant systems, the herbage concentrations of the base cations generally increase while concentrations of the transition metals decrease or show no change.

4.4. Implications to Food Chains

Fluidized bed boiler ash contains heavy metals that may be toxic to animals and humans if they entered into the human food chain. Research on possible effects of FBA application to soils on the nutritional quality of food crops has been conducted with a number of animal species, such as chicks (*Gallus gallus*),[55] rats (*Rattus rattus*),[48,56] pigs (*Sus scrofa*),[28] hamsters (*Mesocricetus auratus*),[56] sheep (*Ovis aries*),[27,31] and cattle (*Bos taurus*).[26] All these studies confirm that treatment of agricultural soils for up to six years with high levels of FBA (up to 9080 kg ha^{-1}) as a liming agent, or a fertilizer source, causes no measurable adverse effects on composition of food crops or herbage, or on the health and productivity of animals consuming diets produced on such soils.

Vona et al.[31] studied the effects of FBA (8960 kg ha^{-1}) on the health and performance of sheep grazing pastures and found that there were no differences in weight gains of either ewes or lambs within years due to treatment and no effect of treatment on blood hematocrit levels. Similarly, there were either no differences or small differences in the concentrations of macro- and micro-elements in the carcass, liver and kidney of ewes and lambs associated with treatment, and tissue concentrations of elements were within normal ranges.[31] Based on this experiment and other relevant studies, including rat (*Rattus rattus*),[48,56] pig (*Sus scrofa*),[28] and hamster (*Mesocricetus auratus*), Vona et al.[31] concluded that treatment of agricultural soils with high levels of FBA causes no adverse effects on the health of animals consuming diets produced on such soils.

Cochran et al.[27] evaluated nutrient utilization by sheep fed forage grown on soil treated with dolomitic limestone (11.26%

Mg, 4540 kg ha^{-1}) and FBA (0.08% Mg, 9080 kg ha^{-1}). Forage analysis indicated that the FBA treatment greatly increased Ca and S concentration but had no effect on Mg concentration in forage, compared with untreated control, whereas the limestone treatment increased Ca and Mg concentration. They found that apparent digestibility of cellulose was higher for forages grown with amendments than for the non-amended control and in addition the apparent digestibility of hemicellulose was higher for forage grown on the limestone treatment than on the FBA treatment. Apparent absorption of S was higher for lambs on the FBA treatment than for those on the limestone treatment. These differences were related to differences in mineral concentrations of the forages. Therefore, Cochran et al.[27] concluded that FBA treatment may have enhanced utilization of Ca and S and improved digestibility of some fiber components by lambs.

4.5. Acid Mine Spoil Reclamation

Large volumes of FBA are required when FBA is used to reclaim acidic surface mine spoils.[3,4,10,23] Many mine spoils are high in pyrite (FeS_2), which oxidizes to form sulfuric acid, resulting in highly acidic (pH $\leq$ 3.5) spoil.[8] Therefore, it can be difficult or impossible to establish vegetation on these sites without altering the strongly acidic environment. Lack of vegetative cover causes these areas to be highly erosive and drainage waters from such sites can cause severe off-site environmental damage because of acidity, high salt content and sedimentation.[57] The application rate of alkaline amendment needed for reclamation of acidic mine spoils can be five to ten times greater than used for agricultural soils.[4] Furthermore, the use of FBA on these sites is attractive because it helps solve the FBA disposal problem while alleviating some of the problems associated with another coal waste product.[8]

Stehouwer et al.[4] investigated the efficacy of FBA as an amendment for acidic mine spoils and the potential for adverse environmental impacts from salts and trace elements when FBA was applied at high rates needed for mine spoil reclamation. In their eight month column leaching and greenhouse study, they found that FBA (60% $CaCO_3$ equivalence) was effective in neutralizing acid conditions in the mine spoils when FBA was added to spoil at 30 to 60 g kg^{-1} spoil weight. With FBA rates at greater than 30 to 60 g kg^{-1} spoil, high leachate pH and Ca and Mg concentrations indicated some potential for amelioration of phytotoxic conditions below the zone of incorporation. In addition, leachate EC, dissolved organic C, and S, tended to increase with increased FBA rate. Fluidized bed boiler ash also reduced leachate concentrations of Al, Fe, Mn and Zn that are frequently phytotoxic in acid soils. Decreased solubility and mobility of these metals would also improve surface and drainage water quality. They found that when FBA was applied at 160 g kg^{-1} spoil or less, leachate concentrations of trace elements of environmental and regulatory concern remained very small. In fact, most were less than primary drinking water standards. Improved plant growth (tall fescue) was observed when FBA was mixed with acid spoils.[10] Therefore, Stehouwer et al.[4] concluded that FBAs are very suitable for mine spoil amendment if amendments are limited to amounts that will not cause excessively high pH, or, phytotoxic salt concentrations. Provided leachates are not highly alkaline, there is little potential for adverse effects on water and soil quality from trace elements. This agrees to the findings by Sidle et al.[23] who used lower rates of FBA in their column leaching experiment.

5. Potential Use of FBA as a Subsoil Acidity Ameliorant

5.1. Background

In many parts of the world acid conditions in subsoil horizons pose a major limitation to the successful production of many important crop species.[38,43,58,59] Furthermore, intensive cropping accompanied by high nitrogen use may lead to subsurface acidification of soils that are not naturally acidic.[60,61] It is becoming more common therefore that subsurface soil acidity constrains plant rooting depth. The problem is often manifested as drought intolerance and/or poor use of nutrients associated with shallow rooting.[43] It is a common practice to incorporate lime into the plough layer to raise the pH of acid soils and neutralize and precipitate plant-toxic Al. Unfortunately, downward movement of base after liming is extremely slow because OH^- and HCO_3^- ions are consumed by acidity in the topsoil and no mobile anions remain. In addition the increased pH generates more negative charge on soil particle surfaces, which acts to retain more Ca^{2+} ions in the surface soil.[62] The most effective procedure for promoting root growth in the acidic subsoil is to mix lime throughout the entire soil volume in which roots are desired.[63] However, in most situations this procedure is not economically feasible.[44]

The expense and difficulty of mixing lime into the subsoil have led researchers to examine surface applied amendments, such as gypsum, in the amelioration of subsoil acidity. Reeve & Sumner[64] found that surface incorporated gypsum moved through a simulated profile of a Natal Oxisol and reduced the level of exchangeable Al in the subsoil more effectively than surface incorporated lime. Since then, a number of studies have been conducted on the surface incorporation of gypsum and gypsiferous waste materials.[42,44] In all the experiments that have been conducted, it is clear from the cumulative responses obtained that very appreciable yield increases of maize and soybeans,[30,65,66] alfalfa,[46] wheat,[67] sorghum,[66] coffee, rice, beans, peaches, cotton[42] can be obtained.

Increased yields of crops grown on acid soils with surface incorporated gypsum have been attributed to increased root length density in the subsoil.[65] Confirmatory evidence for the benefits of deeper rooting following gypsum use is to be found in water extraction pattern down the profile. Sumner & Carter[68] studied the effect of gypsum on alfalfa root distribution in the profile of an Ultisol. Gypsum treatment resulted in substantial penetration of roots to deeper than 1 m, whereas in the control plots few roots penetrated beyond 0.6 m. On both a mass and length basis, the quantity of roots in the subsoil of the gypsum treatment exceeded that of the control three to five fold, which is reflected in the alfalfa yield response.[44] Reports corroborating this type of response have been published for maize, apples, beans and rice.[42] In all cases, the gypsum treatment increased the Ca level in the subsoil, decreased the Al level and promoted root growth, so the roots were able to extract more water from subsoil and had lower leaf water potential than in control treatments. This increased water uptake allows the plant to better overcome drought stress and therefore accumulate dry matter more successfully on the gypsum treated soils.[42]

In a number of cases, however, there is no yield response to gypsum application,[59] and a relative decrease in root growth has been observed at intermediate depths in the subsoil.[65,68] At present there is no entirely satisfactory explanation for this effect.

5.2. Mechanisms of Subsoil Acidity Amelioration by Using Gypsum

Although gypsum does not have the ability to neutralize soil acidity in the same way as limestone, it does produce a number of effects that can partially or fully mitigate the severity of several detrimental aspects of soil acidity. When a $CaSO_4$ front is leached into subsurface soil, the ameliorative effect is due to the increased supply of Ca^{2+} and the decrease in toxic Al^{3+}. Possible mechanisms are: the "self-liming effect",[64] increased subsurface soil Ca^{2+},[42] decreased monomeric Al saturation,[41] increasing the ionic strength of the soil solution,[69] precipitation of basic aluminum sulfate minerals,[70] $AlSO_4^+$ ion pair formation,[69] and displacement and leaching of Al^{3+}.[71]

The "self-liming effect"[64] is based on the adsorbed SO_4^{2-} displacing OH^- from hydrous oxide surfaces, followed by hydrolysis and precipitation of exchangeable Al as in Equations 4 and 5.[43]

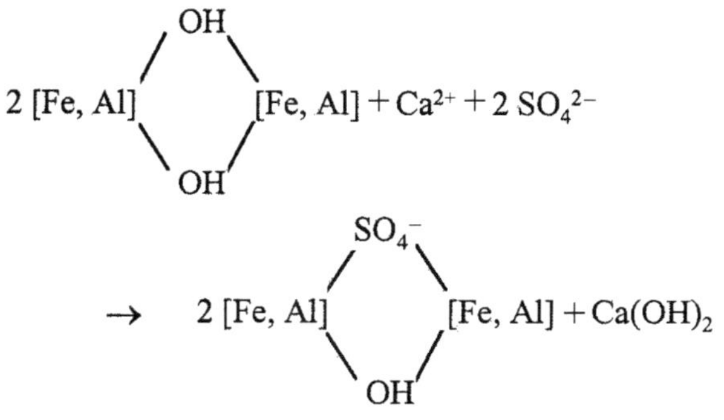

(4)

$$2Al^{3+} + 3Ca(OH)_2 \rightarrow 2Al(OH)_3 + 3Ca^{2+} \quad (5)$$

Sulfate adsorption by soils is associated primarily with Fe- and Al-oxides and allophane.[72] Therefore, the mineralogical composition of a soil will influence the effectiveness of gypsum as an ameliorant for subsurface soil acidity.

5.3. Criteria for Identification of Gypsum-Responsive Soils

In order to make efficient use of gypsum, it is necessary to be able to predict which soils will respond to its application. Characteristics of soil below the plough layer which affect the likelihood of crop rooting improvement include the subsoil exchangeable Ca^{2+} level, the exchangeable Al level and the response of soil minerals to sulfate addition.[41] Where exchangeable Ca^{2+} levels are less than 0.1 cmol $_{(+)}$ kg^{-1}, there is a high probability of root growth responses to gypsum, regardless of Al levels, because of root growth response to Ca^{2+} as a nutrient. At higher levels of Ca^{2+} where the nutritional needs of the plant roots are met, the probability of response to gypsum is controlled more strongly by the percentage Al saturation of effective cation exchange capacity (ECEC) of the soil.[41] Ritchey et al.[66] reported that for low ECEC soils of central Brazil, gypsum responses were usually high for Al saturation levels of 65% or greater and low where Al saturation was less than 35%.

Ritchey et al.[41] and Shainberg et al.[42] found that mineralogy of soils and resulting charge characteristics may influence the reactions of gypsum and, therefore, may vary the effectiveness of gypsum in alleviating acidic soil infertility in various soils. Determining mineralogical composition is, however, more expensive and time-consuming than measuring extractable Ca^{2+} and Al^{3+} levels which are routinely assessed in soil analytical laboratories. To avoid this problem, Sumner[44] published a simple method to predict whether a subsoil will self-lime when gypsum is applied. It is based on the capacity of the soil to sorb SO_4^{2-} and raise the pH through displacement of hydroxyl groups. If the pH of the supernatant of a soil shaken intermittently overnight with 0.005 M $CaSO_4$ (10 g soil to 25 ml solution) is higher than the pH when shaken with 0.005 M $CaCl_2$, and the EC of the 0.005 M $CaSO_4$ solution is lower after shaking than before (indicating sorption of Ca^{2+} and SO_4^{2-}), then the subsoil is considered a good candidate for gypsum treatment.

An alternative test, which encompasses both self-liming and other beneficial soil chemical changes, is the bioassay method that was originally used to detect plant responses to Ca^{2+} deficiency and Al toxicity.[66] In the biological method, four-day root lengths of seedlings of the actual cultivar being used on the farm are compared in subsoil with and without the addition of 1% by volume of gypsum. A leaching treatment can be applied if appropriate. Where growth in the gypsum treated soil is improved by 15% or less, limited benefit is expected from field use of gypsum. Where the improvement is 30% or more, gypsum application may be beneficial.[66]

Soil physical properties can also significantly affect the probability of success for using gypsum to ameliorate subsoil acidity. In soils with impermeable pan layers, root penetration will be limited by physical barriers regardless of gypsum application.[41]

5.4. FBA—A Prospective Ameliorant for Subsoil Acidity

Topsoil incorporated gypsum is able to increase base status and reduce exchangeable Al in the subsoil much more effectively than lime. At the topsoil level, however, gypsum is less effective than lime in totally eliminating exchangeable Al.[73] When gypsum is used as a subsoil acidity amendment, it is necessary to incorporate lime into the plough layer to mitigate the acidity in the topsoil.[42] Because FBA contains both 'lime' and gypsum, it can not only neutralize topsoil acidity, but also act as an effective ameliorant for subsoil acidity. The 'lime' in FBA is CaO or $Ca(OH)_2$ in slaked FBA,[11,21] which has the advantage of being quicker reacting with topsoil acidity than ordinary lime and the Ca^{2+} and $SO_4{}^{2-}$ from gypsum can readily move into subsoil to ameliorate Al toxicity there.

In a leaching study, Wang et al.[45] incorporated lime, FBA and flue gas desulfurization gypsum (FGDG) in the top 50 mm of repacked columns of a variable charge allophanic soil (Patua) in New Zealand. Topsoil incorporated FBA ameliorated top and subsurface soil acidity through liming and the "self-liming effect", respectively. The soil solution pH of the top and subsurface layers of the Patua soil were raised to pH 6.40 and 5.35 respectively, by the FBA treatment, compared with pH 4.80 and 4.65 in the control treatment. Consequently phytotoxic monomeric Al concentrations in soil solution of the FBA treatment were reduced to less than 0.1 μM Al, compared with that of 8–64 μM Al in the untreated control. Although FGDG had a similar "self-liming effect" on the subsurface of the allophanic soil, it had no effect on the topsoil acidity. In contrast, the lime treatment only reduced the topsoil acidity (Figure 2). In a pot experiment, Wang et al.[46] found that the FBA and FGDG treatments (top 50 mm soil incorporated) significantly improved alfalfa (*Medicago sativa.* L) root penetration into the subsurface of the allophonic soil ($P < 0.05$), whereas the lime treatment had no effect on root elongation in the acidic subsurface of the soil. In these experiments, no "self-liming effect" was observed in the subsurface of an ultisol, which is dominated by permanently charged clay minerals.[45,46]

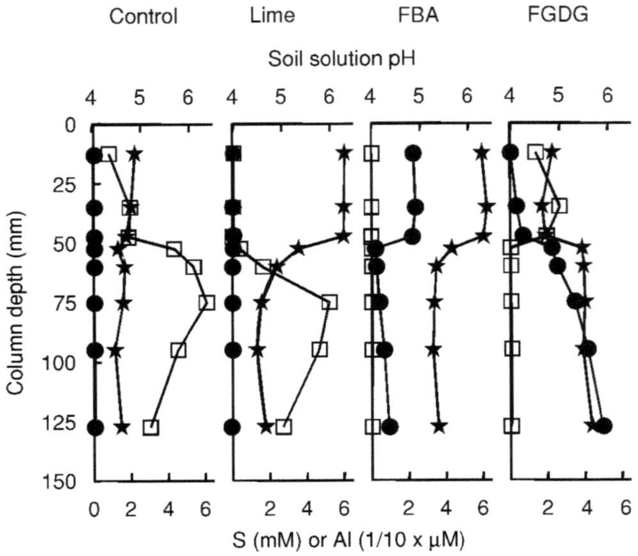

FIGURE 2. Concentrations of labile monomeric Al (□), sulfate (λ), and pH (H) in soil solutions of the sections sampled from leached columns of a variable charged allophanic soil, after lime, fluidized bed boiler ash (FBA) and flue gas desulfurization gypsum (FGDG) were incorporated into the top 50 mm layer (adapted from Wang et al.).[45]

6. FBA as a Sulfur Source

Sulfur is essential to plant nutrition. Sulfur ranks in importance with N as a constituent of the amino acids cysteine, cystine, and methionine in proteins that account for 90% of S in plants. It is also involved in the formation of oil in crops such as peanut (*Arachis hypogaes* L.), soybean [*Glycine max* (L.) Merr.], flax (*Linum usitissimum*), and rapeseed (*Brassica campestris*).[74] Since 1950s, S deficiencies have been reported with increasing frequency throughout the world, including New Zealand.[75,76]

The reasons given for the increasing S deficiencies world wide are: (1) the shift from low analysis to high analysis fertilizers containing little or no S, (2) use of high-yielding crop varieties that remove greater amounts of S from the soil, (3) reduced industrial S emission into the atmosphere due to pollution-control measures (e.g. production of FBA, and decreased use of high S content fossil fuels), (4) decreased use of S in pesticides, and (5) declining S reserves in soil due to erosion, leaching, and crop removal. Increased consumption of sulfur free, high analysis fertilizers is seen as the most important reason for the increasing S deficiency worldwide.[77] Thus, with increasing S deficiencies worldwide, some industrial by-products such as waste gypsum and FBA deserve serious consideration for agricultural applications where mined gypsum has traditionally been used.

Because the S in FBA is mainly in the form of anhydrite ($CaSO_4$) or gypsum ($CaSO_4.2H_2O$) after hydration, FBA and gypsum should have, as an S source, similar effect on crops in S deficient soils. Gypsum, including industrial by-products, such as phosphogypsum, has been successfully used as a S fertilizer to increase the yields of winter wheat,[78] rice[79] and pasture.[52,80] Use of gypsum as a source of S has been extensively reviewed by Alcordo & Rechcigl.[81] To date, data on the use of FBA as a S fertilizer is limited, although Terman et al.[3] demonstrated that FBA is an effective S source for growing corn and peanuts.

Approximately 50% of S in slaked FBA is present as sparingly soluble ettringite, which can act as a slow-release S fertilizer.[52] Therefore, slaked FBA shows potential to be used as a basal S fertilizer in soils prone to severe leaching, which are common in some high rainfall areas of New Zealand and in soils with low anion retention capacities.[52] Results from a field study in the peat soil indicated that ettringite-S in slaked FBA was less affected by leaching loss than gypsum and resulted in longer-term pasture yield S responses to FBA treatments.[52] This was supported by herbage S analysis where, 8 months after application, herbage S contents in the high rates of FBA treatments (6616 and 26462 kg ha^{-1}) were significantly higher than the control.

7. Effect of FBA on Soil Physical Properties

Surface applied phosphogypsum has been successfully used to reduce soil dispersion and erosion.[10] Fluidized bed boiler ash has been utilized to increase water infiltration into soil, reduce soil surface sealing and control soil erosion because gypsum is one of the main components in FBA.[35,36,37] It is possible to use FBA to ameliorate the restrictive nature of fragipan horizons.[41] With dissolution, gypsum increases the electrolyte and Ca concentrations of the soil solution and runoff, thus reducing swelling of soil and dispersion of clay particles.[35] In the case of FBA, which is rich in $CaSO_4$ and CaO, the primary effect will be an increase in soil pH. This could enhance dispersion, particularly of variable-charge soils, and the electrolytes released from the FBA may be insufficient to flocculate clay particles. Therefore FBA may actually increase erosion of variable-charge soils. For soils with a predominance of permanent charge colloids, such as soils with 2:1 type clays, increases in pH have less effect on surface negative charge. Therefore, FBA can be utilized for controlling soil erosion on agricultural land and construction sites, especially those with permanently charged clays.[35,37] Reichert & Norton[35] measured the steady-state infiltration rates (I_s) for five soils, with and without addition of 5000 kg ha^{-1} surface applied FBA (74% CaSO4, 23% CaO, total dissolved salts 4700 mg litre^{-1}). They found that surface sealing and erosion were significantly reduced by the surface applied FBA on all studied soils, but with a lesser effectiveness on the illitic and smectitic soils. For infiltration, FBA was most effective on the smectitic soil, except when swelling is high. The values of I_s for the control ranged from 1.8 to 5.8 mm h^{-1} and were increased 3.6 to 5.0 fold with the FBA. They explained that the increased electrolyte concentration reduced swelling and enhanced flocculation and possibly sedimentation of soil clays, thus decreasing sediment concentration in the runoff and sediment yield. Reichert et al.[37] suggested that FBA has great potential for reducing sealing and erosion on permanent charge soils. The FBA effectiveness on variable charge soils was dependent on soil pH, its buffering capacity and flocculation.[37] Care needs to be taken, however, because compositionally, FBA is similar to Portland cement.[9,11,18] When applied at high rates as a soil amendment, FBA exhibits strong pozzolanic properties and can limit water infiltration and root penetration.[4,10] The pozzolanic nature of FBA can create large cemented 'chunks' of the material in the field. Pozzolans are siliceous materials containing Al that react with CaO to form slow-hardening cement. This pozzolanic property of FBA can benefit apple orchards,[18] where rates of FBA up to 1.12×10^5 ha^{-1} were applied as a within-row cap in an established apple orchard. The surface applied material formed porous cement that prohibited weed growth for up to four years after application, but had no adverse effects on crop yields.

Wang[52] tested the effectiveness of using FBA to ameliorate water repellency in an acidic peat soil, where severe water repellency is a common problem in dry summer and autumn periods. It was found that normal liming and S fertilizer rates of FBA application had negligible effect on water repellency and water infiltration into dry peat soils. However, a high rate (up to 19185 kg ha^{-1}) of surface dressed FBA applied to a peat soil under pasture can reduce the water repellency of the surface soil and increase the rate of water infiltration. This may be caused by the high alkalinity of the applied FBA, which modified the hydrophobic nature of the peat soil by removing fatty acids from the soil particle surface.

8. FBA as a Phosphorus Inactivator

A number of studies were conducted to test the potential of using FBA to reduce solubility and mobility of P in soils with high P concentration and in P-rich organic wastes.[82,83,84] Calcium in FBA acts as a P fixer by reacting with soluble P to form insoluble Ca-phosphates at neutral and alkaline conditions.[85] In a packed column leaching study on dairy manure, Elrashidi et al.[86] reported that P in leachate in FBA treatment was reduced by more than 80% compared with the control treatment containing dairy manure only. In a laboratory incubation and extraction study, Dou et al.[83] found that FBA applied at a rate of 400 g kg^{-1} reduced readily soluble P by 50–60% in P-rich manures including dairy, swine and broiler litter.

Stout et al.[84,87,88] reported that FBA has consistently reduced concentrations of water-soluble phosphorus in a number of soils with high concentration of soluble P. They found that amendment with FBA resulted in a shift from readily available resin P and less available NaOH-extractable Fe- and Al-bound P fractions to HCl-extractable Ca-bound P. They suggest that the neutralizing capacity of FBA is the primary factor in shifting a sizable portion of the soil P to the Ca-bound P fraction.[87] However, considering the high treatment rates (10 g kg^{-1} or higher) required to significantly reduce readily available P in soils, Codling et al.[82] suggest that FBA may not be suitable for field application to sequester P in high P soils, due to the large increase in soil alkalinity.

9. Sustainability of Soil Fertility in Relation to FBA Application

As more stringent emission regulations for coal-fired power plants are enforced, it is likely that the use of fluidized bed combustion and other desulfurization technologies will increase dramatically and will result in the production of increasing amounts of FBA and similar by-products. If vast amounts of FBA and other coal combustion by-products are produced, however, it seems unlikely that there will be sufficient demand for liming materials in agricultural production to consume these amounts and there is considerable interest in the use of land for disposal of these by-products by application at rates much higher than those required to lime soil.[8] Although very high rates of FBA applied (up to 1.12×10^5 kg ha^{-1}) to orchards improved the apple productivity and had no detrimental effect on the soil,[18,20] there are concerns that this use of FBA harms the quality of the soil.[34]

Some reports reflect considerable interest in the concept of soil quality as it relates to sustainable productivity in agricultural systems and in the development of indices to assess changes in such quality.[89] It has been suggested that measurement of changes in soil enzyme activities may provide useful measures of the changes in soil quality.[90] McCarty et al.[33] investigated the influence of FBA (up to 22400 kg ha^{-1}) and $CaCO_3$ on soil pH urease, phosphatase, arylsulfatase and dehydrogenase activities in soil. They found that the effect of both liming materials on the enzyme activities studied was largely due to their influence on soil pH, and FBA functioned as soil liming material in a manner similar to that of $CaCO_3$ and seemed to have little adverse effect on enzyme activities.[33]

The importance of maintaining soil organic matter for the maintenance of good soil quality is well established because of its central role in establishing the physical structure, fertility and biological activity of soil.[89] For example, organic N in soil provides the majority of N for plant growth and it seems obvious that destruction of this organic N would degrade the fertility of soil. In addition, it has been suggested that monitoring the rate of various N transformations in soil can provide useful measures of soil quality and indicate the changes in the quality that can result from environmental disturbances.[90] To determine the influence of FBA on the transformation and fate of soil N and to assess the potential impact of land disposal of FBA on soil quality, Siddaramappa et al.[34] studied the effect of FBA (at the rates of 22.5, 45, 90 and 180 $\times$ 10^3 kg ha^{-1}) on mineralisation and volatile loss of N in soil. The results showed that FBA treatments, except at 22.5×10^3 kg ha^{-1}, caused substantial mineralisation of organic soil N and volatile loss of this N as NH_3 due to the highly alkaline environment (up to pH 12.8). The authors suggested that such extreme alkaline conditions caused chemical degradation and volatile loss of as much as 10% of the organic N in soil. Therefore, improper disposal of FBA on land can have a substantial negative impact on the quality of soil organic matter.[34]

10. Conclusions and Research Priorities

Increasing concern about emissions of SO_2 in flue gas and associated problems with acid rain worldwide has promoted the development of desulfurization technologies that reduce these emissions and result in the production of new by-products, such as FBA. As more stringent emission regulations for high-S coal fired industries are enforced, it is likely that the use of fluidized bed boilers will increase dramatically and will result in the production of increasing amounts of FBA materials and other similar by-products.

This review chapter has indicated that carefully managed FBA application to soils can ameliorate soil acidity and S deficiency without causing adverse effects. Fluidized bed boiler ashes are sufficiently variable in composition and their reaction with different soil types that recommendations for agronomic use of FBA should be material- and site-specific.

Further studies are required to confirm the conditions under which products like FBA, which contain both lime and gypsum, may reach their full potential to ameliorate both surface soil and subsurface soil acidity. These are likely to be on previous unfertilized allophanic soils and those fertilized with S free phosphate fertilizers.

When using gypsiferous materials, however, the soil available Mg and K status should be monitored to avoid a gypsum-Ca induced deficiency of these nutrients. Development of a system for recommendation of FBA application rates to ameliorate top and subsoil acidity plus provide S nutrition would provide farmers with the necessary information to stimulate greater agronomic use of FBA materials.

In the environmental protection area, more information is required on the suitability of FBA as an amendment to reduce the concentration of P in drainage waters from excessively fertilized soils and sewage sludges under field conditions. Similarly more widespread use of FBA to control acid mine drainage will only come if greater information on FBA properties is made available to mining companies.

References

1. Fennelly, P. F., Klemm, H., Hall, R. R. and Durocher, D. F., Coal burns cleaner in a fluid bed. *Environ. Sci. Tech.,* 11, 244, 1977.
2. Lawson, N. W., Fluidized bed boilers. in *Proceedings of coal research conference*, Coal Research Association of New Zealand, Wellington, New Zealand, 1987, 1.
3. Terman, G. L., Kilmer, V. J., Hunt, C. M. and Buchanan, W., Fluidized bed boiler waste as a source of nutrients and lime. *J. Environ. Qual.,* 7, 147, 1978.
4. Stehouwer, R. C., Sutton, P. and Dick, W. A., Minespoil amendment with dry flue gas desulfurization by-products: plant growth. *J. Environ. Qual.,* 24, 861, 1995.
5. Stout, W. L. and Priddy, W. E., Use of flue gas desulphurisation (FGD) by-product gypsum on alfalfa. *Commun. Soil Sci. Plant Anal.,* 27, 2419, 1996.
6. Ruth, L. A., Regeneration of $CaSO_4$ in FBC. in*Proceedings of Fourth International Conference on Bed-bed Combustion*, The MITRE Corporation, McLean, AV, USA. 1975, 425.
7. Stehouwer, R. C., Dick, W. A., Sutton, P., Acidic soil amendment with a magnesium-containing fluidized bed combustion by-product. *Agron. J.,* 91, 24, 1999.

8. Carlson, C. L. and Adriano, D. C., Environmental impacts of coal combustion residues. *J. Environ. Qual.*, 22, 227, 1993.
9. ACI Committee 225, *Guide to the selection and use of hydraulic cements*. ACI Manual of Concrete Practice 1987, Part 1, ACI 225Rangitata-85, 1987.
10. Dick, W.A., Hao, Y., Stehouwer, R.C., Bigham, J.M., Wolfe, W.E., Adriano, D., Beeghly, J.H., and Haefner, R.J., Beneficial uses of flue gas desulfurization by-products: examples and case studies of land application. in *Land application of agricultural, industrial, and municipal by-products*, Fower, J. F. and Dick, W. A. Eds., SSSA Book Series No. 6. SSSA, Madison, WI., 2000, 505.
11. Wang, H. L., Hedley, M. J. and Bolan, N. S., Chemical properties of fluidised bed boiler ash relevant to its use as a liming material and fertiliser. *N. Z. J. Agric. Res.*, 38, 249, 1995.
12. Korcak, R. F., Fluidized bed material as a calcium source for apples. *HortScience* 14, 163, 1979.
13. Korcak, R. F., Effects of applied sewage sludge compost and fluidized bed material on apple seedling growth. *Commun. Soil Sci. Plant Anal.*, 11, 571, 1980.
14. Korcak, R. F., Fluidized bed material as a lime substitute and calcium source for apple seedlings. *J. Environ. Qual.*, 9, 147, 1980.
15. Korcak, R. F., Effectiveness of fluidized bed material as a calcium source for apples. *J. Am. Soc. Hort. Sci.*, 107, 1138, 1982.
16. Korcak, R. F., Utilization of fluidized bed material as a calcium and sulfur source for apples. *Commun. Soil Sci. Plant Anal.*, 15, 879, 1984.
17. Korcak, R. F., Effect of coal combustion waste used as lime substitutes on nutrition of apples on three soils. *Plant Soil*, 85, 437, 1985.
18. Korcak, R. F., Fluidized bed material applied at disposal levels. *J. Environ. Qual.*, 17, 469, 1988.
19. Korcak, R. F., High-gypsum by-products as soil amendments for horticultural crops. *HortTechnolog*, 3, 156, 1993.
20. Korcak, R. F. and Kemper, W. D., Long-term effects of gypsiferous coal combustion ash applied at disposal levels on soil chemical properties. *Plant Soil*, 154, 29, 1993.
21. Stout, W. L., Hern, J. L., Korcak, R. F., Carlson, C. W., Manual for applying fluidized bed combustion residue to agricultural lands. U. S. Department of Agriculture, Agricultural Research Service, ARS-74. Washington, D.C., USA, 1988.
22. Stout, W. L., Sharpley, A. N. and Landa, J., Effectiveness of coal combustion by-products in controlling phosphorus export from soils. *J. Environ. Qual.*, 4, 1239, 2000.
23. Sidle, R. C., Stout, W. L., Hern, J. L. and Bennett, O. L., Solute movement from fluidized bed combustion waste in acid soil and mine spoil columns. *J. Environ. Qual.*, 8, 236, 1979.
24. Edwards, J. H., Horton, B. D., White, A. W. Jr. and Bennett, O. L., Fluidized bed combustion residue as an alternative liming material and Ca source. *Commun. Soil Sci. Plant Anal.*, 16, 621, 1985.
25. Edwards, J. H., White, A. W. Jr. and Bennett, O. L., Effects of fluidized bed combustion residue on pecan seedling growth and nutrient content. *Commun. Soil Sci. Plant Anal.*, 16, 639, 1985.
26. Cochran, M. A., Fontenot, J. P., Perry, H. D. and Allen, V. G., Effect of repeated applications of fluidized-bed combustion residue to reclaimed surface mined pastures, in *Proceedings of the 16th International Grassland Congress*. Versailles, France; Association Francaise pour la Production Fourragere, Nice, France, 1989, 1247.
27. Cochran, M. A., Perry, H. D., Fontenot, J. P. and Allen, V. G., Nutrient utilization by sheep fed forage grown on soil treated with fluidized-bed combustion residue. *J. Anim. Sci.*, 69, 318, 1991.
28. Whitsel, T. J., Reid, R. L., Stout, W. L., Hern, J. L. and Bennett, O. L., Quality of diets with fluidised bed combustion residue treatment. II. Swine trials. *J. Environ. Qual.*, 17, 556, 1988.
29. Marsh, B. H. and Grove, J. H., Plant and soil composition as affected by an alternative lime source containing sulfate. *Soil Sci. Soc. Am. J.*, 56, 1831, 1992.
30. Marsh, B. H. and Grove, J. H., Surface and subsurface soil acidity: soybean root response to bearing-bearing spent lime. *Soil Sci. Soc. Am. J.*, 56, 1837, 1992.
31. Vona, L. C., Meredith, C., Reid, R. L., Hern, J. L., Perry, H. D. and Bennett, O. L., Effects of fluidised bed combustion residue on the health and performance of sheep grazing hill pastures. *J. Environ. Qual.*, 21, 335, 1992.
32. Stehouwer, R. C., Sutton, P, Fowler, R. K. and Dick, W. A., Minespoil amendment with dry flue gas desulfurization by-products: element solubility and mobility. *J. Environ. Qual.*, 24, 165, 1995.
33. McCarty, G. W., Siddaramappa, R., Wright, R. J., Codling, E. E. and Gao, G., Evaluation of coal combustion by-products as soil liming materials: their influence on soil pH and enzyme activities. *Bio. Fertil. Soils.*, 17, 167, 1994.
34. Siddaramappa, R., McCarty, G. W., Wright. R. J. and Codling, E. E., Mineralisation and volatile loss of nitrogen from soils treated with coal combustion by-products. *Biol. Fertil. Soils*, 18, 279, 1994.
35. Reichert, J. M. and Norton, L. D., Fluidized bed bottom-ash effects on infiltration and erosion of swelling soils. *Soil Sci. Soc. Am. J.*, 58, 1483, 1994.
36. Reichert, J. M. and Norton, L. D., Fluidized bed combustion bottom-ash effects on infiltration and erosion of variable-charge soils. *Soil Sci. Soc. Am. J.*, 60, 275, 1996.
37. Reichert, J. M., Norton, L. D. and Huang, C., Sealing, amendment, and rain intensity effects on erosion of high-clay soils. *Soil Sci. Soc. Am. J.*, 58, 1199, 1994.
38. Stout, W. L., Daily, M. R., Nickeson, T. L., Svendsen, R. L. and Thompson, G. P., Agricultural uses of alkaline fluidized bed combustion ash: case studies. in *Proceedings of the 1995 International Ash Utilization Symposium*, Lexington, Kentucky, 1995.
39. Zaifnejad, M., Ritchey, K. D., Clark, R. B., Baligar, V. C. and Martens, D. C., Chemical properties of acid soil treated with coal combustion by-products and leached. *Commun. Soil Sci. Plant Anal.*, 27, 2783, 1996.
40. Crews, J.T., and Dick, W.A., Liming acid forest soils with flue gas desulfurization by-product, growth of Northern red oak and leachate water quality. *Environ. Pollut.*, 103, 55, 1998.
41. Ritchey, K.D., Clark, R.B., Elrashidi, M.A., and Baligar, V.C., Properties and examples of beneficial use of gypsumlike by-products. in *Land application of agricultural, industrial, and municipal by-products*, Fower, J.F. and Dick, W.A. Eds., *SSSA Book Series No. 6.*, SSSA, Madison, WI, 2000, 537.
42. Shainberg, I., Sumner, M. E., Miller, W. P., Farina, M. P. W., Pavan, M. A. and Fey, M. V., Use of gypsum on soils: a review. *Adv. Soil Sci.*, 9, 1, 1989.
43. Sumner, M. E., Gypsum and acid soils: the world scene. *Adv. Agron.*, 51, 1, 1993.

44. Sumner, M. E., Amelioration of subsoil acidity with minimum disturbance. in *Subsoil management techniques*, Jayawardane, N. J. and Steward, B. A. Eds., Lewis Publishers Inc., Boca Raton, FL, USA. 1994, 147.
45. Wang, H., Hedley, M. J., Bolan, N. S., Horne, D. J., The influence of surface incorporated lime and gypsiferous by-products on surface and subsurface soil acidity. I. Soil solution chemistry. *Aust. J. Soil Res.*, 37, 165, 1999.
46. Wang, H., Hedley, M. J., Bolan, N. S., Horne, D. J., The influence of surface incorporated lime and gypsiferous by-products on surface and subsurface soil acidity. II. Root growth and agronomic implications. *Aust. J. Soil Res.*, 37, 181, 1999.
47. Marschner, H., *Mineral nutrition of higher plants*, Academic Press, London, UK, 1986.
48. Cahill, N. J., Reid, R. L., Head, M. K., Hern, J. L. and Bennett, O. L., Quality of diets with fluidized bed combustion residue treatment. I. Rat trials. *J. Environ. Qual.*, 17, 550, 1988.
49. Zaifnejad, M., Ritchey, K. D., Clark, R. B., Baligar, V. C. and Martens, D. C., Fluidized bed combustion by-product treatment and leaching of acid soil affects growth and boron acquisition of maize. *Commun. Soil Sci. Plant Anal.*, 29, 255, 1998.
50. Clark, R. B., Zeto, S. K., Ritchey, K. D. and Baligar, V. C., Mineral acquisition by maize grown in acidic soil amended with coal combustion products. *Commun. Soil Sci. Plant Anal.*, 32, 1861, 2001.
51. Baker, D. E. and Chesnin, L., Chemical monitoring of soils for environmental quality and animal and human health. *Adv. Agron.*, 27, 305, 1975.
52. Wang, H., *Potential uses of fluidized bed boiler ash (FBA) as a liming material, soil conditioner and sulphur fertilizer*, Ph.D. Thesis, Massey University, Palmerston North, New Zealand, 1996.
53. Clark, R. B., Zeto, S. K., Ritchey, K. D., Baligar, V. C., Boron accumulation by maize grown in acidic soil amended with coal combustion products. *Fuel*, 78, 179, 1999.
54. Shear, C. B. and Faust, M., Nutritional ranges in deciduous tree fruits and nuts. *Hortic. Rev.*, 2, 142, 163. 1980.
55. Mitchell, D. M., May, J. D. and Bennett, O. L., Effect of fluidized bed combustion residue on performance and physiology of broilers. *Poultry Sci.*, 62, 2378, 1983.
56. Fashandi, E. L., Reid, R. L., Stout, W. L., Hern, J. L. and Bennett, O. L., The effect of fluidized bed combustion residue on the composition and nutritional quality of food crops for hamsters and rats. *Qual. Planta: Plant Food. Human Nutr.*, 25, 359, 1985.
57. Sutton, P. and Dick, W. A., Reclamation of acidic mined lands in humid areas. *Adv. Agron.*, 41, 377, 1987.
58. Farina, M. P. W. and Channon, P., Acid-subsoil amelioration I. A comparison of several mechanical procedures. *Soil Sci. Soc. Am. J.* 52, 169, 1988.
59. Simpson, J. R.; Pinkerton, A. and Lazdovakis, J., Interaction effects of subsoil acidity and water on the root behaviour and root growth of some genotypes of lucerne. *Aust. J. Agric. Res.*, 30, 609, 1979.
60. Bolan, N. S., Adriano, D. C., and Curtin, D., Soil acidification and liming interactions with nutrient and heavy metal transformation and bioavailability. *Adv. Agron.*, 78, 216, 2003.
61. Bolan, N. S., Hedley, M. J. and White, R. E., Processes of soil acidification during nitrogen cycling with emphasis on legume based pastures. *Plant Soil*, 134, 63, 1991.
62. Alva, A. K., Sumner, M. E. and Miller, W. P., Reactions of gypsum or phosphogypsum in highly weathered acid subsoils. *Soil Sci. Soc. Am. J.*, 54, 993, 1990.
63. Coventry, D. R., Reeves, T. G., Brooke, H. D., Ellington, A. and Slattery, W. J., Increasing wheat yields in north-eastern Victoria by liming and deep ripping. *Aust. J. Exp. Agric.*, 27, 679, 1987.
64. Reeve, N. G. and Sumner, M. E., Amelioration of subsoil acidity in Natal Oxisols by leaching of surface-applied amendments. *Agrochemophysica* 4, 1, 1972.
65. Farina, M. P. W. and Channon, P., Acid-subsoil amelioration II. Gypsum effects on growth and subsoil chemical properties. *Soil Sci. Soc. Am. J.* 52, 175, 1988b.
66. Ritchey, K. D., Feldhake, C. M., Clark, R. B. and Souza, D. M. G., Improved water and nutrient uptake from subsurface layers of gypsum-amended soils. in *Agricultural utilization of urban and industrial by-products, ASA Special Publication No. 58*, Madison, WI, 1995, 157.
67. McLay, C. D. A., Ritchie, G. S. P. and Porter, W. M., Amelioration of subsurface acidity in sandy soils in low rainfall regions. I. Responses of wheat and lupins to surface-applied gypsum and lime. *Aust. J. Soil Res.*, 32, 835, 1994.
68. Sumner, M. E. and Carter, E., Amelioration of subsoil acidity. *Commun. Soil Sci. Plant Anal.*, 19, 1309, 1988.
69. McLay, C. D. A., Ritchie, G. S. P., Porter, W. M. and Cruse, A., Amelioration of subsurface acidity in sandy soils in low rainfall regions. II. Changes to soil solution composition following the surface application of gypsum and lime. *Aust. J. Soil Res.*, 32, 847, 1994.
70. Adams, F. and Rawajfih, Z., Basaluminite and alunite: A possible cause of sulfate retention by acid soils. *Soil Sci. Soc. Am. J.*, 41, 686, 1977.
71. Oates, K. M. and Caldwell, A. G., Use of by-product gypsum to alleviate soil acidity. *Soil Sci. Soc. Am. J.*, 49, 915, 1985.
72. Parfitt, R. L., Anion adsorption by soils and soil materials. *Adv. Agron.*, 30, 1, 1978.
73. Shamshuddin, J., Fauziah, I. C. and Sharifuddin, H. A. H., Effects of limestone and gypsum application to a Malaysian Ultisol on soil solution composition and yields of maize and groundnut. in *Proceedings of International Symposium of Plant—Soil Interactions at Low pH*. Wright, R. J., Baligar, V. C. and Murrmann, R. P. Eds., Kluwer Academic Publishers, Dordrecht, The Netherlands. 1991, 397.
74. Tisdale, S. L., Nelson, W. L. and Beaton, J. D., *Soil fertility and fertilizers*, 4th ed. Macmillan, NY. USA. 1985.
75. The Sulfur Institute., *The fourth major nutrient*, The Sulfur Institute, Washington, D. C. USA. 1982.
76. Tisdale, S. L., Reneau, R. B., Jr. and Platuo, J. S., Atlas of sulfur deficiencies. in *Sulfur in Agriculture*, Tabatabai, M. A. Ed., American Society of Agronomy/Crop Science Society of America/Soil Science Society of America, Madison, WI, USA. 1986, 296.
77. Morris, R. J., The importance of sulphur in agriculture—An overview. in *Proceedings of the International symposium on sulphur in Agricultural Soils*, Portch, S. and Hussain, Sk. G. Eds., The Bangladesh Agricultural Research Council and The Sulphur Institute, Dhaka, Bangladesh, 1986, 1.
78. Oates, K. M. and Kamprath, E. J., Sulfur fertilisation of winter wheat grown on deep sandy soils. *Soil Sci. Soc. Am. J.*, 49, 925, 1985.

79. Chien, S. H., Hellums. D. T. Henao, J., Greenhouse evaluation of elemental sulfur and gypsum for flooded rice. *Soil Sci. Soc. Am. J.*, 51, 120, 1987.
80. Woodhouse, W. W., Long-term fertility requirements of Coastal bermudagrass. III. Sulphur. *Agron. J.*, 61, 705, 1969.
81. Alcordo, I. S. and Rechcigl, J. E., Phosphogypsum in agriculture: a review. *Adv. Agron.*, 49, 55, 1993.
82. Codling, E. E., Mulchi, C. L. and Chaney, R. L., Biomass yield and phosphorus availability to wheat grown on high phosphorus soils amended with phosphate inactivating residues. III. Fluidized bed coal combustion ash. *Commun. Soil Sci. Plant Anal.*, 33, 1085, 2002.
83. Dou, Z., Zhang, G. Y., Stout, W. L., Toth, J. D., Ferguson, J. D., Efficacy of Alum and Coal Combustion By-Products in Stabilizing Manure Phosphorus. *J. Environ. Qual.*, 32, 1490, 2003.
84. Stout, W. L., Sidle, R. C., Hern, J. L. and Bennett, O. L., Effects of fluidized bed combustion waste on the Ca, Mg, S and Zn levels in red clover, tall fescue, oats and buckwheat. *Agron. J.*, 71, 662, 1979.
85. Lindsay, W.L., *Chemical Equilibria in Soils*, Wiley, New York, 1979.
86. Elrashidi, M.A., Baligar, V.C., Korcak, R.F., Persaud, N. and Ritchey, K.D., Chemical composition of leachate of dairy manure mixed with fluidized bed combustion residue. *J. Environ. Qual.*, 28, 1243, 1999.
87. Stout, W.L., Sharpley, A.N., Gburek, W.J., Pionke, H.B., Reducing phosphorus export from croplands with FBC fly ash and FGD gypsum. *Fuel* 78, 175, 1999.
88. Stout, W.L., Sharpley, A.N., Pionke, H.B., Reducing soil phosphorus solubility with coal combustion by-products. *J. Environ. Qual.*, 27, 111, 1998.
89. Parr, J. F., Papendick, R. I., Hornick, S. B. and Meyer, R. E., Soil quality: Attributes and relationships to alternative and sustainable agriculture. *Am. J. Altern. Agric.*, 7, 5, 1992.
90. Visser, S. and Parkinson, D., soil biological criteria as indicators of soil quality. *Am. J Altern. Agric.*, 7, 33, 1992.

24
Potential of Fly Ash and Organic Wastes for Uses as Amendments to Agricultural Soils: A Review

G. S. Ghuman, K. S. Sajwan, and S. Paramasivam
Department of Natural Sciences and Mathematics, Savannah State University, P.O. Box 20600, Savannah, GA 31404, USA

Among the naturally available resources, coal plays a predominant role as a source of energy. It was estimated that by the year 2000, the production of coal in the United States would reach 1.91×10^9 tons[1]. The coal excavated from different parts of the country or of the world such as lignite, bituminous, subbituminous and Australian hard and soft brown coals are different in their chemical composition and physical characteristics[2]. Increased use of coal in power plants has resulted in the accumulation of vast quantities of coal residues of which fly ash is the major fraction. Approximately 63 million tons of fly ash is generated annually in the U.S. Of this amount, about 66% is discarded in lagoons or landfills. Half of the utilized fly ash is used in concrete as pozzolanic material because it contains constituents that combine with lime at ordinary temperatures and in the presence of water form cementitious compounds[3].

A great deal of research has been done to find the proper utilization and disposal of waste products of coal-fired power plants[4,5,6]. Fly ash is commonly used in landfilling, in construction industry, as filler in asphalt mix, stabilizer for road bases, thermal insulation, and deodorization of animal wastes[7]. Fly ash usually contains high concentrations of Ca, K, Na, Mg, B, Ni and Mn. Earlier reports indicated that the elements present in crude coal are partitioned not only among the different types of coal ash residues but also among particles of different sizes in fly ash[8,9]. Fly ash consists of particles of all sizes ranging from 2 to 1000 μm in diameter. Much of the earlier work conducted to utilize fly ash was centered on efforts to amend the soil characteristics for proper vegetation growth. Fly ash usually contains high concentrations of Ca, K, Na, Mg, B, Ni and Mn. Alkaline fly ash due to the presence of oxides has been used to neutralize acidity[10]. Furr et al.[11] studied the uptake of several elements by vegetables grown in potted soil amended with 10% by weight of fly ash and found that As, B, Ca, Cu, Fe, Hg, I, K, Mg, Mo, Ni, Sb and Se were higher in concentration in the edible portions of at least three of the crops grown on fly ash amended soil as compared to control crops. Massive applications are usually associated with adverse effects to soils and growing plants.

The effect of fly ash amended compost application to agricultural crops was not investigated in earlier studies. Adriano et al.[12] suggested that the use of coal residues in conjunction with other materials such as sewage sludge, peat and animal manures needs to be explored. Menon et al.[13] studied the utilization of fly ash to amend the characteristics of organic compost for agricultural use in acid soil. Menon et al.[14] studied the effects of coal fly ash-amended composts on the yield and elemental uptake by plants. Ghuman et al.[15] studied the up take of multielements by corn from fly ash-compost amended soil. Compost contains relatively lower concentrations of several elements present in fly ash particularly the Ca, B and Cd.

Coal Combustion

Over the past few decades, the American Public has become increasingly wary of nuclear power because of concern about radiation releases from normal plant operations, plant accidents and nuclear waste. For containing it, the American public and electric utilities have preferred coal combustion as a power source. United States is fueled by coal compared with 14.8% for nuclear energy[16].

Ghuman et al.[15] compared the composition of crude coal and various fractions of fly ash including the weathered ash from SRS-484 (Savannah River Site) power plant. Selected results are included in the following Tables:

Results in Tables 1 and 2 show increased concentrations of K, Na, Ca, Mg and Fe in fly ash fractions as compared to those in crude coal. Weathered ash contained greater K and Na and less Ca, Mg and Fe than the fresh ash. Manganese, Zn, Ni and Cr increased in fly ash fractions relative to crude coal but Cu remained nearly the same in both categories. Manganese, and Cu decreased in weathered ash compared to fresh ash, but Zn, Ni and Cr were nearly equal in both fractions. Lithium, Sr, Ba, Ti, As and P levels increased in fly ash over those of crude coal. Titanium and As concentrations

TABLE 1. Major and some minor metals in coal and fly ash from SRS-484.

Sample	K	Na	Ca	Mg	Fe	Mn	Zn	Cu
			ppm					
I Crude Coal	405	142	1984	272	9607	19	34	69
II Fly Ash Boiler 1–4	3125	306	3158	874	18264	49	13	28
III Fly Ash Boiler 1–13	5861	853	4055	1389	15376	55	48	60
IV Fly Ash Boiler 1–16	5843	842	4166	1409	15828	56	44	62
V Fresh Ash	1409	164	4588	1126	107749	927	22	47
VI Weathered Ash	1884	281	1012	322	38383	89	23	26

markedly increased in weathered ash over those in fresh ash but P levels showed substantial decrease from 775 ppm to 145 ppm[17].

Composition of Coal Pile Leachate

Ghuman[15a] studied the composition and metal species in coal pile leachate affected groundwater at Savannah River Site (SRS), South Carolina. At SRS D-Area a coal-fired electric power and steam generating plant was built in 1952. To provide fuel for this plant a 9-acre unlined coal storage pile had been present since that time. In 1978 a 12.5-acre Coal Pile Runoff Basin (CPRB) (Figure 1.) was built adjacent to the coal storage pile to contain runoff from the pile. Seepage from the CPRB and the ground surrounding the coal pile had contaminated the water table aquifer in an area of 15–20 acres. The basin consists mostly of unconsolidated sands and clays. The sands are predominantly quartz sands with minor amounts of feldspars, micas and heavy minerals. Kaolinite with minor amounts of smectite is the dominant clay mineral[18].

TABLE 2. Trace metals and phosphorus in coal and fly ash from SRS-484.

Sample	Ni	Cr	Li	Sr	Ba	Ti	As	P
				ppm				
I Crude Coal	12	4	3	122	73	179	20	32
II Fly Ash Boiler 1–4	16	13	30	198	286	497	109	116
III Fly Ash Boiler 1–13	31	34	75	272	336	785	294	260
IV Fly Ash Boiler 1–16	33	32	74	275	340	774	295	248
V Fresh Ash	17	13	23	151	273	370	59	775
VI Weathered Ash	17	16	10	135	218	447	91	145

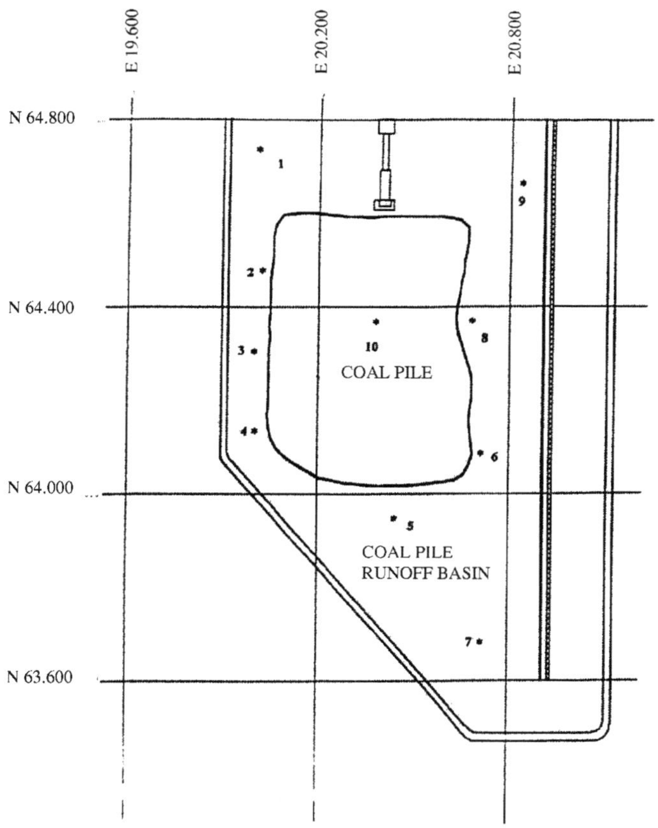

FIGURE 1. Locations on D-Area coal pile of Hydropunch and Hydrocone samples.

The groundwater contaminated by acidic Coal Pile Leachate (CPL) contains predominantly sulfate and metals such as Fe, Al, Cr, Cd and Pb. The groundwater also contains halogenated volatile compounds and elevated concentrations of organic pollutants such as trichloroethylene (TCE). Conventional approaches to environmental remediation of such problems have relied primarily on physical isolation or removal technologies including groundwater pumping. In most cases, pump and treat has not succeeded in restoring aquifers to regulatory standards. This may be due to desorption of contaminant constituents from the soil particles in aquifers and to channeling effects associated with variations in hydraulic conductivity. Understanding the mobility of contaminating metals and other constituents in the form of various species in groundwater was important for designing the remedial systems.

Analysis of speciation includes an estimation of the distribution of a metal among various dissolved aqueous species and an estimation of the amount of metal that is associated with mobile colloids. In addition, the reactions between the contaminant plume and aquifer minerals are important to understand the transport of contaminant metals. Analytical data for the chemical composition of 18 groundwater samples collected from seven monitoring wells around the coal pile in D-Area

TABLE 3. Analyses of groundwater samples of D-Area taken with the Hydropunch.

Sample & Depth (ft.)	pH	Al	Ba	Ca	Mg	Na	K
	mg L^{-1}						
D8: 16.5–17.5	4.6	0.121	0.0152	0.885	0.336	3.390	0.00
D8: 22–23	5.0	0.478	0.0242	0.924	0.426	3.780	0.00
D8: 25–26	5.1	0.121	0.0152	0.885	0.336	3.390	0.00
D7: 19–20	3.8	20.700	0.0345	79.700	46.100	14.900	0.00
D7: 24–25	4.1	3.580	0.0277	37.000	16.400	12.700	0.00
D7: 30–31	5.6	3.310	0.0656	26.200	55.200	32.900	4.76
D6: 22–23	5.9	0.077	0.1590	6.940	1.400	9.280	0.00
D6: 26–27	6.0	0.173	0.1510	6.960	1.400	10.800	0.00
D5: 17–18	4.8	0.074	0.0401	2.710	1.082	5.235	0.00
D5: 21–22	5.0	0.138	0.0189	2.310	0.475	3.800	0.00
D5: 26–27	5.1	0.076	0.0171	4.030	0.557	3.400	0.00
D4: 17–18	4.3	0.000	0.1810	36.000	29.900	14.600	3.74
D4: 21–22	3.6	20.100	0.0279	218.000	118.000	11.800	4.85
D4: 27–28	4.5	2.630	0.1230	92.200	34.300	11.100	0.00
D3: 17–18	4.1	7.450	0.0272	58.800	35.400	15.200	4.04
D3: 21–22	3.7	19.900	0.0157	84.100	61.800	15.800	4.46
D3: 27–28	3.5	11.400	0.0152	387.000	90.700	29.700	8.50
D2: 26–27	4.4	0.881	0.0152	156.000	34.600	6.000	0.00

Sample & Depth (ft.)	Fe	Mn	Zn	Ni	Si	Cl	SO^4
	mg L^{-1}						
D8: 16.5–17.5	2.360	0.020	0.060	0.000	11.00	7.08	84.60
D8: 22–23	8.130	0.089	0.060	0.041	11.60	5.96	2.08
D8: 25–26	2.360	0.020	0.060	0.000	11.00	0.01	0.00
D7: 19–20	2.880	0.735	0.641	0.242	13.70	3.80	602.00
D7: 24–25	0.872	0.390	0.383	0.238	11.80	3.80	243.00
D7: 30–31	27.000	10.300	2.030	0.298	19.30	39.30	1591.00
D6: 22–23	17.60	0.357	0.081	0.041	9.54	7.59	55.30
D6: 26–27	18.80	0.351	0.302	0.039	10.30	7.80	46.20
D5: 17–18	1.05	0.085	0.041	0.029	6.11	12.60	1.16
D5: 21–22	0.38	0.042	0.108	0.000	7.79	6.03	3.33
D5: 26–27	0.58	0.052	0.130	0.000	9.70	4.05	6.93
D4: 17–18	2.49	0.421	0.126	0.274	15.80	10.40	980.00
D4: 21–22	23.00	41.500	1.120	0.483	12.70	3.36	2130.00
D4: 27–28	68.50	3.950	0.374	0.288	7.94	15.30	1120.00
D3: 17–18	218.00	5.100	0.535	0.535	12.30	6.63	1180.00
D3: 21–22	273.00	20.000	0.780	0.000	11.40	4.52	1680.0
D3: 26–27	3.02	10.000	0.398	0.083	7.28	7.75	1700.00
D2: 26–27	83.10	10.900	0.960	0.284	11.70	3.81	1120.00

(Fig. 1) are listed in Table 3. For simplicity, all the available data were not included in this Table.

Chemical composition data in Table 3 show high concentrations of Fe, Ca, Mg, Si, Cl and SO_4in groundwater samples of all sites. Maximum concentrations observed were 273 mg L^{-1} Fe (D3: 21–22), and 218 mg L^{-1} Ca, 118 mg L^{-1} Mg, 2130 mg L^{-1} SO_4 (D4: 21–22). Relatively high concentration of Al was found in sample D4: 21–22, but Al concentrations were quite variable in samples of different sites and at different depths. Silicon concentration was consistently high in all the samples analyzed, which indicates its uniform dissolution in a low pH soil. In general, the chemistry of coal pile leachate affected groundwater is vertically stratified, but the order of stratification of various constituents is not the same. For instance, groundwater in the lower zones of sites D: 3, D: 4 and D: 5 (26–27 ft. depth) contained elevated concentrations of Ca, Mg and sulfate. In groundwater sample of D: 3 and D:5 sites, the concentration of Fe decreased with depth. This variation was probably due to the short distance migration of coal pile runoff (CPR) into the ground and also due to the direction of flow of contaminant plume from the western to the eastern side of the coal pile (Fig. 1).

Speciation of Aluminum

Percent distribution of Al among different species possibly present in groundwater is depicted in Figure 2. The species have been characterized into groups of three pH ranges encountered in the study samples. The delineation of pH ranges is as follows: pH range 3.5–4.6 (10 samples); pH range 4.8–5.1 (5 samples); and pH range 5.6–6.0 (3 samples). The direction of groundwater flow is from sites 8 and 6 part of coal pile toward the sites 2, 3 and 4 (Fig. 1). This direction of flow of coal pile leachate from the vicinity of the pile and the adjacent coal pile runoff basin (CPRB) seems to have caused low pH in groundwater sample of sites D2, D3, D4 and shallow-depth samples of sites D7 and D8. At low pH, the dissolution of kaolinite mineral in the soil consumes acid and provides additional Al to precipitate as sulfate. For this reason, MINTEQ/A2 Model predicts the presence of $Al(SO_4)^+$ and $Al(SO_4)_2$ in these low-pH groundwater samples. The ionic species of Al^{+3} is prevalent up to 38% in low pH ranges (3.5–4.6) and (4.8–5.1), but decreased markedly at a minimum value of 4% in the groundwater samples of pH range 5.6–6.0. The formation of $AlOH^{+2}$ and $Al(OH)_2{}^+$ species were negligible (up to 3%) in low pH range (pH 3.5–4.6), but increased markedly in both pH groups above a pH of 4.8. Only a small percentage of Al $(OH)_3$ AQ species was present in the groundwater of pH range (4.8–5.1), but increased significantly to an average value of 34% in the groundwater samples of pH range (5.6–6.0). It is evident that the formation of different species of Al in the groundwater is pH dependent and each species is limited to a certain pH range.

Speciation of Fe, Mn, Zn and Sulfate

In Table 4, the average percent distribution of Fe, Mn, Zn and sulfate has been characterized into four pH-range groups. The limits of these pH ranges were arbitrarily chosen on the basis of similarities of species in the study samples as revealed by MINTEQ/A2 calculations. The pH ranges included in Table 4 are: pH range: 3.5–4.1 (6 samples), pH range 4.3–4.5 (3 samples), pH range 4.6–5.1 (6 samples), and pH range 5.6–6.0 (3 samples). The ionic species of Fe^{++} is found in high percentages in the groundwater samples of all pH values and is apparently the highest in the pH range of 4.6–5.1. The $FeSO_4$ (AQ) species is adequately formed in the groundwater samples of pH below 4.5 and it is quite low in the samples with pH above 4.6. Ferric iron (Fe^{+3}) species is only 2.8% in the samples of pH range 3.5–4.1, which may indicate that at higher pH the ferric iron is precipitated in the form of insoluble hydroxide and other minerals. The formation of $FeOH^{+2}$ decreases with an increase in pH while the formation of $Fe(OH)_2{}^+$ increases with rising pH (Table 4). A small percentage (1.1%) of $Fe(OH)_3$ is formed only in the pH range 5.6–6.0. $FeSO_4{}^+$ is mainly formed in the pH range 3.5–4.1.

Manganese, another transition metal showed relatively small variation of species as compared to the species of Fe. The presence of ionic Mn^{+2} increased from 70.7% in the low pH-range (3.5–4.1) samples to 98.0% in the high pH-range (4.6–5.1) samples. On the other hand, $MnSO_4$ (AQ) was formed in appreciable quantities in low-pH groundwater (pH ranges: 3.5–4.1 and 4.3–4.5) and decreased markedly in high-pH samples. The distribution of ionic zinc (Zn^{++}) is less variable than Mn and is appreciably present in groundwater of all pH values.

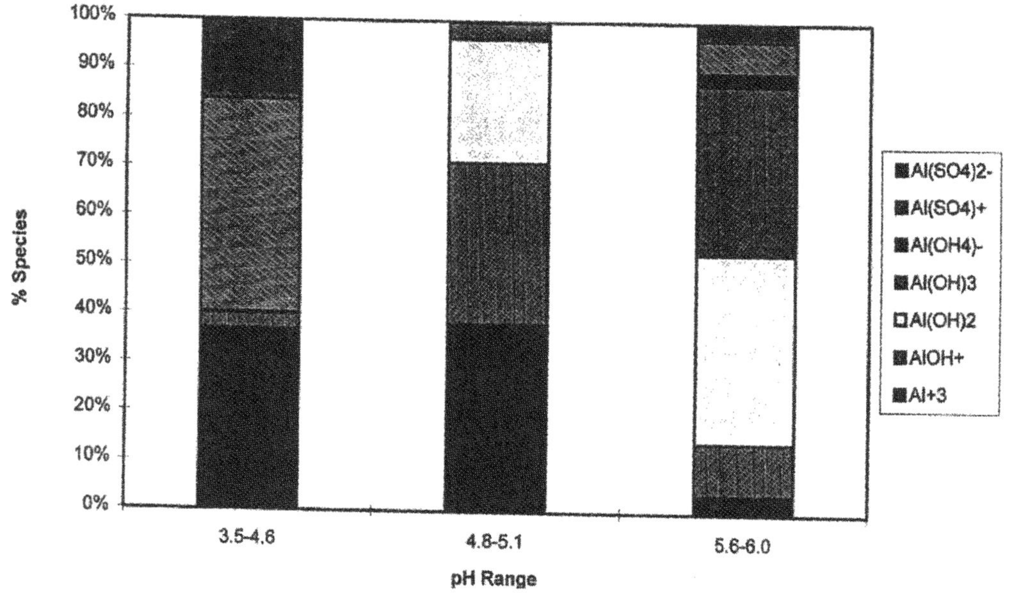

FIGURE 2. Percent distribution of different species of Al metal in D-Area groundwater taken with Hydropunch (Sampled April–May, 1990).

TABLE 4. Percent distribution of different species of metal and sulfate in D-Area Groundwater taken with Hydropunch (Sampled April–May, 1990).

Species	pH 3.5–4.1	pH 4.3–4.5	pH 4.6–5.1	pH 5.6–6.0
	Percent distribution in each pH range			
Fe^{++}	71.2	67.4	98.1	83.4
$FeSO_4$(AQ)	28.8	32.6	1.6	16.6
Fe^{+++}	2.8	–	–	–
$FeOH^{+2}$	30.3	15.6	4.4	–
$Fe(OH)^3$ (AQ)	–	–	–	1.1
$FeSO_4$	25.8	2.7	–	–
Mn^{+2}	70.7	66.9	98.0	83.1
$MnSO_4$ (AQ)	29.3	33.1	1.8	16.9
Zn^{+2}	63.5	59.3	97.6	79.0
$ZnSO_4$(AQ)	33.4	37.8	2.2	19.1
$Zn(SO_4)_2{}^{-2}$	3.1	2.9	–	1.8
$SO_4{}^{-2}$	77.0	85.0	97.8	92.7
$MgSO_4$(AQ)	5.7	4.0	–	2.0
$CaSO_4$(AQ)	8.6	7.3	1.4	1.9
$FeSO_4$(AQ)	4.4	3.7	0.6	2.6
$Al(SO_4)^+$	2.1	–	–	–
$Al(SO_4)_2{}^-$	2.2	–	–	–

An additional species Zn $(SO_4)_2{}^{-2}$ was formed in minor quantities in low-pH solutions (pH 3.5–4.5) probably due to the abundance of sulfate ions and their affinity to Zn.

Sulfate ion is the predominant constituent in the coal pile leachate and persists in high percentages during its mobility through an acid soil. For this reason, the MINTEQ/A2 calculation model predicts an abundance of this free ion in the groundwater of all pH ranges. In addition to other chemical combinations of sulfate, there is a possibility of formation of $MgSO_4$ (AQ), $CaSO_4$ (AQ) and $FeSO_4$ (AQ) in small percentages (3.6–8.6%) in acidic groundwaters within a pH rang of 3.5–4.5. As indicated in the speciation of Al (Fig. 2.), there is a possibility of about two percent of sulfate ion forming the species Al $(SO_4)^+$ and Al $(SO_4)_2{}^-$ in highly acidic groundwater in the pH range of 3.5–4.1.

Utilization of Fly Ash-Amended Compost

To determine the agricultural use of coal fly ash, Menon et al.[19] prepared fly ash-amended compost by using different fly ash: compost ratios. Different grades of fly ash-amended composts were prepared by mixing and equilibrating fly ash and home-made organic compost. The mixtures used were: fly ash only (100% FA), compost only (0% FA), 15% FA, 20% FA, 30% FA, 40% FA, 50% FA and 60% FA. Water-extractable constituents of fly ash-amended compost samples were analyzed and the results are included in Table 5.

Results presented in Table 5 show that there was a significantly higher level of nutrients in the 20–40% fly ash-amended composts compared to their concentrations in fly ash alone. The essential nutrients such as N, P, K and S that are low in fly ash,

TABLE 5. Water-extractable constituents of fly ash-amended compost Treatments.

Treatments	K	Na	Ca	Mg	Zn	B	P	S	N
	Elements (ppm)								
Fly ash (FA)	184	145	1895	41	0.2	29.3	ND[a]	683	7.9
Home-made Compost only	3197	2199	294	175	2.6	23.3	560	4951	893.0
Comp. 2 15% FA	2575	1162	274	159	1.9	29.1	269	3986	913.0
Comp. 3 20% FA	2952	1404	544	385	1.3	18.1	367	4308	1320
Comp. 4 30% FA	2059	1586	343	187	0.7	32.4	133	4044	761
Comp. 5 40% FA	1501	1299	347	155	0.6	30.9	69	3469	1238
Comp. 6 50% FA	1501	852	363	163	0.5	40.3	73	3760	1624
Comp. 7	1189	2033	528	157	0.2	41.5	24	3292	1077

[a] Not detected

TABLE 6. Analysis and yield of sorghum plants grown in sifted soil mixed with fly ash-amended "home made" compost of different compositions.

	Elements (ppm) and dry wt. Yield of plants (g)									
Treatments	K[a]	Ca[a]	Mg[a]	Zn	Mn	B	P[b]	S[c]	N[c]	Yield
0A-DControl	1.2	11.1	7.4	14.1	37.9	23.7	17.8	6.2	2.0	16.0
1A-D 0%FA	4.0	6.7	4.7	19.4	25.3	17.4	41.9	8.0	2.7	27.0
2A-D 15%FA	4.1	8.0	5.0	19.2	26.6	18.5	40.0	6.8	2.2	30.0
3A-D 20%FA	4.6	7.0	5.5	26.9	22.8	17.1	41.0	7.3	1.9	31.0
4A-D 30%FA	3.8	9.5	7.3	19.3	28.2	21.1	31.6	6.1	2.0	33.0
5A-D 40%FA	4.0	8.7	6.3	20.4	28.3	26.9	38.1	5.4	1.9	32.0
6A-D 50%FA	3.8	7.2	4.9	17.9	32.0	18.7	31.9	6.0	2.0	34.0
7A-D 60%FA	3.4	10.0	6.7	18,5	26.0	61.0	26.4	6.0	1.9	27.0

[a]Concentration (10^{-4}); [b]Concentration (10^{-3}); [c]Concentration (10^{-2})

were considerably higher in fly ash-amended composts. All fly ash-amended composts except the one with 30% fly ash, have a higher N content than either the pure organic compost or pure fly ash. This indicates that chemical reaction and mineralization of N were probably taking place in the fly ash-amended mixture during composting.

Optimization of the Fly Ash-to-Compost Ratio for Plant Growth

Menon et al. (1992) performed a set of greenhouse experiments to optimize the fly ash-amended compost—to—soil ratio. Pots lined with plastic bag were filled with various proportions of fly ash-amended compost (20% FA) and soil making the total weight 7 kg in each pot. The soil and compost were mixed thoroughly with a twin-shell blender. Corn seeds were planted and then harvested after 8 weeks. Plant shoots were washed and dried for 3–5 days at 70 °C to obtain dry shoot yield. The results of this study were as follows:

FA-compost/soil ratio:	1:19	1:9	1:6	1:4	1:3
Dry shoot yield/pot (g):	55.4	58.2	67.7	74.9	79.4

These data indicated that the plants grown using FA compost-to-soil ratio of 1:3 gave the maximum yield. Subsequent experiments conducted with 20% FA-amended composts using three different organic composts revealed that the "home-made" compost was the best in terms of the utilization of coal fly ash. Only this compost, mixed with 20% FA, released the additional quantities of Ca and Mg present in the system. Additionally, boron, which may be detrimental for plant growth when present in large quantities, is released in smaller quantities in this specimen than in others.

The optimum composition of FA-amended compost (20–40% FA) was established by determining the yield of two plants, corn and sorghum, grown using different compositions of fly ash and "home-made" compost at the constant FA compost-to-soil ratio of 1:3. The results elemental analysis of dry shoots of sorghum plants grown in sifted sandy loam soil mixed with fly ash-amended "home-made" compost of different compositions are included in Table 6.

It can be seen from Table 6 that sorghum plants grown in soil treated with fly ash-amended compost in the 20–40% FA range, had significant amounts of the essential nutrients including N, P, and K and gave higher dry shoot yields than the remaining treatments. It may be concluded that during composting, the most important function of "home made" compost was to release the nutrients such as N, P and K that were present in only small amounts in the fly ash. Data indicate that plant yield decreased as the boron levels in the plant increased. It is clear that boron uptake by sorghum plants was lower with the treatments at the 20–40% FA level.

Uptake of Multielements by Corn from Fly Ash-Compost Amended Soil

Ghuman et al.[15] studied the feasibility of application of fly ash-compost mixture to soils for the availability and uptake of elements by corn (Zea mays L.). The crop was grown in potted Ogeechee sandy loam soil using eight treatments: soil alone, soil amended with 15% compost, and soil amended with 2, 5, 10, 15, 20 and 25% of fly ash-amended compost. Fly ash for this study was sampled from a coal-fired power plant (484-D) at the Department of Energy Savannah River Site. For the compost material, one bag of 'compost toast' was purchased from Bricker's Farm, Augusta, Georgia. The Ap horizon of Ogeechee sandy loam soil (Siliceous Thermic Typic Ochragults) was collected from the field of Savannah State University, Savannah, Georgia. A homogeneous fly ash-compost mixture (20% and 80% Compost) was prepared by equilibrating the two components for two weeks in a plastic container. The mixture was kept moist by adding deionized distilled water (Purity 1.2 μmho) and the contents were stirred periodically with a spatula. Greenhouse pot culture experiment was conducted growing corn with eight treatments. Eight treatments included: soil alone, soil amended with 15% compost, soil amended with 2, 5, 10, 15, 20 and 25% of fly ash-amended compost using four replicates of each. Supplemental application of

TABLE 7. Water-soluble elements in experimental soil, fly ash, compost and FA-compost.

Element	Soil alone	Fly Ash (484-D SRS)	Compost	FA-Compost
		mg kg^{-1}		
Ca	222	1895	411	149
Mg	22	41	67	42
Na	3269	145	1376	790
K	261	284	783	657
Zn	0.04	2.2	3.7	0.53
Mn	0.23	4.6	6.1	2.30
Cd	ND	0.07	ND	ND
Cu	ND	0.7	2.9	3.10
Ni	1.90	3.0	ND	ND
PH	5.50	6.90	7.0	7.2

ND: Not detected

N, P and K was not made in any of the treatments. Corn plants (Zea mays L.), were grown in pots for a period of six weeks and then harvested and dried to constant weight.

Elemental Analysis of Soil, Fly Ash, Compost and Corn Plants

Water-soluble elements in the soil, fly ash, compost and fly ash-amended compost were determined by equilibrating 10 g of air dry sample with 100 mL of deionized distilled water (1:10 dilution), agitating and obtaining the abstract. For corn tissue analysis, one-gram ground samples from each treatment were ashed in a muffle furnace at 500° C. The ashed sample was extracted with 10 mL of 6N HCl. The clear extract was used for the estimation of various elements by atomic absorption emission. Results are included in Tables 7 and 8.

Results in Table 7 indicate that SRS 484-D fly ash is relatively high in Ca [13] but contains smaller quantities of Mg, K, Zn, Mn, Cd, Cu and Ni when compared to compost and FA-compost. Water-soluble elements were determined to find their rapid availability to crop. The pH of SRS 484-D was similar to that of compost and FA-compost. Elemental concentrations in corn plants are displayed in Table 8. It was observed that Ca, Zn, Mn, Cu and Mg concentrations were higher in plants grown in soil without any amendments compared to soils either amended with compost or FA-compost. This may be due to stunted growth of the plants, which resulted in low dry matter yield, and high concentrations of these elements. The concentrations of K, Zn and Ni were particularly high in corn plants grown in soil amended with 15% compost (Treatment 1). With 20% FA-compost application (Treatment 2), the concentrations of Ca, Mg, Zn, Cu and Ni in corn tissues were fairly high, but K concentrations decreased to a minimum as compared to all other treatments. In general, Ca and Mg concentrations in corn plants showed consistent decrease with the increasing percentage of FA-compost mixture in soil matrix (Treatments 3–7). Cadmium concentrations were observed to be low, varying within a narrow range of 0.36 to 1.07 mg kg^{-1} dry weight in the corn plants of all treatments. This observation was in agreement with the earlier study of Furr et al [11], who found no increase of Cd in the crops as a result of fly ash treatment.

Dry matter yield of corn and total uptake of various elements per pot are presented in Table 9. Corn plants grown in soil without any amendment had low corn yield and minimum uptake of all elements. Amending soil with 15% compost improved the corn yield and the uptake of all elements, especially that of Zn (3.1 mg pot^{-1}). Corn yields improved over the control in all FA-compost treatments, but the maximum yield was obtained in the 20% and 25% treatments. Page et al. [5] produced higher yields of several agronomic crops by the application of fly ash to either calcareous or acid soils at rates ranging up to 8% (by weight). The increase in dry matter yield of corn when grown in soil amended with FA-compost appears to be associated with the supply and uptake of macronutrients K, Ca and Mg, micronutrients Zn, Mn and Cu (Tables 8 and 9),

TABLE 8. Concentrations of elements in corn grown in fly ash-compost mixture- amended soil in different ratios[a].

	0	1	2	3	4	5	6	7
Treatment	Control Soil	Soil+15% Compost	Soil+2% FA-Comp	Soil+5% FA-Comp	Soil+10% FA-Comp	Soil+15% FA-Comp	Soil+20% FA-Comp	Soil+25% FA-Comp
Element				mg kg^{-1}				
Ca	8844 [b]	4156	6792	5406	5965	4788	4044	3788
Mg	5853	4353	6123	5894	5138	4395	3741	3704
Na	26187	29813	18188	28250	32969	42219	32750	36812
K	9631	13004	5575	7700	7638	9538	11775	13562
Zn	57	43	38	37	40	44	41	31
Mn	51	27	36	32	29	33	33	33
Cd	0.80	0.89	0.81	0.98	0.72	0.36	1.07	0.36
Cu	4.20	3.50	4.00	4.20	3.30	3.30	3.80	3.80
Ni	2.70	4.70	3.70	4.10	0.80	4.60	1.00	4.20

[a] Fly ash-compost mixture was prepared with 20% FA.
[b] Each value is the mean of four replicates.

TABLE 9. Concentrations of elements in corn grown in fly ash-compost mixture-amended soil in different ratios[a].

	0	1	2	3	4	5	6	7
Treatment	Control Soil	Soil+15% Compost	Soil+2% FA-Comp	Soil+5% FA-Comp	Soil+10% FA-Comp	Soil+15% FA-Comp	Soil+20% FA-Comp	Soil+25% FA-Comp.
Yield				g dry wt. Pot-1				
	15.8[a]	72.5	55.4	57.6	58.2	67.6	74.9	79.4
Element				mg kg^{-1}				
Ca	140	300	376	312	345	324	303	299
Mg	99	315	339	336	298	297	280	293
Na	414	2151	1006	1634	1914	2860	2450	3086
K	161	941	310	444	443	645	882	1076
Zn	0.78	3.15	2.08	2.14	2.32	2.97	3.06	2.500
Mn	0.77	1.94	2.01	1.84	1.70	2.20	2.44	2.600
Cd	0.014	0.064	0.045	0.057	0.042	0.024	0.080	0.028
Cu	0.067	0.257	0.223	0.240	0.194	0.225	0.280	0.298
Ni	0.039	0.336	0.202	0.236	0.045	0.312	0.077	0.220

[a] Each value is the mean of four replicates

and perhaps some other essential elements. Previous studies reported that Ca and/or Mg could reduce K uptake by plants grown in fly ash treated soils [16,17]. Plant availability of many other nutrient elements may also be affected by the shifting of soil pH.

Reduced plant uptake or deficiency symptoms of Zn have been observed in fly ash treated soils[16,17,18,19]. Data in Table 9 show a consistent increase in the uptake of essential elements by corn with the application of increasing amounts of FA-Compost mixture to the potted soil. It appears that the greater availability of most elements to corn plants occurred from the components of both the fly ash and the compost in the medium. This may lead to the conclusion that the blending of compost with fly ash in small amounts, improved the bioavailability of essential and non-essential plant food elements from fly ash, which resulted in better growth as compared to control without any detrimental effects.

Plant nutrient concentrations were correlated with the composition of fly ash-amended compost for corn to determine (a) if plant nutrient contents of the FA-amended composts were related to plant yield and (b) if there was any relation between the elemental content of the plant and its concentration in the FA-amended compost. The linear equations corresponding to these plots and 'r' values are given in Figures 3 and 4.

Figure 3 shows a positive correlation (r = 0.86) between the dry matter yield of corn and the uptake of K pot^{-1}. It is evident that among other factors, greater availability of K from soil and FA-Compost mixture (Table 7) was responsible for the improved growth of corn. Pietz et al.[22] also showed that the concentration of water-soluble K was the highest in sludge + lime-amended treatments of coal refuse material. In their study, potential toxicity from metals in the coal refuse material was most likely reduced by complexing and neutralizing reactions with the applied sewage sludge and lime.

Figure 4 shows a significant correlation between the total uptake of essential micronutrients Mn and Cu by corn (r = 0.95). This seems to be a synergistic interaction between these nutrients. It appears that positive and meaningful correlations depend on the nature of plant species. It is sufficient to report that at least some nutrients in fly ash have a direct or indirect impact on their assimilation in plants.

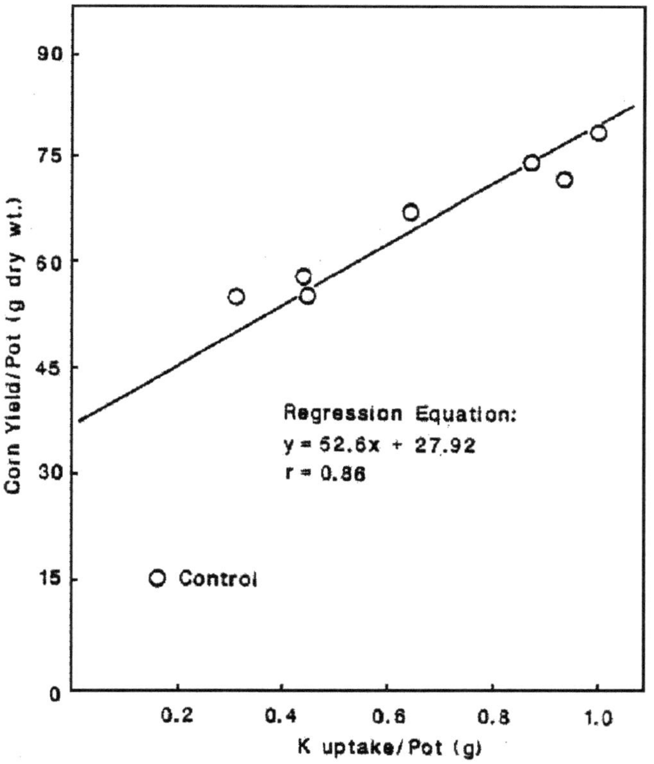

FIGURE 3. Correlation between corn yield and K uptake from soil treated with Fly Ash-Compost mixture in different ratios.

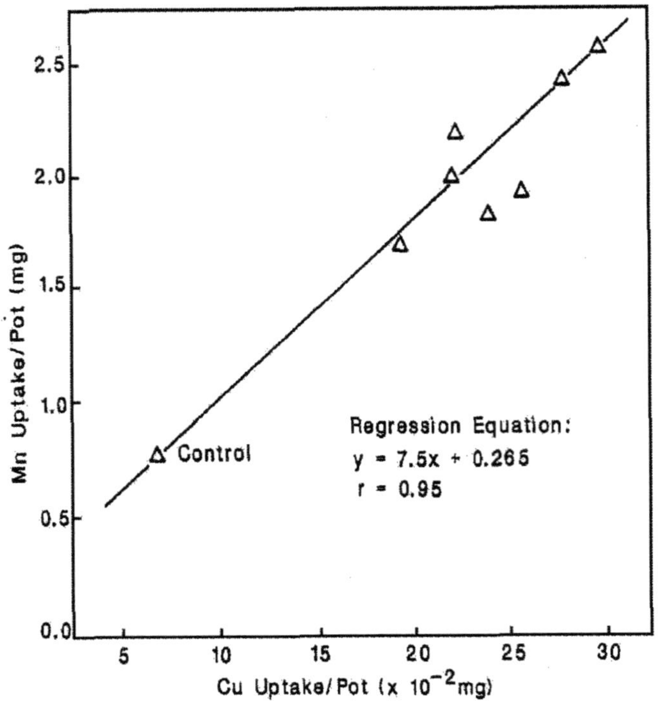

FIGURE 4. Relationship between Mn and Cu uptake by corn from soil treated with Fly Ash-Compost mixture in different ratios.

Summary

Vast quantities of coal residues and fly ash require proper disposal and utilization. Fly ash contains high concentrations of various useful and toxic elements. Runoff from a coal pile and its leachate splits into various ionic species as it enters the soil and ultimately contaminates the groundwater. A portion of the accumulating fly ash is used in concrete as pozzolanic material. Studies reveal that amending the soil with 20–25% fly ash-compost mixture proves beneficial for the growth of certain crops, particularly corn and sorghum without any deleterious effects. Fly ash-compost (from organic wastes) mixture applied to the soil minimizes the uptake of certain toxic elements.

References

1. Gordon, R.L.: 1978, Science **200**, 153.
2. Yudovich, Ya.E., Korycheva, A.A., Obruchnikov, A.S., and Stepanov, Yu.V.1972, Geochem. Internl. 712 (Trans. from Gookhimiya, No. 8, 1023, 1972).
3. Jones, L.W.: 1990, Interference Mechanisms in Waste Stabilization/Solidification Processes, R&D Project Summary EPA/600/S2-89/067.
4. Adriano, D.C., Woodard, T.A., and Ciravolo, T.G.: 1978, J. Environ. Qual. **7**, 416.
5. Elseewi, A.A., Bingham F.T., and Page, A.L.: 1978a, Growth and Mineral Composition of Lettuce and Swiss Chard Grown on Fly Ash Amended Soils, in D.C. Adriano and I.L. Brisbin (eds.), Environmental Chemistry and Cycling Processes CONF-760429. U.S. Dept. Commerce, Springfield, VA p. 568.
6. Phung, H.T., Lund, L.J., and Page, A.L.: 1978, Potential Use of Fly Ash as a Liming Material, in D.C. Adriano and I.L. Brisbin (eds.), Environmental and Cycling Processes CONF-760429, Technical Information Center, U.S. Dept. Energy.
7. Bolch, Jr., W.E.: 1980, 'Solid Waste and Trace Element Impact' in A.E.S. Green, (ed.), Coal Burning Issues, Univ. Press of Florida, Gainesville, Chapter 12:231.
8. Page, A.L., Elseewi, A.A., and Straughan, I.R.: 1979, Residue Reviews **71**, 103.
9. Davison, R.L., Nausch, D.F.S., Wallace, J.R., and Evans, Jr., C.A.: 1974, Environ. Sci. Technol. **8**, 1107.
10. Fail, Jr., J.L., and Wochok, Z.S.: 1977, Plant and Soil **48**, 472.
11. Furr, A.K., Kelly, W.C., Bache, C.A., Gutenmann, W.H., and Lisk, D.J.: 1976, J. Agric. Food Chem. **24**, 885.
12. Adriano, D.C., Page, A.L., Elseewi, A.A., Chang, A.C., and Straughan, I.: 1980, J. Environ. Qual. **9**, 333.
13. Menon, M.P., Ghuman, G.S., James, J., Chandra, K., and Adriano, D.C.: 1990, Water, Air and Soil Pollution **50**, 343.
14. Menon, M.P., Sajwan, K.S., Ghuman, G.S., James, J., Chandra, K, and Bacon, B.: 1993, Water, Air and Soil Pollution **69**, 209.

15a. Ghuman, G.S., Menon, M.P., Chandra, K, James, J., Adriano, D.C., and Sajwan, K.S.: 1994, Water, Air and Soil Pollution **72**, 285.

15b. Ghuman, G.S.: 1997, Modeling of Metal Species in Coal Pile Leachate Affected Groundwater. Summer Research Report submitted to Savannah River Site, 11p.

16. Yahoo: 2004, Search Results for Coal: Coal Combustion, www.ornl.gov/ORNL Review/rev26-34/text/colmain.html-
17. Ghuman, G. S.: 1994, Georgia J. Sci. **52**, 85.
18. Anderson, M.A., Bertsch, P.M., Feldman, S.B., and Zelazny, L.W.: 1991, Interactions of acidic metal-rich coal pile runoff with a subsoil, Environ. Sci. Technol. **25**, 2038.
19. Menon, M.P., Ghuman, G.S., James, J., and Chandra, K.: 1992, J. Environ. Sci. Health **27 (4)**, 1127.
20. Mulford, F.R., and Martens, D.C.: 1971, Soil Sci. Soc. Am. Proc. **35**, 296.
21. Schnappinger, Jr., M.G., Martens, D.C., and Plant, C.O.: 1975, Environ. Sci. Technol. **9**, 258.
22. Pietz, R.I., Carlson, Jr., C.R., Peterson, J.R., Zeng, D.R., and Lue-Hing, C: 1989, J. Environ. Qual. **18**, 169.

25
Evaluation of Bahiagrass (*Paspalum notatum*) as a Vegetative Cover for a Landfill Containing Coal Combustion Waste

Bon-Jun Koo[1], Christopher Barton[1], and Domy Adriano[2]

[1]*University of Kentucky, Department of Forestry, Thomas Poe Cooper Bldg., Lexington, KY 40546-0073, USA*

[2]*University of Georgia, Savannah River Ecology Laboratory, Drawer E, Aiken, SC 29802, USA*

1. Abstract

A vegetative cover is a remedial technique utilized on landfills and waste sites for soil stabilization and for the physical and/or chemical immobilization of contaminants. Many herbaceous plants, primarily grasses, exhibit rapid growth, are moderately resistance to environmental stress, and are therefore often used as cover crops in environmental restoration and remediation projects. Use of bahiagrass (*Paspalum notatum*) was examined as a potential cover species and phytostabilizer on an unlined landfill (488-D Ash Basin, 488-DAB) containing approximately one million Mg of coal combustion wastes (CCWs) at the U.S. Department of Energy's Savannah River Site (SRS) in South Carolina. Use of soil amendments and treatments to relieve physical limitations at the site (compaction) and promote vegetation success were implemented and assessed. The influence of these treatments on metal uptake by bahiagrass was also evaluated. Results indicated that the survival of bahiagrass growing in plots treated with a surface amendment (15 cm layer of material applied over the CCWs) was the highest in those containing a topsoil cover and followed the order: topsoil > biosolid > ash > apatite > control. Ripping of the landfill prior to planting also resulted in increased survival for the bahiagrass. Significant differences with respect to survival and metal uptake were not observed in plots that were inoculated with vesicular-arbuscular mycorrhizae (VAM) over those not inoculated. However, significant differences ($p < 0.05$) were observed in plant tissue concentrations of Al, Cr, Fe, Ni, and Zn in plots treated with ash over those of the topsoil and biosolid treatments. Results indicated that the use of soil amendments and subsurface (physical) treatments were essential for plant survival and that periodic monitoring of plant species should be continued to ensure that metal toxicity and secondary contaminant problems do not arise with time.

2. Introduction

Contamination of soils and groundwater from industrial activities such as mining, smelting, electroplating, and ore refining is a problem of global significance. Phytoremediation is a low-cost alternative to traditional environmental remediation techniques, where vegetation and associated microfauna are used to remove, contain, or render harmless environmental contaminants[1]. A vegetative cover is one such technique that may be utilized for the physical and/or chemical immobilization of contaminants (i.e., phytostabilization). The vegetation provides a network of roots to stabilize surface soils, as well as above ground biomass to absorb energy from wind and water forces that would otherwise lead to erosion and bulk contaminant transport[2]. However, plants grown on these sites must be able to survive in the presence of the contaminant of concern and not contribute to the problem through uptake and potential transfer to other trophic groups.

Many herbaceous plants (primarily grasses) exhibit rapid growth, are moderately resistance to environmental stress, and are effective stabilizers of soils. As such, environmental restoration and remediation of degraded lands such as mines, brownfields and landfills often rely upon the use of grasses for revegetation[3,4,5]. Successful restoration, however, depends on the selection of the best plant species and their successful establishment as well as the nutrient uptake characteristics of the root system and application of soil amendments[6,7].

Bahiagrass (*Paspalum notatum*), a C4 creeping perennial plant that produces both rhizomes and stolons can from a dense sod and is common species used for remediation of heavy metal contaminated sites[8,9]. Xia (2004) reported that vetiver and bahiagrasses showed good potential for phytoremediation of a metal contaminated site in China by exhibiting high survival rates, up to 99% and 96%, respectively, and high uptakes of Pb and Cd[10].

Establishment of pine trees and bahiagrass on an unlined landfill (488-D Ash Basin, 488-DAB) containing approximately one million Mg of coal combustion wastes (CCWs) that has resulted in the formation of acidic drainage (AD) was examined at the U.S. Department of Energy's Savannah River Site (SRS) in South Carolina as a potential method for phytostabilizing the contaminated area. Pyrite (FeS_2), found within the CCWs, undergoes oxidation in the presence of air and water and produces an acidic water enriched with iron, aluminum, sulfate, heavy metals and trace elements (e.g., lead, selenium, arsenic, mercury, zinc) that has contributed to groundwater deterioration in the area and poses a threat to biota on, and adjacent to, the basin[11,12]. Use of a vegetative cover to control AD via enhanced evapotranspiration has been hypothesized by Barton et al.[12]. Not only would a vegetative cover influence the redox conditions and AD generation of a waste site, but enhanced buffering capacity due to organic metabolites from root exudates and plant decay could help to break the acid production cycle[13]. Biosolids applied for a vegetative cover, such as municipal sewage sludge and animal waste, are an important group of soil amendments that are increasingly being used in agro-forestry and reclaimed lands[14]. As well as supplying plant nutrients, the organic matter (OM) in biosolids enhances aeration, porosity, tilth, and water retention capacity of soils[14].

A review on pyrite oxidation and subsequent implications on the 488-DAB and tree vegetation on the 488-DAB to assess the potential for phytoavailability of trace elements were reported by Barton et al.[12] and Maharaj et al.[15] but not for grass vegetation. As such a study was performed to obtain information to assess the potential for plant growth and metal uptake by the bahiagrass growing with different subsurface (ripped + compost, ripped, and control) and surface treatments (topsoil, biosolid, ash, apatite, and control) on the 488-DAB landfill.

3. Materials and Methods

3.1. Site Description and Background

The 488-DAB is an unlined, earthen containment basin located on the Savannah River Site, SC that received sluiced fly ash, dry fly ash and coal reject material from the early 1950's to the mid 1990's. Non hazardous wastes deposited in the basin contain metals typical of fly ash and coal (As, Co, Cr, Cu, Fe, Mn, Ni and Zn). The ash basin is $\sim 550 \times 180 \times 5.5$ m in size and contains $\sim 5.4 \times 10^5$ m^3 of waste material. Based on limited soil borings, the waste seems to be variably saturated, having both wet and dry zones that vary with depth. Hydraulic anisotropies may have been created when various materials were introduced to the basin. Alternatively, post-depositional features such as diagenetic "hard pans" may have developed as materials weathered over time, creating local anisotropies. Heavy equipment use on the basin and excessive grading has also resulted in a highly compacted surface.

3.2. Greenhouse Study

Greenhouse studies were initiated to evaluate potential amendments and/or treatments required for plant growth in the extremely acidic (1:1 pH 2.0) basin material. Treatments included the use of passivating agents to inhibit pyrite oxidation; and mycorrhizal fungi to enhance plant nutrient and water absorption, and promote feeder root health and longevity. Forty-eight gravity lysimeters were packed with 4 Kg of air-dried coal reject material and amended with 4 Kg of either coal reject material (control); 1:1 coal reject/fly ash mixture; or 2:1:1 coal reject/topsoil/compost mixture. Eight lysimeters from each group were leached with one pore volume of either a dilute H_2O_2/KH_2PO_4/NaOAc solution or distilled water. Thereafter, all lysimeters were leached weekly with 0.25 pore volumes of distilled water. After leaching was initiated, one-half of the lysimeters were inoculated with vesicular-arbuscular mycorrhizae (VAM) (four per treatment/amendment pair). Subsequently, all lysimeters were planted with a 5×5 cm plug of bahiagrass (*Paspalum notatum*). Plant survival and growth characteristics were monitored for a period of nine months following transplanting.

3.3. Field Study

A randomized block design consisting of three subsurface treatments (blocks) and five duplicated surface amendments (treatments) was developed for the 488-D basin (Fig. 1). Subsurface blocks were established to evaluate the influence of compaction and nutrient loading on plant survival and root penetration in the basin material. The three blocks included: 1) disturbed (ripped and disked) and compost amended (150 wet

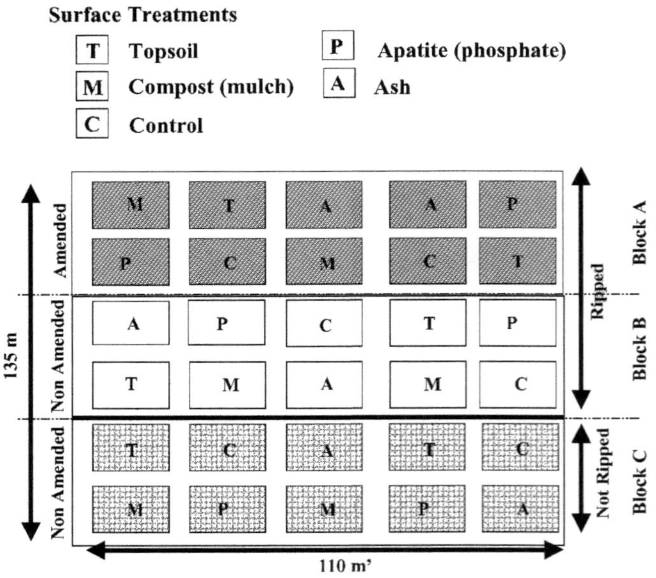

FIGURE 1. Randomized plot layout consisting of three subsurface treatments (blocks) and five duplicated surface amendments (treatments) on the 488-D basin at the Savannah River Site, SC.

Mg/ha Carolina compost), 2) disturbed (ripped and disked) with no compost amendment and 3) undisturbed (no ripping, or disking).

The five surface treatments outlined below were applied randomly to 2 of the 10 plots within each of the above three blocks: 1) control (no additional amendments or mechanical treatment applied), 2) apatite and super triple phosphate at a rate of 3.4 Kg of P per ha, 3) topsoil cover (10–15 cm), 4) biosolid cover (10-15 cm), and 5) ash cover (10–15 cm) (from A-area powerhouse), Each block was approximately 0.5 ha in size. Each treatment covered a surface area of 232 m^2. Five meter buffer strips were situated between each treatment cell.

Site preparation began in December, 1999 with the application of sixty wet tons per acre of Carolina compost to Block A. Subsequently, Blocks A and B were ripped to a depth of nearly one meter, and then plowed with a cutting harrow. All tillage implements were pulled with dozers to minimize additional compaction. Ten tons per acre of agricultural lime ($CaCO_3$) was broadcast over the entire study area in October of 2000. Surface amendments were applied from November 2000 thru January 2001.

Prior to the cover application, samples were collected from each of the materials and the basin (Table 1). The samples were analyzed for pH, EC, total recoverable metals[16]; total exchangeable bases and nitrate[17]; and total N, and C following methods outlined in the Methods of Soil Analysis[18]. Samples for metal analysis were digested using the HCl/HNO_3-based microwave digestion method[19] followed by inductively coupled plasma-mass spectrometry or optical emission spectrometry (ICP-MS, ICP-OES). Quality assurance-quality control protocols were followed for all analytical procedures as outlined in USEPA Method 6020[16].

Containerized loblolly (*Pinus taeda*) and Virginia (*Pinus virginiana*) pine seedlings were transplanted on the basin in February 2001. Each plot received 100 trees on 1.5-m centers. Information pertaining to the forest species study and results is described by Barton et al[20]. Bahiagrass for use on the 488-D basin was grown in the greenhouse. The grass was seeded in flats that contained either a commercially available potting soil, or a potting soil inoculated with vesicular-arbuscular mycorrhizae (VAM). Six 1 × 1 meter subplots were established on one-half of the treatment plots on the basin, representing each treatment of each block. Each subplot received twenty 5 × 5 cm plugs of a grass with and without VAM in May 2001. The transplants were watered immediately after planting, but received no additional irrigation afterwards. Survival measurements were performed on an annual basis for bahiagrass. By the second growing season, naturally seeded grasses and weeds had become a nuisance and posed a significant threat to the planted seedlings by competing for resources (water and nutrients). As such, all plots were mowed in May 2002 between the seedling rows and growth measurements of the bahiagrass ceased.

Tissue samples were collected from the grass plots in July, 2004 for analysis. The plots were not mowed in 2004, so samples were representative of fully grown species. The plant specimens were washed with distilled water to remove dust and particulate material, and then dried at 60°C. The dried plant tissue was ground using a Tekmar Analytical Mill and digested following USEPA methodology[19]. Metal analysis was performed by ICP-OES. The accuracy and precision of the plant tissue digestion and analysis was determined by the analysis of a spiked NIST standard reference material (SRM 1515-apple leaves).

TABLE 1. Chemical characteristics of substrate materials prior to application.*

Parameter	Soil	Biosolid/compost	Ash	Coal reject material
pH (1:1)	5.07 (0.62)	6.83 (1.03)	5.25 (0.35)	1.72 (0.75)
EC (1:5)	0.07 (0.08)	2.66 (0.53)	1.2 (0.28)	5.92 (1.86)
NO_3-N	10.3 (3.0)	31.0 (14.7)	BDL	BDL
P†	6.8 (2.9)	BDL	BDL	BDL
K†	17.2 (10.0)	79.6 (19.2)	8.5 (0.7)	2.3 (1.5)
Mg†	48.7 (34.3)	146.0 (107.6)	7.5 (4.9)	174.3 (82.2)
Ca†	208.2 (118.3)	766.3 (64.5)	272 (67.8)	526.0 (47.0)
Al‡	4193.8 (146.3)	430	3960	1767.9 (495)
Fe‡	3510.5 (195.5)	351	8220	20476.4 (13827.1)
Mn‡	222.3 (8.89)	2.33	42.5	7.18 (12.3)
Zn‡	14.3 (11.1)	4.95	11.5	1.96 (6.00)
Cd‡	BDL	0.13	BDL	0.1 (0.58)
Pb‡	8.15 (1.1)		18.9	28.4 (12.4)
As‡	9.88 (0.72)	BDL	33.4	64.7 (43.0)
Se‡	BDL	BDL	7.77	8.88 (5.80)
OM (%)		31.9		
C (%)	1.29 (1.22)	13.3 (1.0)	26.4 (1.2)	11.6 (2.25)
N (%)	0.08 (0.05)	0.41 (0.6)	0.48 (0.8)	0.21 (0.04)

* mg Kg^{-1} except where noted otherwise.
† Mehlich-1 method (HNO_3-H_2SO_4).
‡ U.S. EPA method 200.2 (HNO_3-HCl).
BDL = below detection limit.

3.4. Statistical Analysis

General mean and standard error calculations for the replicate analyses were performed using Microsoft Excel. Analysis of variance (ANOVA), general linear models (PROC GLMs), and PROC Univariate were performed to determine significant differences in elemental composition of the harvested bahiagrass. Due to the high mortality in control and apatite plots, and the loss of replication, statistical evaluation of the surface treatments was performed for the topsoil, ash and biosolid treatments only. All statistical models were performed using SAS[21]. Plant tissue concentrations are considered statistically different at $p < 0.05$.

4. Results and Discussion

4.1. Substrate Material and Soil Amendments

Chemical characteristics of the basin materials and soil amendments are presented in Table 1 and are detailed in the previous papers[12,15]. The pH of the soil, biosolid and ash materials were neutral to slightly acidic, while the basin material was highly acidic and potentially phytotoxic. The coal reject material and the A-area ash were greatly limited with respect to nutrients. The 488-DAB material was also characterized by elevated Fe and Al concentrations, which were over those of background levels for Coastal Plain soils in South Carolina[22], but not in the critical range for plant growth and survival. The biosolid data also indicate the presence of primary and secondary plant nutrients that, although initially low, may serve as a slow release fertilizer in addition to serving as a buffering agent.

4.2. Greenhouse Study

Within two weeks after transplanting, all bahiagrass died in the non-amended (control) lysimeters. Grasses in the ash and topsoil amended lysimeters, leached with only distilled water, exhibited 100% survival and showed good growth characteristics over the nine-month period. Not only was the grass capable of growing in the cover material, but root growth into the reject material was clearly exhibited by the end of the study. The ash (A-Area power plant) amendment significantly ($p < 0.05$) enhanced plant growth over all other treatments (Fig. 2). The use of VAM inoculant showed some outward benefits to plant growth with respect to color, however, the inoculum did not significantly enhance plant growth.

The use of $H_2O_2/KH_2PO_4/NaOAc$ as a pyrite microencapsulator showed some benefit from the standpoint of eluent water quality (Table 2), but was deleterious to plant growth and survival. Nearly all of the grass sprigs died in these pots. Even though the overall pyrite oxidation rate was lowered in the $H_2O_2/KH_2PO_4/NaOAc$ pots, extracts from soil samples collected in the upper 10cm showed elevated sulfate and EC concentrations over that of the non treated pots. The H_2O_2 may have contributed to heightened oxidation of pyrite at the surface, and oxidation of the naturally occurring passivating agents of OM within the amended layer. As such, maintenance of a lower oxidation potential within the surface layer of pots leached only with distilled water may have resulted in lower plant mortality.

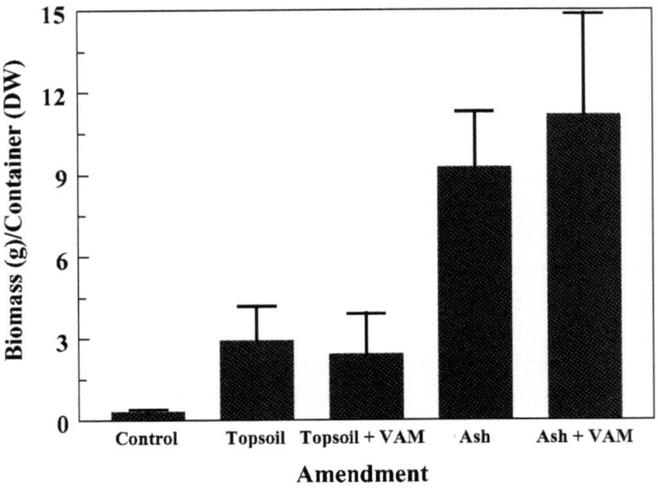

FIGURE 2. Biomass of bahiagrass per container from greenhouse study (n = 4 pots).

4.3. Field Study

For the discussion of vegetation survival, data obtained during comprehensive surveys performed in September, 2001 and May, 2002 were utilized. The seedlings had been in the field for one and two seasons, respectively. Examination of the blocking effect on seedling survival in the first year revealed that ripped + compost block exhibited the highest survival followed by ripped only and control (Table 3). In the second season ripped + compost still exhibited the highest survival followed by ripped only and control but ripped only and control treatments were not different. For 2002, survival in ripped + compost as a subsurface treatment was significantly greater than those in ripped only and control while there were no significant differences concerning application of inoculation with VAM for survival of bahiagrass (Table 3). This response is likely the result of enhanced physicochemical characteristics of the plots due to ripping and compost incorporation. Ripped

TABLE 2. Eluent chemistry from pots with and without $H_2O_2/KH_2PO_4/NaOAc$ added as a potential pyrite microencapsulator.

Parameter	Concentration with coating	Concentration without coating
pH (su)	2.49 (0.12)	2.32 (0.2)
EC (mS cm^{-1})	4150 (249)	4850 (197)
Fe (mg L^{-1})	395 (47)	555 (62)
SO_4(mg L^{-1})	1580 (101)	2140 (163)
Al (mg L^{-1})	131 (29)	124 (21)
Acidity ($CaCO_3$ eq.)	1085 (97)	1230 (144)

TABLE 3. Survival of bahiagrass (*Paspalum notatum*) on 488-DAB in percentage. Twenty springs were planted per plot in May 2001.*

Treatment	Ripped + compost + VAM**	Ripped + compost	Ripped + VAM	Ripped	Control + VAM	Control	Date
Topsoil	100	100	100	100	100	100	9/12/2001
Ash	100	90	100	100	30	0	9/12/2001
Biosolid	80	85	50	40	100	85	9/12/2001
Apatite	100	100	5	0	0	0	9/12/2001
Control	15	15	0	0	0	0	9/12/2001
Topsoil	100	100	100	100	100	100	5/24/2002
Ash	30	10	5	0	20	0	5/24/2002
Biosolid	50	75	10	0	100	50	5/24/2002
Apatite	25	100	0	0	0	0	5/24/2002
Control	0	0	0	0	0	0	5/24/2002

* Plots were mowed after 5/24/2002 collection and plants in ash treated plot were dead because boron toxicity.
** VAM = vesicular-arbuscular mycorrhizae.

+ compost subsurface treatment exhibited lower compaction which aids the system by allowing for root penetration, gas exchange and water infiltration[23]. The deep incorporation of the biosolids also aided the system by providing a fertilizer for the grasses and by buffering acidity in the rooting zone. Average soil pH (1:1) for the 10 cm depth during the second year was 5.2 for the ripped + compost treatment versus 4.2 and 4.4 for the ripped and control treatments, respectively.

The effect of the surface treatment on grass survival indicated that the topsoil and biosolid amendments greatly improved survivability on the basin (Table 3). The topsoil surface treatment exhibited 100% survival of grasses on all plots regardless of subsurface treatment. At least 50% survival of grasses planted in the ripped + compost and control plots containing a biosolid cover was observed. Oddly, only 10% survival was shown in the biosolid plot on the ripped only treatment. Examination of soil samples collected in 2002 revealed that the biosolid plot in the ripped only treatment contained a lower pH, C and N content and a higher EC concentration than that observed on the biosolid plot in the control subsurface treatment (Table 4). This information, linked with observations in the field, led us to believe that the biosolid cover in the ripped only plot had experienced some erosion, which reduced the effect of the amendment and heightened the expression of the underlying CCWs. The ash treatment appeared to be performing well during the first growing season, as anticipated from the greenhouse study, but showed a dramatic decline during year two. Toxicity effects shown to the grasses in the field may not have been expressed in the greenhouse due to the limited time of the greenhouse experiment and/or due to a dilution of a salt effect by periodic watering of the greenhouse pots[11,12]. The apatite treatment was impressive (100% survival) in one plot on the ripped + compost treatment; however, complete mortality was observed on the ripped only and control. The limited effect of the apatite treatment is not surprising given the response of the microencapsulation study in

TABLE 4. Average characteristics of soil samples collected during the second growing season at the 0-10 cm depth on the 488-DAB.

Subsurface treatment	Surface treatment	pH (1:1)	EC (1:3)†	N (%)	C (%)
Ripped + compost	Biosolid	6.42	2.05	0.45	19.45
Ripped + compost	Control	4.51	2.20	0.49	19.93
Ripped + compost	Topsoil	4.55	0.30	0.37	16.46
Ripped + compost	Ash	5.79	0.05	0.42	24.48
Ripped + compost	Apatite	4.13	2.35	0.43	18.52
Ripped	Biosolid	5.20	2.70	0.48	18.45
Ripped	Control	2.53	3.69	0.41	17.87
Ripped	Topsoil	5.12	0.19	0.31	13.58
Ripped	Ash	4.54	0.39	0.40	19.18
Ripped	Apatite	3.73	2.54	0.45	21.46
Control	Biosolid	6.54	2.39	0.55	19.53
Control	Control	3.68	2.78	0.37	19.20
Control	Topsoil	5.31	0.22	0.31	15.53
Control	Ash	4.44	0.81	0.47	21.21
Control	Apatite	2.34	4.99	0.52	22.39

† Units of measurement = mS cm^{-1}.

the greenhouse (Table 2). Complete mortality was observed in all control plots regardless of the subsurface treatment. In all, the treatment cover applications tended to enhance the buffering capacity of the substrate, and provided a more favorable rooting environment for the survival of bahiagrass.

Statistically, the ripped + compost and topsoil treatment exhibited a significantly higher survival than those with the same cover in ripped only and control. The ash treatment in ripped + compost and ripped only was significantly higher than that observed in control. Although statistical analyses were not possible for the apatite and control treatments due to complete mortality in some plots, the importance of the amendments on plant survival and growth is clearly evident (Table 3).

Total elemental composition of the bahiagrass did reveal significantly higher tissue concentrations of Al, B, Cr, Fe, Ni, and Zn in the ash treatment over those of the other surface treatments ($p < 0.05$) (Table 5) but not in the range of critical plant toxicity[14]. Ash treatment B concentrations were near the level of critical toxicity for grasses (25 – 50 mg Kg^{-1}) and may have contributed to the high mortality observed on those plots. Metal concentrations were not "significantly" different among subsurface treatments except for Al. Results indicated that Mn, B, Al, and Zn were accumulated by bahiagrass growing on the 488-DAB, however, Ni, Cd and Cr were not accumulated at levels above that of the CCWs. Similar instances have been reported elsewhere by Maharaj et al.[15]. In all, metal concentrations of bahiagrass examined were found to be within the regulatory limits for solid waste disposal established by the U. S. Environmental Protection Agency[25] and therefore do not pose contamination risk at this time. However, the periodic monitoring is suggested to ensure that toxicity and secondary contaminant problems do not arise.

5. Conclusions

The results indicated that deep ripping, subsurface treatment, and surface amendments were needed for the successful establishment of bahiagrass on the 488-DAB. Bahiagrass located on ripped + compost plots containing a topsoil cover exhibited the highest survival rates. Bahiagrass on ripped only and control subsurface treatments, regardless of the surface treatment, were likely influenced by the lack of subsurface amendments and subsequent benefits it provided, which resulted in high mortality. There were no significant differences concerning application of inoculation with vesicular-arbuscular mycorrhizae (VAM) for survival and metal concentrations of bahiagrass. There were also no significant differences ($p < 0.05$) concerning plant tissue concentrations of metals among the subsurface treatments but Al, Cr, Fe, Ni, and Zn concentrations in the tissue treated with ash as a surface treatment significantly ($p < 0.05$) differed from those of topsoil and biosolid treatments. Based upon these results, establishment of a vegetative cover on the 488-D basin with bahiagrass is feasible pending that the site is properly prepared using treatments and amendments as outline above. Given the chemical characteristics of the CCWs in the basin, however, periodic monitoring of the grass should be routinely performed to ensure that metal toxicity and/or secondary contaminant problems do not arise.

Acknowledgements. This work was funded in part by the U.S. Department of Energy—Savannah River Operations Office through the U.S. Forest Service-Savannah River (Interagency Agreement DE-IA09-00SR22188) and the Savannah River Ecology Laboratory through the University of Georgia Research Foundation, Inc (Financial Assistance Award

TABLE 5. Effect of subsurface and surface treatments on elemental concentrations for bahiagrass tissue collected on the 488-DAB.*

	Elements (mg kg^{-1})								
Treatment	Al	B	Cd	Cr	Cu	Fe	Mn	Ni	Zn
				Subsurface treatment effect					
Ripped + compost	88.81	23.27	0.24	3.57	3.13	121.56	54.89	1.07	14.33
	(A)	(A)	(A)	(A)	(A)	(A)	(A)	(A)	(A)
Ripped	55.42	10.63	0.04	3.52	2.92	65.46	36.18	0.87	12.70
	(AB)	(A)	(A)	(A)	(A)	(A)	(A)	(A)	(A)
Control	41.80	15.19	0.14	3.89	2.42	68.20	96.67	1.03	18.32
	(B)	(A)	(A)	(A)	(A)	(A)	(A)	(A)	(A)
				Surface treatment effect					
Topsoil	55.58	11.06	0.34	3.58	3.02	66.80	107.22	1.27	13.96
	(B)	(B)	(A)	(B)	(A)	(B)	(A)	(AB)	(B)
Biosolid	46.85	18.64	0.07	3.15	2.49	70.28	39.63	0.37	9.22
	(B)	(AB)	(A)	(B)	(A)	(B)	(A)	(B)	(B)
Ash	97.69	23.76	BDL	4.61	2.97	144.50	43.26	1.55	26.92
	(A)	(A)	(A)	(A)	(A)	(A)	(A)	(A)	(A)

* Bonferroni's Test using the GLM Procedure in the SAS System done to determine significant differences between species ($p = 0.05$); means with the same letter are not significantly different.
BDL = below detection limit.

DE-FC09-96SR18546). Support was also provided by the U.S. Forest Service—Southern Research Station, Center for Forested Wetlands Research; and the Environmental Research and Training Laboratory (ERTL) at the University of Kentucky.

References

1. Cunningham, S. D., and Ow, D. W., Promises and prospects of phytoremediation-update on biotechnology, *Plant Physiol.*, 110, 715, 1996.
2. Brown, S. L., Henry, C. L., Chaney, R. L., Compton, H., and DeVolder, R. S., Using municipal biosolids in combination with other residuals to restore metal-contaminated mining areas, *Plant Soil*, 249, 203, 2003.
3. Entry, J. A., Watrud, L. S., and Reeves, M., Accumulation of ^{137}Cs and ^{90}Sr from contaminated soil by three grass species inoculated with mycorrhizal fungi, *Environ. Pollut.*, 104, 449, 1999.
4. Gillbert, M., Minesite rehabilitation, *Trop. Grassland*, 34, 147, 2000.
5. Ye, Z. H., Wong, J. W. C., and Wong, M. H., Vegetation response to lime and manure compost amendments on acid lead/zinc mine tailings: a greenhouse study, *Rest. Ecol.*, 8, 289, 2000.
6. Bradshaw, A. D., and Huttl, R. F., Future minesite restoration involves a broader approach, *Ecol. Eng.*, 17, 87, 2001.
7. Koo, B-J., Adriano, D. C., Bolan, N. S., and Barton, C. D., Soil biology: root exudates and microorganisms, in *Encyclopedia of Soils in the Environment*, Vol. 3, Hillel, D., Ed., Elsevier Ltd., Oxford, U.K., 2005, 421–428.
8. Kalmbacher, R. S., and Martin, F. G., Effect of flooding on seed germination and emergence of three pasture grasses, *Soil Crop Sci. Soc. Fl. Proc.*, 57, 73, 1998.
9. Xia, H. P., and Shu, W. S., Resistance to and uptake of heavy metals by *Vetiveria zizanioides* and *Paspalum notatum* from lead/zinc mine tailings, *Acta Ecol. Sinica*, 21, 1121, 2001. (In Chinese).
10. Xia, H. P., Ecological rehabilitation and phytoremediation with four grasses in an oil shale mined land, *Chemosphere*, 54, 345, 2004.
11. Danker, R. M., Adriano, D. C., Koo, B-J., Barton, C. D., and Punshon, T., Soil amendments promote vegetation establishment and control acidity in coal combustion waste., in *Chemistry of Trace Elements in Fly Ash*, Sajwan, K. S., Alva, A. K., and Keefer, R. F., Eds., Kluwer Academic/ Plenum Publishers, New York, 2003, 235–257.
12. Barton, C. D., Paddock, L., Romanek, C., Maharaj, S., and Seaman, J., Phytostabilization of a Landfill Containing Coal Combustion Waste I. Geochemical Characterization and Implications for Remediation. *Environ. Geosci.*, 11(2), 2005. (In press).
13. Kleinman, R. L. P., and Rastogi, V., Reducing acid mine drainage liabilities using bactericides & other control technologies, in 13th Annual National Meeting American Society for Surface Mining and Reclamation Workshop #8, 1996.
14. Adriano, D. C., *Trace Elements in Terrestrial Environments*, 2nd Edition, Springer, New York, 2001.
15. Maharaj, S., Barton, C. D., Koo, B-J., and Newman, L. A., Phytoavailability of trace elements from a landfill containing coal combustion waste, in *Chemistry of Trace Elements in Fly Ash*, Sajwan, K. S., Alva, A. K., and Keefer, R. F., Eds., Kluwer Academic/ Plenum Publishers, New York, 2005. (Accepted).
16. USEPA, *Methods for the determination of metals in environmental samples*, Method 200.2 EPA/600/R-94/111, Washington, DC, United States Environmental Protection Agency, 1994.
17. NRCS, Natural Resources Conservation Service, *Soil Survey Laboratory Methods Manual*, Soil Survey Investigations, Report No. 42, United States Department of Agriculture, Washington, DC, U.S.A. 1996.
18. Sparks, D. L., Page, A. L., Helmke, P. A., Loeppert, R. H., Soltanpour, P. N., and Tabatabai, M. A., Johnson, C. T., and Sumner, M. N., *Methods of Soil Analysis: Chemical Methods*, Milwaukee, WI: SSSA Publications, 1996.
19. USEPA, *Microwave assisted acid digestion of siliceous and organically based Matrice,*. Method 3052, Rev 0, In SW-846: Test Methods for Evaluating Solid Waste, Physical/Chemical Methods, Washington, DC: USEPA Office of Solid Waste, 1996.
20. Barton, C. D., Marx, D., Adriano, D., Koo, B-J., Newman, L., Czapka, S., and Blake, J., Phytostabilization of a Landfill Containing Coal Combustion Waste II. Field Evaluation, *Environ. Geosci.*, 11(3), 2005. (In press).
21. SAS, *SAS/STAT User's Guide*, Version 8, SAS Institute Inc., Cary, NC, 1999.
22. Canova, J. L., Elements in South Carolina inferred background soil and sediment samples, *South Carolina Geol.*, 41, 11, 1999.
23. Bledsoe, L., Varsa, E. C., Chong, S. K., Olsen, F. J., Klubek, B. P., and Stucky, D. J., The effects of deep tillage on reclaimed mine spoils, in *Prime Farmland Reclamation*, Dunker, R. E. et al., Eds., University of Illinois at Urbana-Champaign, Department of Agronomy, 1992, 51–58.
24. Vogel, W. G. Revegetating surface mined lands with herbaceous and woody species together, in *Trees for Reclamation*, USDA Forest Service and Interstate Mining Compact Commission Technical Report, 1980.
25. U.S. EPA, Appendix A: Generic SSLs for the residential and commercial/industrial scenarios, 2001, http://www.epa.gov/superfund/resources/soil/ssgmarch01.pdf.

About the Editors

Kenneth S. Sajwan is a Professor and the Director of the Environmental Science Program in the Department of Natural Sciences and Mathematics at Savannah State University, Savannah, Georgia, USA. Dr. Sajwan earned a B.S. in Agriculture and Animal Husbandry, an M.S. in Agronomy from India, and Ph.D.s in Post Harvest Technology from the Indian Institute of Technology, Kharagpur, India, and in Soil Chemistry and Plant Nutrition from Colorado State University. After completing his Ph.D. from Colorado State University he devoted nine months as a Reader to teaching and research at the graduate level in Water Use Management at the Indian Institute of Technology-Roorkee (formerly University of Roorkee) in India. Dr. Sajwan joined Savannah State University during the fall of 1992 as an Associate Professor and was promoted to full Professor in 1996. Prior to coming to Savannah State University, he worked as an Assistant Professor at the University of Georgia's Savannah River Ecology Laboratory in Aiken, South Carolina. His previous work experience also includes a World Bank consultancy to Colombia, South America, and research associateships at the University of Wisconsin and the University of Kentucky. Dr. Sajwan also holds an Adjunct Professorship appointment at Alabama Agricultural & Mechanical University and the University of South Carolina at Aiken, and an Faculty Affiliate appointment at the Institute of Ecology, University of Georgia.

Dr. Sajwan has taught number of courses during his career at Savannah State University. These courses include Introduction to Environmental science, limnology. Contaminant Hydrology, Environmental Ethics, Environmental Law, Environmental Chemistry and Analysis, Environmental Health Safety and risk Assessment, Environmental and Evolutionary Issues, Ecology, Hazardous Waste management, Principles of Ecotoxicology, Environmental Impact Assessment, Special Topics, Senior Seminar, Senior Research, and Internship. Dr. Sajwan has been recognized as a devoted and talented teacher and his accomplishments are reflected in the success of his students and his uncanny ability to motivate, challenge and inspire his students to excel academically both in the classroom and beyond. Dr. Sajwan has received several awards for his outstanding contribution to teaching and research. Dr. Sajwan was a recipient of the Richard Nicholson Award for Excellence in Science Teaching (2005), the National Science Teachers Association Distinguished Science Teacher Award (2004), the Ernest L. Boyer International Award for Teaching, Learning and Technology (2003), Board of Regents' University system of Georgia Teaching Excellence Award (2002), and the White House Millennium Award for Teaching and Research Excellence (2001). In addition, he is the recipient of the Board of Regents' of the University System of Georgia Distinguished Professor of teaching and Learning Award for the 1998–1999 academic year at savannah State University. Dr. Sajwan is also the recipient of the 1999 International Award for Innovative excellence in Teaching, Learning and Technology for the year 1999.

Dr. Sajwan has edited four books, *Trace Elements in the Environment, Chemistry of Trace Elements in Fly Ash, Trace Elements in Coal and Coal Combustion Residues, and Biogeochemistry of Trace elements in Coal and Coal Combustion Byproducts*. In addition, he has published two laboratory manuals, *Environmental Science Laboratory Manual and General Biology Laboratory Manual*, and has published over hundred (100) articles in peer reviewed journals, serials, conference proceedings, and symposia. Dr. Sajwan is very well known nationally and internationally in his area of research. His primary research includes biogeochemistry of trace elements, environmental chemistry, ecotoxicology, and chemical equiliribia in soils. Currently, Dr. Sajwan in investigating the potential benefits and environmental impact of the application of coal ash and organic waste mixtures to agricultural lands for crop production.

Irena Twardowska is a Research Hydrogeochemist and a Head of the Laboratory of Non-point Contamination of the Terrestrial and Aquatic Environment at the Institute of Environmental Engineering of the Polish Academy of Science in Zabrze, Silesia, Poland. She is also Professor of Environmental Engineering at the Jan Dlugosz University in Czestochowa, Poland. Dr. Twardowska received her D.Sc. in Hydrogeology from the University of Mining and Metallurgy in Krakow in 1986, her Ph.D. in Environmental Engineering from Silesia Technical University in Gliwice in 1965, and her M.S. in Sanitary Engineering from the same University in 1960. Her entire carrier has been spent with the Polish Academy of Sciences, Institute of Environmental Engineering in Zabrze, Poland, where she held the positions of senior scientist and research group leader, and since 1987 of associate Professor and Laboratory head.

Dr. Twardowska has published three monographs, more than 200 papers and chapters in books, presented papers at more than 90 international symposia, is an author of 4 patents on industrial waste dumps construction. She has been leading numerous multidisciplinary national and international research projects, recently and currently in collaboration with Germany, India, Israel, Norway and Greece. Her research interests concern the generation, transformations, release, migration and immobilization of contaminants in solid waste disposal sites, soil and vadose zone, effect of these processes on groundwater quality and development of the pollution prevention and control measures.

Dr. Twardowska is a member of 10 scientific societies and a Council Member and past–president of SECOTOX—International Society of Ecotoxicology and Environmental Safety, and co-organizer of its Central and Eastern European Section. From 1993 to 1996 she was a vice-chair of the Subcommittee for Waste Examination of the Committee of Analytical Chemistry of the Polish Academy of Sciences. Since 1996 she has been a member, and since 2003 a vice-chair of Technical Committee No. 216 on Solid Wastes of the Polish Standardization Committee and since 1998 a representative of Polish Standardization Committee in CEN/TC 262 on Waste Characterization in European Standardization Committee. She is a member of Editorial Board and Editor-in-Chief of two scientific journals

Dr. Twardowska is a recipient of three Awards of the Secretary of the Polish Academy of Sciences, Golden Award of the Polish Ministry of Environment, Silver Cross from the State Council of Poland and Silver and Golden Awards from the Regional Council of Silesia for achievements in environmental research. She has been a certified expert of the Polish Ministry of Environment and the Regional authorities, and a recognized frequent advisor to the industry in the field of the environment protection.

Tracy Punshon is a Program Coordinator in the Advanced Analytical Center for Environmental Sciences at the University of Georgia's Savannah River Ecology Laboratory, located in Aiken, South Carolina, USA. Dr. Punshon earned a Bachelor of Science degree in Applied Biology in 1992, majoring in pollution ecology, and a Ph.D. in 1996—both from Liverpool John Moore's University (LJMU). The subject of her Ph.D. was the resistance of willows (*Salix* spp.) to heavy metal pollution in the soil. The response of plants to soil metal contamination has been a central theme throughout her research career. After completing her Ph.D. Dr Punshon worked for the Ministry of Agriculture, Fisheries and Food in collaboration with the University of Reading and LJMU evaluating the ability of soil amendments to reduce bioavailable cadmium and arsenic concentrations in contaminated agricultural soils.

In the spring of 1998, Dr. Punshon began working with Dr. D.C. Adriano at the Savannah River Ecology Laboratory. Her research focused on the effect of coal combustion by-products on the health and elemental composition of various crop plants. In the fall of 2000, Dr. Punshon joined CRESP (Consortium for Risk Evaluation with Stakeholder Participation) at the University of Rutgers, and studied the bioavailability and trophic transfer of metal and radionuclides, based at SREL. During this time, Dr. Punshon conducted research at the National Synchotron Light Source at the Brookhaven National Laboratory, using x-ray spectroscopy to elucidate metal distribution and speciation in plants growing in heavily contaminated environments, and showed a contaminant signature within the annual rings of willows growing on a metal-contaminated former radiological settling pond. At the same time, Dr. Punshon's research on coal combustion by-products developed into a collaboration with other SREL scientists to study the speciation and bioavailability of arsenic and selenium in plants growing on flue-gas desulfurization residue amended soils using size-exclusion chromatography coupled to ICP-MS.

Dr. Punshon has authored and co-authored 22 papers in refereed journals, conference proceedings, and special reports; with a further 5 invited publications in books and peer-reviewed journals. Her key research on metal distribution in willows has been featured in local newspapers, the UGA news website and the website of the Office of Biological and Environmental Research (Department of Energy). Dr. Punshon is a member of the editorial committee for the International Conference on the Biogeochemistry of Trace Elements, and a regular reviewer for the Journal of Environmental Quality, The Science of the Total Environment and Environmental Pollution.

Ashok K. Alva is currently a Research Leader at the USDA-Forage and Vegetable Crops Laboratory, Prosser, Washington, USA. He received his Ph.D. degree in Soil Chemistry and Plant Nutrition from the Pennsylvania State University in 1983. He has conducted a wide range of collaborative research in Universities in the United States, Australia, Denmark, Thailand, and India. In addition to several invitational visits to international Universities and Research Institutes, he has been involved in cooperative research projects in Brazil, and China.

He has conducted extensive research on the use of coal combustion byproducts, including fly ash and the gas desulfurization gypsum, as agricultural soil amendments. His research expertise includes: reaction and transport of agricultural chemicals in soils; heavy metal toxicity to plants and techniques to alleviate such toxicities; soil solution chemistry and speciation of metals using thermodynamic chemical speciation models; sequential fractionation of metals and relationship between the metal concentrations in sequential fractionation and single extractions; chemistry of acid soils and factors affecting plant growth; application of tissue analysis for diagnosis and correction of nutritional status of plants. His current research also include development of irrigation and nutrient best management practices in sandy soils to increase the nutrient uptake efficiency, and minimize non-point source pollution of groundwater.

He has authored and co-authored 23 invited book chapters; and over 252 papers in refereed journals, conference proceedings, and extension publications. Dr. Alva has made over 105 presentations at various international, national, and local meetings. He has supervised 13 postdoctoral fellows, 7 visiting scientist, and several graduate students. Dr. Alva is a "Fellow" of the "American Society of Agronomy" and the "Soil Science Society of America". He served as an Associate Editor of the Journal of Environmental Quality, and a Co-Editor for the Florida State Horticultural Society Proceedings. Currently, he serves in the Editorial Board for the Journal of Crop Improvement, and is a technical reviewer for 14 international journals, and several competitive grant proposals. He received the University of Florida Meritorious Faculty Performance Awards in 1997 and 1998, and USDA-ARS—Superior Research Leadership Awards for 2000 through 2004.

Index

A

488-DAB (488-D Ash Basin) CCWs, 195, 196, 199, 197. *See also* phytoavailability
  As concentration, 197
Abelmoschus esculentus, 172
ACAA. *See* American Coal Ash Association
acid leachate, 195
acid mine drainage, 50, 52, 54, 142
  acidity, 146
  Ca level, 145
  coal ash leaching behavior, 50
  fluidised bed boiler ash (FBA), 203
  fluidized bed combustion (FBC), 50
  fly ash (FA), 6
  Mn level, 145
  wetland substrate
    solution geochemistry gradients, 142–149
acid neutralization, 50, 190
acid rain, 24, 212. *See also* flue gas desulfurization
acid rock drainage, 33–39
  oxygen as regent, 34
  prevention and control, 33
AD (acidic drainage), 195, 226
adsorbtion
  effect of adsorbate [trace element] concentration, 151, 153
    Cr, Pb and Cu removal, 153
    rate in mixed metal solution, 151
  effect of adsorbent concentration, 153
    McInt-FA, 152
    PWW-FA, 152
  effect of contact time, 151
  mercury, 99–104
  rate of, 151
Agrostis palustris, 125
Agrostis stolonifera, 125
air penetration resistance, 35–36. *See also* fly ash
  lime enrichment, 36
  solidification, 36
alkaline ash leachate, 96
alkaline CCBs
  FBC, 51
  FGD, 51
aluminum speciation, 219
AMD. *See* acid mine drainage
American Coal Ash Association, 6, 13, 19
ammonium sulfate, 26
  as fertilizer, 26
  from flue gas, 26
    dry regenerable sodium process, 26
    Krupps Kopper Walther process, 26
    wet ammonia process, 26
Andropogon gerardii, 199
anthropogenic impact, 41
  on elemental distribution, 46
Arachis hypogaea, 159, 160, 210
ARD. *See* acid rock drainage
ARW (artificial rainwater), 135, 136, 137
  column experiments, 135
As and Se speciation, 115
  by ion chromatography ICP-MS, 117
  in soil and plant tissues, 116
ash and sludge application
  plant metal uptake, 160
ash leachate, alkaline, 96
ash pond effluent characteristics
  enrichment of elements, 48
  pH and concentrations of elements, 43, 48
ASW (available soil water), 159, 160
Avena sativa, 207

B

BA. *See* bottom ash
Baccharis halimifolia, 200
BAF. *See* bioaccumulation factor
BAFA root mix, 128
bahiagrass (*Paspalum notatum*)
  vegetative cover for a landfill, 225–231
BARC. *See* bottom ash in road constructions
basal substrate
  effect of pH, 180
bed boiler ash, fluidised. *See* fluidised bed boiler ash
bermudagrass, 191
bioaccumulation factor, 90, 92, 95
  average metal bioaccumulation, 92, 93
boiler slag, 3, 4, 22
Bos taurus, 207
bottom ash in road constructions
  grassed swales, 87
  heavy metals leaching, 87
  metal retention sustainability, 96
  pollutant trap in, 88
  road leachate, 87, 88
  soil in swales, 92–93
  vegetation in roadside swales, 88–96
  vegetation samples, 90
bottom ash, 3, 4, 22, 124, 125, 132, 190. *See also* coal combustion by-products; coal combustion products
  boiler slag, 4, 22
  from coal, 4
  from municipal solid waste (MSW), 4
  in road constructions (BARC), 87–96
  MSWI, 87
  re-use potential of, 4
  silica blasting grit, 22
  wet-bottom furnaces, 22
box-and-whisker plots, 91
Braun-Blanquet scale, 90
by-product fertilizer
  N-fertilizer, 28
  vegetable culture test, 27, 28
by-product gypsum, 25

C

CAAA. *See* Clean Air Act Amendments of 1990
calcareous soil
  leaching
    of soil nutrients and metals, 171–176

calcium carbonate equivalency, 190
PL, 192
cation exchange capacity, 48, 69, 161
CCBs, alkaline, 51
CCBs. *See* coal combustion by-products
CCE. *See* calcium carbonate equivalency
CCP. *See* coal combustion products
CCW. *See* coal combustion wastes
Cd removal
adsorbent FA concentration, 152
contact time, 152
CEC. *See* cation exchange capacity
centrifuge method, 135, 136
class-C FA
acid-neutralizing efficacy, 190, 191, 193
amelioration of soil acidity, 190
CCE, 190
in liming soil, 193
lime substitution, 191
pulverized limestone, 190
turfgrass, 193
utility turves, 191
Clean Air Act Amendments of 1990, 4
Clean Air Act, 6, 114
NSR, 6
CO_2 hydrate technique, 30
coal
combustion by-products (CCBs), 21–22, 150, 171, 184
combustion products (CCP), 3–9, 15–18, 124, 190
combustion wastes (CCW), 195, 197, 225
electricity generation, 13
IEA, 14
Kyoto Protocol, 13, 14
low rank coal, 21, 24
production, 13–14
spontaneous combustion, 33, 34
coal and ash piles, 87–150
transport and leachability of metals
acid mine drainage wetland substrate, 142
arsenic and selenium speciation, 114
fly ash materials comparison, 150
grassed swales, 87
leachability of trace metals from sandy or rocky soils, 105
municipal solid waste incineration bottom ashes, 87
putting green root mixes, 124
removal of mercury, 99
solute leaching, 134
coal ash
as agriculture soil amendment, 159–225
chemical properties, 161
element uptake by maize (*Zea mays* L.), 184
evaluation of bahiagrass (*Paspalum notatum*), 225
fluidised bed boiler ash (FBA), 202
leaching of soil nutrients and metals, 50–58, 171
ornamental plant production, 177
Oxyaquic Haplustalf appended with, 159
phytoavailability, 195
potential of fly ash and organic wastes, 216
sandy or rock soils, 105
soil acidity with class-C fly ash, 190
transport and plant uptake of Zn, 159
utilization rate of, 22
coal combustion by-products, 21–22, 150, 171, 184
acid mine drainage, 51, 184
boiler slag, 50
bottom ash (BA), 22
coal ash, 50
coal mine reclamation, 51
FBC, 51
flue gas, 50
flue gas desulfurization (FGD), 51
gypsum, 22
fly ash (FA), 22, 150
in construction applications, 171
leaching behavior of, 50, 56
physicochemical properties of, 125
recycling, 114–115
structural fills, 51
surface reclamation, 52
utilization of, 22–24
coal combustion flue gas desulfurization waste
on element uptake by maize (*Zea Mays* L.), 184–189
coal combustion products, 3–9, 15–18, 124, 190
amended root mixes
chromium concentrations, 130
American Coal Ash Association (ACAA) report, 13, 15
and biosolids, 17
as soil amendments, 115
boiler slag (BS), 3, 15
bottom ash (BA), 3, 15, 16
disposal and utilization, 18, 191
environmental hazards associated with, 3, 13–20
FBC ash, 15
finance transfer, 19
flue gas desulfurization (FGD) residues, 3, 6, 15
fly ash (FA), 3, 4, 6, 15, 16, 17, 114
generation, use and disposal, 15
in India, 17
in Poland, 19
in USA, 16
Hg input source, 3
metal(loid) content of, 3
micron-scale spatial metal analysis techniques, 4
MWLP, 52
production of, 3
radionuclides, 3
recycling policy, 3, 13, 18
application to soil, 114
regulatory instruments, 18–19
in India, 18
in Poland, 18
in USA, 19
root mix columns, 130
sand amendment, 125
soil application, 3
utilization, 6, 15, 16
concrete/cement and construction materials production, 17
wallboard production from FGD gypsum, 17
waste management enterprise, 18
weathering behavior, 52
coal combustion residues, 13–66
environmental impact of, 11
alternative by-products of, 21
coal and coal combustion products, 13
coal ash leaching behavior in acid mine water, 50
delineation of water and sediment contamination, 41
fly ash as sealing material, 33
heavy metals adsorption, 66
simultaneous $SO_2/SO_3/NO_x$ treatment of coal-fired flue gas, 21
trace elements in fly ashes, 61
coal combustion waste, 195, 197, 225
acidic drainage (AD), 195, 197
and pyrite, 195
boiler slag (BS), 195
bottom ash (BA), 195
flue gas desulfurization residue (FGD), 195
fly ash (FA), 195
landfill, 225
MCL criteria for radionuclide concentrations, 62
phytoavailability of trace elements from, 195–201
radioactivity of, 62
sequential extraction procedure, 196
coal fly ash, 66–83
as amendment for Florida's sandy or coarse soils, 105
as soil amendment, 105
Alfisols, 105
Entisols, 105
Spodosols, 105
trace metals, 105

TCLP, 106
trace elements, 105
coal pile leachate
aluminum speciation, 219
composition of, 217–218
Fe, Mn, Zn and sulfate speciation, 219–220
fly ash-amended compost, 220
optimization of fly ash-to-compost ratio for plant growth, 221
uptake of multielements by corn from, 221
ground water contamination, 217
coal pile runoff basin, 217, 218, 219
coal pile runoff, 218
coal-based electricity generation
IEA, 14
Kyoto Protocol, 13, 14
coal-fired flue gas, 21–32
ammonium sulfate, 26
by-product fertilizer, 29
environmental problems
acid rain, 24
gaseous by-products, 29
pollutant in, 24
SO_3 concentration, 24–25
solid by-products, 29
coal-fired plants
combustion waste, 61–65
FGD system
wet scrubber process, 24
mercury removal, 99
column experiments. *See also* solute leaching
saturated, 135
unsaturated, 135
column leachings, 126
combustion products, coal, 3–9
compost, 216, 222
fly ash-amended, 220
optimization of fly ash-to-compost ratio for plant growth, 221
uptake of multielements by corn from, 221
CON. *See* control rooting mix
container substrate
fly ash, dolomite, and components of, 178
ornamental plant production, 178
control rooting mix, 125
core aerification, 191
CPL. *See* coal pile leachate
CPR. *See* coal pile runoff
CPRB. *See* coal pile runoff basin
Cr removal, 150. *See also* PWW-FA
adsorbent source, 152, 153, 154
contact time, 152, 153, 154
creeping bentgrass, 125
Cynodon dactylon, 105

D

delineation
of water and sediment contamination, 41
desulfurization residues, flue gas. *See* flue gas desulfurization residues
distribution coefficient, 100, 101
drinking water radioactivity
maximum concentration limits (MCL), 61
dry regenerable sodium process, 26
dry-bottom boilers, 22

E

EBRD. *See* European Bank for Reconstruction and Development
EC (electric conductivity), 96
ECEC (effective cation exchange capacity), 209
EGR. *See* enhanced gas recovery
electron beam flue gas treatment, 26
cost and plant capacity, 29
SO_2, SO_3 and NO_x removal, 26
vs. wet limestone FGD, 29
enhanced gas recovery, 30
enhanced oil recovery, 30
Entisols, 105
environmental risks
acid rain, 24
As and Se concentration, 122
fly ash (FA), 63, 172
gypsum disposal, 25
landfilling, 177
landscape, 24
road construction with MSWI, 87
solute leaching, 134
EOR. *See* enhanced oil recovery
Eupatorium compositifolium, 200
European Bank for Reconstruction and Development, 25
EXC. *See* exchangeable fraction

F

FA. *See* fly ash
Fagopyrum sagittatum, 207
FBA. *See* fluidised bed boiler ash
FBC. *See* fluidized bed combustion
fertilizer
agricultural effect of, 27
ammonia nitrate, 27
ammonium sulfate, 27
sulphur, 203
Festuca arundinacea, 7, 207
FGD. *See* flue gas desulfurization residues
FGDG. *See* flue gas desulfurization gypsum
flue gas
ammonium sulfate, 26
by low-rank coal, 24
electron beam process, 26
Marista II TPP, 27
SO_2 transformation into ammonium sulfate, 26
flue gas desulfurization, 21, 114, 185
amended soil
arsenic and selenium speciation, 114–123
coal-fired flue gas, 23
disposal by-product, 21
green house study, 184
on element uptake by maize (*Zea Mays* L.), 184–189
SO_x content, 21
sulfur dioxide emission, 23, 24
acid rain, 24
wet limestone process, 23
flue gas desulfurization residue, 3–7, 51, 210
agronomic application, 7
alkaline nature, 6
amended layer
As concentration, 117, 121
Se concentration, 117
bioavailability of P, 7
by-product, 22–25, 28, 30
ammonium sulfate, 25
economic evaluation of, 25
gypsum, 23, 25
comparison with FA, 7
effect on maize growth and element uptake, 186–188
effect on soil leachate, 186
on electric conductivity, 186
on pH, 186
FA impurities, 115
with FA20, 6
landfilling, 6
metal concentration impact, 115
metalloid concentration in, 117
as neutralizing agents, 7
with organic waste material, 7
without organic waste material, 7
pH balance restoration, 7
reuse of CO_2, 21
as soil amendment, 185
SO_2 removal, 21, 26
wallboard material production, 6
wet limestone process, 21, 24, 28
flue gas scrubbing, 114
fluidised bed boiler ash
as liming material, 205–208
acid mine spoil reclamation, 208
food chain implications, 207–208
plant elemental composition, 207
soil acidity, 205–208
soil chemical properties, 206–207
as liming material, soil conditioner and sulfur fertilizer, 202–215
as mine spoil amendment, 207
as phosphorus inactivator, 211
as S fertilizer, 203, 210

fluidised bed (*cont.*)
as subsoil acidity ameliorant, 208–210
by using gypsum, 209
gypsum-responsive soils, 209
B concentrations, 204
bioassay method, 209
Ca content, 203
chemical characterization of, 203
crop responses to, 205
effect of pH, 207
effect on plant elemental composition, 204
effect on soil physical properties, 211
flue gas desulfurization gypsum (FGDG), 210
fuel and the limestone source, 203
improper disposal of, 212
leachate concentrations, 208
magnesium content, 203
metallic concentration, 207
physical and chemical properties, 203
phytotoxic elements, 204
selenium content, 204
SO_2 removal, 202
soil acidity, 203
soil erosion control, 211
soil fertility sustainability, 211–212
topsoil acidity, 210
fluidized bed combustion, 50, 51
fly ash, 3, 4, 6, 16, 22, 124, 125, 190, 216, 222. *See also* coal combustion by-products; coal combustion products
acid mine drainage, 6
adsorbate, 151
adsorbent, 151–156
agricultural land application, 6
agricultural lime, 193
air penetration resistance, 35
amended compost application, 216, 220
water-extractable constituents of, 220
amended container substrates, 177
chemical properties, 179
for plant production, 179, 180
amendments to agricultural soils, 216–224
and sewage sludge (SS), 6
as sealing agent, 33–40, 37
dense FA:water mixtures for coal seams fire control in mining areas, 37
preventive sealing of mining waste dumps for ARD, 38
spontaneous combustion control, 38
bermudagrass sod production management, 6
in calcareous soil, 174
chemical properties of, 178
class-C
amelioration of soil acidity, 190–194
dry-bottom boilers, 22
environmental risks minimization, 6
green root mix, 132
heavy metal content of, 27
leachate volumes, 173
leaching potential, 6
materials, 150–157
metal(loid) enrichment and toxicity, 6
metals and metalloids, 115
molybdenum concentration, 129
neutralize acidity, 216
ornamental plant production, 177–183
plant nutrient imbalance, 6
Portland cement replacement, 22
pozzolan, 22
restoration, 6
severely eroded land remediation, 6
solute leaching, 134–141
spoil pH, 51
trace elements, 59–83
As, 134
Se, 134
water-holding capacity, 134
XAFS, 4
XANES, 4
fly ash-amended compost. *See* compost, fly ash-amended
fossil fuel power plants
removal of mercury, 99–104
Freundlich distribution coefficients, 66
functionalized nanoporous sorbents, 99–104. *See also* mercury removal

G
GAC. *See* granulated activated carbon
Gallus gallus, 207
gas desulfurization gypsum, 210
GEESI (General Electric Environmental Services), 26
GHG. *See* greenhouse gas
Golf course putting green, 124, 191
Gossypius hirsutus, 121
granulated activated carbon (GAC), 103
grassed swales
as pollutant trap, 87
metal translocation, 87
green leachate collection, 125
green root mixes, 124, 126
greenhouse columns, 127
greenhouse gas, 14
gypsum, 23, 209, 210
as source of S, 210
as subsoil acidity amendment, 210
ECEC, 209
environmental problems, 23
gas desulfurization, 210
with fly ash, 24

H
hard and soft acid base principle, 102
HDPE. *See* high density polyethylene
heavy metal adsorption, 66–83
batch equilibrium technique, 68
calcium nitrate, 68
heavy metals content
in soil, coal ash and sewage sludge, 70–71
heavy metals extraction, 69
HG-AAS. *See* hydride generation atomic absorption spectrometry
high density polyethylene, 88, 90
Hordeum vulgare, 105, 177
HSAB. *See* hard and soft acid base principle
hydraulic conductivity, 36–37
hydride generation atomic absorption spectrometry, 116
Hypoestes phyllostachya, 177, 178, 180, 182

I
ICP-MS. *See* inductively coupled plasma-mass spectrometry
ICP-OES. *See* inductively coupled plasma optical emission spectroscopy
IEA. *See* International Energy Agency
inductively coupled plasma optical emission spectroscopy, 90, 116
inductively coupled plasma-mass spectrometry, 135, 90, 100, 116, 191
standard operating conditions for, 116
influent wastewater
trace element, 150
International Energy Agency, 14
electrification rates, 15
energy development index, 14
ion exchange resin, 102
ion selective electrode, 42
iron leaching, 107
ISE. *See* ion selective electrode

K
Krome gravelly loam soil, 172
Krupps Kopper Walther process, 26
Kyoto Protocol, 13
in Japan, 14
refusal by USA, 14

L
landfill
bahiagrass (*Paspalum notatum*), 225
fly ash erosion, 177, 216
leachate generation, 177
phytoavailability of trace elements, 195–201
Langmuir isotherm, 101
for methyl mercury adsorption, 101
leachate, 105
alkaline ash, 96
coal pile, 217–219

concentration of trace metals, 107
Cd, 108, 109, 112
Cu, 108
Fe, 107, 108
Mn, 107
Mn, 108
Pb, 107
Pb, 108
Zn, 108
metal concentrations in, 131
quality and trace element transport, 129–131
volumes, 173
leaching, 50
of Ca and trace metals, 174–175
CCB interaction, 51
column, 126
concentrations of heavy metals, 171
effect of fly ash, 171–176
FBC ash with Site M and P, 53, 54
of inorganic P, 171, 174
leaching front, 134
leaching solutions
SPLP, 50
toxicity characteristic leaching procedure (TCLP), 50, 106
mine water, 51, 52, 56
of NO_3-N and NH-N, 173–174
residence time, 140
soil amendments, 172
solute, 134–141
toxicity characteristic, 50, 101
fly and bottom ash, 50
rainfall simulation, 106
least significant difference, 90
Leymus arenarius, 90, 93
lignite combustion
flue gas, 26
limestone FGD
CO_2 emission, 23
limestone gravel, 146
liming FBA, 205–208. *See also* fluidised bed boiler ash
acid mine spoil reclamation, 208
food chain implications, 207–208
plant elemental composition, 207
soil acidity, 205–208
Linaria vulgaris, 90, 93
low rank coal, 21, 24
LSD. *See* least significant difference
Lycopersicon esculentum, 105, 177, 180

M
MAD. *See* maximum allowable depletion
Malus domestica, 207
matric potential, 138
maximum allowable depletion, 159, 160
maximum concentration limits, 61, 105, 107
McInt-FA, 152, 153
Cu removal, 153, 155
Mn removal, 154
Pb removal, 152, 155
Zn removal, 153, 155
MCL. *See* maximum concentration limits
Medicago sativa, 7, 105, 177, 210
mercury adsorption, 99–104
by thiol-SAMMS, 101
effect of cation mixtures, 102
effect of complexing ligands, 103
effect of pH, 100
functionalized nanoporous sorbents, 99–104
granulated activated carbon (GAC), 103
ion exchange resin, 102
ionic strength, 100
resin, 103
thiol-SAMMS, 99, 103
mesocosm soil, 116
As concentrations, 117
Se concentrations, 117, 118
Mesocricetus auratus, 207
metal adsorption, heavy, 66–83
metal and metalloid
bioavailability
CCP disposal and reuse, 3
speciation, 115
As and Se, 115
in plant tissues, 121
soils, 120
metal contamination, 41. *See also* sediments and water
metalloid concentration
in plant tissues, 120
mine spoil, 51
mine water leaching procedure, 51, 52, 56
Mn and Pb removal
adsorbent FA concentration, 152, 153, 154
adsorbent source, 152, 153, 154
contact time, 152, 153, 154
molecular self-assembly, 99
MSW. *See* municipal solid waste
MSWI. *See* municipal solid waste incineration
mullite, 46
municipal solid waste incineration, 4, 87, 96. *See also* bottom ash
BARC, 87
roads construction, 87
MWLP. *See* mine water leaching procedure

N
nanoporous sorbent adsorbed
molecular self-assembly, 99
surfactant molecules, 99
thiol-SAMMS, 99, 100, 104
nanoporous sorbents, functionalized, 99–104. *See also* mercury removal
nanoporousthiol-SAMMS sorbent, 103
Nephrolepis exaltata, 180
new source review
ACAA, 6
pollution control devices, 6
N-fertilizer, 26, 29
ammonium sulfate, 30
consumption, 28
residual fertilization effects, 28
NSR. *See* new source review

O
Okra biomass, 173
Orangeburg soil, 117, 118
organic wastes
amendments to agricultural soils, 216–224
ornamental plant production, 177. *See also* fly ash
Oryza sativa, 105, 177
Ovis aries, 207
oxidation basins, 142
Oxyaquic Haplustalf
with coal ash and sewage sludge, 159–170
Zn movement and uptake, 159–170

P
Panicum capillarye, 200
Panicum milaceum, 200
particle size distribution, 124
Paspalum notatum, 225, 226
Pb removal
contact time, 152
PFA. *See* pulverized fuel ash
PG root mixes, 124
hydraulic conductivity, 124
sand-sized bottom ash, 124
silt-sized fly ash, 124
PG. *See* putting green
Phalaris arundinacea, 90, 93, 95, 96
Phaseolus vulgaris, 105, 177
phosphogypsum, 211
phosphorus inactivator, 211. *See also* fluidised bed boiler ash
physicochemical properties
soil, coal ash and sewage sludge, 69–70
phytoavailability
488-DAB (488-D Ash Basin) CCWs, 195, 196, 199, 197
of trace elements from landfill, 195–201
phytoremediation, 225
phytotoxicity
sewage sludge, 159
Pinus taeda, 227
Pinus virginiana, 227
PL. *See* pulverized limestone
plant metal uptake, 160
Plantago maritima, 90, 93, 95

pollution monitoring
  total heavy metal concentrations in sediments, 41
Potentilla (Argentina) anserina, 90, 93
PRB (powder river basin), 34
PSA (pressure swing adsorption), 30
PSD. *See* particle size distribution
PTSA (pressure temperature swing adsorption), 30
pulverized fuel ash, 171
pulverized limestone, 190, 191
putting green, 124
  columns, 126
  leachate, 126
  root mix
    CCP bottom and fly ash as inorganic amendments/sand-substitutes, 125
PWW-FA, 153. *See also* trace element removal
  carbon concentration, 152
  Cd removal, 152, 154
  contact time, 152
  Cr removal, 152, 154
  Mn removal, 152, 154
  Pb removal, 152, 154, 155
pyrite, 195

R
radioactive mine water
  A type, 61
  B type, 61
  Ra removal, 62
radioactivity
  coal combustion waste, 62
  drinking water, 61
  fly ash (FA), 63
radionuclides
  occurrence and sorption of, 61–65
Rattus rattus, 207
removal mechanism
  Cd removal, 152, 154
  Cr removal, 150, 152, 154
  Cu removal, 153, 155
  Fe removal, 142
  Hg removal, 99–104
  Mn removal, 152, 154
  NO_x removal, 27
  Pb removal, 152, 154, 155
  Ra removal, 62
  SO_2 removal, 26, 27, 202
  Zn removal, 153, 155
river sediment characteristics
  element concentrations, 42, 43, 45, 46
  pH and element concentrations, 43
road construction material. *See also* bottom ash in road constructions
  MSWI bottom ash, 87
road leachate
  HDPE, 88
roadside swales, 88–90
  vegetation of, 90–92
root mix, 124, 125
  columns
    BAFA, 126
    BASFA, 126
  effects on PSD, 128
  hydraulic conductivity, 126–127
  hydrology, 127–129
  leachate
    trace element concentration, 124
  in sand, 124

S
SAMMS. *See* self-assembled monolayers on mesoporous silica
sand sized particles, 125
saturated experiments, 137–138
Savannah River Site, 195, 196, 217
SCR. *See* selective catalytic removal
scrubbing system
  dry, 6
  wet, 6
sediments
  metal contamination, 41
  physicochemical properties
    CEC, 48
    clay content, 47
    mineralogy, 46
    specific surface area, 48
selective catalytic removal, 26
selenium concentrations, 115, 134
selenium mobility and bioavailability, 121
selenomethionine, 115
self-assembled monolayers on mesoporous silica, 99. *See also* mercury removal
sewage sludge amendment, 6, 66, 68, 159
  amendment, 66–83
  chemical properties, 161
  crop production, 67
  land application of, 66
SFA (sand sized particles) PG mix, 130
site-specific ash leaching procedure, 50
sludge
  amendment, 66
  temporal variation of water extractable Zn, 163
SO_2 emissions. *See also* flue gas desulfurization, residues
  Clean Air Act Amendments of 1990 (CAAA), 4
soil
  amendments
    biosolids, 172
    compost, 172
    leaching of soil nutrients and metals, 172
  chemical properties, 161
  contamination
    road material, 96
  fractionation, 197–199
  metal concentration, 96
  metal fixation in, 97
  profile
    cation exchange capacity (CEC), 161
    trace metal movement, 160
    Zn concentration distribution, 163, 165
  quality
    metal translocation, 87
    road material impact, 87
  types, 66–83
solid waste
  coal ash, 66
  sewage sludge, 66
solute leaching, 134–141
  matric potential, 135–136
    centrifuge method, 135–136
    saturated experiments, 137–138
    unsaturated experiments, 138
  pore solution residence time, 134
  UFA, 134
solution geochemistry gradients, 142–149
sorbents, functionalized nanoporous, 99–104. *See also* mercury removal
Sorghum vulgaris, 6
specific surface area, 48
Sphagnum peat moss, 125
SPLP, 50
Spodosols, 105
SRS. *See* Savannah River Site
SS. *See* sewage sludge
subsoil acidity ameliorant, 208–210. *See also* fluidised bed boiler ash
  gypsum, 208
  plant rooting depth, 208
sulfur content
  removal
    flue gas desulfurization residue (FGD), 114
  SO_2, 24–25
  SO_3, 24–25
swales
  metal concentration, 92, 93
  soil in, 92–93
  vegetation in, 93

T
TCE. *See* trichloroethylene
TCLP. *See* toxicity characteristic leaching procedure
Thiobacillus ferrooxidans, 33
thiol-SAMMS, 100, 101, 102, 103
  in adsorbing mercury, 101
  leachability of, 101
  mercury adsorption, 101

trace element
effect of adsorbent concentration, 152
leachability of, 106
TCLP, 106
removal from aqueous system, 150–157
adsorption, 152–155
from sandy or rocky soils amended with coal fly ash, 105–113
sorptive behaviour
effect of pH, 151
transport, 124
trichloroethylene, 217
Trifolium pratense, 207
Trifolium repens, 7
Triticum aestivum, 105, 177
turfgrass site, 191
root mix, 124
Typha latifolia, 90, 93, 96, 143

U
UFA. *See* unsaturated flow apparatus
UNFCC. *See* United Nations Framework Convention on Climate Change
United Nations Framework Convention on Climate Change
greenhouse gas, 14
unsaturated flow apparatus, 134
unsaturated fly ash amended columns
ionic strength, 139

V
VAM. *See* vesicular-arbuscular mycorrhizae
vegetative cover, 225
AD control, 226
vesicular-arbuscular mycorrhizae, 225, 226, 227
Vicia cracca, 114, 116, 121
selenium concentrations in, 120

W
water and sediment contamination, delineation of, 41
water extractable concentration, 159
of Zn, 159
WCI. *See* World Coal Institute
WEC. *See* water extractable concentration
wet scrubber (limestone) process, 21, 23, 29, 24. *See also* FGD
by-product utilization rate, 21
CO_2 emissions, 24
limestone slurry, 28
wet-bottom furnaces, 22
wetland substrate depth, 145
aerobic basins and limestone substrates, 142
alkalinity, 147
anaerobic sequences, 143
concentration gradients, 144
depth zones, 145
limestone gravel layer, 145
organic layer, 145
surface zone, 145
water column, 145
effect of pH, 143
gypsum precipitation, 148
interstitial solution compositions, 145
limestone substrate, 142
manganese and Al levels, 146, 147
metal retention, 142
mobilization of Fe and SO_{2-}, 147
overload, 144
oxidation basins, 142
pH level, 146
precipitation, 147
redox level, 143, 145
removal of Mn, 143
saturation indices, 145, 146, 147, 148
seasonal effects, 147
solution speciation analysis, 146, 147
spring rainfall, 147
substrate depth effects, 145–147
sulfide levels, 143
surface or subsurface water flow, 142
wet-type ESP
SO_3 removal, 26
wet-type process, 26
GEESI process, 26
Krupps Kopper Walther process, 26
World Coal Institute, 14

X
XAFS (X-ray absorption fine structure spectroscopy), 4
XANES (X-ray absorption near-edge structure spectroscopy), 4

Z
Zea mays, 105, 121, 177, 184, 188, 190
Zn concentration
adsorbent FA concentration, 153, 154
contact time, 153, 154
distribution in plant parts, 165
by ash, 165
by mixture, 165
effect of irrigation schedule, 166
in plant parts, 165
in kernel (storage organs), 165
in shoot (vegetative tissues), 165
in soil, 165
movement and uptake
ASW, 159
MAD, 159
WEC, 159
solid waste application, 166
transfer efficiency, 165
transport, 159–170
WEC temporal variation, 162–163

MIX
Papier aus verantwortungsvollen Quellen
Paper from responsible sources
FSC® C105338

If you have any concerns about our products,
you can contact us on
ProductSafety@springernature.com

In case Publisher is established outside the EU,
the EU authorized representative is:
Springer Nature Customer Service Center GmbH
Europaplatz 3, 69115 Heidelberg, Germany

Printed by Libri Plureos GmbH
in Hamburg, Germany